AF839332

Current Topics in Membranes, Volume 66

Structure and Function of Calcium Release Channels

Current Topics in Membranes, Volume 66

Series Editors

Dale J. Benos
Department of Physiology and Biophysics
University of Alabama
Birmingham, Alabama

Sidney A. Simon
Department of Neurobiology
Duke University Medical Centre
Durham, North Carolina

Current Topics in Membranes, Volume 66

Structure and Function of Calcium Release Channels

Edited by

Irina I. Serysheva
Department of Biochemistry and Molecular Biology
The University of Texas Medical School
Houston, TX, USA

ELSEVIER

AMSTERDAM • BOSTON • HEIDELBERG • LONDON
NEW YORK • OXFORD • PARIS • SAN DIEGO
SAN FRANCISCO • SINGAPORE • SYDNEY • TOKYO
Academic Press is an imprint of Elsevier

Academic Press is an imprint of Elsevier
30 Corporate Drive, Suite 400, Burlington, MA 01803, USA
525 B Street, Suite 1900, San Diego, CA 92101-4495, USA

First edition 2010

Copyright © 2010 Elsevier Inc. All rights reserved

No part of this publication may be reproduced, stored in a retrieval system or transmitted in any form or by any means electronic, mechanical, photocopying, recording or otherwise without the prior written permission of the publisher

Permissions may be sought directly from Elsevier's Science & Technology Rights Department in Oxford, UK: phone: (+44) (0) 1865 843830; fax: (+44) (0) 1865 853333, email: permissions@elsevier.com. Alternatively you can submit your request online by visiting the Elsevier web site at http://elsevier.com/locate/permissions, and selecting *Obtaining permission to use Elsevier material*

Notice
No responsibility is assumed by the publisher for any injury and/or damage to persons or property as a matter of products liability, negligence or otherwise, or from any use or operation of any methods, products, instructions or ideas contained in the material herein. Because of rapid advances in the medical sciences, in particular, independent verification of diagnoses and drug dosages should be made

ISBN: 978-0-12-381037-3
ISSN: 1063-5823

For information on all Academic Press publications
visit our website at elsevierdirect.com

Printed and bound by CPI Group (UK) Ltd, Croydon, CR0 4YY

Transferred to Digital Print 2012

Working together to grow
libraries in developing countries

www.elsevier.com | www.bookaid.org | www.sabre.org

ELSEVIER BOOK AID International Sabre Foundation

Contents

Contributors

Numbers in parentheses indicate the pages on which the authors' contributions begin.

Matthew J. Betzenhauser (273) Department of Physiology and Cellular Biophysics, Columbia University Medical School, New York City, New York, USA

Ilya Bezprozvanny (323) Department of Physiology, University of Texas Southwestern Medical Center at Dallas, Dallas, Texas, USA

Darren F. Boehning (191) Department of Neuroscience and Cell Biology, University of Texas Medical Branch, Galveston, Texas, USA

Joanne Carney (49) Department of Cardiology, Wales Heart Research Institute, School of Medicine, Cardiff University, Heath Park, Cardiff, United Kingdom

Don-On Daniel Mak (235) Department of Physiology, University of Pennsylvania, Philadelphia, Pennsylvania, USA

Anamika Dayal (115) Department of Medical Genetics, Molecular and Clinical Pharmacology, Innsbruck Medical University, Innsbruck, Austria

Robert T. Dirksen (139) Department of Pharmacology and Physiology, University of Rochester Medical Center, Rochester, New York, USA

Clara Franzini-Armstrong (1) Department of Cell and Developmental Biology, University of Pennsylvania School of Medicine, Philadelphia, Pennsylvania, USA

Manfred Grabner (115) Department of Medical Genetics, Molecular and Clinical Pharmacology, Innsbruck Medical University, Innsbruck, Austria

Suresh K. Joseph (299) Department of Pathology and Cell Biology, Thomas Jefferson University, Philadelphia, Pennsylvania, USA

J. Kevin Foskett (235) Department of Physiology; Department of Cell and Developmental Biology, University of Pennsylvania, Philadelphia, Pennsylvania, USA

Derek R. Laver (69) School of Biomedical Sciences, University of Newcastle; Hunter Medical Research Institute, Callaghan, New South Wales, Australia

Zheng Liu (27) Wadsworth Center, New York State Department of Health, Albany, New York, USA

Steven J. Ludtke (171) National Center for Macromolecular Imaging, Verna and Marrs McLean Department of Biochemistry and Molecular Biology, Baylor College of Medicine, Houston, Texas, USA

Sammy A. Mason (49) Department of Cardiology, Wales Heart Research Institute, School of Medicine, Cardiff University, Heath Park, Cardiff, United Kingdom

Gerhard Meissner (91) Department of Biochemistry and Biophysics, University of North Carolina, Chapel Hill, North Carolina, USA

Barry V. L. Potter (209) Department of Pharmacy and Pharmacology, Wolfson Laboratory of Medicinal Chemistry, University of Bath, Claverton Down, Bath, United Kingdom

Andrew M. Riley (209) Department of Pharmacy and Pharmacology, Wolfson Laboratory of Medicinal Chemistry, University of Bath, Claverton Down, Bath, United Kingdom

Ana M. Rossi (209) Department of Pharmacology, University of Cambridge, Cambridge, United Kingdom

Irina I. Serysheva (171) Department of Biochemistry and Molecular Biology, The University of Texas Medical School, Houston, Texas, USA

Colin W. Taylor (209) Department of Pharmacology, University of Cambridge, Cambridge, United Kingdom

Cedric Viero (49) Department of Cardiology, Wales Heart Research Institute, School of Medicine, Cardiff University, Heath Park, Cardiff, United Kingdom

Terence C. Wagenknecht (27) Wadsworth Center, New York State Department of Health, Albany; Department of Biomedical Sciences, School of Public Health, State University of New York at Albany, Albany, New York, USA

Lan Wei (139) Department of Pharmacology and Physiology, University of Rochester Medical Center, Rochester, New York, USA

Alan J. Williams (49) Department of Cardiology, Wales Heart Research Institute, School of Medicine, Cardiff University, Heath Park, Cardiff, United Kingdom

David I. Yule (273) Department of Pharmacology and Physiology, University of Rochester Medical School, Rochester, New York, USA

Preface

Irina I. Serysheva

Intracellular Ca^{2+} signaling is a strictly controlled spatial and temporal event caused by the orchestrated mobilization of Ca^{2+} into the cytoplasm via Ca^{2+} channels, from the extracellular milieu (Ca^{2+} influx) or from the intracellular stores (Ca^{2+} release). The focus of this book is on ligand-gated Ca^{2+} channels that mediate release of Ca^{2+} from the intracellular stores such as the endoplasmic and sarcoplasmic (ER/SR) reticulum into the cytoplasm in response to appropriate messenger or protein interaction signals. These channels are essential to a wide variety of physiological processes, including muscle contraction, heartbeat, and brain function. Since the discovery of Ca^{2+} release channels about 30 years ago, intensive research efforts of numerous laboratories around the world were focused on studies of these channels to understand molecular mechanisms underlying their function. There have been major advances in the field of structure and function of Ca^{2+} release channels over the recent years. This book aims to provide a concise and informative overview of these developments in one volume. To achieve this goal, contributions from scientists at the forefront of their respective fields outline the most exciting discoveries, provide critical analysis of their implications, and give perspectives for future research on Ca^{2+} release channels.

Two families of intracellular Ca^{2+} release channels have been identified: the ryanodine receptor (RyR) and the inositol 1,4,5-trisphosphate receptor (IP_3R), that are reviewed in Part I and II of this book, respectively. RyRs are primary Ca^{2+} release channels in muscle cells and are key players in the generation of Ca^{2+} signals triggering muscle contraction. IP_3Rs are detected in all tissues and cell types with the highest density in the Purkinje cells of cerebellum. IP_3, primary cellular agonist of IP_3R channels, is generated in the cell as a result of phosphoinositide metabolism in response to numerous stimuli, such as hormones, growth factors, neurotransmitters, odorants, and light. Localized in the ER/SR membranes, Ca^{2+} release channels are the largest tetrameric ion channels, known to date, with a megadalton molecular mass. Both channel families share significant sequence homology that accounts for the many functional similarities between RyRs and IP_3Rs. The chapter by Clara Franzini-Armstrong, whose studies have provided the first glimpses of RyRs in the native membrane, discusses the ultrastructure of

RyRs in a variety of muscles. Current advances in the 3D structure determination of Ca^{2+} release channels are discussed by Terry Wagenknecht and Zheng Liu with the focus on RyR, and Steven Ludtke and Irina Serysheva provide the chapter on the structure of IP_3R. The chapters by Alan Williams and colleagues and by Darren Boehning are focused on molecular basis for the gating of Ca^{2+} release channels and review proposed mechanisms underlying the ion translocation through Ca^{2+} release channels across the cell membrane. The chapters by Derek Laver and Kevin Foskett and Don-On Daniel Mak are dedicated to the regulation of ion conduction properties of Ca^{2+} release channels by different intracellular modulators, while Colin Taylor and colleagues analyze the structural basis for the interaction between the IP_3 receptor and its naturally occurring high-affinity agonists, adenophostins, in their chapter. Regulation of Ca^{2+} release channels by protein kinases and redox active species is critically reviewed in the chapters by Gerhard Meissner, David Yule, and Suresh Joseph. The chapter by Manfred Grabner and Anamika Dayal is on the structure–functional relationships between the RyRs and their molecular partners in muscle cells, the voltage-gated L-type Ca^{2+} channels. Finally, the overview of human diseases related to the defects in Ca^{2+} release channels or to the modulation of their activity is given in two chapters by Robert Dirksen and Lan Wei (RyR-linked diseases) and Ilya Bezprozvany (IP_3R-linked diseases).

In attempt to cover the most topics in this active field of research within a single volume, it is inevitable that some areas are not presented in great detail or omitted due to the lack of space. Thus, we apologize in advance if some opinions and views are not discussed. Overall, this book should serve as a useful and an informative resource to a large group of research scientists including new researchers and students who are just entering the rapidly growing field of ion channels. Finally, it should provide an important reference for research-oriented clinicians aiding in novel drug discovery.

I am grateful to all of the authors for their keen interest in this publication and for taking time to provide their chapters to this book, to the Elsevier Editors for their efficient support during the book preparation. I also would like to thank my family for their invaluable assistance and patience throughout my work on this book.

Previous Volumes in Series

Current Topics in Membranes and Transport

Volume 23 Genes and Membranes: Transport Proteins and Receptors* (1985)
Edited by Edward A. Adelberg and Carolyn W. Slayman

Volume 24 Membrane Protein Biosynthesis and Turnover (1985)
Edited by Philip A. Knauf and John S. Cook

Volume 25 Regulation of Calcium Transport across Muscle Membranes (1985)
Edited by Adil E. Shamoo

Volume 26 Na^{+}–H^{+} Exchange, Intracellular pH, and Cell Function* (1986)
Edited by Peter S. Aronson and Walter F. Boron

Volume 27 The Role of Membranes in Cell Growth and Differentiation (1986)
Edited by Lazaro J. Mandel and Dale J. Benos

Volume 28 Potassium Transport: Physiology and Pathophysiology* (1987)
Edited by Gerhard Giebisch

Volume 29 Membrane Structure and Function (1987)
Edited by Richard D. Klausner, Christoph Kempf, and Jos van Renswoude

Volume 30 Cell Volume Control: Fundamental and Comparative Aspects in Animal Cells (1987)
Edited by R. Gilles, Arnost Kleinzeller, and L. Bolis

Volume 31 Molecular Neurobiology: Endocrine Approaches (1987)
Edited by Jerome F. Strauss, III, and Donald W. Pfaff

**Part of the series from the Yale Department of Cellular and Molecular Physiology.*

Volume 32 Membrane Fusion in Fertilization, Cellular Transport, and Viral Infection (1988)
Edited by Nejat Düzgünes and Felix Bronner

Volume 33 Molecular Biology of Ionic Channels* (1988)
Edited by William S. Agnew, Toni Claudio, and Frederick J. Sigworth

Volume 34 Cellular and Molecular Biology of Sodium Transport* (1989)
Edited by Stanley G. Schultz

Volume 35 Mechanisms of Leukocyte Activation (1990)
Edited by Sergio Grinstein and Ori D. Rotstein

Volume 36 Protein–Membrane Interactions* (1990)
Edited by Toni Claudio

Volume 37 Channels and Noise in Epithelial Tissues (1990)
Edited by Sandy I. Helman and Willy Van Driessche

Current Topics in Membranes

Volume 38 Ordering the Membrane Cytoskeleton Trilayer* (1991)
Edited by Mark S. Mooseker and Jon S. Morrow

Volume 39 Developmental Biology of Membrane Transport Systems (1991)
Edited by Dale J. Benos

Volume 40 Cell Lipids (1994)
Edited by Dick Hoekstra

Volume 41 Cell Biology and Membrane Transport Processes* (1994)
Edited by Michael Caplan

Volume 42 Chloride Channels (1994)
Edited by William B. Guggino

Volume 43 Membrane Protein–Cytoskeleton Interactions (1996)
Edited by W. James Nelson

Volume 44 Lipid Polymorphism and Membrane Properties (1997)
Edited by Richard Epand

Volume 45 The Eye's Aqueous Humor: From Secretion to Glaucoma (1998)
Edited by Mortimer M. Civan

Volume 46 Potassium Ion Channels: Molecular Structure, Function, and Diseases (1999)
Edited by Yoshihisa Kurachi, Lily Yeh Jan, and Michel Lazdunski

Volume 47 Amiloride-Sensitive Sodium Channels: Physiology and Functional Diversity (1999)
Edited by Dale J. Benos

Volume 48 Membrane Permeability: 100 Years since Ernest Overton (1999)
Edited by David W. Deamer, Arnost Kleinzeller, and Douglas M. Fambrough

Volume 49 Gap Junctions: Molecular Basis of Cell Communication in Health and Disease
Edited by Camillo Peracchia

Volume 50 Gastrointestinal Transport: Molecular Physiology
Edited by Kim E. Barrett and Mark Donowitz

Volume 51 Aquaporins
Edited by Stefan Hohmann, Søren Nielsen and Peter Agre

Volume 52 Peptide–Lipid Interactions
Edited by Sidney A. Simon and Thomas J. McIntosh

Volume 53 Calcium-Activated Chloride Channels
Edited by Catherine Mary Fuller

Volume 54 Extracellular Nucleotides and Nucleosides: Release, Receptors, and Physiological and Pathophysiological Effects
Edited by Erik M. Schwiebert

Volume 55 Chemokines, Chemokine Receptors, and Disease
Edited by Lisa M. Schwiebert

Volume 56 Basement Membranes: Cell and Molecular Biology
Edited by Nicholas A. Kefalides and Jacques P. Borel

Volume 57 The Nociceptive Membrane
Edited by Uhtaek Oh

Volume 58 Mechanosensitive Ion Channels, Part A
Edited by Owen P. Hamill

Volume 59 Mechanosensitive Ion Channels, Part B
Edited by Owen P. Hamill

Volume 60 Computational Modelling of Membrane Bilayers
Edited by Scott E. Feller

Volume 61 Free Radical Effects on Membranes
Edited by Sadis Matalon

Volume 62 The Eye's Aqueous Humor
Edited by Mortimer M. Civan

Volume 63 Membrane Protein Crystallization
Edited by Larry DeLucas

Volume 64 Leukocyte Adhesion
Edited by Klaus Ley

Volume 65 Claudins
Edited by Alan S. L. Yu

SECTION 1

RYR Ca^{2+} RELEASE CHANNELS

CHAPTER 1

RyRs: Their Disposition, Frequency, and Relationships with Other Proteins of Calcium Release Units

Clara Franzini-Armstrong
Department of Cell and Developmental Biology, University of Pennsylvania School of Medicine, Philadelphia, Pennsylvania, USA

I. OVERVIEW

The location, distribution, size, and molecular architecture of calcium release units in a variety of muscles are illustrated and discussed in terms of the phylogenetic, developmental, and molecular factors that affect them.

Current Topics in Membranes, Volume 66 1063-5823/10 $35.00
Copyright 2010, Elsevier Inc. All right reserved.
DOI: 10.1016/S1063-5823(10)66001-2

II. INTRODUCTION

The focus of this chapter is on locations, positions, and interactions of ryanodine receptors (RyRs). RyRs are located in junctional domains of the sarcoplasmic reticulum (jSR), where they form part of a large supramolecular complex and also interact with the voltage-dependent Ca^{2+} channel (CaV) of transverse (T) tubules and plasmalemma. The cytoplasmic domains of RyRs are seen as feet in thin sections for electron microscopy (EM) and the position of CaV channels is detected by freeze-fracture.

The jSR is located in close apposition to either plasmalemma or T tubules. Closely spaced feet (RyR) occupy the junctional gap between the jSR and the surface membranes and these sites release Ca^{2+}, hence the common appellation of Ca^{2+} release units (CRUs). Similarities and differences between cardiac and skeletal muscle CRUs abound and are particularly useful in unraveling structure–function correlations.

III. CARDIAC CRUs

A. The Classical Description of Dyads and Peripheral Couplings

CRUs of myocardial cells (peripheral couplings and dyads) are junctions of flattened jSR cisternae with surface membranes, separated from each other by small, slightly variable distances that are significantly larger than the distance between adjacent RyRs in the junction. The general appearance and the molecular composition of peripheral couplings and dyads is essentially the same and the two are functionally equivalent. In normal hearts, the jSR cisterna of CRUs is a single structure, marked by the content of clustered calsequestrin (CASQ) in the lumen.

In cardiac CRUs, the appearance of feet in the junctional gap varies from distinct densities to blurred and indistinct profiles even within the same junction. The average thickness of thin sections for EM is in the range of 50–60 nm and RyRs are separated by ~30 nm. Thus, feet images in EM represent the superimposed profiles of two RyRs. Visibility of feet in an ordered array is strongly affected by the orientation of the section plane relative to the array axis. Until very recently (see below), the disposition of RyRs in cardiac myocytes was not directly visualized (with a single exception; Tijskens, Meissner, & Franzini-Armstrong, 2003) and the question of whether a single coherent group, or array, of RyR covers the entire junctional gap of a cardiac CRU was not explored in detail. Instead, estimates of CRU feet content were based on the approximate size of the jSR–surface membrane contacts, resulting in numbers as high as ~30–270 RyRs/CRU from EMs and of 80–140 from light microscopy (Chen-Izu et al., 2006).

B. New Views of Dyads and Peripheral Couplings

1. New Findings

Apparently minor but actually highly significant modifications of current views have been introduced by two recent works that explore cardiac CRUs by electron tomography (Hayashi et al., 2009) or by sophisticated light microscopy (Baddeley et al., 2009).

Single tomographic images of embedded tissue afford views of CRUs that are equivalent to those obtained in very thin sections, providing direct identification of CRU components and the ability of reconstructing 3D images. An extensive tomographic reconstruction of CRUs in mouse ventricle finds a discrepancy between coverage of T tubules surface by SR, and the extent of sites containing identifiable RyR profiles. The latter are smaller and less continuous than the former, so that a classical dyad must be redefined as a collection of smaller junctional areas. Once alerted by tomography to the discontinuity in the disposition of RyRs at the sites of SR–T tubule association, the discontinuity can be directly observed in thin sections. Examples (among many) are in Fig. 1A of Asghari, Schulson, Scriven, Martens, and Moore (2009). The result of this changed view is that coherent RyR clusters for mice are smaller than previously estimated, being on the average composed of ~43 elements, and often containing as few as 15 RyRs. The classical dyads contain more than one group of these smaller coherent aggregates. The myocardium of a *pan* triadin null mouse model (*Trdn–/–*) in which all proteins of the junctional supramolecular complex, including RyRs, are downregulated offers a striking example of this discontinuity by showing that T–SR contacts in each dyad are discontinuous (Chopra et al., 2009).

While the revised view of dyads as composed of smaller subcomponents that are held under the umbrella of a common jSR is in easy agreement with the thin section images, some details of the tomography-derived descriptions are not consistent with previous views of dyads. In the mesh models of the T tubules and associated structures of Hayashi et al. (2009), the filled-in outlines of the jSR, colored in yellow, form an almost continuous coverage of the T tubules. If this is correct, then thin sections should reveal jSR profiles (defined by their periodic dense content of CASQ) covering most of T tubules' surface. Instead, the jSR profiles are discrete entities, separated by well-defined spaces. The discrepancy can easily be explained by the inclusion of some SR elements that are not junctional, but are adjacent to T tubules, into the 3D reconstruction of the jSR.

A striking new publication with somewhat similar, but even more precise, results regarding the size, arrangement, and complex geometry of RyR clusters in cardiac muscle involves a close look at peripheral couplings of rat myocardium using an advanced optical technique that allows detection of single RyR positioning (Baddeley et al., 2009). The new results show that a classical peripheral coupling is constituted of several smaller RyR clusters, that the individual clusters are quite variable in size and shape, and that in most cases noticeable gaps in the RyR array within each smaller cluster exist, so that on the average the cluster content of RyRs is as low as ~14 RyRs (Fig. 1A and B). However, at least one-third of the clusters reside within small groups at short distances from each other (edge-to-edge distances of

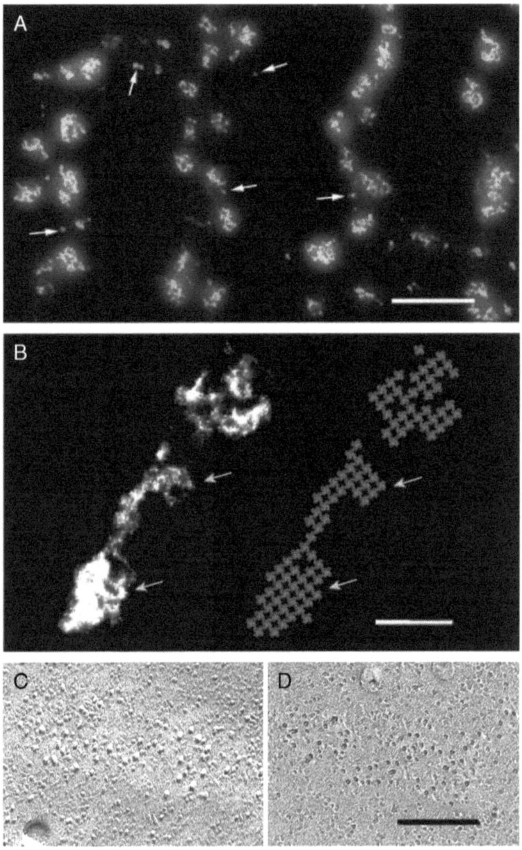

FIGURE 1 (A, B) TIRF microscopy images of antibody labeled RyR clusters in peripheral couplings of rat cardiac myocytes. (A) Diffraction-limited fluorescence image (red) marking the positions of individual peripheral couplings, is overlaid over the images of RyR clusters at single protein resolution. (B) Reconstruction of two clusters shows the distribution of RyR arrays described in the text. Reprinted with permission from Baddeley et al. (2009). (C, D) Clusters of CaV1.2 channels (large particles) in the plasmalemma of finch cardiac cells mimic the disposition of RyR clusters in (A) and (B). Scale bars: (A) 1 μm; (B, C) 200 nm. (See Color Insert.)

50 nm or less). A comparison between the diffraction-limited image of fluorescent spots and the high-resolution view of clusters could be interpreted to indicate that each diffraction-limited spot, containing more than one RyR cluster, corresponds to the classically described peripheral coupling.

Data from this work are consistent with a disposition of RyR in an orthogonal lattice, apparently similar to that exhibited by RyR1 in skeletal muscle (Fig. 1B). However, the orthogonal positioning of RyRs centers is not sufficient to specify the actual RyR interactions in the cluster. Body muscles of arthropods (Loesser, Castellani, & Franzini-Armstrong, 1992), skeletal muscle of primitive fish (Di Biase & Franzini-Armstrong, 2005), and muscles of higher vertebrates (Block, Imagawa, Campbell, & Franzini-Armstrong, 1988) all have RyRs disposed in orthogonal lattices, but details of intermolecular interactions differ. The precise 2D arrangement of RyR2 in cardiac muscle is not yet completely elucidated.

A second new finding is that individual RyR clusters have extremely variable sizes and have lattices that differ in orientation even when adjacent to each other, as if they have formed independently by accrual around stochastically generated crystallization centers. Given the presence of a protein that insures cluster location, appropriately sized RyR clusters may form without requiring an explicit scaffold or other process that tightly controls cluster size (Baddeley et al., 2009). The third finding is that the shape of RyR arrays is quite different from circular or slightly elongated ellipses, and that the arrays are often quite incomplete, missing individual elements.

CaV1.2 are clustered in correspondence of RyR groups within CRUs of cardiac muscle (Sun et al., 1995), though, differently from skeletal muscle, there is no apparent direct link between the two molecules. Nonetheless, a specific rapport is established during cardiac muscle differentiation between clusters of CaV1.2 and RyR2 (Protasi, Sun, & Franzini-Armstrong, 1996). Figure 1C and D shows two freeze-fracture images of the plasmalemma in finch heart (Bossen, Sommer, & Waugh, 1978), showing clusters of large particles (CaV1.2 channels), whose disposition mimics remarkably well that of RyR2 in Fig. 1B. The connection that allows such a correspondence in the disposition of RyR and Cav1 in cardiac muscle is not known.

2. Functional Implications

Presumably the groups of SR/surface contacts and/or individual RyR clusters that are grouped within a super cluster in the definition by Baddeley et al. (2009) correspond to the classically described CRUs. The term couplon, first introduced in skeletal muscle (Stern, Pizarro, & Rios, 1997), can be used to indicate a single SR/surface membrane contact containing a single coherent cluster of RyRs within a cardiac CRU. The relationship between couplon (RyR cluster) and CRU (RyR super cluster) is quite clear in Fig. 1, reproduced from Baddeley et al. (2009), in which each diffraction-limited spot corresponding to a peripheral coupling (shown as a fuzzy outline in "A") is composed of several couplons (the brightest areas, better resolved in "B").

The main question is whether the partial fragmentation of the classical CRU into smaller couplons requires a major revision of spark generation and/or overall Ca^{2+} release models in cardiac muscle. Baddeley et al. (2009) discuss this point arguing that $\sim$30% of couplons (RyR clusters) have edge-to-edge distances $\leq$50 nm and are likely to trigger Ca^{2+} release from each other. The RyRs in closely spaced couplons that are part of the same CRU may behave, in respect to activation by Ca^{2+}, in the same manner as RyRs that are in close contact to each other within each couplon, so that the fragmentation of a CRU into smaller, closely spaced couplons may not have a direct functional implication. If, on the other hand, an additional functional interaction between cardiac RyRs requires actual physical contact between the proteins, then the fragmentation of a CRU into smaller couplons may have a significant effect.

IV. CRUs IN SKELETAL AND INVERTEBRATE BODY MUSCLES

A. Complete RyR Clusters

The new findings in cardiac muscle show that the classically described CRUs are composed of smaller, closely spaced contacts containing independently generated, often incomplete, RyR clusters or couplons with different orientations. This raises the question whether a similar revision of CRU structure is necessary for other muscles. In the classical descriptions of couplons in skeletal and invertebrate body muscles each couplon forms a single, internally coherent, contact with surface membranes, and contains a single RyR cluster. A revision of this basic description is not necessary: a single, continuous 2D crystal, occupies the whole junctional gap of each couplon, without any fragmentations and/or changes in the orientation of the crystalline axis within the couplon (Fig. 2). In this respect skeletal and body muscles differ from vertebrate cardiac muscle.

B. Is Clustering of RyRs in Skeletal and Body Muscles a Stochastic Event?

If the formation of ordered RyR arrays within skeletal muscle couplon is purely a stochastic event, as proposed for cardiac muscle (Baddeley et al., 2009), orientations of RyR clusters in individual couplons should be random.

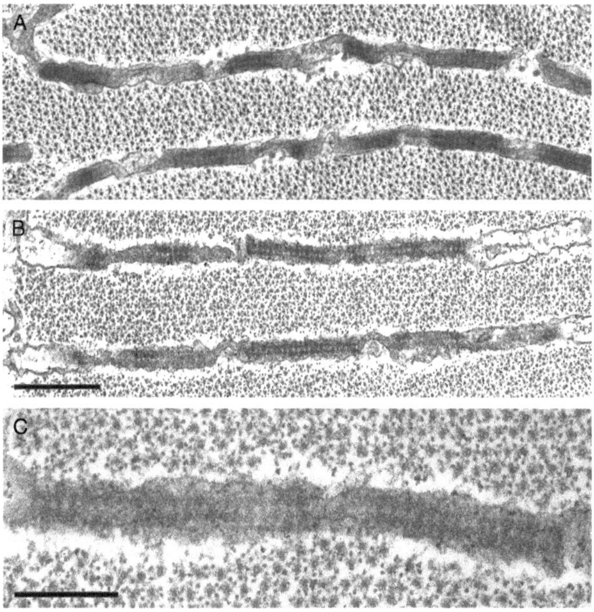

FIGURE 2 Cross sections of muscle fibers from the toadfish swimbladder. Short segments containing double rows of feet represent individual couplons. Couplons are shorter in male (A) than in female (B) muscles. Interruptions in the long rows of feet occur only where the couplon ends or leaves the section (A–C). Scale bars: (A, B) 480 nm; (C) 240 nm.

In most skeletal muscle, on the contrary, feet form two (occasionally three) rows, which have the same, constant orientation relative to the long axis of T tubules in couplons throughout the whole fiber (Fig. 2), indicating that the orientation of RyR arrays is restrained. The simplest explanation is that the constant orientation is imposed by the long and narrow dimensions of the junctional SR membrane, which must fit between the myofibrils and thus restrict the direction of growth of 2D RyR crystals to one direction.

If the above consideration is correct, the orientation of feet clusters should be variable where no such restraints are imposed. This is the case for SR/T tubule junctions that are longitudinal and for peripheral couplings, both of which are under no geometric restraint. Figure 3 illustrates longitudinally oriented triads/dyads in adult EDL fibers from a mouse null for the CASQ1 gene (A–C); in normal tonic fibers from the frog (D–F); and in the muscle of an arthropod (a spider, G–I)). In all three types of CRUs, each couplons is filled by a single, coherent array of feet, but the orientations of the arrays in different couplons are variable. A second example is shown in Fig. 4 where arrays of CaV1.1 tetrads are imaged in freeze-fracture replicas of the

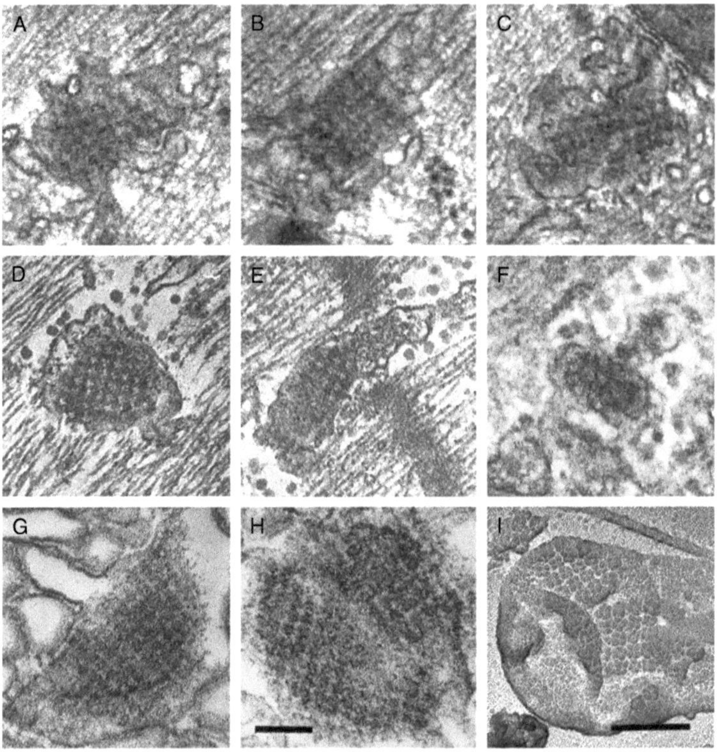

FIGURE 3 Images of feet arrays in longitudinally oriented couplons from the EDL of a CASQ1 null mouse (A–C courtesy of Paolini et al., 2007); from slow tonic fibers of the frog (D–F); and from muscles of a scorpion (G–I courtesy of Loesser et al., 1992). The orientation of feet arrays is variable relative to the fiber long axis in different couplons. Scale bars: (A–H) 150 nm; (I) 150 nm.

plasmalemma from a cultured cell line of skeletal muscle origin (A) and from zebra fish larvae (B). The disposition of feet arrays is indirectly indicated by the arrays of CaV1.1 channels in the surface membrane. Again note variations in the orientation of adjacent clusters fitting a classical stochastic hypothesis of their independent generation.

C. Size Variations of Couplons Are Limited: A Modified Stochastic Model

If the assembly of RyR clusters is not affected by restraining factors, the size of couplons should vary within a continuum that includes very small and fairly large elements. Instead, the size range is limited, and specified for each

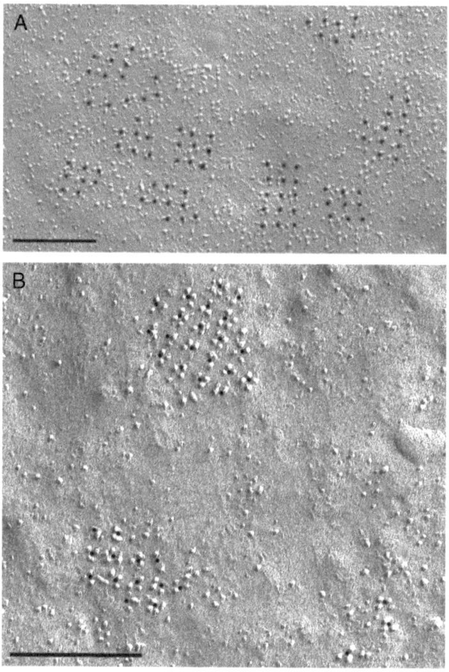

FIGURE 4 Freeze fracture of the plasmalemma in BC3H1 cells (A; Protasi et al., 1997) and a zebra fish larva (B; Schredelseker et al., 2005) showing assemblies of CaV1.1. The position of tetrads (marked by central dots) reflects the position of underlying assemblies of RyRs. Note variable orientations and sizes of the clusters. Scale bars: 240 nm.

fiber type. The segments of T tubule covered by each couplons are overall significantly shorter in slow twitch fibers than in fast twitch fibers (Franzini-Armstrong, Ferguson, & Champ, 1988). The very fast swimbladder muscles of male and female toadfish illustrate this point well. The overall content of RyR is basically the same for the two sexes, but muscles from males have more frequent and significantly smaller couplons than from females (Fig. 2). An easy way of obtaining this difference is to restrict the sites at which feet aggregate to those that contain another protein, for example, the docking protein junctophilin, and to fix the ratio of RyRs/docking protein. At equal concentrations of RyRs, a higher concentration of junctophilin would result in smaller but more frequent couplons (male vs. female toadfish swimbladder). At lower concentration of junctophilin and RyRs, the couplons would be both less frequent and small (slow twitch fibers), while at higher, but not limiting, concentrations of junctophilins and higher concentrations of RyRs, the junctions would be more frequent and also larger (fast twitch fibers),

V. FACTORS AFFECTING CRU ASSEMBLY IN SKELETAL AND CARDIAC MUSCLES

A. *Several Steps in CRU Formation Affect the Assembly and Final Structure of CRUs*

The difference between single coherent arrays of RyRs in CRUs of skeletal (and body) muscles and the incomplete and fragmented arrays of RyRs in cardiac muscle may simply be due different stoichiometry of components. However, assembly of CRUs is a complex phenomenon and several factors may affect the outcome.

1. Docking

The formation of CRUs requires the segregation of the jSR proteins constituting the couplon's supramolecular complex (RyR, CASQ, triadin, junctin, among others) and RyR associated proteins (homer, calmodulin, FKBP12) to the jSR cisternae, and the association of the jSR with the surface membranes. On the surface membranes' side of the junctions, CaV1 channels also cluster in correspondence of CRUs.

A close association between SR and surface membranes, appropriately called docking is the first event in CRU's formation. In developing skeletal and cardiac muscle small SR cisternae are docked at the plasmalemma before RyR and CaV channels are detectable (Franzini-Armstrong, Pincon-Raymond, & Rieger, 1991). Evidence for a direct role of junctophilins, first identified in skeletal muscle triads, in the docking is quite clear (Takeshima, Komazaki, Nishi, Iino, & Kangawa, 2000). Lack of junctophilin type 1 results in a significant decrease in the number of triads in the developing jaw muscle, which is normally rich in this protein isoform (Ito et al., 2001). Whether complete silencing of both junctophilin types present in skeletal muscle completely inhibits triad formation is not yet fully established (Hirata et al., 2006).

Docking precedes the other steps and thus, it is not surprising it can occur in the absence of other major components of CRUs. In skeletal muscle, docking occurs in paralytic "dyspedic" mouse muscles lacking either only RyR1 (Takekura, Nishi, Noda, Takeshima, & Franzini-Armstrong, 1995) or both RyR1 and RyR3 (Ikemoto et al., 1997) and also in paralytic muscles from avian "crooked neck dwarf" mutation lacking aRyR (Ivanenko, McKemy, Kenyon, Airey, & Sutko, 1995). In frog ventricle, expressing basically no RyR2, SR docks to the plasmalemma to form peripheral couplings containing clusters of CaV1.2 (Tijskens et al., 2003). In skeletal muscle, docking occurs in the absence of CaV1.1 (Flucher et al., 1993), and of both RyR1 and Cavl.1 (Felder, Protasi, Hirsch, Franzini-Armstrong, & Allen, 2002).

Other proteins of the junctional complex are also not intrinsically necessary for docking. In both skeletal and cardiac muscle, SR docks to surface membranes in the absence of CASQ (Knollmann et al., 2006), triadin (Shen et al., 2007), and junctin (C. Franzini-Armstrong et al., unpublished observations).

2. Is Docking Essential?

Docking to surface membrane seems to be an essential step for the assembly of CRUs in skeletal muscle. In the absence of junctophilin expression and appropriate targeting to ER and plasmalemma of RyR1 and CaV1 is not sufficient to form CRUs (Takekura et al., 1995).

Adult cardiac muscle, on the other hand, contains assemblies of RyR clusters associated with CASQ even in the absence of docking. Corbular SR and extended junctional SR (ejSR) are present in mammalian atrium, in all parts of the avian hearts, and occasionally in mammalian ventricle. These are functional CRUs that are not associated either with the plasmalemma or with T tubules (Bossen et al., 1978). However, docked CRUs are of great importance in cardiac muscle function, because they are the only sites where depolarization of the surface membrane can initiate localized Ca^{2+} releases. The absence of junctophilin 2, the cardiac isoform, results in arrested cardiac development and embryonic lethality (Takeshima et al., 2000).

3. Trapping of Proteins Within CRU Proteins

RyR of SR couplons and CaV of surface membranes interact with each other in excitation–contraction coupling. In skeletal muscle, they control each other's gating in a bidirectional fashion by direct molecular interaction (Grabner, Dirksen, Suda, & Beam, 1999), while in cardiac muscle they are though to affect each other's function indirectly. In both muscle types, RyR and CaV channels must be in close proximity to each other in order to interact, and yet they do not require each other's presence in order to be retained within CRUs. So some other protein, perhaps junctophilin itself, must trap the two channels within a CRU, once they reach the site. Trapping of the two sets of molecules within the limited space of a CRU increases their effective local concentration, so that despite their low affinity there is a high probability that at any instant an interaction may occur between some of the components of the two apposed clusters of channels. In skeletal muscle, a tetrad constituted of four CaV1.1 is linked to the four subunits of a RyR. It is not known whether all four components of a tetrad need to be linked to a single RyR in order to activate it, but it is noticeable that even when a patch of Cav1.1 is poorly populated as in differentiating cells at least one complete tetrads is usually present in each patch (Protasi et al., 1997).

B. Coclustering of CRU Proteins

Once within a CRU, the component proteins cocluster into functionally and structurally related assemblies. In skeletal muscle, RyR1 of CRUs self-assemble into orderly orthogonal arrays even in the absence of CaV1.1. The latter, on the other hand, even if clustered at a CRU, strictly require the presence of and anchorage to RyR1 in order to assemble into an orthogonal lattice of tetrads (Protasi, Franzini-Armstrong, & Allen, 1998). Conformational coupling (Paolini, Fessenden, Pessah, & Franzini-Armstrong, 2004) and functional bidirectional interaction between the CaV1.1 and RyR1 (Grabner et al., 1999) require a link between four CaV1.1 and the four subunits of the RyRs (Protasi et al., 2002). The tetrad is a structural expression of this link. Cardiac RyR2 and the second skeletal isoform, RyR3, are not appropriate substitutes for this interaction. Vice versa, cardiac CaV1.2 cannot substitute for skeletal CaV1.1 in forming tetrads with RyR1. In cardiac muscle, RyR2 and CaV1.2 cluster at CRUs, but while RyR2, similarly to RyR1, form orderly clusters (Baddeley et al., 2009). CaV1.2 retain a random disposition within CRU-associated clusters. Despite the apparent lack of direct coupling, the two cardiac channels aggregate within CRUs in a concerted fashion (Fig. 1; Protasi et al., 1996).

An extensive search has been made for the domains of RyR1 and of the a1 CaV1.1 that are necessary for their specific structural association (the formation of tetrads) as well as for the bidirectional talk. Tetrad association is necessary but not sufficient for a skeletal muscle type interaction, but the domains necessary for the two channels to link hands are closely related but not identical to those that allows cross talk between them. Surprisingly, the small bCaV1 subunit is also essential for the RyR1–Cav1.1 association (Schredelseker et al., 2005). In the absence of a1, the channel region of the complex, a2 fails to target to CRUs (Flucher, Morton, Froehner, & Daniels, 1990). The reverse is not true: in the absence of the a2 subunit, CaV1.1 are present in CRUs and they link to RyR1 forming tetrads (Gach et al., 2008).

CASQ interacts with two other components of couplons, junctin and triadin, and through them with RyRs. However, neither the presence of RyRs, nor that of Cav1.1, is necessary for the association of the triadin–CASQ complex with couplons (Flucher et al., 1993). The converse is also true: CASQ, triadin, and junctin are not needed to the assembly of RyR and CaV1 clusters within CRUs in both skeletal and cardiac muscle.

Triadin and junctin strongly affect the disposition of CASQ within couplons. CASQ exists mostly as a long linear polymers randomly folded within in the jSR lumen (Fig. 5). It is likely that monomeric CASQ, being very small is simply not detected in the EM. In close proximity of the jSR membrane CASQ is condensed into denser spots due to its link with triadin and junctin.

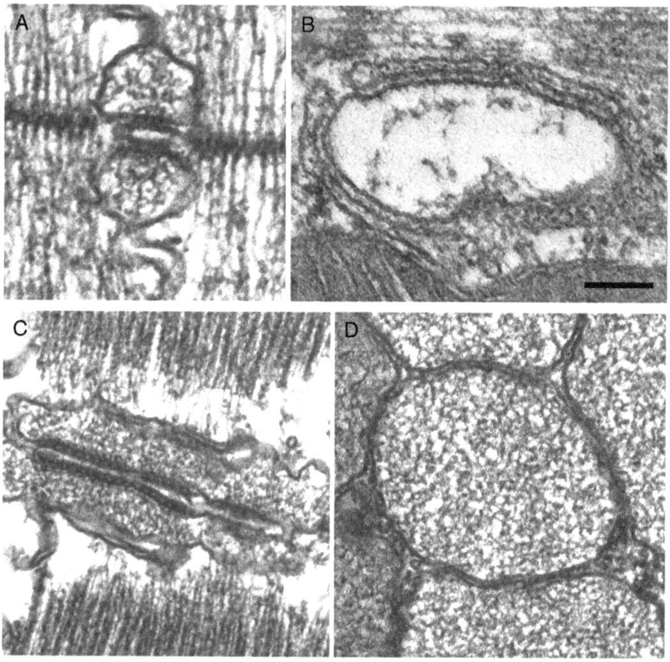

FIGURE 5 The CASQ content of jSR in CRUs of frog skeletal (A, C) and rat cardiac (B, rat) jSR appear different. However, when CASQ is overexpressed of cardiac SR (D), the structure is remarkably similar to skeletal (Tijskens et al., 2003). Scale bar: 480 nm.

The effect is much stronger in cardiac muscle, where CASQ/triadin/junctin ratio is lower and CASQ appears as a string of beads that fully occupy the very narrow couplon lumen (Fig. 5). In skeletal muscle and in cardiac muscle overexpressing CASQ, the clustering effect of triadin and junctin is diluted by the large amount of CASQ.

The absence of CASQ has an apparently opposite effect on the structure of cardiac and skeletal jSR. While in the former the CASQ-null cisternae are slightly enlarged (Knollmann et al., 2006) in the latter they are greatly diminished in size (Paolini et al., 2007). The likely explanation is that in cardiac muscle, the absence of CASQ releases the tight of CSAQ/Tr/Jct interaction, allowing the SR membrane to relax into a slightly wider cisterna. Indeed, in triadin null myocardium where CASQ expression is reduced, the jSR cisternae also become wider (Chopra et al., 2009). In skeletal muscle and in CASQ overexpressing myocardium, the abundant luminal CASQ gel imposes a wide shape to the jSR cisternae. When CASQ is reduced, either directly in CASQ1 null (Paolini et al., 2007) or indirectly in triadin null mice (Shen et al., 2007) the SR cisternae collapse into a smaller size.

Mitsugumin 29 is a protein of the triad, which may be associated with the other components of the jSR. However, contribution of mg29 to CRU architecture is unclear, since there is no apparent change in the close SR–T tubule association in the absence of this protein (Pan et al., 2002).

C. Corbular SR: Structural and Functional Considerations

"Corbular" and "ejSR" are equivalent, specialized CRUs of cardiac SR that contain RyR and associated proteins, but are not "docked." These domains are frequent in myocytes that lack T tubules, but have a relatively large diameter such as most atrial myocytes in mammals, and all chambers of the avian heart (Jewett, Sommer, & Johnson, 1971). Myocardium, including that of birds, expresses a single RyR isoform (Jorgensen, Shen, Arnold, McPherson, & Campbell, 1993) located at all types of CRUs (Junker, Sommer, Sar, & Meissner, 1994), so ejSR is equivalent to couplons in other CRUs. The question is whether ejSR participates in e–c coupling under normal conditions. In the case of frog atrial cells where the fiber diameter is smaller than 10 μm (Page & Niedergerke, 1972) and no CRUs of any sort are present inside the cell, it clear that Ca^{2+} liberated at the cell surface and/or entering from the extracellular space must be sufficient to activate contraction. Mammalian atrium and avian myocardium, which have larger cells that contain ejSR and frequent peripheral coupling with docked jSR. Does release from internal ejSR also come into play? Evidence for triggered internal Ca^{2+} release in mammalian atrial myocytes (Woo, Cleemann, & Morad, 2005) gives a positive answer. Presumably release at centrally located ejSR is triggered via diffusion of Ca^{2+} initially released at peripheral couplings, and then amplified by sequential release from ejSR sites. In this respect, it is noticeable that the average edge-to-edge distance between ejSR is very small: 135 $\pm$ 15 nm in chicken and $<$20 nm in the fast beating finch myocardium (Bossen et al., 1978) and comparable to the distances between couplons within dyads/peripheral couplings (Hayashi et al., 2009). Peripheral couplings, on the other hand are separated by larger distances ($\sim$470 nm in chicken and finch; Franzini-Armstrong, Protasi, & Ramesh, 1999).

An unanswered question regarding corbular and ejSR is the clustering mechanism for junctional proteins in these SR domains that are not "docked."

VI. ISOFORM-SPECIFIC FEATURES OF RYR DISTRIBUTION

The three known RyR isoforms have different distributions (Sutko & Airey, 1996) and contribute in different ways to the assembly of CRUs. Skeletal muscle fibers contain RyR1 and RyR3 and equivalent isoforms in variable ratios. RyR1 is essential to e–c coupling: spontaneous or induced RyR1 null mutations, and mutations resulting in nonfunctional RyR1, are birth lethal because the muscles are paralyzed and the fibers develop poorly and/or degenerate. RyR1 are always strictly located within the junctional gap of CRUs and assemble into semicrystalline arrays, although parameters of the array are slightly different in primitive fish versus more evolved fish and higher vertebrates (Di Biase & Franzini-Armstrong, 2005).

RyR3 are located at CRUs, but they are expressed at low levels in mammalian muscles (Flucher, Conti, Takeshima, & Sorrentino, 1999), and are not essential to e–c coupling and/or muscle differentiation (Takeshima et al., 1996). In bird, amphibia, and some fish muscles RyR1/RyR3 ratio is approximately 1. In this case RyR3 remains segregated into a parajunctional position within CRUs, so that they face the myofibrils and not the T tubules (Felder & Franzini-Armstrong, 2002). However, in the absence of RyR1, RyR3 can take their place within the junctional gap. This is rarely seen in myotubes of RyR1 (or αRyR) null mutants, and more clearly when RyR3 are expressed in the 1B5 dyspedic cell line (Protasi et al., 2000). In these cells, RyR3 are inserted within the CRUs and colocalize with triadin and CaV1.1, although they fail to form a structural and functional link with CaV1.1.

Differently from RyR1, RyR2 of vertebrate myocardium, and the single RyR gene product of invertebrate muscles (Takeshima et al., 1994) are found in clusters that are either extrajunctional and/or parajunctional. In cardiac muscle, ejSR and corbular SR are examples of extra junctional RyR2. Evidence for additional extrajunctional SR, either in proximity of caveolae (Scriven, Klimek, Asghari, Bellve, & Moore, 2005) or positioned in the SR opposite the middle of the sarcomere (Lukyanenko, Ziman, Lukyanenko, Salnikov, & Lederer, 2007) has not been directly confirmed by EM. Several possibilities exist. One is that some of the RyR detected by gold labeling at the A band level is simply RyR that belongs to dyads associated with a longitudinal segment of the T tubule (Asghari et al., 2009). A second possibility is that additional RyRs are individual molecules not clustered and thus not visible in EM images. These may be RyRs in the process of diffusing along the membrane on their way from RER to dyads.

In muscles of invertebrate, the array of feet often extends farther than the area of membrane directly apposed to T tubules, so that a number of feet are parajunctional (see example in Fig. 3). Thus, differently from RyR1, the single RyR isoform of arthropods (Takeshima et al., 1994) can take both a junctional and a parajunctional position.

VII. ARCHITECTURE OF SR AND T TUBULE MEMBRANES IS MUSCLE- AND FIBER-TYPE SPECIFIC

Both SR and T tubules have shapes that repeat at every sarcomere with some degree of regularity, but vary greatly from one type of muscle (and organism) to another. As noted by Porter and Palade (1957), "the sarcoplasm of the striated muscle fiber contains an equivalent of the endoplasmic reticulum which bears some unique and interesting structural relations to the myofibrils. This leads to the reasonable deduction that in its function the system may represent an important factor in the contractile phenomena these cells display." More than 50 years and thousands of publications later, we know very little about what actually determines the final SR/T tubule distribution. The question has more than academic interest, because the two membrane systems are subject to considerable alterations in experimental and/or pathological conditions.

A. Varieties of T Tubule Networks

1. Embryology and Phylogeny

When all types of muscles at all ages and under different conditions are considered, the T tubule networks are considerably more variable and intricate than originally described (Andersson-Cedregren, 1959). The almost perfect transverse orientation of T tubules' networks in adult vertebrate muscles is a relatively novel development in terms of evolution times. The T tubules in body muscles of arthropods have all possible dispositions, ranging from mostly transverse to mostly longitudinal, with numerous cross connections (Veratti, 1961).

During skeletal muscle differentiation T tubules initially invade the space between the myofibrils, possibly the path of least resistance, and are mostly longitudinal. Transition to the precise, sarcomere-related transverse orientation of the adult muscle can be quite rapid (a matter of hours in zebrafish tail muscles) or very slow (1 month in mouse), but in both cases it requires the previous assembly of myofibrils and the formation of CRUs.

Even in the adult, the T tubules are not exclusively transverse. Longitudinal extensions (Veratti, 1961) connect adjacent transverse networks, but do not associate with the jSR. Figure shows a cascade of longitudinal tubules that marks the sites where the striations of adjacent myofibrils are mismatched, at a Vernier dislocations (Peachey & Eisenberg, 1978). These longitudinal extensions are presumably remnants of the original orientation of T tubules during muscle differentiation (Flucher, Takekura, & Franzini-Armstrong, 1993) (Fig. 6).

Differentiation of T tubules in ventricular myocardium also involves an initial, random distribution of membrane invaginations and a subsequent change into a predominantly transverse network, but with extensive longitudinal extensions (Forbes, Hawkey, & Sperelakis, 1984). However, in the case of adult myocardium longitudinal extensions play a more significant role than in skeletal muscle. Once seen in its entirety the cardiac network is

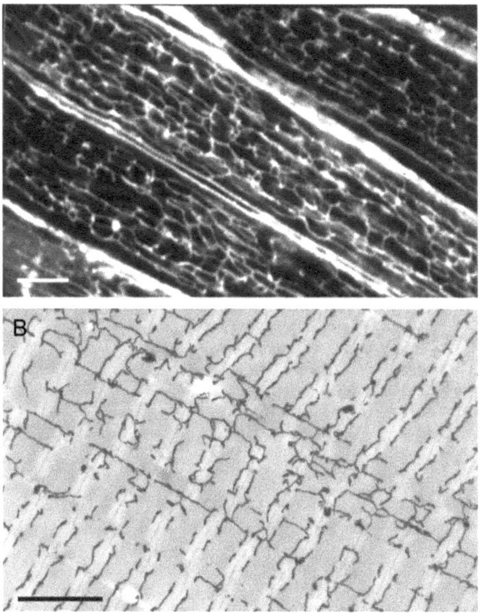

FIGURE 6 (A) T tubules (labeled with a lipid soluble fluorescent dye) are mostly longitudinal in this fiber from a chick muscle at ~E18. (B) A ripple of longitudinal T tubules remains in adult muscle fibers at Vernier displacements of the cross striation. Scale bars: 5 μm.

remarkably disordered and variable (Soeller & Cannell, 1999) and the transverse network (the "Z rete") comprises a net that is less complete and has larger spacings compared to skeletal muscle (Leeson, 1978). Another, significant detail is that CRUs in cardiac muscle are formed by association of jSR with both longitudinally and axially oriented tubules, resulting in location of some CRUs away from the Z line (Asghari et al., 2009).

2. T Tubules and CRUs Revert to More Primitive Forms Under Pathological Conditions

A disordered arrangement of T tubules, with loss of the precise transverse orientation of the network is a common result of a variety of muscle pathologies. Cardiac muscle function is extremely dependent on well-regulated Ca^{2+} movements, and thus it is highly sensitive to any alterations in the architecture of its SR/T tubule membrane system. Minor as well as profound alterations in Ca^{2+} regulation and T tubule architecture have been detected in cardiac hypertrophy and failure (Bers, 2006), reviewed by Brette and Orchard (2007). In view of these changes in the relationship between T tubules, CRUs and the myofibrils, several questions need to be addressed. One is whether changes in T tubule architecture are the cause of the result of the pathology; the second is whether T tubule disorder signals a disconnection of T tubules from CRUs or of CRUs from the myofibrils. Since CRU anchor T tubules to the myofibrils, both would result in T network disarrangement. Further, is the disorder due to a regression from a fully differentiated architecture to a more primitive one, or is the expression of an ongoing process of degeneration and regeneration, where the latter does not achieve a final result? A partial answer to some of these questions comes from the observation that reduced e–c coupling effectiveness in experimental cardiac hypertrophy may be interpreted as a reduction in functional coupling of CaV1.2 and RyR2 channels (Gomez et al., 1997) and traced to an apparent disconnection between RyR and CaV1.2 clusters in the failing heart (Song et al., 2006). A similar, although more limited, separation of RyR and CaV1.2 clusters is found in a *pan* triadin null mouse model, where expression levels of all junctional proteins are decreased, with an impairment of e–c coupling and susceptibility to ventricular arrhythmia (Chopra et al., 2009).

T tubules in fast fibers of skeletal muscles respond in a very sensitive manner to a number of interventions that also influence the relationship of T tubules to SR and the frequency of CRUs. Disordered T networks with frequent longitudinal extensions; evidence of new T tubule formation and multiple SR–T tubule contacts (pentads, heptads) are found in denervated (Takekura, Tamaki, Nishizawa, & Kasuga, 2003); eccentrically exercised (Takekura, Fujinami, Nishizawa, Ogasawara, & Kasuga, 2001) and, on a slower time scale, in immobilized muscles (Takekura, Kasuga, Kitada, & Yoshioka, 1996)

B. The Position and Orientation of CRUs

1. Position of CRUs in the Fiber is Determined by Links to Myofibrils

CRUs in the fiber interior are located along T tubules (except for corbular SR in cardiac muscle) giving the impression that location of T tubules dictates that of CRUs. On the contrary, it is the position of CRUs relative to the myofibrils that determines the location of the T tubules' networks. The specific position of CRUs relative to the bands of the myofibrils is achieved before T tubules acquire their transverse orientation during skeletal muscle differentiation (Flucher et al., 1993). Additionally, CRUs' specific proteins coassemble into CRUs before the T network is fully organized in both skeletal and cardiac muscle and completion of longitudinal SR protein organization is important in the positioning of triads (Cusimano, Pampinella, Giacomello, & Sorrentino, 2009). Finally, corbular and ejSR of myocardium are located at the Z lines, even though they are not associated with T tubules. A link between ankyrin, 1.5, an intrinsic protein of the SR membrane, and obscurin, a giant protein of the myofibrils is thought to provide the necessary anchorage (Kontrogianni-Konstantopoulos et al., 2006). Transitions in the location of these proteins during differentiation may reflect changes in the arrangement of membranes (Giacomello & Sorrentino, 2009) but the story is complex and not yet fully clarified.

The positioning of peripheral couplings raises an additional question. These mostly have a very specific location on the plasmalemma, usually along one to three rows that encircle the fibers at the level of Z line-I bands of the underlying myofibrils (Fig. 1), see Di Biase and Franzini-Armstrong (2005) for amphioxus and fish muscle; Tijskens et al. (2003) for cardiac myocytes. It is not clear whether this position is also determined by a link to the myofibrils.

2. Orientation of CRUs Depends on Fiber Type, Age Origin, and Genotype

Orientation of CRUs depends on the muscle and fiber type of origin. In twitch fibers of adult vertebrate muscles the majority of triads have an elongated shape and are oriented with their long axis perpendicular to the fiber's axis and thus in a transverse orientation (Fig. 2). In slow tonic fibers of frog, however, the predominant orientation is longitudinal (Fig. 3). In vertebrate cardiac muscle arrays of RyRs in the dyads have variable dimensions and mostly combine longitudinal and transversely oriented regions as they wrap around the T tubule. In muscles of most arthropods, dyads and triads have a predominantly longitudinal orientation (Fig. 3).

During differentiation of twitch fibers in skeletal muscle CRUs (a mixture of dyads and triads) initially have a longitudinal orientation and slowly rotate into a transverse position during late embryonic and early postnatal periods (Flucher, Takekura, et al., 1993). In the fast pectoralis muscle of chicken triads maintain a longitudinal orientation for long periods after hatching and then change orientation while also transitioning between A–I band junction to the final Z line level (Takekura, Shuman, & Franzini-Armstrong, 1993).

The transverse versus longitudinal orientation of the CRU axis is not likely to have an effect of e–c coupling, but embryology and phylogeny concur in showing that a longitudinal orientation of the CRU axis, like that of T tubules, is more primitive. It is interesting that reversal to a longitudinal orientation is the common result of various genetic manipulations and/or alterations of normal muscle function, but in mammalian muscles it is mostly restricted to some types of fast twitch fibers. In the absence of CASQ (Paolini et al., 2007) and triadin (Shen et al., 2007) a relatively high percentage of triads have a longitudinal position. In these two engineered muscles, it is possible that the longitudinal CRU is simply the remnant of an incomplete differentiation.

A clear, completely reversible shift from transverse to longitudinal CRU orientation of the triads occurs within a very brief period of time following denervation and subsequent reinnervation (Takekura et al., 1996).

Acknowledgments

Supported by NIH RO1 HL-48093.

References

Andersson-Cedregren, E. (1959). Ultratsructure of motor end plate and sarcoplasmic creticulum of mouse skeletal muscle fibers. *Journal of Ultrastructure Research, 1*(Suppl.), 5–100.

Asghari, P., Schulson, M., Scriven, D. R., Martens, G., & Moore, E. D. (2009). Axial tubules of rat ventricular myocytes form multiple junctions with the sarcoplasmic reticulum. *Biophysical Journal, 96*(11), 4651–4660.

Baddeley, D., Jayasinghe, I. D., Lam, L., Rossberger, S., Cannell, M. B., & Soeller, C. (2009). Optical single-channel resolution imaging of the ryanodine receptor distribution in rat cardiac myocytes. *Proceedings of the National Academy of Sciences of the United States of America, 106*(52), 22275–22280.

Bers, D. M. (2006). Altered cardiac myocyte Ca regulation in heart failure. *Physiology (Bethesda), 21*, 380–387.

Block, B. A., Imagawa, T., Campbell, K. P., & Franzini-Armstrong, C. (1988). Structural evidence for direct interaction between the molecular components of the transverse tubule/sarcoplasmic reticulum junction in skeletal muscle. *The Journal of Cell Biology, 107*, 2587–2600.

Bossen, E. H., Sommer, J. R., & Waugh, R. A. (1978). Comparative stereology of the mouse and finch left ventricle. *Tissue Cell, 10*, 773–784.

Brette, F., & Orchard, C. (2007). Resurgence of cardiac t-tubule research. *Physiology (Bethesda), 22*, 167–173.

Chen-Izu, Y., McCulle, S. L., Ward, C. W., Soeller, C., Allen, B. M., Rabang, C., et al. (2006). Three-dimensional distribution of ryanodine receptor clusters in cardiac myocytes. *Biophysical Journal, 91*, 1–13.

Chopra, N., Yang, T., Asghari, P., Moore, E. D., Huke, S., Akin, B., et al. (2009). Ablation of triadin causes loss of cardiac Ca^{2+} release units, impaired excitation-contraction coupling, and cardiac arrhythmias. *Proceedings of the National Academy of Sciences of the United States of America, 106*, 7636–7641.

Cusimano, V., Pampinella, F., Giacomello, E., & Sorrentino, V. (2009). Assembly and dynamics of proteins of the longitudinal and junctional sarcoplasmic reticulum in skeletal muscle cells. *Proceedings of the National Academy of Sciences of the United States of America, 106*, 4695–4700.

Di Biase, V., & Franzini-Armstrong, C. (2005). Evolution of skeletal type e-c coupling: A novel means of controlling calcium delivery. *The Journal of Cell Biology, 171*, 695–704.

Felder, E., & Franzini-Armstrong, C. (2002). Type 3 ryanodine receptors of skeletal muscle are segregated in a parajunctional position. *Proceedings of the National Academy of Sciences of the United States of America, 99*, 1695–1700.

Felder, E., Protasi, F., Hirsch, R., Franzini-Armstrong, C., & Allen, P. D. (2002). Morphology and molecular composition of sarcoplasmic reticulum surface junctions in the absence of DHPR and RyR in mouse skeletal muscle. *Biophysical Journal, 82*, 3144–3149.

Flucher, B. E., Andrews, S. B., Fleischer, S., Marks, A. R., Caswell, A., & Powell, J. A. (1993). Triad formation: Organization and function of the sarcoplasmic reticulum calcium release channel and triadin in normal and dysgenic muscle in vitro. *The Journal of Cell Biology, 123*, 1161–1174.

Flucher, B. E., Conti, A., Takeshima, H., & Sorrentino, V. (1999). Type 3 and type 1 ryanodine receptors are localized in triads of the same mammalian skeletal muscle fibers. *The Journal of Cell Biology, 146*, 621–630.

Flucher, B. E., Morton, M. E., Froehner, S. C., & Daniels, M. P. (1990). Localization of the $\alpha 1$ and $\alpha 2$ subunits of the dihydropyridine receptor and ankyrin in skeletal muscle triads. *Neuron, 5*, 339–351.

Flucher, B. E., Takekura, H., & Franzini-Armstrong, C. (1993). Development of the excitation-contraction coupling apparatus in skeletal muscle: Association of sarcoplasmic reticulum and transverse tubules with myofibrils. *Developmental Biology, 160*, 135–147.

Forbes, M. S., Hawkey, L. A., & Sperelakis, N. (1984). The transverse-axial tubular system (TATS) of mouse myocardium: Its morphology in the developing and adult animal. *The American Journal of Anatomy, 170*, 143–162.

Franzini-Armstrong, C., Ferguson, D. G., & Champ, C. (1988). Discrimination between fast- and slow-twitch fibres of guinea pig skeletal muscle using the relative surface density of junctional transverse tubule membrane. *Journal of Muscle Research and Cell Motility, 9*, 403–414.

Franzini-Armstrong, C., Pincon-Raymond, M., & Rieger, F. (1991). Muscle fibers from dysgenic mouse in vivo lack a surface component of peripheral couplings. *Developmental Biology, 146*, 364–376.

Franzini-Armstrong, C., Protasi, F., & Ramesh, V. (1999). Shape, size, and distribution of Ca(2+) release units and couplons in skeletal and cardiac muscles. *Biophysical Journal, 77*, 1528–1539.

Gach, M. P., Cherednichenko, G., Haarmann, C., Lopez, J. R., Beam, K. G., Pessah, I. N., et al. (2008). $\alpha 2\delta 1$ dihydropyridine receptor subunit is a critical element for excitation-coupled calcium entry but not for formation of tetrads in skeletal myotubes. *Biophysical Journal, 94*, 3023–3034.

Giacomello, E., & Sorrentino, V. (2009). Localization of ank1.5 in the sarcoplasmic reticulum precedes that of SERCA and RyR: Relationship with the organization of obscurin in developing sarcomeres. *Histochemistry and Cell Biology, 131*, 371–382.

Gomez, A. M., Valdivia, H. H., Cheng, H., Lederer, M. R., Santana, L. F., Cannell, M. B., et al. (1997). Defective excitation-contraction coupling in experimental cardiac hypertrophy and heart failure. *Science, 276*, 800–806.

Grabner, M., Dirksen, R. T., Suda, N., & Beam, K. G. (1999). The II-III loop of the skeletal muscle dihydropyridine receptor is responsible for the Bi-directional coupling with the ryanodine receptor. *The Journal of Biological Chemistry, 274*, 21913–21919.

Hayashi, T., Martone, M. E., Yu, Z., Thor, A., Doi, M., Holst, M. J., et al. (2009). Three-dimensional electron microscopy reveals new details of membrane systems for Ca^{2+} signaling in the heart. *Journal of Cell Science, 122*, 1005–1013.

Hirata, Y., Brotto, M., Weisleder, N., Chu, Y., Lin, P., Zhao, X., et al. (2006). Uncoupling store-operated Ca^{2+} entry and altered Ca^{2+} release from sarcoplasmic reticulum through silencing of junctophilin genes. *Biophysical Journal, 90*, 4418–4427.

Ikemoto, T., Komazaki, S., Takeshima, H., Nishi, M., Noda, T., Iino, M., et al. (1997). Functional and morphological features of skeletal muscle from mutant mice lacking both type 1 and type 3 ryanodine receptors. *The Journal of Physiology, 501*(Pt 2), 305–312.

Ito, K., Komazaki, S., Sasamoto, K., Yoshida, M., Nishi, M., Kitamura, K., et al. (2001). Deficiency of triad junction and contraction in mutant skeletal muscle lacking junctophilin type 1. *The Journal of Cell Biology, 154*, 1059–1067.

Ivanenko, A., McKemy, D. D., Kenyon, J. L., Airey, J. A., & Sutko, J. L. (1995). Embryonic chicken skeletal muscle cells fail to develop normal excitation-contraction coupling in the absence of the α ryanodine receptor. Implications for a two-ryanodine receptor system. *The Journal of Biological Chemistry, 270*, 4220–4223.

Jewett, P. H., Sommer, J. R., & Johnson, E. A. (1971). Cardiac muscle. Its ultrastructure in the finch and hummingbird with special reference to the sarcoplasmic reticulum. *The Journal of Cell Biology, 49*, 50–65.

Jorgensen, A. O., Shen, A. C., Arnold, W., McPherson, P. S., & Campbell, K. P. (1993). The Ca^{2+}-release channel/ryanodine receptor is localized in junctional and corbular sarcoplasmic reticulum in cardiac muscle. *The Journal of Cell Biology, 120*, 969–980.

Junker, J., Sommer, J. R., Sar, M., & Meissner, G. (1994). Extended junctional sarcoplasmic reticulum of avian cardiac muscle contains functional ryanodine receptors. *The Journal of Biological Chemistry, 269*, 1627–1634.

Knollmann, B. C., Chopra, N., Hlaing, T., Akin, B., Yang, T., Ettensohn, K., et al. (2006). Casq2 deletion causes sarcoplasmic reticulum volume increase, premature Ca^{2+} release, and catecholaminergic polymorphic ventricular tachycardia. *The Journal of Clinical Investigation, 116*, 2510–2520.

Kontrogianni-Konstantopoulos, A., Catino, D. H., Strong, J. C., Sutter, S., Borisov, A. B., Pumplin, D. W., et al. (2006). Obscurin modulates the assembly and organization of sarcomeres and the sarcoplasmic reticulum. *The FASEB Journal: Official Publication of the Federation of American Societies for Experimental Biology, 20*, 2102–2111.

Leeson, T. S. (1978). The transverse tubular (T) system of rat cardiac muscle fibers as demonstrated by tannic acid mordanting. *Canadian Journal of Zoology, 56*, 1906–1916.

Loesser, K. E., Castellani, L., & Franzini-Armstrong, C. (1992). Dispositions of junctional feet in muscles of invertebrates. *Journal of Muscle Research and Cell Motility, 13*, 161–173.

Lukyanenko, V., Ziman, A., Lukyanenko, A., Salnikov, V., & Lederer, W. J. (2007). Functional groups of ryanodine receptors in rat ventricular cells. *The Journal of Physiology, 583*, 251–269.

Page, S. G., & Niedergerke, R. (1972). Structures of physiological interest in the frog heart ventricle. *Journal of Cell Science, 11*, 179–203.

Pan, Z., Yang, D., Nagaraj, R. Y., Nosek, T. A., Nishi, M., Takeshima, H., et al. (2002). Dysfunction of store-operated calcium channel in muscle cells lacking mg29. *Nature Cell Biology, 4*, 379–383.

Paolini, C., Fessenden, J. D., Pessah, I. N., & Franzini-Armstrong, C. (2004). Evidence for conformational coupling between two calcium channels. *Proceedings of the National Academy of Sciences of the United States of America, 101*, 12748–12752.

Paolini, C., Quarta, M., Nori, A., Boncompagni, S., Canato, M., Volpe, P., et al. (2007). Reorganized stores and impaired calcium handling in skeletal muscle of mice lacking calsequestrin-1. *The Journal of Physiology, 583*, 767–784.

Peachey, L. D., & Eisenberg, B. R. (1978). Helicoids in the T system and striations of frog skeletal muscle fibers seen by high voltage electron microscopy. *Biophysical Journal, 22*, 145–154.

Porter, K. R., & Palade, G. E. (1957). Studies on the endoplasmic reticulum. III. Its form and distribution in striated muscle cells. *The Journal of Biophysical and Biochemical Cytology, 3*, 269–300.

Protasi, F., Franzini-Armstrong, C., & Allen, P. D. (1998). Role of ryanodine receptors in the assembly of calcium release units in skeletal muscle. *The Journal of Cell Biology, 140*, 831–842.

Protasi, F., Franzini-Armstrong, C., & Flucher, B. E. (1997). Coordinated incorporation of skeletal muscle dihydropyridine receptors and ryanodine receptors in peripheral couplings of BC3H1 cells. *The Journal of Cell Biology, 137*, 859–870.

Protasi, F., Paolini, C., Nakai, J., Beam, K. G., Franzini-Armstrong, C., & Allen, P. D. (2002). Multiple regions of RyR1 mediate functional and structural interactions with α(1S)-dihydropyridine receptors in skeletal muscle. *Biophysical Journal, 83*, 3230–3244.

Protasi, F., Sun, X. H., & Franzini-Armstrong, C. (1996). Formation and maturation of the calcium release apparatus in developing and adult avian myocardium. *Developmental Biology, 173*, 265–278.

Protasi, F., Takekura, H., Wang, Y., Chen, S. R., Meissner, G., Allen, P. D., et al. (2000). RYR1 and RYR3 have different roles in the assembly of calcium release units of skeletal muscle. *Biophysical Journal, 79*, 2494–2508.

Schredelseker, J., Di Biase, V., Obermair, G. J., Felder, E. T., Flucher, B. E., Franzini-Armstrong, C., et al. (2005). The β1a subunit is essential for the assembly of dihydropyridine-receptor arrays in skeletal muscle. *Proceedings of the National Academy of Sciences of the United States of America, 102*, 17219–17224.

Scriven, D. R., Klimek, A., Asghari, P., Bellve, K., & Moore, E. D. (2005). Caveolin-3 is adjacent to a group of extradyadic ryanodine receptors. *Biophysical Journal, 89*, 1893–1901.

Shen, X., Franzini-Armstrong, C., Lopez, J. R., Jones, L. R., Kobayashi, Y. M., Wang, Y., et al. (2007). Triadins modulate intracellular Ca(2+) homeostasis but are not essential for excitation-contraction coupling in skeletal muscle. *The Journal of Biological Chemistry, 282*, 37864–37874.

Soeller, C., & Cannell, M. B. (1999). Examination of the transverse tubular system in living cardiac rat myocytes by 2-photon microscopy and digital image-processing techniques. *Circulation Research, 84*, 266–275.

Song, L. S., Sobie, E. A., McCulle, S., Lederer, W. J., Balke, C. W., & Cheng, H. (2006). Orphaned ryanodine receptors in the failing heart. *Proceedings of the National Academy of Sciences of the United States of America, 103*, 4305–4310.

Stern, M. D., Pizarro, G., & Rios, E. (1997). Local control model of excitation-contraction coupling in skeletal muscle. *The Journal of General Physiology, 110*, 415–440.

Sun, X. H., Protasi, F., Takahashi, M., Takeshima, H., Ferguson, D. G., & Franzini-Armstrong, C. (1995). Molecular architecture of membranes involved in excitation-contraction coupling of cardiac muscle. *The Journal of Cell Biology, 129*, 659–671.

Sutko, J. L., & Airey, J. A. (1996). Ryanodine receptor Ca^{2+} release channels: Does diversity in form equal diversity in function? *Physiological Reviews, 76*, 1027–1071.

Takekura, H., Fujinami, N., Nishizawa, T., Ogasawara, H., & Kasuga, N. (2001). Eccentric exercise-induced morphological changes in the membrane systems involved in excitation-contraction coupling in rat skeletal muscle. *The Journal of Physiology, 533*, 571–583.

Takekura, H., Kasuga, N., Kitada, K., & Yoshioka, T. (1996). Morphological changes in the triads and sarcoplasmic reticulum of rat slow and fast muscle fibres following denervation and immobilization. *Journal of Muscle Research and Cell Motility, 17*, 391–400.

Takekura, H., Nishi, M., Noda, T., Takeshima, H., & Franzini-Armstrong, C. (1995). Abnormal junctions between surface membrane and sarcoplasmic reticulum in skeletal muscle with a mutation targeted to the ryanodine receptor. *Proceedings of the National Academy of Sciences of the United States of America, 92*, 3381–3385.

Takekura, H., Shuman, H., & Franzini-Armstrong, C. (1993). Differentiation of membrane systems during development of slow and fast skeletal muscle fibres in chicken. *Journal of Muscle Research and Cell Motility, 14*, 633–645.

Takekura, H., Takeshima, H., Nishimura, S., Takahashi, M., Tanabe, T., Flockerzi, V., et al. (1995). Co-expression in CHO cells of two muscle proteins involved in excitation-contraction coupling. *Journal of Muscle Research and Cell Motility, 16*, 465–480.

Takekura, H., Tamaki, H., Nishizawa, T., & Kasuga, N. (2003). Plasticity of the transverse tubules following denervation and subsequent reinnervation in rat slow and fast muscle fibres. *Journal of Muscle Research and Cell Motility, 24*, 439–451.

Takeshima, H., Ikemoto, T., Nishi, M., Nishiyama, N., Shimuta, M., Sugitani, Y., et al. (1996). Generation and characterization of mutant mice lacking ryanodine receptor type 3. *The Journal of Biological Chemistry, 271*, 19649–19652.

Takeshima, H., Komazaki, S., Nishi, M., Iino, M., & Kangawa, K. (2000). Junctophilins: A novel family of junctional membrane complex proteins. *Molecular Cell, 6*, 11–22.

Takeshima, H., Nishi, M., Iwabe, N., Miyata, T., Hosoya, T., Masai, I., et al. (1994). Isolation and characterization of a gene for a ryanodine receptor/calcium release channel in *Drosophila melanogaster*. *FEBS Letters, 337*, 81–87.

Tijskens, P., Meissner, G., & Franzini-Armstrong, C. (2003). Location of ryanodine and dihydropyridine receptors in frog myocardium. *Biophysical Journal, 84*, 1079–1092.

Veratti, E. (1961). Investigations on the fine structure of striated muscle fiber read before the Reale Istituto Lombardo, 13 March 1902. *The Journal of Biophysical and Biochemical Cytology, 10*(4 Suppl.), 1–59.

Woo, S. H., Cleemann, L., & Morad, M. (2005). Diversity of atrial local Ca^{2+} signalling: Evidence from 2-D confocal imaging in Ca^{2+}-buffered rat atrial myocytes. *The Journal of Physiology, 567*, 905–921.

CHAPTER 2

Electron Microscopy of Ryanodine Receptors

Terence C. Wagenknecht*,† and Zheng Liu*

*Wadsworth Center, New York State Department of Health, Albany, New York, USA
†Department of Biomedical Sciences, School of Public Health, State University of New York at Albany, Albany, New York, USA

I. OVERVIEW

This review summarizes the contributions of cryo-electron microscopy (cryo-EM) and single-particle three-dimensional reconstruction (3D cryo-EM) to understanding the structural architecture of ryanodine receptors (RyRs). 3D reconstructions at about 1-nm resolution have been determined for RyRs in putatively open and closed states, and are beginning to reveal the

Current Topics in Membranes, Volume 66
1063-5823/10 $35.00
Copyright 2010, Elsevier Inc. All right reserved.
DOI: 10.1016/S1063-5823(10)66002-4

mechanism of calcium release channel gating and of allosteric communication over distance greater than 10 nm between the cytoplasmic and transmembrane (TM) regions of the receptor. The information content of 3D cryo-EM is greatly increased when combined with biochemical and molecular biological labeling, and strategies for docking atomic structures of subcomponents into cryo-EM density maps or RyRs. In the future, electron microscopy will produce more detailed structures for RyRs and allow the characterization of ever more complex macromolecular assemblies in which RyRs participate such as the excitation–contraction coupling machinery in skeletal and cardiac muscle.

II. INTRODUCTION

RyRs, a class of intracellular calcium release channels, are the largest known ion channels, being homotetramers of net molecular weight ~2.3 MDa (Fill & Copello, 2002; Hamilton & Serysheva, 2009). All three of the mammalian isoforms—RyR1, RyR2, RyR3—have been characterized by cryo-EM and 3D reconstruction and, as expected based upon their high-sequence similarity, shown to have nearly identical 3D structures. For simplicity, the amino acid sequence numbers cited in this review will refer to the RyR1 isoform unless stated otherwise. RyR1 and RyR2 are highly expressed in skeletal and cardiac muscle, respectively, where they have prominent roles in excitation–contraction coupling (see Chapter 1 by Franzini-Armstrong). This review will summarize the contributions of cryo-EM in elucidating the structure of RyRs, and will surmise on the future role that EM will play in this endeavor.

III. CRYO-EM OF MACROMOLECULAR COMPLEXES

Most current knowledge of the 3D architecture of RyRs has been obtained by cryo-EM of isolated, solubilized receptors combined with "single-particle" image processing, an approach we will refer to here as 3D cryo-EM (Hamilton & Serysheva, 2009; Liu & Wagenknecht, 2005). Indeed, the RyR was among the first biological macromolecules to which this technology was applied (Radermacher et al., 1992, 1994). More generally, 3D cryo-EM has evolved as a powerful method for elucidating the 3D architecture of large protein and ribonucleo–protein complexes, albeit at less than atomic resolution. Typically, for this technique, a few microliters of aqueous solution of purified macromolecules are deposited onto a standard EM grid containing a thin specimen support film of carbon (or holey carbon). The specimen is then blotted, and rapidly frozen (< 1 ms) by plunging into a cryogen, usually liquid ethane. Grids so prepared contain regions of thin (~100 nm) vitreous (amorphous) ice in which are embedded well-preserved macromolecules. The amount of RyR protein required per grid ranges from a few tenths of a microgram (when RyRs adsorb to a carbon support) to several micrograms (e.g., to obtain RyRs suspended in solution over the holes in a carbon support).

Electron micrographs collected using suitably low electron doses and containing thousands of images of individual complexes are processed by computer. Sometimes micrographs need to be recorded with the specimen stage tilted so as to achieve a sufficiently diverse range of particle orientations to compute a 3D reconstruction. The processing includes particle selection, sorting of the images according to orientation of the particles (and conformation or composition if heterogeneity in these properties is present in the population). Finally, structurally homogeneous sets of variously oriented particles are combined to produce 3D reconstructions. For more details on the methodology and recent applications of this technology, including some in which near-atomic resolution has been achieved, see the review by Cheng and Walz (2009). This review focuses on recent accomplishments of cryo-EM in elucidating the structure of RyRs. As will become apparent, the efficacy of 3D cryo-EM is greatly enhanced when it is combined with other biophysical techniques (Ludtke, Serysheva, Hamilton, & Chiu, 2005; Samso, Wagenknecht, & Allen, 2005; Serysheva et al., 2008).

IV. THREE-DIMENSIONAL ARCHITECTURE OF RYR AS DETERMINED BY CRYO-EM

Since 1994, numerous 3D reconstructions at modest resolutions (2–4 nm) have been determined from micrographs of vitreously frozen RyRs for all three isoforms and of various RyR–ligand complexes. Since 2005, two independent groups have described the most detailed structures currently available (~1 nm resolution) of RyR1, presumably in the closed state (Ludtke et al., 2005; Samso et al., 2005; Serysheva et al., 2008). Better resolutions should be forthcoming, but certain technical difficulties, discussed in Section VII.A, are slowing the progress compared to what has been achieved for other macromolecular complexes of similar size.

An overview of one of the 1-nm resolution structures is shown in Fig. 1. The overall shape of RyR has been described as resembling that of a mushroom, with the stem corresponding to the transmembrane (TM) region, and the cap, which is overall square (~280 Å edge length by 115 Å thickness) in shape when viewed along the long axis of the stem corresponding to the

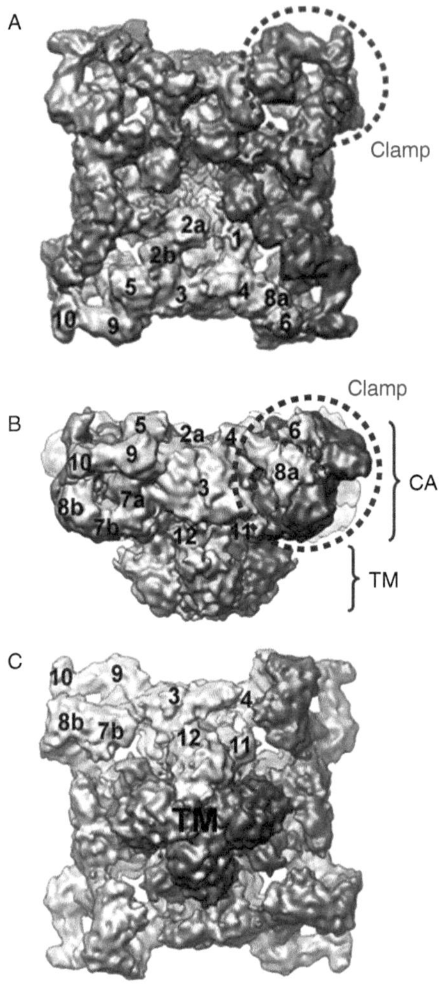

FIGURE 1 Surface representation from 3D cryo-microscopy of skeletal muscle RyR. Three orientations are shown: (A) "top" view along the fourfold symmetry axis; (B) "side" view, normal to fourfold symmetry axis, showing the transmembrane (TM) and cytoplasmic region/assembly (CA); (C) "bottom" view along the fourfold symmetry axis showing the surface that contains the TM region. The proposed subunit boundaries are indicated by showing each subunit in a different color. The length of one edge of the CA is 28 nm. One of the "clamp" structures that form the corners of the CA are indicated in (A) and (B). The numerals indicate distinctive subregions that are reproducibly resolved in 3D reconstructions. Resolution is estimated at 1 nm. Figure modified and adapted from Serysheva et al. (2008) with permission. (See Color Insert.)

cytoplasmic region (CA, cytoplasmic assembly). The TM region of the RyR subunit is composed of ~700 amino acids located at the carboxy terminus of the RyR polypeptide, and most or all of the remaining ~4300 amino acids comprise the CA (Du & MacLennan, 2005; Du, Sandhu, Khanna, Guo, & MacLennan, 2002).

The CA consists of distinctive interconnected globular *subregions* (likely corresponding to one or several protein folding domains) arranged such that a substantial fraction of the volume enclosed by the CA is accessible to solvent. These subregions are indicated by the numerals (1–11) superimposed on the model depicted in Fig. 1. Because of the "loose" connection of the subregions, the surface of the CA contains many cavities and crevices, including two solvent-filled channels per subunit that lead from one face of the CA to the other. One particularly conspicuous crevice exists between the large subregion 3 (also called the "handle" domain) and subregions 8, 8a; as discussed later, this crevice contains a binding site for the RyR-regulatory protein calmodulin (CaM).

Distinctive structures termed "clamps" form the corners of the CA. The clamps comprise subregions 5–10, and contain one of the holes mentioned earlier. In skeletal muscle, the clamp structures appear to interact with dihydropyridine receptors (Paolini, Protasi, and Franzini-Armstrong, 2004), which sense depolarization of the plasma membrane and control the calcium channel activity of RyR1 during excitation–contraction coupling.

The differently colored regions in Fig. 1 are hypothesized to represent the shape of the four identical subunits. This assignment is based on the assumption that interdomain interactions that appear weakest (based upon surface area of contact regions) represent intersubunit boundaries, which may not always be the case, and so the subunit boundaries depicted in Fig. 1 might change somewhat in the future.

The best resolutions achieved to date (1 nm) from cryo-EM of RyR are just sufficient to identify *some* α-helices and, less reliably, β-sheets, but tracing of the polypeptide chain is not yet possible. Serysheva et al. (2008) have used automated as well as direct visual analysis of the density maps to predict more than 40 α-helices and 7 β-sheets. The helices in the TM region are of particular interest, and are discussed further later. Why such a large (about 90% of the RyR total mass) and structurally complex CA is present in RyRs has been the subject of speculation. Numerous-binding sites for regulators of RyR calcium channel activity are known to be present in the CA, as is discussed later, but it is not clear that the degree of regulation of RyRs is greater than that of other ligand-gated channels for which the extramembrane mass is far less (e.g., the neurotransmitter gated ion channels).

V. α-HELICES IN THE TM REGION AND THE MECHANISM OF CALCIUM CHANNEL GATING

Both of the independently determined 1-nm resolution 3D reconstructions of RyR in the closed state resolve α-helices in the TM regions (Ludtke et al., 2005; Samso et al., 2005). Two of these helices are of particular interest because they can be related to the mechanism of the calcium channel activity of RyR. Sequence analysis and biophysical characterization of wild-type and modified RyRs has led to a structural model of the ion-conducting pore of RyR that is analogous to that of the potassium channels, such as KscA, whose atomic structure is known (Balshaw, Gao, & Meissner, 1999; Welch, Rheault, West, & Williams, 2004; Zhao et al., 1999). KcsA, like RyR, is a tetramer, and the ion-conducting pore comprises four sets of three characteristic α-helices, each set contributed by an individual subunit. Two of the α-helices (termed "inner" and "outer") are TM and the third (the "pore" helix) traverses only part of the way through the bilayer. The pore helix is connected to an extended nonhelical loop that lines the narrowest portion of the channel pore and functions as the "selectivity filter" to establish ion specificity (Doyle et al., 1998). Both of the reported 1-nm cryo-EM reconstructions of RyR resolve α-helices that are analogous to the inner and pore helices of KcsA (Fig. 2). Besides the pore and inner helices, additional α-helices have been tentatively identified in the TM region but their precise number and nature remain to be definitively established (Ludtke et al., 2005; Samso et al., 2005, 2009).

One puzzling difference between the two independently determined reconstructions, both of which were determined under conditions expected to favor the closed state of the channel, is that Ludtke et al. (2005) found the inner helix to be bent whereas Samso et al. (2005) observed a nearly straight conformation. Based upon comparisons to atomic structures of potassium channels, the straight conformation is to be expected for a closed channel and the bent helix is predicted for an open channel (Jiang et al., 2002). Ludtke et al. therefore have proposed that RyR's mechanism of gating differs from that of potassium channels whereas the results of Samso et al. do not demand a difference in the gating mechanism. An explanation for this discordance is lacking. However, a recent report (Samso et al., 2009) describes a 3D reconstruction of RyR determined under conditions favoring the open state of the receptor, and in that reconstruction, the inner helix assumes the bent configuration (Fig. 2), lending support for the idea that the mechanism of channel gating in RyR is similar to that employed by the KcsA potassium channel.

Figure 2 illustrates the structural differences between the open and closes states in the TM assembly that were observed in the recent study by Samso et al. (2009). Strong elongate densities in the RyR1 density map match quite

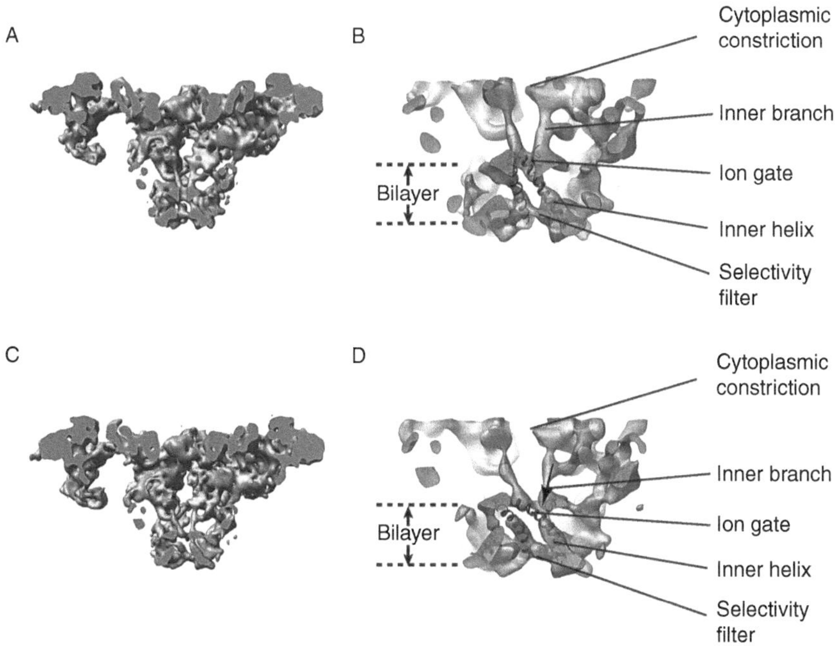

FIGURE 2 Comparison of 3D reconstructions determined for skeletal RyRs under conditions that favor open and closed states. (A) Side view of "closed" state sliced along the fourfold axis at an angle 11° from the diagonal of the square defined by the CA. (B) Close-up of the TM region for the "closed" state and showing the inner α-helices (ribbon) from the crystal structure of the KcsA channel docked into the cryo-EM density map. (C) "Open" state represented as described for (A). (D) Close-up of TM region of open state with α-helices from the inner helices of the KvAP crystal structure docked into the cryo-EM map. Structural features that are described in the text are shown at right (B) and (D). The arrow in (D) indicates an interruption in the otherwise continuous density of the inner helix which corresponds to a "glycine hinge" in the KvAP channel. Figure modified from Samso et al. (2009). (See Color Insert.)

well the configurations of the inner helices observed in X-ray structures of potassium channels, both for the open and closed states of the channels. In the open state, the shorter pore helices are also resolved near the luminal mouth of the channel. The ion gate is assigned to the region where the inner helices, which are tilted, come closest to one another, similar to the ion gate in the KcsA potassium channel (Doyle et al., 1998). Intriguingly, situated above the putative ion gate (i.e., toward the cytoplasmic side of the TM region) are another set of elongate densities, termed "inner branches," which may or may not correspond to α-helices; the inner branches, which move apart radially when the RyR goes from open to closed states, appear to provide a physical linkage between the ion gate and the CA of RyR, where several modulator ligands are known to bind (e.g., FKBP12, CaM) and to influence channel gating. Thus, the inner branches are likely important in the conformational coupling that is known to occur between the massive CA and the TM region. However, the mechanisms of channel regulation by modulators that bind to the CA certainly involve complexities yet to be discovered. The conformational differences in the CA between the open and closed states of RyR that were observed by Samso et al. (2009) are smaller than might have been expected; although most of the subregions undergo slight movements, the largest are less than 1 nm (e.g., subregion 10, located at the corners of the CA, appears to move about 0.8 nm downward (toward the TM-containing face) when RyR goes from closed to open states. Four previously determined 3D reconstructions of RyRs in putatively open states show substantially larger differences from the closed states (Orlova, Serysheva, van Heel, Hamilton, & Chiu, 1996; Serysheva, Schatz, van Heel, Chiu, & Hamilton, 1999; Sharma, Jeyakumar, Fleischer, & Wagenknecht, 2000, 2006). Although the resolution of these earlier reconstructions is severalfold worse than 1 nm, it is not clear why this factor should produce larger apparent conformational changes in the CA than the 1-nm resolution reconstruction. One possible explanation is that not all RyR activators promote the same conformational coupling between the CA and TM regions. Clearly, additional study, including determination of higher resolution structural models of RyRs under more native and varied conditions, is needed to characterize the mechanism(s) of channel gating and conformational coupling of the CA to the TM region.

VI. SYNERGISM OF 3D CRYO-EM AND OTHER BIOPHYSICAL/ BIOCHEMICAL TECHNIQUES

A. *Cryo-EM of Chemically Modified RyRs to Determine Modulator-Binding Sites and Surface-Exposed Sequences*

Once initial 3D reconstructions of RyRs became available it was quite easy to determine additional reconstructions at moderate resolutions (2–4 nm) of RyRs that had been chemically modified by various ligands, mainly for the purpose of determining the locations of the ligands. The main requirements are that the ligand be sufficiently large to be detectable at the anticipated resolution, which is generally not a limiting factor for protein ligands, and the ligand must bind with sufficient affinity to survive dilution (typically to nanomolar concentration) and adsorption to an EM specimen grid. Indeed, most published reconstructions of RyRs have been of various RyR–ligand complexes, the results of which are summarized in Fig. 3. Typically, a reconstruction of the RyR–ligand complex is determined and a difference density

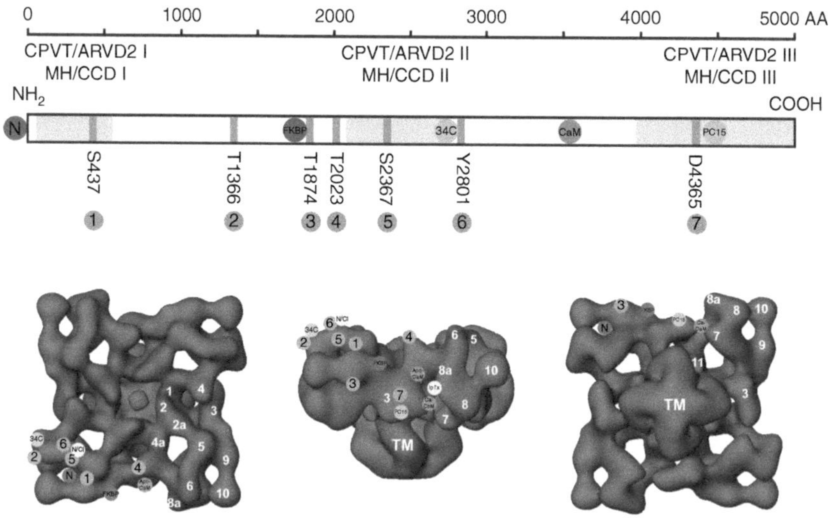

FIGURE 3 Summary of surface-exposed amino acid residues and functional sites that have been mapped on the surface of RyR by 3D cryo-microscopy. *Top*: schematic of the amino acid sequence of RyR2. The green bars indicate amino acid locations that were mapped by the GFP-insertion method. Also superimposed on the sequence are subsequences proposed to be involved in binding modulatory ligands. Pink bars indicate regions that are frequently mutated in diseases of skeletal (malignant hyperthermia (MH), central-core disease (CCD)) and heart (catecholaminergic polymorphic ventricular tachycardia (CPVT) and arrhythmogenic right ventricular dysplasia (ARVD)) muscle. (*Bottom* three panels: surface representations of RyR2 (Meng et al., 2007) in top (cytoplasmic-facing), side, and bottom (SR-lumen-facing) orientations with labels indicating sites that have been localized (each site occurs at four symmetrically related positions in the tetrameric RyR2, but only one site is shown for each localization). Numbers in white indicate landmark subregions. For technical reasons, it was easier to perform the GFP insertions using the cardiac isoform RyR2, but the results are directly transferable to skeletal RyR1 owing to the high-sequence conservation between the two isoforms and their essentially identical 3D structures (Sharma et al., 1998, 2006). The length of one edge of the RyR (lower left image) is 28 nm. Abbreviations: CaM, calmodulin; F, FKBP; 34C, monoclonal antibody 34C; PC15, monoclonal antibody PC15, also known as I29 (Benacquista et al., 2000); N, amino-terminus. (See Color Insert.)

map is formed by subtraction of the map obtained for RyR lacking the ligand from the map of the RyR–ligand complex. Usually, a strong, statistically significant, positive difference dominates the difference density map, and this is attributed to the ligand. Commonly, the difference maps show additional weaker positive and negative densities which can be due to noise or, more interestingly, to conformational alterations induced by the ligand.

Among the first modulatory ligands to be mapped were the immunophilins, FKBP12 and FKBP12.6, one or both of which bind with high affinity to all three RyR isoforms and modulate channel gating (Chelu, Danila, Gilman, & Hamilton, 2004; Marks, 1996). FKBP12/FKBP12.6 bind to the CA at the intersection of regions 9 and 3 (Fig. 3), about 13 nm from the ion-conducting pore (Samso, Shen, & Allen, 2006; Sharma et al., 2006; Wagenknecht et al., 1997). This finding was one of the first indications that long-range communication (> 10 nm) occurs between CA and TM regions in RyRs.

CaM is another regulator that, in both its Ca^{2+}-bound and Ca^{2+}-free (apo-CaM) forms, constitutively associates with RyRs (Aracena, Hidalgo, & Hamilton, 2005; Meissner, 2004). 3D cryo-EM has shown that both CaM forms interact with region 3 of the skeletal RyR, but only Ca^{2+}–CaM is wedged in the deep cleft formed by regions 3 and 8 (Fig. 3) whereas apo-CaM is displaced by about 3.3 nm out of the cleft onto the lateral face of subregion 3 (Samso & Wagenknecht, 2002; Wagenknecht et al., 1997). At present it not clear whether the different binding locations of Ca^{2+}–CaM and apo-CaM are due to a conformational change in RyR that is triggered by Ca^{2+} regulatory sites on the RyR that causes a common CaM-binding site to move, or to chemically distinct interaction sites on RyR for the two forms of CaM. As is the case for FKBP12, the binding locations of Ca^{2+}–CaM and apo-CaM are greater than 10 nm from the ion-conducting pore in the TM region. Higher resolution reconstructions are needed to determine the nature of the conformational changes that occur in RyRs consequent to regulation by CaM.

Figure 3 also shows a cryo-EM localization of the protein CLIC2, a more recently identified potential modulator of skeletal and cardiac RyRs (Board, Coggan, Watson, Gage, & Dulhunty, 2004; Dulhunty, Pouliquin, Coggan, Gage, & Board, 2005; Meng et al., 2009). CLIC2 is an unusual protein that exists in two structural states: as an integral membrane protein functioning as an intracellular chloride channel and as a soluble protein, one of whose functions seems to be to regulate RyRs by interacting with the CA region at a location that bridges regions 5 and 6.

3D cryo-EM has also been used to map the locations of two toxins, Imperatoxin A and natrin, that affect the calcium channel activity of RyRs. Imperatoxin A, a 33-amino acid peptide produced in the venom of the scorpion *Pandinus imperator*, binds with nanomolar affinity to skeletal and cardiac RyR, and enhances channel activity. Imperatoxin binds at the base of the cleft formed by subregions 3, 4, 7, and 8 (Fig. 1), quite close to where Ca^{2+}–CaM binds (Samso, Trujillo, Gurrola, Valdivia, & Wagenknecht, 1999). Intriguing experimental findings have implicated Imperatoxin A as binding to a site that interacts with the cytoplasmic II–III cytoplasmic loop of the dihydropyridine receptor at the triad junctions of skeletal muscle (Dulhunty, Curtis, Watson, Cengia, & Casarotto, 2004; Gurrola et al., 1999). The II–III loop plays a key role in the conformational coupling between the voltage-sensing dihydropyridine receptor and RyR that occurs during skeletal excitation–contraction coupling (Tanabe, Beam, Adams, Nicodome, & Numa, 1990).

Natrin, a toxin from a snake venom has been shown to inhibit skeletal RyR activity, and 3D cryo-EM has been used to localize this toxin's binding site to a location between subregions 5 and 6 on the CA of RyR (Fig. 3; Zhou et al., 2008). This site is likely an important one for understanding regulation of RyR activity. First, it is located at an intersubunit boundary (see Fig. 1) in the model proposed by Serysheva et al. (1998). Second, at least some of the mutations that are responsible for malignant hyperthermia and central-core disease in skeletal muscle and certain forms of sudden cardiac death map to regions 5 and 6 (discussed in the following section). Further, according to a hypothesis for which supporting evidence is accumulating (Ikemoto, 2005; Ikemoto & Yamamoto, 2000), the disease mutations that lie in these two regions operate by a common mechanism involving the weakening of a critical interdomain (also intersubunit) interaction that results in appropriate gating of RyR to the open state. Thus, natrin may well inhibit the transition of RyR from a closed to an open state by strengthening the interaction between subregions represented by or contained within regions 5 and 6. CLIC2, discussed earlier, binds in a mode that is nearly indistinguishable from natrin, and therefore might inhibit RyR in the same manner as natrin.

B. Cryo-EM of Modified RyRs to Determine Surface-Exposed Amino Acid Residues

Current 3D reconstructions from cryo-EM are inadequate for resolving amino acid side chains, and it is unclear when reconstructions containing sufficient detail for this purpose will be achieved. One experimental approach to partially address this problem involves cryo-EM of recombinant chimeric RyRs in which a molecule of green fluorescent protein (GFP) has been inserted at a precise location within RyR's sequence that is judged likely to be exposed on the surface of RyR. 3D reconstructions at moderate resolution ($\sim$3 nm) of such chimeras easily resolve the covalently linked GFP whose center-of-mass can be determined with a precision estimated at 1 nm. Results from such studies are useful for correlating specific amino acid residues that have been implicated in particular functions (e.g., regulatory phosphorylation sites) with their location on the 3D architecture and, in some instances, for testing specific hypotheses regarding domain–domain interactions and dynamics involving amino acid sequences within the RyR subunit. Following are some selected examples of results for 3D cryo-EM of GFP-RyR chimeras. The amino acid residue numbers in this section refer to the cardiac isoform of RyR.

1. Divergent Regions

Among the first amino acids to be localized making use of GFP insertions were at positions Asp-4365, Thr-1366, Thr-1874 (Liu, Zhang, Li, Chen, & Wagenknecht, 2002; Liu, Zhang, Wang, Chen, & Wagenknecht, 2004; Zhang et al., 2003); these amino acids are located, respectively, within regions of RyR's sequence known as divergent regions 1, 2, and 3, and their locations are indicated in Fig. 3 by the green circles labeled "7," "2," and "3," respectively (Sorrentino & Volpe, 1993). The divergent regions are highly variable among the three RyR isoforms, and are likely to be responsible in large part for the different roles played by RyRs in various tissues and subcellular locations. All three divergent regions map to the CA, with DR1 in subregion 3 and DR2 and DR3 in subregions 6 and 9, respectively, which lie in the "clamp" structures.

2. Amino-Terminal and Central "Disease-Causing Mutation" Regions

GFP insertions at Ser-437 and Ser-2367 map near to one another on the cytoplasmic region of the 3D structure of RyR2 (green dots numbered 1 and 5 in Fig. 3) (Liu, Wang, Zhang, Chen, & Wagenknecht, 2005; Wang et al., 2007). Ser-437 and Ser-2367 are located in the amino-terminal and central regions (pink regions at top of Fig. 3), respectively, of RyR's amino acid sequence. These regions have been found to be frequently mutated in the skeletal diseases malignant hyperthermia and central-core disease, and in certain forms of sudden cardiac death (Cerrone, Napolitano, & Priori, 2009; Dirksen & Avila, 2005; Dulhunty, Beard, Pouliquin, & Kimura, 2006; Jurkat-Rott, McCarthy, & Lehmann-Horn, 2000; Priori & Napolitano, 2005; Robinson, Carpenter, Shaw, Halsall, & Hopkins, 2006; Stowell, 2008). Ikemoto and colleagues have presented evidence supporting their hypothesis that mutations lying in these two regions interfere with an interaction between two structural domains in the 3D structure of RyRs (Ikemoto & Yamamoto, 2002; Murayama et al., 2007; Oda et al., 2005; Priori & Napolitano, 2005; Yamamoto et al., 2008). The result from cryo-EM that Ser-437 and Ser-2367 are indeed adjacent to one another in the 3D structure, despite their large separation in the sequence, is consistent with the Ikemoto hypothesis.

3. Phosphorylation Sites on RyR

The role of phosphorylation in regulating RyRs is of current interest. Three sites that have come under intense study in RyR2 are Ser2030 (2065 in RyR1) and two closely associated sties Ser-2808 (2843 in RyR1) and Ser-2814 (Thr-2848 in RyR1). Amino acid residues that are *near* to these sites have been localized using the GFP insertion technique (see Fig. 3 green circles 4 (Thr-2023) and 6 (Tyr-2801)) (Jones et al., 2008; Meng et al.,

2007). The two GFP sites lie on different domains in the cytoplasmic region of the receptor with Thr-2023 in subregion 4 and Tyr-2801 on subregion 6. According to an hypothesis proposed by Marks and colleagues (Reiken et al., 2003; Wehrens et al., 2006) hyperphosphorylation at these sites occurs during heart failure and causes the dissociation of FK506-binding protein (FKBP12.6) from RyR2, which leads to a RyR2 that is "leaky" to Ca^{2+}. One might therefore expect that at least one of the phosphorylation sites would be near to the receptor-bound FKBP12.6, but, as is apparent from Fig. 3, this is not the case—both subregions 4 and 6, which contain the sites of phosphorylation, are distant from the FKBP12.6 site ("FKBP" in Fig. 3) and so neither of these sites appears to be *directly* involved in FKBP12.6 binding. It should be pointed out that the hypothesis is controversial (e.g., Xiao, Sutherland, Walsh, & Chen, 2004).

C. Pseudo-Atomic Structures of RyR by Integration of Atomic Structures into Cryo-EM Density Maps

An atomic model for RyR, most likely determined by either X-ray crystallography or cryo-EM, will hopefully become available in the near future. In the meantime, atomic models of fragments of RyR (Amador et al., 2009; Kim, Shin, Kim, & Eom, 1999; Lobo and Van Petegem, 2009) or of its macromolecular ligands such as CaM (Maximciuc, Putkey, Shamoo, & MacKenzie, 2006) and FKBP12/FKBP12.6 (Deivanayagam, Carson, Thotakura, Narayana, & Chodavarapu, 2000; Griffith et al., 1995) have already been described. A common application of cryo-EM is to dock such high-resolution structural models into lower resolution density maps from cryo-EM of larger complexes that bind the ligand. Sometimes the docking is done manually, but with increasing frequency, automated algorithms are applied to improve the accuracy of the docking (Chacon & Wriggers, 2002; Hinsen, Reuter, Navaza, Stokes, & Lacapere, 2005; Topf et al., 2008; Volkmann, 2009; Volkmann & Hanein, 1999; Wriggers, Milligan, & McCammon, 1999). Of course, the accuracy depends upon the resolution of the cryo-EM density map. When density maps are determined at 1-nm resolution or better, the fitting can be sufficiently accurate that "pseudoatomic models" can be plausibly proposed and, to the extent they are correct, these models greatly increase the information content of cryo-EM reconstructions. An example of docking an atomic structure into a cryo-EM map of RyR is illustrated in Fig. 2B and D where the inner helices of the X-ray models of the potassium channels KcsA and KvAP could be directly fitted without modification into the density maps for RyR in its closed and open states, respectively (Samso et al., 2009). Other examples are the docking of an atomic structure of FKBP12 into a cryo-EM density map of an RyR–FKBP complex (Samso et al., 2006), docking of an homology-modeled atomic structure of RyR1 amino acid residues 12–565 into a cryo-EM derived 3D reconstruction of skeletal RyR (Serysheva et al., 2008), and docking of CLIC2 and natrin in their binding complexes with RyR1 (Zhou et al., 2008).

VII. OUTLOOK AND PERSPECTIVES

A. *Pathways to Higher Resolution*

Clearly, improved resolution of 3D reconstructions from cryo-EM of RyRs is needed. Conceivably, a sufficient number of fragments of RyRs could be crystallized, solved to atomic resolution, and docked into moderate resolution models of RyR derived from cryo-EM so as to achieve a pseudo-atomic model of the complete RyR, but this would be a tedious, long-term endeavor. We think it more likely that an atomic structure for a RyR will be determined by X-ray crystallography or 3D cryo-microscopy. As has been mentioned, progress in attaining subnanometer resolution for RyRs by cryo-EM has been slower than for other macromolecular complexes, such as the ribosome which is of comparable size and complexity to RyR. Heterogeneity within populations of isolated RyRs is suspected to be the reason that higher resolutions have not been achieved. Improved image processing techniques for dealing with heterogeneity are under development in many laboratories (Spahn & Penczek, 2009), but it is unlikely that improvements in this area can completely compensate for structural heterogeneity. A reduction in the heterogeneity that is present in RyR populations will likely be required to attain atomic or near-atomic resolution. If the heterogeneity is due to partial depletion of RyR-associated proteins, then improved purification methods or biochemical manipulations of the purified receptors (e.g., addition of missing components) should lead to more homogenous image populations and better resolution. However, heterogeneity could also be due to variability in the conformations of RyR molecules, and ameliorating this situation could be difficult. Various conditions have been tried to "lock" RyRs in defined conformational states, but with limited or uncertain success (Ludtke et al., 2005; Orlova et al., 1996; Samso et al., 2009). Quite possibly, the detergents that are required to maintain the solubility of RyRs cause partial destabilization of native structure, particularly within the CA, which was shown in the earliest studies to exhibit variability among 3D reconstructions (Radermacher et al., 1994). If milder detergents cannot be found, then methods to remove detergents and to reconstitute RyRs into a more native lipid environment may be required to achieve improved resolution in RyR reconstructions. Briefly, three types of specimens could be useful for this purpose: (1) 2D arrays of RyRs embedded in lipid bilayers, (2) RyRs incorporated into lipid vesicles, (3) RyRs incorporated into nanodiscs. Attempts to induce RyRs to form ordered 2D arrays have been reported (Yin, Blayney, & Lai, 2005; Yin, D'Cruz, & Lai, 2008; Yin, Han, Wei, & Lai, 2005; Yin & Lai, 2000), and methods for determining high-resolution reconstructions from 2D crystals are well developed, but thus far the arrays that have been obtained for RyRs are not well ordered. Single-particle image processing methodology has recently been described and applied to a membrane protein that was reconstituted into lipid vesicles (Wang & Sigworth, 2009); large membrane complexes such as RyRs would appear to be good candidates for application of this technology. A more limited association of lipids with RyR might be achieved by incorporating RyR's TM region into a nanodisc. Nanodiscs are disc-shaped elements of lipid bilayer that are bounded by a scaffold protein such as apolipoprotein A-I (Denisov, Grinkova, Lazarides, & Sligar, 2004; Nath, Atkins, & Sligar, 2007). Nanodisc-associated RyRs would not be much larger than detergent-solubilized RyRs, and should be tractable by existing single-particle image processing methods, as was demonstrated in a recent electron microscopy study of nanodiscs containing single molecules of integrin (Ye et al., 2010).

B. Supramolecular Structures Involving RyRs

Following the determination of atomic models for RyRs, electron microscopy will continue to make important contributions to understanding RyRs' roles in cellular Ca^{2+} signaling. The availability of atomic level information for RyRs will greatly enhance the interpretability of results from cryo-EM studies.

RyRs exist in multiple copies within the cellular machinery that mediates excitation–contraction coupling, and are but one of many proteins that contribute to signal transduction (Dulhunty, 2006; Treves et al., 2009). Many of the non-RyR proteins interact directly with RyR, and reconstitution of RyRs associated to these proteins can produce specimens might be more readily characterized by cryo-EM than X-ray crystallography (e.g., the RyR–FKBP12 complex discussed in Section VI.A; Samso et al., 2006).

An emerging and potentially powerful technique for characterizing the organization of macromolecular complexes in their cellular context is cryo-electron tomography (CET) (Hoenger & Mcintosh, 2009; Koning & Koster, 2009; Leis, Rockel, Andrees, & Baumeister, 2009; Milne & Subramaniam, 2009). Specimens for CET can be thin regions (up to $\sim$200 nm) of intact cells, cryo-sections of tissue or cells (Hsieh, Leith, Mannella, Frank, & Marko,

2006; Marko, Hsieh, Schalek, Frank, & Mannella, 2007; Norlen, Oktem, & Skoglund, 2009), or isolated subcellular organelles or microsomal fractions. 3D reconstructions are obtained from a series of tilted images of the specimen, similar to what is done in medical tomography. Current limitations of CET are that the 3D density maps are low in contrast, of limited resolution (usually worse than 3 nm), and effective labeling strategies for specific proteins of interest are not currently available. However, in those instances where large complexes (e.g., RyRs) are detectable and present in multiple occurrences in the tomograms, it is feasible to apply single-particle techniques and thereby obtain highly detailed structural models of complex subcellular machinery. Initial attempts employing CET to determine density maps of RyRs and their surroundings in isolated triad junctions from skeletal muscle have been described) and some new structural features have been resolved in the tomograms (Renken et al., 2009; Wagenknecht, Hsieh, Rath, Fleischer, & Marko, 2002). In the near future improved resolution and other technological advances promise to greatly expand our knowledge of the architecture of the excitation–contraction coupling apparatus.

References

Amador, F. J., Liu, S., Ishiyama, N., Plevin, M. J., Wilson, A., MacLennan, D. H., et al. (2009). Crystal structure of type I ryanodine receptor amino-terminal β-trefoil domain reveals a disease-associated mutation "hot spot" loop. *Proceedings of the National Academy of Sciences of the United States of America, 106*, 11040–11044.

Aracena, P., Hidalgo, C., & Hamilton, S. L. (2005). RyR1 modulation by calmodulin. In X.D.H.T. Wehrens & A. R. Marks (Eds.), *Ryanodine receptors—Structure, function and dysfunction in clinical diseases* (pp. 163–168). New York: Springer.

Balshaw, D., Gao, L., & Meissner, G. (1999). Luminal loop of the ryanodine receptor: A pore-forming segment? *Proceedings of the National Academy of Sciences of the United States of America, 96*, 3345–3347.

Benacquista, B. L., Sharma, M. R., Samso, M., Zorzato, F., Treves, S., & Wagenknecht, T. (2000). Amino acid residues 4425-4621 localized on the three-dimensional structure of the skeletal muscle ryanodine receptor. *Biophysical Journal, 78*, 1349–1358.

Board, P. G., Coggan, M., Watson, S., Gage, P. W., & Dulhunty, A. F. (2004). CLIC-2 modulates cardiac ryanodine receptor Ca^{2+} release channels. *The International Journal of Biochemistry and Cell Biology, 36*, 1599–1612.

Cerrone, M., Napolitano, C., & Priori, S. G. (2009). Catecholaminergic polymorphic ventricular tachycardia: A paradigm to understand mechanisms of arrhythmias associated to impaired Ca^{2+} regulation. *Heart Rhythm, 6*, 1652–1659.

Chacon, P., & Wriggers, W. (2002). Multi-resolution contour-based fitting of macromolecular structures. *Journal of Molecular Biology, 317*, 375–384.

Chelu, M. G., Danila, C. I., Gilman, C. P., & Hamilton, S. L. (2004). Regulation of ryanodine receptors by FK506 binding proteins. *Trends in Cardiovascular Medicine, 14*, 227–234.

Cheng, Y., & Walz, T. (2009). The Advent of near-atomic resolution in single-particle electron microscopy. *Annual Review of Biochemistry, 78*, 723–742.

Deivanayagam, C. C. S., Carson, M., Thotakura, A., Narayana, S. V. L., & Chodavarapu, R. S. (2000). Structure of FKBP12.6 in complex with rapamycin. *Acta Crystallographica D, 56*, 266–271.

Denisov, I. G., Grinkova, Y. V., Lazarides, A. A., & Sligar, S. G. (2004). Directed self-assembly of monodisperse phospholipid bilayer nanodiscs with controlled size. *Journal of the American Chemical Society, 126*, 3477–3487.

Dirksen, R. T., & Avila, G. (2005). Pathophysiology of muscle disorders linked to mutations in the skeletal muscle ryanodine receptor. In X.D.H.T. Wehrens & A. R. Marks (Eds.), *Ryanodine receptor: Structure, function and dysfunction in clinical disease* (pp. 229–242). New York: Springer.

Doyle, D. A., Cabral, J. M., Pfuetzner, R. A., Kuo, A., Gulbis, J. M., Cohen, S. L., et al. (1998). The structure of the potassium channel: Molecular basis of K^+ conduction and selectivity. *Science, 280*, 69–77.

Du, G. G., & MacLennan, D. H. (2005). Topology and transmembrane organization of ryanodine receptors. In X.D.H.T. Wehrens & A. R. Marks (Eds.), *Ryanodine receptor: Structure, function and dysfunction in clinical disease* (pp. 9–23). New York: Springer.

Du, G. G., Sandhu, B., Khanna, V. K., Guo, X. H., & MacLennan, D. H. (2002). Topology of the Ca^{2+} release channel of skeletal muscle sarcoplasmic reticulum (RyR1). *Proceedings of the National Academy of Sciences of the United States of America, 99*, 16725–16730.

Dulhunty, A. F. (2006). Excitation-contraction coupling from the 1950s into the new millennium. *Clinical and Experimental Pharmacology and Physiology, 33*, 763–772.

Dulhunty, A. F., Beard, N. A., Pouliquin, P., & Kimura, T. (2006). Novel regulators of RyR Ca^{2+} release channels: Insight into molecular changes in genetically-linked myopathies. *Journal of Muscle Research and Cell Motility, 27*, 351–365.

Dulhunty, A. F., Curtis, S. M., Watson, S., Cengia, L., & Casarotto, M. G. (2004). Multiple actions of imperatoxin A on ryanodine receptors: Interactions with the II-III loop "A" fragment. *The Journal of Biological Chemistry, 279*, 11853–11862.

Dulhunty, A. F., Pouliquin, P., Coggan, M., Gage, P. W., & Board, P. G. (2005). A recently identified member of the glutathione transferase structural family modifies cardiac RyR2 substate activity, coupled gating and activation by Ca^{2+} and ATP. *The Biochemical Journal, 390*, 333–343.

Fill, M., & Copello, J. A. (2002). Ryanodine receptor calcium release channels. *Physiological Reviews, 82*, 893–922.

Griffith, J. P., Kim, J. L., Kim, E. E., Sintchak, M. D., Thomson, J. A., Fitzgibbon, J., et al. (1995). X-ray structure of calcineurin inhibited by the immunophilin-immunosuppressant FKBP12-FK506 complex. *Cell, 82*, 507–522.

Gurrola, G. B., Arevalo, C., Sreekkumar, R., Lokuta, A. J., Walker, J. W., & Valdivia, H. H. (1999). Activation of ryanodine receptors by imperatoxin A and a peptide segment of the II-III loop of the dihydropyridine receptor. *The Journal of Biological Chemistry, 274*, 7879–7886.

Hamilton, S. L., & Serysheva, I. I. (2009). Ryanodine receptor structure: Progress and challenges. *The Journal of Biological Chemistry, 284*, 4047–4051.

Hinsen, K., Reuter, N., Navaza, J., Stokes, D. L., & Lacapere, J. J. (2005). Normal mode-based fitting of atomic structure into electron density maps: Application to sarcoplasmic reticulum Ca-ATPase. *Biophysical Journal, 88*, 818–827.

Hoenger, A., & Mcintosh, J. R. (2009). Probing the macromolecular organization of cells by electron tomography. *Current Opinion in Cell Biology, 21*, 89–96.

Hsieh, C. E., Leith, A., Mannella, C. A., Frank, J., & Marko, M. (2006). Towards high-resolution three-dimensional imaging of native mammalian tissue: Electron tomography of frozen-hydrated rat liver sections. *Journal of Structural Biology, 153*, 1–13.

Ikemoto, N. (2005). Intra-molecular domain-domain interaction: A key mechanism for calcium channel regulation of ryanodine receptors. In X.D.H.T. Wehrens & A. R. Marks (Eds.), *Ryanodine receptor: Structure, function and dysfunction in clinical disease* (pp. 53–65). New York: Springer.

Ikemoto, N., & Yamamoto, T. (2000). Postulated role of inter-domain interaction within the ryanodine receptor in Ca^{2+} channel regulation. *Trends in Cardiovascular Medicine, 10*, 310–316.

Ikemoto, N., & Yamamoto, T. (2002). Regulation of calcium release by interdomain interaction within ryanodine receptors. *Frontiers in Bioscience, 7*, D671–D683.

Jiang, Y., Lee, A., Chen, J., Cadene, M., Chait, B. T., & MacKinnon, R. (2002). Crystal structure and mechanism of a calcium-gated potassium channel. *Nature, 417*, 515–522.

Jones, P. P., Meng, X., Xao, B., Cai, S., Bolstad, J., Wagenknecht, T., et al. (2008). Localization of PKA phosphorylation site, Ser(2030), in the three-dimensional structure of cardiac ryanodine receptor. *The Biochemical Journal, 410*, 261–270.

Jurkat-Rott, K., McCarthy, T., & Lehmann-Horn, F. (2000). Genetics and pathogenesis of malignant hyperthermia. *Muscle and Nerve, 23*, 4–17.

Kim, S., Shin, D. W., Kim, D. H., & Eom, S. H. (1999). Crystallization and preliminary X-ray crystallographic studies of the D2 region of the skeletal muscle ryanodine receptor. *Acta Crystallographica D, 55*, 1601–1603.

Koning, R. I., & Koster, A. J. (2009). Cryo-electron tomography in biology and medicine. *Annals of Anatomy, 191*, 427–445.

Leis, A., Rockel, B., Andrees, L., & Baumeister, W. (2009). Visualizing cells at the nanoscale. *Trends in Biochemical Sciences, 34*, 60–70.

Liu, Z., & Wagenknecht, T. (2005). Three-dimensional reconstruction of ryanodine receptors. In X.D.H.T. Wehrens & A. R. Marks (Eds.), *Ryanodine receptor: Structure, function and dysfunction in clinical disease* (pp. 25–34). New York: Springer.

Liu, Z., Wang, R. W., Zhang, J., Chen, S. R. W., & Wagenknecht, T. (2005). Localization of a disease-associated mutation site in the three-dimensional structure of the cardiac muscle ryanodine receptor. *The Journal of Biological Chemistry, 280*, 37941–37947.

Liu, Z., Zhang, J., Li, P., Chen, S. R. W., & Wagenknecht, T. (2002). Three-dimensional reconstruction of the recombinant type 2 ryanodine receptor and localization of its divergent region 1. *The Journal of Biological Chemistry, 277*, 46712–46719.

Liu, Z., Zhang, J., Wang, R. W., Chen, S. R. W., & Wagenknecht, T. (2004). Location of divergent region 2 on the three-dimensional structure of cardiac muscle ryanodine receptor/calcium release channel. *Journal of Molecular Biology, 338*, 533–545.

Lobo, P. A., & Van Petegem, F. (2009). Crystal structures of the N-terminal domains of cardiac and skeletal muscle ryanodine receptors: Insights into disease mutations. *Structure, 17*, 1505–1514.

Ludtke, S. J., Serysheva, I. I., Hamilton, S. L., & Chiu, W. (2005). The pore structure of the closed RyR1 channel. *Structure, 13*, 1203–1211.

Marko, M., Hsieh, C., Schalek, R., Frank, J., & Mannella, C. (2007). Focused-ion-beam thinning of frozen-hydrated biological specimens for cryo-electron microscopy. *Nature Methods, 4*, 215–217.

Marks, A. R. (1996). Cellular functions of immunophilins. *Physiological Reviews, 76*, 631–649.

Maximciuc, A. A., Putkey, J. A., Shamoo, Y., & MacKenzie, K. R. (2006). Complex of calmodulin with a ryanodine receptor target reveals a novel, flexible binding mode. *Structure, 14*, 1547–1556.

Meissner, G. (2004). Molecular regulation of cardiac ryanodine receptor ion channel. *Cell Calcium, 35*, 621–628.

Meng, X., Wang, G. L., Viero, C., Wang, Q. L., Mi, W., Su, X. D., et al. (2009). CLIC2-RyR1 interaction and structural characterization by cryo-electron microscopy. *Journal of Molecular Biology, 387*, 320–334.

Meng, X., Xiao, B. L., Cai, S. T., Huang, X. J., Li, F., Bolstad, J., et al. (2007). Three-dimensional localization of serine 2808, a phosphorylation site in cardiac ryanodine receptor. *The Journal of Biological Chemistry, 282*, 25929–25939.

Milne, J. L. S., & Subramaniam, S. (2009). Cryo-electron tomography of bacteria: Progress, challenges and future prospects. *Nature Reviews. Microbiology, 7*, 666–675.

Murayama, T., Oba, T., Hara, H., Wakebe, K., Ikemoto, N., & Ogawa, Y. (2007). Postulated role of interdomain interaction between regions 1 and 2 within type 1 ryanodine receptor in the pathogenesis of porcine malignant hyperthermia. *The Biochemical Journal, 402*, 349–357.

Nath, A., Atkins, W. M., & Sligar, S. G. (2007). Applications of phospholipid bilayer nanodiscs in the study of membranes and membrane proteins. *Biochemistry, 46*, 2059–2069.

Norlen, L., Oktem, O., & Skoglund, U. (2009). Molecular cryo-electron tomography of vitreous tissue sections: Current challenges. *Journal of Microscopy, 235*, 293–307.

Oda, T., Yano, M., Yamamoto, T., Tokuhisa, T., Okuda, S., Doi, M., et al. (2005). Defective regulation of interdomain interactions within the ryanodine receptor plays a key role in the pathogenesis of heart failure. *Circulation, 111*, 3400–3410.

Orlova, E. V., Serysheva, I. I., van Heel, M., Hamilton, S. L., & Chiu, W. (1996). Two structural configurations of the skeletal muscle calcium release channel. *Nature Structural Biology, 3*, 547–552.

Paolini, C., Protasi, F., & Franzini-Armstrong, C. (2004). The relative position of RyR feet and DHPR tetrads in skeletal muscle. *Journal of Molecular Biology, 342*, 145–153.

Priori, S. G., & Napolitano, C. (2005). Cardiac and skeletal muscle disorders caused by mutations in the intracellular Ca^{2+} release channels. *The Journal of Clinical Investigation, 115*, 2033–2038.

Radermacher, M., Rao, V., Grassucci, R., Frank, J., Timerman, A. P., Fleischer, S., et al. (1994). Cryo-electron microscopy and three-dimensional reconstruction of the calcium release channel ryanodine receptor from skeletal muscle. *The Journal of Cell Biology, 127*, 411–423.

Radermacher, M., Wagenknecht, T., Grassucci, R., Frank, J., Inui, M., Chadwick, C., et al. (1992). Cryo-EM of the native structure of the calcium release channel/ryanodine receptor from sarcoplasmic reticulum. *Biophysical Journal, 61*, 936–940.

Reiken, S., Lacampagne, A., Zhou, H., Kherani, A., Lehnart, S. E., Ward, C., et al. (2003). PKA phosphorylation activates the calcium release channel (ryanodine receptor) in skeletal muscle: Defective regulation in heart failure. *The Journal of Cell Biology, 160*, 919–928.

Renken, C., Hsieh, C. E., Marko, M., Rath, B., Leith, A., Wagenknecht, T., et al. (2009). Structure of frozen-hydrated triad junctions: A case study in motif searching inside tomograms. *Journal of Structural Biology, 165*, 53–63.

Robinson, R., Carpenter, D., Shaw, M. A., Halsall, J., & Hopkins, P. (2006). Mutations in RYR1 in malignant hyperthermia and central core disease. *Human Mutation, 27*, 977–989.

Samso, M., Feng, W., Pessah, I. N., & Allen, P. D. (2009). Coordinated movement of cytoplasmic and transmembrane domains of RyR1 upon gating. *PLoS Biology, 7*, 980–995.

Samso, M., Shen, X. H., & Allen, P. D. (2006). Structural characterization of the RyR1-FKBP12 interaction. *Journal of Molecular Biology, 356*, 917–927.

Samso, M., Trujillo, R., Gurrola, G. B., Valdivia, H. H., & Wagenknecht, T. (1999). Three-dimensional location of the imperatoxin A binding site on the ryanodine receptor. *The Journal of Cell Biology, 146*, 493–499.

Samso, M., & Wagenknecht, T. (2002). Apocalmodulin and Ca^{2+}-calmodulin bind to neighboring locations on the ryanodine receptor. *The Journal of Biological Chemistry, 277*, 1349–1353.

Samso, M., Wagenknecht, T., & Allen, P. D. (2005). Internal structure and visualization of transmembrane domains of the RyR1 calcium release channel by cryo-EM. *Nature Structural and Molecular Biology, 12*, 539–544.

Serysheva, I. I., Ludtke, S. J., Baker, M. L., Cong, Y., Topf, M., Eramian, D., et al. (2008). Subnanometer-resolution electron cryomicroscopy-based domain models for the cytoplasmic region of skeletal muscle RyR channel. *Proceedings of the National Academy of Sciences of the United States of America, 105*, 9610–9615.

Serysheva, I. I., Schatz, M., van Heel, M., Chiu, W., & Hamilton, S. L. (1999). Structure of the skeletal muscle calcium release channel activated with Ca^{2+} and AMP-PCP. *Biophysical Journal, 77*, 1936–1944.

Sharma, M. R., Jeyakumar, L. H., Fleischer, S., & Wagenknecht, T. (2000). Three-dimensional structure of ryanodine receptor isoform three in two conformational states as visualized by cryo-electron microscopy. *The Journal of Biological Chemistry, 275*, 9485–9491.

Sharma, M. R., Jeyakumar, L. H., Fleischer, S., & Wagenknecht, T. (2006). Three-dimensional visualization of FKBP12.6 binding to an open conformation of cardiac ryanodine receptor. *Biophysical Journal, 90*, 164–172.

Sharma, M. R., Penczek, P., Grassucci, R., Xin, H.-B., Fleischer, S., & Wagenknecht, T. (1998). Cryoelectron microscopy and image analysis of the cardiac ryanodine receptor. *The Journal of Biological Chemistry, 273*, 18429–18434.

Sorrentino, V., & Volpe, P. (1993). Ryanodine receptors—How many, where and why. *Trends in Pharmacological Sciences, 14*, 98–103.

Spahn, C. M. T., & Penczek, P. A. (2009). Exploring conformational modes of macromolecular assemblies by multiparticle cryo-EM. *Current Opinion in Structural Biology, 19*, 623–631.

Stowell, K. M. (2008). Malignant hyperthermia: A pharmacogenetic disorder. *Pharmacogenomics, 9*, 1657–1672.

Tanabe, T., Beam, K. G., Adams, B. A., Nicodome, T., & Numa, S. (1990). Regions of the skeletal muscle dihydropyridine receptor critical for excitation-contraction coupling. *Nature, 346*, 567–569.

Topf, M., Lasker, K., Webb, B., Wolfson, H., Chiu, W., & Sali, A. (2008). Protein structure fitting and refinement guided by cryo-EM density. *Structure, 16*, 295–307.

Treves, S., Vukcevic, M., Maj, M., Thurnheer, R., Mosca, B., & Zorzato, F. (2009). Minor sarcoplasmic reticulum membrane components that modulate excitation-contraction coupling in striated muscles. *The Journal of physiology, 587*, 3071–3079.

Volkmann, N. (2009). Confidence intervals for fitting of atomic models into low-resolution densities. *Acta Crystallographica D, 65*, 679–689.

Volkmann, N., & Hanein, D. (1999). Quantitative fitting of atomic models into observed densities derived by electron microscopy. *Journal of Structural Biology, 125*, 176–184.

Wagenknecht, T., Hsieh, C. E., Rath, B. K., Fleischer, S., & Marko, M. (2002). Electron tomography of frozen-hydrated isolated triad junctions. *Biophysical Journal, 83*, 2491–2501.

Wagenknecht, T., Radermacher, M., Grassucci, R., Berkowitz, J., Xin, H.-B., & Fleischer, S. (1997). Locations of calmodulin and FK506-binding protein on the three-dimensional architecture of the skeletal muscle ryanodine receptor. *The Journal of Biological Chemistry, 272*, 32463–32471.

Wang, L., & Sigworth, F. J. (2009). Structure of the BK potassium channel in a lipid membrane from electron cryomicroscopy. *Nature, 461*, 292–295.

Wang, R. W., Chen, W. Q., Cai, S. T., Zhang, J., Bolstad, J., Wagenknecht, T., et al. (2007). Localization of an NH_2-terminal disease-causing mutation hot spot to the "clamp" region in the three-dimensional structure of the cardiac ryanodine receptor. *The Journal of Biological Chemistry, 282*, 17785–17793.

Wehrens, X. H. T., Lehnart, S. E., Reiken, S., Vest, J. A., Wronska, A., & Marks, A. R. (2006). Ryanodine receptor/calcium release channel PKA phosphorylation: A critical mediator of heart failure progression. *Proceedings of the National Academy of Sciences of the United States of America, 103*, 511–518.

Welch, W., Rheault, S., West, D. J., & Williams, A. J. (2004). A model of the putative pore region of the cardiac ryanodine receptor channel. *Biophysical Journal, 87*, 2335–2351.

Wriggers, W., Milligan, R. A., & McCammon, J. A. (1999). Situs: A package for docking crystal structures into low-resolution maps from electron microscopy. *Journal of Structural Biology, 125*, 185–195.

Xiao, B. L., Sutherland, C., Walsh, M. P., & Chen, S. R. W. (2004). Protein kinase A phosphorylation at serine-2808 of the cardiac Ca^{2+}-release channel (ryanodine receptor) does not dissociate 12.6-kDa FK506-binding protein (FKBP12.6). *Circulation Research, 94*, 487–495.

Yamamoto, T., Yano, M., Xu, X. J., Uchinoumi, H., Tateishi, H., Mochizuki, M., et al. (2008). Identification of target domains of the cardiac ryanodine receptor to correct channel disorder in failing hearts. *Circulation, 117*, 762–772.

Ye, F., Hu, G., Taylor, D., Ratnikov, B., Bobkov, A. A., McLean, M. A., et al. (2010). Recreation of the terminal events in physiological integrin activation. *The Journal of Cell Biology, 188*, 157–173.

Yin, C. C., Blayney, L. M., & Lai, F. A. (2005). Physical coupling between ryanodine receptor-calcium release channels. *Journal of Molecular Biology, 349*, 538–546.

Yin, C. C., D'Cruz, L. G., & Lai, F. A. (2008). Ryanodine receptor arrays: Not just a pretty pattern? *Trends in Cell Biology, 18*, 149–156.

Yin, C. C., Han, H. M., Wei, R. S., & Lai, F. A. (2005). Two-dimensional crystallization of the ryanodine receptor Ca^{2+} release channel on lipid membranes. *Journal of Structural Biology, 149*, 219–224.

Yin, C. C., & Lai, F. A. (2000). Intrinsic lattice formation by the ryanodine receptor calcium-release channel. *Nature Cell Biology, 2*, 669–671.

Zhang, J., Liu, Z., Masumiya, H., Wang, R., Jiang, D. W., Li, F., et al. (2003). Three-dimensional localization of divergent region 3 of the ryanodine receptor to the clamp-shaped structures adjacent to the FKBP binding sites. *The Journal of Biological Chemistry, 278*, 14211.

Zhao, M., Li, P., Li, X., Zhang, L., Winkfein, R. J., & Chen, S. R. W. (1999). Molecular identification of the ryanodine receptor pore-forming segment. *The Journal of Biological Chemistry, 274*, 25971–25974.

Zhou, Q., Wang, Q. L., Meng, X., Shu, Y. Y., Jiang, T., Wagenknecht, T., et al. (2008). Structural and functional characterization of ryanodine receptor-natrin toxin interaction. *Biophysical Journal, 95*, 4289–4299.

CHAPTER 3

The Ryanodine Receptor Pore: Is There a Consensus View?

Joanne Carney, Sammy A. Mason, Cedric Viero, and Alan J. Williams
Department of Cardiology, Wales Heart Research Institute, School of Medicine, Cardiff University, Heath Park, Cardiff, United Kingdom

I. OVERVIEW

The intracellular Ca^{2+}-release channel referred to as the ryanodine receptor (RyR) is a key factor in a plethora of biological and pathophysiological processes and is therefore the focus of interest of many scientists and

Current Topics in Membranes, Volume 66
Copyright 2010, Elsevier Inc. All right reserved.
1063-5823/10 $35.00
DOI: 10.1016/S1063-5823(10)66003-6

clinicians. In recent years, an ever-growing body of evidence has emerged detailing the ligands and mechanisms involved in the regulation of RyR activity, and how this might be altered in disease. This information is reviewed in other articles in this publication. Here we review a fundamental aspect of RyR function, namely the structures and mechanisms that control which, and how many, ions can flow through the open channel and underpin the unusual ion handling characteristics of RyR and its great efficiency as a Ca^{2+}-release channel. Therefore, we will limit our discussion to a relatively small region of the channel that provides the pathway for ion movement across the membrane: the pore-forming region (PFR). Our present understanding of these processes has been informed by a wide range of approaches. Information on the likely structure of the PFR has been provided by high-resolution cryo-electron microscopy (cryo-EM) and by molecular modeling, following the identification of structural analogies between RyR and other ion channels. The mechanisms involved in discrimination and translocation have emerged from detailed characterization of channel function, theoretical models, and the interpretation of the consequences of residue mutation in the PFR. Together, these approaches have been used to identify regions of RyR that are critical for ion movement and may contribute to the binding site for ryanodine.

II. INTRODUCTION

Endo/sarcoplasmic reticulum (ER/SR) intracellular membrane systems contain two species of Ca^{2+}-release channel, the inositol trisphosphate receptor ($InsP_3R$) and the RyR, so named because this channel possesses a high-affinity binding site for the plant alkaloid ryanodine (Sutko, Airey, Welch, & Ruest, 1997). Three mammalian isoforms of RyR have been identified: RyR1 found predominantly in skeletal muscle, RyR2 found in both cardiac muscle and the brain, and RyR3 which is expressed at low levels in a wide range of tissues (Fleischer, 2008). Functional RyR channels are homotetramers with each monomer containing approximately 5000 amino acids. Detailed information on the functional properties of RyR channels has been obtained following the incorporation of individual channels into artificial phospholipid bilayers which permits the measurement of single channel gating and ion handling (Williams, West, & Sitsapesan, 2001). RyR channels are extremely efficient Ca^{2+}-release channels and this efficiency is underpinned by the structure and mechanisms of ion handling of the PFR of the molecule. Our present understanding of these parameters will be reviewed here.

III. ION HANDLING IN RYR

A. Basic Properties of Ion Handling in the RyR Isoforms

The first characterization of ion handling in RyR was carried out by Meissner and colleagues using purified rabbit RyR1 (Smith et al., 1988). The studies demonstrated that RyR1 (a) is cation selective, (b) is divalent selective, (c) has a higher unitary conductance for monovalent than divalent cations, and (d) is essentially nonselective between group 1a monovalent cations. These basic features were confirmed in detailed characterizations of monovalent (Lindsay, Manning, & Williams, 1991) and divalent ion handling (Tinker & Williams, 1992) in purified sheep RyR2 (see Table I).

In addition to the above, RyR2 was found to be relatively nonselective between individual mono- or divalent cations, somewhat selective for divalents ($P_{X^{2+}}/P_{K^+} \sim 6$), while exhibiting a high affinity (μM) for divalents (Tinker, Lindsay, & Williams, 1992b) compared to a relatively low affinity (mM) for monovalent cations. The different experimental conditions that have been used to examine ion handling in the RyR isoforms makes a direct comparison of properties difficult; however, basic features are compared in Table II.

TABLE I
Ion Handling in RyR2

X^+	pX^+/pK^+	γ (pS)	K_D (mM)	γ_{max} (pS)
K^+	1.00	723	19.9	900
Na^+	1.15	446	17.8	516
Cs^+	0.61	440	34.0	588
Rb^+	0.87	621	n.d.	n.d.
Li^+	0.99	215	9.1	248
X^{2+}	pX^{2+}/pK^+	γ (pS)	*K_D (mM)	pX^{2+}/pBa^{2+}
Ba^{2+}	5.8	202 ± 5	0.165	1.0
Sr^{2+}	6.7	166 ± 4	0.123	1.1
Ca^{2+}	6.5	135 ± 5	0.116	1.1
Mg^{2+}	5.9	89 ± 4	0.116	1.1

pX^+/pK^+: bi-ionic 210 mM.
γ (pS): determined at symmetrical 210 mM [X^+] or bi-ionic with [K^+/X^{2+}].
K_D and γ_{max} from Michaelis–Menten conductance–activity plots. [X^+] or *predicted from RyR Eyring rate theory model (Tinker et al., 1992b).
pX^{2+}/pK^+: bi-ionic 210 mM.
pX^{2+}/pBa^{2+}: bi-ionic with 210 mM [Ba^{2+}/X^{2+}].
n.d.: not determined.

TABLE II

Comparison of Ion Handling in RyR Subtypes

	RyR1	RyR2	RyR3
$[K^+]$ γ (pS)	772	723	777
$[Ca^{2+}]$ γ (pS)	123	135	137
pCa^{2+}/pK^+	6.6	6.5	6.3
Permeability sequence	$Li^+>Na^+>K^+>Rb^+>Cs^+$	$Na^+>K^+>Li^+>Rb^+>Cs^+$	n.d.
Conductance sequence	$K^+>Cs^+>Rb^+>Na^+>Li^+$	$K^+>Rb^+>Na^+>Cs^+>Li^+$	n.d.

RyR1, RyR2 and RyR3 K^+ slope conductances were examined in symmetrical $[K^+]$ 250 mM/210 mM/250 mM, respectively. Divalent permeability ratios were examined in bi-ionic $[K^+/Ca^{2+}]$ at 50/50 mM, 210/210 mM and 250/250 mM, respectively. Monovalent permeability ratios were examined in bi-ionic $[K^+/X^+]$ 210/200–210 mM, 210/210 mM and 250/250 mM, respectively. Conductance sequences were compared in symmetrical $[X^+]$ 200–215 mM (RyR1) and in symmetrical 210 mM (RyR2). n.d. : not determined. Readers may refer to the following publications for further information: RyR1 (Shomer et al., 1994; Smith et al., 1988; Xu et al., 1993), RyR2 (Lindsay et al., 1991; Tinker & Williams, 1992) and RyR3 (Chen et al., 1997b).

Differences between the RyR isoforms are minimal; however, there are some limited differences in the relative conductance of monovalent cations and permeability sequences of RyR1 and RyR2 that may suggest some minor differences in the mechanisms governing ion handling in these channels. The sequence for monovalent and divalent cation binding in RyR2 correspond to Eisenman (1962) and Sherry (1969) sequences XI and VII, respectively, indicating the involvement of high field strength sites in the process of discrimination (Tinker et al., 1992b). Based on the structure of Ca^{2+} binding proteins, it was predicted that cation binding at such sites could be coordinated by carboxyl oxygens. To account for the higher relative permeability of divalents over monovalents, it was proposed that the RyR PFR was likely to contain a high density of such sites (Tinker et al., 1992b).

B. What is Responsible for the High Rate of Ion Translocation in RyR?

The maximum rate of cation translocation in RyR is phenomenal, with saturating unitary conductance for monovalent cations approaching 1 nS (approximately twice the theoretical limit for a selective channel based on the laws of diffusion; Hille, 1991). Clearly an enormous unitary conductance is a useful property for a Ca^{2+}-release channel, but how is this achieved? A range of potential contributing mechanisms have been proposed, for example, rates of delivery of cations could be maximized by a high density of acidic residues at the entrance of the PFR (Tinker et al., 1992b) or the RyR PFR could be occupied by more than one ion at a time leading to increased rates of ion exit by ion–ion repulsion (Smith et al., 1988). Another suggestion, based on earlier predictions for large conductance K^+ channels (Latorre & Miller, 1983), was that the PFR of RyR, in comparison with other ion channels, was both wide and short so maximizing both rates of ion entry and exit (Williams, 1992).

The dimensions of some aspects of the RyR PFR have been investigated. The minimum radius of the RyR PFR has been estimated as 3.5 Å based on the relative permeability of organic monovalent cations of known dimensions (Tinker & Williams, 1993). However, subsequent experiments have shown that this region of the PFR is likely to be flexible. With a high driving force neomycin (minimum radius 5 Å), a normally impermeant blocking polycation, can move through the PFR (Mead & Williams, 2002a). The binding of ryanodine to RyR prevents this translocation presumably by decreasing the flexibility of the PFR (Mead & Williams, 2002b). Information on aspects of the length of the PFR is also available.

The length of the voltage drop across the channel has been measured at approximately 10 Å by monitoring blocking parameters of bis-quaternary ammonium ions of varying length (Tinker & Williams, 1995) and the region of RyR in which single-file diffusion occurs has been estimated as 9 Å from measurements of streaming potentials (Tu, Velez, Brodwick, & Fill, 1994). Taken together these parameters are consistent with the proposal that the PFR of RyR is, in comparison with more selective, lower conductance, channels such as the bacterial K^+ channel KcsA, relatively wide and short (Williams et al., 2001).

IV. WHERE IS THE PFR IN THE RYR CHANNEL?

A. *Evidence from Cryo-Electron Microscopy*

Three-dimensional structures of RyR tetramers have been obtained by reconstruction of images gathered from electron micrographs of frozen-hydrated single isolated channels. Of the three RyR isoforms, the most extensive structure–function analysis has been carried out with RyR1 due to its high abundance in skeletal muscle and the relative ease with which it can be purified. Until recently, the most detailed structures available were at resolutions of only 25–30 Å (for a review see Serysheva, Chiu, & Ludtke, 2007). Although no secondary structure elements can be identified at this resolution, it is clear that RyRs have an overall mushroom shape with a large cytoplasmic foot assembly connected to a transmembrane (TM) stalk-like structure (Radermacher et al., 1994; Serysheva et al., 1995). This technique has been used to identify domains in the cytoplasmic foot which are binding sites for various effectors (Liu, Zhang, Wang, Wayne Chen, & Wagenknecht, 2004; Samso & Wagenknecht, 2002; Wagenknecht et al., 1997). By analogy with other ion channels, the PFR of RyR is likely to be located in the TM portion of the structure and an apparent opening running axially into the TM domain has been identified in what is believed to be an open conformation of RyR1 (Orlova, Serysheva, van Heel, Hamilton, & Chiu, 1996). This opening is not visible in the closed conformation of the channel and these studies have also highlighted other structural alterations in RyR that may occur during the transition from a closed state to an open state of the channel. Recent technical advances have yielded structural information on the TM domain of RyR1 at considerably improved resolution and this will be discussed in Section V.B.

B. Analogies with K^+ Channels and Evidence from Functional Studies of Mutant RyR Channels

In 1999, Balshaw, Gao, and Meissner (1999) identified a motif located in a luminal loop between the last two membrane spanning helices in the primary structure of RyR (GGGIGD) (see Fig. 1) as being analogous to the signature selectivity sequence of K^+ channels and suggested that this region was likely to be a component of a PFR in RyR. Consistent with this proposal, mutation of various residues within this motif, and of residues flanking this motif, were shown to alter ion handing properties of both RyR1 and RyR2 (Fig. 1).

1. RyR1

Meissner's laboratory expressed rabbit RyR1 in HEK cells and screened amino acids at the end of the predicted "loop 2" of the PFR (see Fig. 1). R4913E RyR1 channels released Ca^{2+} after caffeine stimulation *in situ*, but [^{3}H]-ryanodine binding was lost and unitary K^+ conductance was decreased by 60% while Ca^{2+} current was also reduced. In contrast, D4917A mutants showed no caffeine-induced Ca^{2+} transients, no [^{3}H]-ryanodine binding, no detectable Ca^{2+} current, but retained 60% of the unitary K^+ conductance (Gao et al., 2000). Of particular interest are mutations of two residues conserved in all RyR isoforms: D4903A and D4907A. Both mutants retained Ca^{2+}-dependent [^{3}H]-ryanodine binding and had unitary conductance indistinguishable from wild-type (WT) channels. Caffeine-evoked Ca^{2+} transients were seen in cells expressing both mutants, but were increased by the D4907A substitution (Gao et al., 2000). The mutant G4894A induced dramatic changes in ion conductance. Maximum unitary K^+ conductance was severely decreased and Ca^{2+} current was abolished, while responses to caffeine and [^{3}H]-ryanodine binding were preserved (Gao et al., 2000).

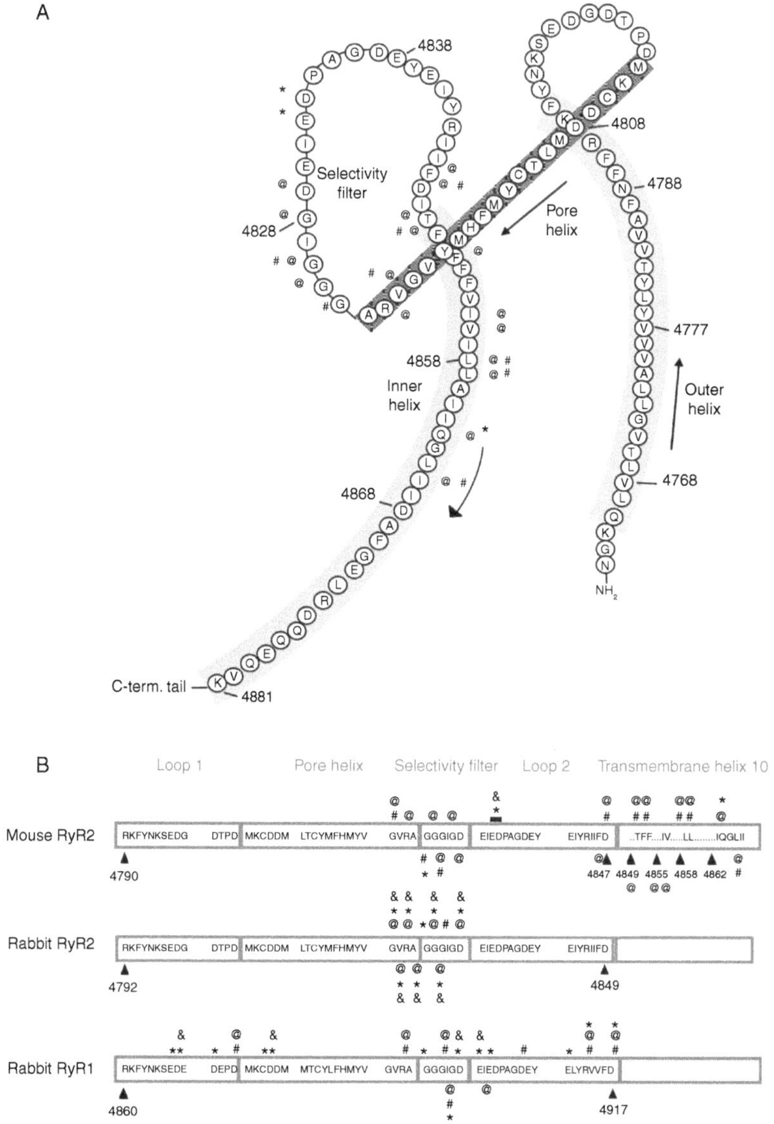

FIGURE 1 Mutagenesis characterization and sequence alignments of the putative pore-forming regions of RyR Ca^{2+}-release channels. (A) Tube map of our model of the mouse RyR2 pore. Meaning of the symbols can be found below. (B) This schematic representation depicts the localization of various residues of the different predicted pore regions (rabbit RyR1, rabbit and mouse RyR2, according to available published data) that underwent substitution into another residue or a series of diverse amino acids. The arrowheads specify the residue numbering according to each isoform of each species. The consequences of the mutations on channel function are indicated as follows: *, ion translocation impairment (permeation, conductance); #, Ca^{2+} release is compromised (in microsomes or cells; stimulation by caffeine); @, decrease or abolition of [^{3}H]-ryanodine binding; &, disruption of the normal ion discrimination (selectivity).

Recently, MacLennan's laboratory characterized the I4898T substitution in the human RyR1 sequence (corresponding to I4897T in rabbits and I4895T in mice), a mutation in the putative selectivity filter which occurs in central-core disease (CCD) patients (Zvaritch et al., 2007). Transient expression of the rabbit ortholog in HEK-293 cells led to the identification of leaky channels (Lynch et al., 1999). However, homozygous I4895T mice displayed altered developmental features with a suppressed RyR1-mediated Ca^{2+} release while the morphology of *in situ* RyR1 clusters was preserved (Zvaritch et al., 2007).

In addition, previous investigations from the same group provided evidence that R4892W, I4897T, and G4898E (three CCD mutations) resulted in reduced Ca^{2+} sensitivity and amplitude of Ca^{2+}-dependent Ca^{2+} release, together with a decrease of [^{3}H]-ryanodine binding, when rabbit channels were coexpressed with SERCA1a in HEK-293 cells (Du, Khanna, Guo, & MacLennan, 2004).

2. RyR2

The involvement of residues in and around the putative selectivity filter in RyR2 has also been investigated. Residue substitutions in the rabbit motif, GVRAGGGIGD (Fig. 1), were studied by MacLennan and colleagues (Du, Guo, Khanna, & MacLennan, 2001). With the exception of I4829A and I4829T (corresponding to the CCD mutant I4897T in RyR1), *in situ* caffeine-evoked Ca^{2+} transients were observed in all mutants (either Ala substitution or Ala into Val). Moreover, there was a complete loss of [^{3}H]-ryanodine binding in all mutants except G4826A, I4829V, and G4830A. A high concentration of ryanodine (10 μM) induced *in situ* Ca^{2+} release in all Ala mutants from 4823 to 4827. Ryanodine also raised the amplitude of caffeine-evoked Ca^{2+} transients in G4828A and restored caffeine sensitivity/responsiveness in I4829A and I4829T. Single-channel recording was possible for all mutants except G4822A and A4825V; however, the unitary K^+ conductance of all mutants within this group except I4829V and G4830A was reduced; in G4824A (corresponding to G4894 in RyR1) conductance was reduced by 97%. The ability to discriminate between Ca^{2+} and K^+ was lost in all of these mutants, except G4826A, I4829V, and G4830A (Du et al., 2001). These findings confirmed the results of earlier investigations reported by Chen and colleagues (Zhao et al., 1999) who characterized residues in the equivalent motif (GVRAGGGIGD) in mouse RyR2 channels. These studies established that residue substitutions at R4822, G4825, G4828, and D4829 reduced or abolished high-affinity [^{3}H]-ryanodine binding, while *in situ* caffeine-induced Ca^{2+} release in cells expressing these mutant RyR2s was comparable with that of WT channels. In contrast, substitution at G4824 yielded RyR2 channels that retained [^{3}H]-ryanodine binding, but displayed decreased caffeine-induced Ca^{2+} release. The K^+ unitary conductance of these channels was reduced drastically (22 pS as compared with 798 pS for the WT). Mutations at G4820 and G4826 reduced caffeine-induced Ca^{2+} release and abolished [^{3}H]-ryanodine binding.

C. Further Analogies Between the PFR of K^+ Channels and RyR

A breakthrough in our understanding of the mechanisms involved in the function of ion channels came about when the crystal structure of the bacterial K^+ channel, KcsA, was solved by MacKinnon and coworkers (Doyle et al., 1998). This prototypical ion channel structure revolutionized our understanding of the structures involved in ion discrimination, ion translocation, and the gating process of K^+ channels, allowing direct comparisons with other cation channels. The pore of KcsA is formed by four identical subunits arranged around a fourfold symmetry axis to form a functional tetramer, each consisting of two TM helices (Fig. 2). An extracellular loop, which folds back into the membrane, connects the helices and is composed of a short pore helix and a sequence of amino acids that contain a selectivity filter incorporating the K^+ channel signature selectivity sequence. The pore forms an inverted tepee with a gate at the cytosolic entrance. The transition from a closed to an open state occurs by the flexing of a

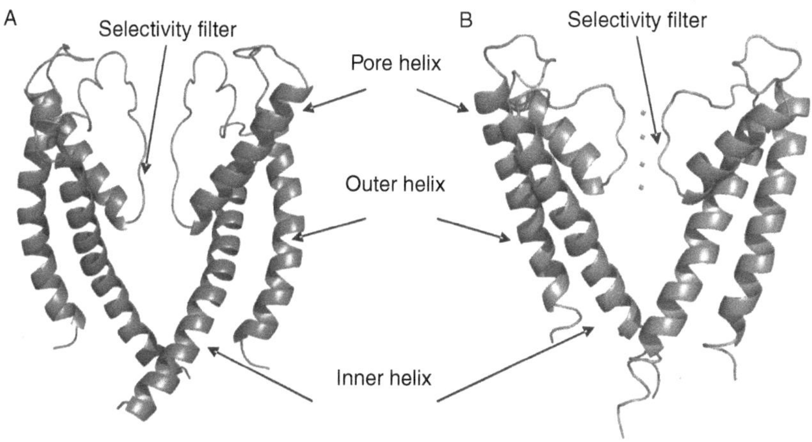

FIGURE 2 (A) RyR pore (analogy model from Welch et al., 2004) and (B) KcsA with four potassium ions shown in the four binding sites of the selectivity filter (1BL8.pdb, Doyle et al., 1998). Two opposing monomers of each channel are shown for clarity.

glycine hinge located approximately halfway along the inner helix (Jiang et al., 2002b). The relatively rigid structure of the selectivity filter of KcsA allows precise coordination of K^+ ions with backbone carbonyl atoms of residues. The hydration state of the permeating ion is also a crucial factor in selecting K^+ over Na^+ (Dudev & Lim, 2009).

V. ATTEMPTS TO IDENTIFY THE STRUCTURE OF THE RYR PFR

A. A Model of the RyR Pore Using KcsA as a Template

Following the identification of a presumed selectivity filter, secondary structure predictions for the loop linking the last two TM helices in RyR indicated that this region was likely to contain structural elements equivalent to those found in the PFR of K^+ channels (Williams et al., 2001). To further test possible structural similarities between the PFRs of RyR and K^+ channels Welch, Rheault, West, and Williams (2004) constructed of a model of the putative RyR PFR using the recently solved crystal structure of KcsA as a template (see Fig. 2).

The model comprises just 2.4% of the total RyR2 monomer incorporating the last two TM helices and the luminal loop of each monomer. By combining information from secondary structure prediction programs and comparing sequence analogies with KcsA, corresponding structural elements in RyR were identified and, using molecular modeling software, a stable tetrameric model for RyR was constructed.

By making quantitative comparisons of the physicochemical properties of KcsA and the RyR homology model, Welch et al. (2004) were able to demonstrate that their model represents a highly plausible model for the PFR region of RyR. While the mechanisms governing ion handling must be vastly different in KcsA and RyR the actual structural elements of the PFRs and their organization are likely to be very similar with both structures having a cytosolic cavity lined by the inner helices that taper inward to form a gate. The most obvious differences between the two structures are the arrangements of the loop between the outer helix and the pore helix and the width and shape of the selectivity filter, RyR being much wider than that of KcsA. There are also marked differences in the solvent accessibility in this region with water penetrating all the way across the selectivity filter in the RyR2 model. Rings of negative charge are located at both cytosolic and luminal ends of both structures although the charge density is much higher in RyR2, a factor that could be important in concentrating cations around the entrance to the pore (see above). The much greater degree of flexibility of the selectivity filter observed in the RyR model, compared to

KcsA, and the higher density of negative charge around the mouths are likely to be key determinants in the observed differences in ion handling between RyRs and KcsA. Molecular dynamics (MD) simulations performed in the RyR model correlate extremely well with previous experimental data with regards to selective ion permeation (Williams et al., 2001) and block of K^+ translocation by tetraethyl ammonium ions (Tinker, Lindsay, & Williams, 1992a). The reader is referred to the original paper for more detailed and quantitative comparisons and assessments of structural components, physicochemical comparisons, and ion handling capabilities of the model (Welch et al., 2004). The RyR pore analogy model contains a glycine residue in the putative inner helix (GLIIDA) (Fig. 1) that could act as a hinge in a gating motif (GXXXXA), as originally identified by Jiang et al. (2002b) in K^+ channels. This raises the possibility that the conformational changes underlying gating in RyRs could be similar to those of K^+ channels.

B. High-Resolution Images of the PFR of RyR

Two reports in 2005 detailed the structures of presumed closed states of RyR1 obtained using cryo-EM at resolutions of around 10 Å. The first structure was compared directly to the atomic model of KcsA, docking this structure into the 3D map of RyR1 (Samso, Wagenknecht, & Allen, 2005).

Apart from the outer helices, the overall orientation and arrangement of structural elements display a high degree of architectural compatibility, with the selectivity filter orientated toward the lumen and the constriction of the inner helices forming a gate at the cytoplasmic face of the structure. The second structure, published by Ludtke, Serysheva, Hamilton, and Chiu (2005) is modeled independently of any known structures and identifies five α-helices in the RyR1 TM domain of which two are identified as the inner helix and pore helix. In contrast to the cryo-EM structure obtained by Samso et al. (2005) the inner helix is modeled with a kink at the putative glycine gating hinge and is therefore compared to the open structure of the K^+ channel MthK (Jiang et al., 2002a; Samso et al., 2005).

The different conformations of the inner helices in these two structures mean that disagreement exists regarding the closed structure of RyR and the theory that kinking of the inner helices is required to open the channel. This may be due to the relatively low level of resolution and the ambiguity as to which conformational state the protein is in during the imaging process (Hamilton & Serysheva, 2009). What is clear from these cryo-EM structures is that the RyR PFR orientates around a central fourfold axis of symmetry and that a funnel like cavity is formed with a wide entrance at the luminal side tapering in towards the cytoplasmic side. These observations strongly correlate with the earlier analogy model (Welch et al., 2004) and indicate that the PFR of RyRs are indeed similar in architecture to K^+ channels and could gate using similar elements.

C. A RyR PFR Model Based on a High-Resolution Cryo-EM Structure

In an attempt to bridge the gap between the low resolution limits of cryo-EM and the previously published RyR PFR model, Ramachandran, Serohijos, Xu, Meissner, and Dokholyan (2009) recently built a molecular model of the putative PFR of RyR1 based on the cryo-EM structure of Ludtke et al. (2005). The unambiguous assignment of several helix-like densities gave a model with remarkable similarities to the MthK K^+ channel, crystallized in the open state. The two main helices, inner helix and pore-forming helix, were found to face each other and line the channel pore, as evidenced in all previous cryo-EM structures. As in previous assignments of the pore structure, acidic residues are located at the mouth of the pore with residues D4899 and E4900, postulated to be important for channel function in RyR1, positioned within the selectivity filter (Wang, Xu, Pasek, Gillespie, & Meissner, 2005; Xu, Wang, Gillespie, & Meissner, 2006).

The low resolution limits of cryo-EM do not allow for modeling of amino acid side chains. However, by replacing the missing side chains and adding the luminal loop, Ramachandran et al. (2009) were able to perform MD simulations and, by taking into consideration data from previously published single channel experiments, were able to examine interactions of the PFR with mono- and divalent ions known to permeate the channel (Lindsay et al., 1991; Tinker & Williams, 1992). MD simulations of selectivity filter mutants known to alter ion handling correlate well with experimental data with ion occupancies observed in the model agreeing well with the charge space competition (CSC) theory, as discussed below. Specific residues located near the selectivity filter show preferential affinity for Ca^{2+} over K^+, supporting the accuracy of the model as a Ca^{2+}-selective channel. The proposed glycine gating hinge occurs at residue 4934 in RyR1, analogous to G4864 modeled in RyR2 by Welch et al. (2004). That recent structures resemble the open state of MthK even when solved in conditions apparently favoring the closed state of the channel indicate that mechanisms other than flexing at the inner helix glycine hinge may be responsible for the transition between closed and open states in RyR. In order to ascertain the correct conformation of the inner helix in the closed state, an improvement in cryo-specimen presentation and image processing is needed.

VI. THEORETICAL APPROACHES TO UNDERSTANDING THE MECHANISMS UNDERLYING ION TRANSLOCATION AND DISCRIMINATION IN RYR

In the absence of detailed information on the structure of the RyR PFR, potential mechanisms underlying ion handling have been explored using theoretical models. A major area of debate that has arisen from these studies is whether the high unitary conductance of RyR is achieved as the result of ion–ion repulsion in a multi-ion pore. It is frequently acknowledged within the field that the interpretation of experimental ion channel permeation phenomena and ion channel occupancy are model dependent. For example, the saturation of conductance in RyR suggests that ions do not move independently of one another (as in solution) with a fit consistent with a single occupancy pore (Lindsay et al., 1991; Tinker & Williams, 1992); however, multi-ion models can show saturation (Gillespie & Boda, 2008; Gillespie & Eisenberg, 2002).

For this reason the ion occupancy state of RyR has been a matter of strong debate with no clear conclusion made due to the presence of two opposing theories; rate theories such as that of Eyring and continuum models such as Poisson–Nernst–Planck (PNP) (Levitt, 1986; Miller, 1999; Nonner, Chen & Eisenberg, 1999) which both describe the patterns of ion permeation in RyR (Chen, Xu, Tripathy, Meissner, & Eisenberg, 1997, 1999; Tinker et al., 1992b). Although there are multiple lines of "model-dependent" experimental evidence supporting multi- (Smith et al., 1988) or single-ion occupancy (Lindsay et al., 1991; Tinker & Williams, 1992; Tinker et al., 1992b) one phenomenon has become central to the argument of ion occupancy in RyR, the anomalous mole fraction effect (AMFE). AMFE is observed as a minima in current, conductance or Erev when the concentration of ions within a bi-ionic mixture varies to produce different ratios of ions (mole fractions). AMFE can be interpreted in terms of Eyring rate theory with a single filing multi-ion pore (Lindsay et al., 1991; Tinker & Williams, 1992) or in terms of resistors in series (PNP—continuum model) which can be shown when (on average) < 1 ion is in the pore (Nonner, Chen, & Eisenberg, 1998). It has long been established that multi-ion channels may not necessarily display AMFE (Hille & Schwarz, 1978) and that single filing is not necessary to observe this phenomenon (Gillespie, Boda, He, Apel, & Siwy, 2008). In RyR, the presence of AMFE is highly dependent on the choice of cation, ion activities, and the method of detecting AMFE (Tomaskova & Gaburjakova, 2008). Therefore, advocates of rate theory models of RyR will be satisfied that single occupancy exists in some ion mixtures (Lindsay et al., 1991; Tinker & Williams, 1992) whereas in others it does not (Gillespie, Giri, & Fill, 2009;

Tomaskova & Gaburjakova, 2008). Recent PNP theories adopt density function theories (DFT) to improve on the previous flawed models which do not describe ions of a finite size and do not consider interactions between ions and channel structure. New models describe selectivity in RyR by CSC where ions can interact with flexible, time-averaged, charged residues where "charge space competition created by ions in a confined geometry results in significant selectivity" (Gillespie & Eisenberg, 2002). In this model, RyR, on average, is multiply occupied in various mole fraction mixtures of monovalent and divalent cations while one, approaching two, divalent cation(s) are present in the pore in pure divalent solutions (Gillespie, Xu, Wang, & Meissner, 2005). However, this model assumes that ions are dehydrated once reaching the pore, whereas an alternative model of RyR demonstrates that ions are partially hydrated within the pore (Welch et al., 2004). In summary, there is agreement within the literature that there is multi-ion occupancy in RyR and therefore this could serve as a mechanism, along with pore dimensions (Section III.B) and negative charge (Sections V.A and VII.B) to maximize ion conduction in the channel.

VII. TESTING PHYSICAL AND THEORETICAL MODELS OF THE RYR PFR BY RESIDUE SUBSTITUTION

A. RyR1

Meissner's group identified potentially important residues by using mutagenesis in conjunction with a theoretical PNP–DFT model and MD simulations in their model of the PFR of RyR1. D4899 and E4900 were shown to be critical for both conductance (ion permeation) and selectivity (high ion binding) (Wang et al., 2005). Simulations and electrophysiological experiments revealed that the mutant D4899Q exhibited a loss of Ca^{2+} preference in the selectivity filter (Ramachandran et al., 2009). Furthermore, analysis of G4898R, a CCD mutant showed a decrease of Ca^{2+} conductance experimentally, whereas K^+ conductance was constitutively high. These data were confirmed by simulations with a decrease in the preference of Ca^{2+} over K^+ in the selectivity filter (Ramachandran et al., 2009). Noteworthy is that both CCD mutations G4898E and G4898R can form homotetramers and heterotetramers in combination with WT channels when expressed in HEK-293 cells. In terms of function, while heterotetrameric mutants can maintain their Ca^{2+}-dependent activity and K^+ conductance, but display decreased Ca^{2+} selectivity (depending on the type of mutants and the combination of subunits), homotetrameric mutants have negligible Ca^{2+}-dependent activity and Ca^{2+} permeation and reduced K^+ conductance (Xu, Wang, Yamaguchi, Pasek, & Meissner, 2008).

B. RyR2

In agreement with earlier experimental data (Mead, Sullivan & Williams, 1998), the RyR PFR analogy model based on KcsA demonstrates the existence of a high density of negatively charged residues at both the luminal and cytosolic mouths of RyR2 pore (Welch et al., 2004). To investigate the significance of this charge, substitutions of acidic residues at the luminal side of the channel were performed. Earlier studies in RyR1 had demonstrated that the neutralization of either the glutamic acid residue equivalent to E4832 or the aspartic acid residue equivalent to D4833 in RyR2 produced no significant change in K^+ conductance (Wang et al., 2005). However, introducing the double substitution ED4832AA in mouse RyR2 channels led to very significant changes in function (Mead-Savery et al., 2009). In comparison with WT RyR2, the open probability of ED4832AA RyR2 is greatly increased and a wide variety of subconductance states are observed. ED4832AA discriminates ideally between cations and anions but discrimination between divalent and monovalent cations is lost.

At low ionic activities unitary conductance of ED4832AA is reduced but conductance increases to values above those seen in WT channels at high activities. MD simulations provide some insights into the consequences of this double substitution suggesting that (a) interactions of E4832 and D4833 with other residues in the PFR stabilize the selectivity filter, (b) E4832 and D4833 neutralization reduces the electric field at the luminal face that in turn could reduce cation delivery to the pore at low activities, and (c) neutralization lowers the electric field in the selectivity filter which, together with destabilization of the filter could underlie enhanced conductance at high ionic activity. In addition to its role as a Ca^{2+}-release channel, RyR also contains a high-affinity binding site for the plant alkaloid ryanodine. Interaction of ryanodine or its derivatives and congeners (ryanoids) results in alterations in both channel gating and ion handling (Sutko et al., 1997; Tinker et al., 1996). As outlined above, residue substitution in and around the proposed selectivity filter of the RyR channels can reduce or abolish the high-affinity interaction of [^{3}H]-ryanodine with its receptor. Another region of the channel that may contribute to the site of ryanoid interaction was uncovered by the investigation of residue substitution in the predicted pore-lining TM helix (TM 10). Mutations D4847A, F4850A, F4851A, L4858A, L4859A, and I4866A severely reduced the release of Ca^{2+} from the ER in HEK-293 cells in response to caffeine and diminished [^{3}H]-ryanodine binding to cell lysates. Mutations F4846A, T4849A, I4855A, V4856A, and Q4863A eliminated or markedly reduced [^{3}H]-ryanodine binding, but cells expressing these mutants responded to caffeine by releasing Ca^{2+} from intracellular stores (Wang et al., 2003). Investigations of single RyR2 channels containing one of these mutants; Q4863A, revealed that ryanodine and other ryanoids bind to the channel as readily as to the WT RyR2. The absence of measureable equilibrium binding of [^{3}H]-ryanodine to Q4863A RyR2 reflects an almost 600-fold increase in the rate of dissociation of the bound ryanoid (Ranatunga, Wayne Chen, Ruest, Welch, & Williams, 2007).

VIII. CONCLUDING REMARKS

In this contribution we have attempted to demonstrate how recent developments in structural biology, electrophysiology and molecular modeling have contributed to an improved understanding of the architecture of the PFR and the mechanisms governing the way in which the RyR channel translocates and discriminates between ions. Together these properties allow SR Ca^{2+} to be released at phenomenal rates through the open RyR channel during excitation–contraction coupling.

Acknowledgment

The work in our laboratory is supported by the British Heart Foundation.

References

Balshaw, D., Gao, L., & Meissner, G. (1999). Luminal loop of the ryanodine receptor: A pore forming segment? *Proceedings of the National Academy of Sciences of the United States of America, 96*, 3345–3347.

Chen, D., Xu, L., Tripathy, A., Meissner, G., & Eisenberg, B. (1997). Permeation through the calcium release channel of cardiac muscle. *Biophysical Journal, 73*, 1337–1354.

Chen, D. P., Xu, L., Tripathy, A., Meissner, G., & Eisenberg, B. (1999). Selectivity and permeation in calcium release channel of cardiac muscle: Alkali metal ions. *Biophysical Journal, 76*, 1346–1366.

Chen, S. R. W., Li, X., Ebisawa, K., & Zhang, L. (1997). Functional characterization of the recombinant type 3 Ca^{2+} release channel (ryanodine receptor) expressed in HEK293 cells. *The Journal of Biological Chemistry, 272*, 24234–24246.

Doyle, D. A., Morais Cabral, J., Pfuetzner, R. A., Kuo, A., Gulbis, J. M., Cohen, S. L., et al. (1998). The structure of the potassium channel: Molecular basis of K^{+} conduction and selectivity. *Science, 280*, 69–77.

Du, G. G., Guo, X., Khanna, V. K., & MacLennan, D. H. (2001). Functional characterization of mutants in the predicted pore region of the rabbit cardiac muscle Ca^{2+} release channel (ryanodine receptor isoform 2). *The Journal of Biological Chemistry, 276*, 31760–31771.

Du, G. G., Khanna, V. K., Guo, X., & MacLennan, D. H. (2004). Central core disease mutations R4892W, I4897T and G4898E in the ryanodine receptor isoform 1 reduce the Ca^{2+} sensitivity and amplitude of Ca^{2+}-dependent Ca^{2+} release. *The Biochemical Journal, 382*, 557–564.

Dudev, T., & Lim, C. (2009). Determinants of K^{+} vs Na^{+} selectivity in potassium channels. *Journal of the American Chemical Society, 131*, 8092–8101.

Eisenman, G. (1962). Cation selective glass electrodes and their mode of operation. *Biophysical Journal, 2*, 259–323.

Fleischer, S. (2008). Personal recollections on the discovery of the ryanodine receptors of muscle. *Biochemical and Biophysical Research Communications, 369*, 195–207.

Gao, L., Balshaw, D., Xu, L., Tripathy, A., Xin, C., & Meissner, G. (2000). Evidence for a role of the lumenal M3-M4 loop in skeletal muscle Ca(2+) release channel (ryanodine receptor) activity and conductance. *Biophysical Journal, 79*, 828–840.

Gillespie, D., & Boda, D. (2008). The anomalous mole fraction effect in calcium channels: A measure of preferential selectivity. *Biophysical Journal, 95*, 2658–2672.

Gillespie, D., Boda, D., He, Y., Apel, P., & Siwy, Z. S. (2008). Synthetic nanopores as a test case for ion channel theories: The anomalous mole fraction effect without single filing. *Biophysical Journal, 95*, 609–619.

Gillespie, D., & Eisenberg, R. S. (2002). Physical descriptions of experimental selectivity measurements in ion channels. *European Biophysics Journal, 31*, 454–466.

Gillespie, D., Giri, J., & Fill, M. (2009). Reinterpreting the anomalous mole fraction effect: The ryanodine receptor case study. *Biophysical Journal, 97*, 2212–2221.

Gillespie, D., Xu, L., Wang, Y., & Meissner, G. (2005). (De)constructing the ryanodine receptor: Modeling ion permeation and selectivity of the calcium release channel. *The Journal of Physical Chemistry B, 109*, 15598–15610.

Hamilton, S. L., & Serysheva, I. I. (2009). Ryanodine receptor structure: Progress and challenges. *The Journal of Biological Chemistry, 284*, 4047–4051.

Hille, B. (1991). *Ionic channels of excitable membranes.* Sunderland, MA: Sinauer Associates Inc.

Hille, B., & Schwarz, W. (1978). Potassium channels as multi-ion single-file pores. *The Journal of General Physiology, 72*, 409–442.

Jiang, Y., Lee, A., Chen, J., Cadene, M., Chait, B. T., & MacKinnon, R. (2002a). Crystal structure and mechanism of a calcium-gated potassium channel. *Nature, 417*, 515–522.

Jiang, Y., Lee, A., Chen, J., Cadene, M., Chait, B. T., & MacKinnon, R. (2002b). The open pore conformation of potassium channels. *Nature, 417*, 523–526.

Latorre, R., & Miller, C. (1983). Conduction and selectivity in potassium channels. *The Journal of Membrane Biology, 71*, 11–30.

Levitt, D. (1986). Interpretation of biological ion channel flux data-reaction-rate versus continuum theory. *Annual Review of Biophysics and Biophysical Chemistry, 15*, 29–57.

Lindsay, A. R., Manning, S. D., & Williams, A. J. (1991). Monovalent cation conductance in the ryanodine receptor-channel of sheep cardiac muscle sarcoplasmic reticulum. *Journal de Physiologie, 439*, 463–480.

Liu, Z., Zhang, J., Wang, R., Wayne Chen, S. R., & Wagenknecht, T. (2004). Location of divergent region 2 on the three-dimensional structure of cardiac muscle ryanodine receptor/calcium release channel. *Journal of Molecular Biology, 338*, 533–545.

Ludtke, S., Serysheva, I., Hamilton, S., & Chiu, W. (2005). The pore structure of the closed RyR1 channel. *Structure, 13*, 1203–1211.

Lynch, P. J., Tong, J., Lehane, M., Mallet, A., Giblin, L., Heffron, J. J., et al. (1999). A mutation in the transmembrane/luminal domain of the ryanodine receptor is associated with abnormal Ca^{2+} release channel function and severe central core disease. *Proceedings of the National Academy of Sciences of the United States of America, 96*, 4164–4169.

Mead, F., & Williams, A. J. (2002a). Block of the ryanodine receptor channel by neomycin is relieved at high holding potentials. *Biophysical Journal, 82*, 1953–1963.

Mead, F., & Williams, A. J. (2002b). Ryanodine-induced structural alterations in the RyR channel suggested by neomycin block. *Biophysical Journal, 82*, 1964–1974.

Mead, F. C., Sullivan, D., & Williams, A. J. (1998). Evidence for negative charge in the conduction pathway of the cardiac ryanodine receptor channel provided by the interaction of K^+ channel N-type inactivation peptides. *The Journal of Membrane Biology, 163*, 225–234.

Mead-Savery, F. C., Wang, R., Tanna-Topan, B., Chen, S. R., Welch, W., & Williams, A. J. (2009). Changes in negative charge at the luminal mouth of the pore alter ion handling and gating in the cardiac ryanodine-receptor. *Biophysical Journal, 96*, 1374–1387.

Miller, C. (1999). Ionic hopping defended. *The Journal of General Physiology, 113*, 783–787.

Nonner, W., Chen, D. P., & Eisenberg, B. (1998). Anomalous mole fraction effect, electrostatics, and binding in ionic channels. *Biophysical Journal, 74*, 2327–2334.

Nonner, W., Chen, D. P., & Eisenberg, B. (1999). Progress and prospects in permeation. *The Journal of General Physiology, 113*, 773–782.

Orlova, E. V., Serysheva, I. I., van Heel, M., Hamilton, S. L., & Chiu, W. (1996). Two structural configurations of t he skeletal muscle calcium release channel. *Nature Structural Biology, 3*, 547–552.

Radermacher, M., Rao, V., Grassucci, R., Frank, J., Timerman, A. P., Fleischer, S., et al. (1994). Cryo-electron microscopy and three-dimensional reconstruction of the calcium release channel/ryanodine receptor from skeletal muscle. *The Journal of Cell Biology, 127*, 411–423.

Ramachandran, S., Serohijos, A. W., Xu, L., Meissner, G., & Dokholyan, N. V. (2009). A structural model of the pore-forming region of the skeletal muscle ryanodine receptor (RyR1). *PLoS Computational Biology, 5*, e1000367.

Ranatunga, K. R., Wayne Chen, S. R., Ruest, L., Welch, W., & Williams, A. J. (2007). Quantification of the effects of a ryanodine receptor channel mutation on interaction with a ryanoid. *Molecular Membrane Biology, 24*, 185–193.

Samso, M., & Wagenknecht, T. (2002). Apocalmodulin and Ca^{2+}-calmodulin bind to neighboring locations on the ryanodine receptor. *The Journal of Biological Chemistry, 277*, 1349–1353.

Samso, M., Wagenknecht, T., & Allen, P. D. (2005). Internal structure and visualization of transmembrane domains of the RyR1 calcium release channel by cryo-EM. *Nature Structural and Molecular Biology, 12*, 539–544.

Serysheva, I. I., Chiu, W., & Ludtke, S. (2007). Single-particle electron cryomicroscopy of the ion channels in the excitation-contraction coupling junction. *Methods in Cell Biology, 79*, 407–435.

Serysheva, I. I., Orlova, E. V., Chiu, W., Sherman, M. B., Hamilton, S. L., & van Heel, M. (1995). Electron cryomicroscopy and angular reconstitution used to visualize the skeletal muscle calcium release channel. *Nature Structural Biology, 2*, 18–24.

Sherry, H. S. (1969). The ion-exchange properties of zeolites. In J. Marinsky (Ed.), *Ion exchange* (pp. 89–133). New York: Marcel Dekker, Inc..

Shomer, N. H., Mickelson, J. R., & Louis, C. F. (1994). Ion selectivity of porcine skeletal muscle Ca^{2+} release channels is unaffected by the Arg615 to Cys615 mutation. *Biophysical Journal, 67*, 641–646.

Smith, J. S., Imagawa, T., Ma, J., Fill, M., Campbell, K. P., & Coronado, R. (1988). Purified ryanodine receptor from rabbit skeletal muscle is the calcium- release channel of sarcoplasmic reticulum. *The Journal of General Physiology, 92*, 1–26.

Sutko, J. L., Airey, J. A., Welch, W., & Ruest, L. (1997). The pharmacology of ryanodine and related compounds. *Pharmacological Reviews, 49*, 53–98.

Tinker, A., Lindsay, A. R., & Williams, A. J. (1992a). Block of the sheep cardiac sarcoplasmic reticulum Ca(2+)-release channel by tetra-alkyl ammonium cations. *The Journal of Membrane Biology, 127*, 149–159.

Tinker, A., Lindsay, A. R., & Williams, A. J. (1992b). A model for ionic conduction in the ryanodine receptor channel of sheep cardiac muscle sarcoplasmic reticulum. *The Journal of General Physiology, 100*, 495–517.

Tinker, A., Sutko, J., Ruest, L., Deslongchamps, P., Welch, W., Airey, J., et al. (1996). Electrophysiological effects of ryanodine derivatives on the sheep cardiac sarcoplasmic reticulum calcium-release channel. *Biophysical Journal, 70*, 2110–2119.

Tinker, A., & Williams, A. J. (1992). Divalent cation conduction in the ryanodine receptor channel of sheep cardiac muscle sarcoplasmic reticulum. *The Journal of General Physiology, 100*, 479–493.
Tinker, A., & Williams, A. J. (1993). Probing the structure of the conduction pathway of the sheep cardiac sarcoplasmic reticulum calcium-release channel with permeant and impermeant organic cations. *The Journal of General Physiology, 102*, 1107–1129.
Tinker, A., & Williams, A. J. (1995). Measuring the length of the pore of the sheep cardiac sarcoplasmic reticulum calcium-release channel using related trimethylammonium ions as molecular calipers. *Biophysical Journal, 68*, 111–120.
Tomaskova, Z., & Gaburjakova, M. (2008). The cardiac ryanodine receptor: Looking for anomalies in permeation properties. *Biochimica et Biophysica Acta, 1778*, 2564–2572.
Tu, Q., Velez, P., Brodwick, M., & Fill, M. (1994). Streaming potentials reveal a short ryanodine-sensitive selectivity filter in cardiac Ca^{2+} release channel. *Biophysical Journal, 67*, 2280–2285.
Wagenknecht, T., Radermacher, M., Grassucci, R., Berkowitz, J., Xin, H. B., & Fleischer, S. (1997). Locations of calmodulin and FK506-binding protein on the three-dimensional architecture of the skeletal muscle ryanodine receptor. *The Journal of Biological Chemistry, 272*, 32463–32471.
Wang, R., Zhang, L., Bolstad, J., Diao, N., Brown, C., Ruest, L., et al. (2003). Residue Gln4863 within a predicted transmembrane sequence of the Ca^{2+} release channel (ryanodine receptor) is critical for ryanodine interaction. *The Journal of Biological Chemistry, 278*, 51557–51565.
Wang, Y., Xu, L., Pasek, D. A., Gillespie, D., & Meissner, G. (2005). Probing the role of negatively charged amino acid residues in ion permeation of skeletal muscle ryanodine receptor. *Biophysical Journal, 89*, 256–265.
Welch, W., Rheault, S., West, D. J., & Williams, A. J. (2004). A model of the putative pore region of the cardiac ryanodine receptor channel. *Biophysical Journal, 87*, 2335–2351.
Williams, A. (1992). Ion conduction and discrimination in the sarcoplasmic reticulum ryanodine receptor/calcium-release channel. *Journal of Muscle Research and Cell Motility, 13*, 7–26.
Williams, A. J., West, D. J., & Sitsapesan, R. (2001). Light at the end of the Ca(2+)-release channel tunnel: Structures and mechanisms involved in ion translocation in ryanodine receptor channels. *Quarterly Reviews of Biophysics, 34*, 61–104.
Xu, L., Jones, R., & Meissner, G. (1993). Effects of local anesthetics on single channel behavior of skeletal muscle calcium release channel. *The Journal of General Physiology, 101*, 207–233.
Xu, L., Wang, Y., Gillespie, D., & Meissner, G. (2006). Two rings of negative charges in the cytosolic vestibule of type-1 ryanodine receptor modulate ion fluxes. *Biophysical Journal, 90*, 443–453.
Xu, L., Wang, Y., Yamaguchi, N., Pasek, D. A., & Meissner, G. (2008). Single channel properties of heterotetrameric mutant RyR1 ion channels linked to core myopathies. *The Journal of Biological Chemistry, 283*, 6321–6329.
Zhao, M., Li, P., Li, X., Zhang, L., Winkfein, R. J., & Chen, S. R. (1999). Molecular identification of the ryanodine receptor pore-forming segment. *The Journal of Biological Chemistry, 274*, 25971–25974.
Zvaritch, E., Depreux, F., Kraeva, N., Loy, R. E., Goonasekera, S. A., Boncompagni, S., et al. (2007). An Ryr1I4895T mutation abolishes Ca^{2+} release channel function and delays development in homozygous offspring of a mutant mouse line. *Proceedings of the National Academy of Sciences of the United States of America, 104*, 18537–18542.

CHAPTER 4

Regulation of RyR Channel Gating by Ca^{2+}, Mg^{2+} and ATP

Derek R. Laver[*,†]
*School of Biomedical Sciences, University of Newcastle, Callaghan, New South Wales, Australia
†Hunter Medical Research Institute, Callaghan, New South Wales, Australia

I. OVERVIEW

This chapter gives the current picture of how calcium release channels (ryanodine receptors, RyRs) in the sarcoplasmic reticulum (SR) are regulated by intracellular ligands. The review focuses on Ca^{2+}, Mg^{2+}, and ATP (adenosine triphosphate), which are key regulators of the RyR. Different RyR isoforms are expressed in skeletal and cardiac muscle. These isoforms are modulated differently by Ca^{2+}, Mg^{2+}, and ATP, leading to the different characteristics of E–C coupling in the two muscle types.

Current Topics in Membranes, Volume 66
1063-5823/10 $35.00
Copyright 2010, Elsevier Inc. All right reserved.
DOI: 10.1016/S1063-5823(10)66004-8

Single channel recordings have revealed a wealth of information about ligand regulation mechanisms. Experimental observations are considered here in the framework of a kinetic model of RyR gating which unifies luminal and cytoplasmic regulation mechanisms. The model is used to gain insight into the role of Ca^{2+}, Mg^{2+}, and ATP in controlling Ca^{2+} release during E–C coupling.

Current evidence indicates the presence of four types of Ca^{2+} sensing site on RyRs. They have two Ca^{2+}-activation sites located in the luminal and cytoplasmic domains of the RyR. These sites feed into a common gating mechanism and produce synergistic activation by luminal and cytoplasmic Ca^{2+}. The cytoplasmic domains also possess two inhibitory sites with Ca^{2+} affinities of $\sim$1 μM and 1 mM. Magnesium, which competes with Ca^{2+}, plays an important role in inhibiting RyRs and shaping the Ca^{2+}-dependent activation of RyRs in muscle. ATP stimulates RyR activation by luminal and cytoplasmic Ca^{2+}. These actions of ATP and other stimulators may share a common mechanism by which cytoplasmic domains modulate RyR activation by luminal and cytoplasmic Ca^{2+}.

II. INTRODUCTION

Key regulators of the RyR are Ca^{2+}, Mg^{2+}, and ATP (Coronado, Morrissette, Sukhareva, & Vaughan, 1994; Meissner, 1994). RyRs are activated by $\sim$1 μM $[Ca^{2+}]_C$ and mM ATP and they inhibited by mM levels of Mg^{2+} (Laver, Baynes, & Dulhunty, 1997; Meissner, 1994). The different isoforms are modulated differently by these ligands and this probably underlies the different characteristics of E–C coupling in different muscle types (Lamb, 2000; Laver, Lenz, & Lamb, 2001). This chapter will concentrate on ligand regulation of RyR2 and its role in cardiac E–C coupling. The focus here is on ligand regulation of RyR2 because cardiac E–C coupling depends entirely on ligand regulation of RyR2 (Bers, 2002), whereas the primary regulator of RyR1isoform in skeletal muscle is the 1,4-dihydropyridine receptor (DHPR) (Lamb & Stephenson, 1991, 1994). This chapter gives the current picture of RyR2 regulation by Ca^{2+}, Mg^{2+}, and ATP. The experimental observations from single channel recording of RyRs are considered in the framework of a functional model of RyR2 gating. The model is used to understand the mechanisms underlying the action of RyR activators such as ATP and to gain insight into the role of Ca^{2+} and Mg^{2+} in controlling Ca^{2+} release during E–C coupling.

III. RYR2 IN CARDIAC CONTRACTION AND PACEMAKING

The cardiac action potential triggers DHPRs causing a rise in $[Ca^{2+}]_C$ and activation of RyR2 channels in the SR via their cytoplasmic facing Ca^{2+}-activation sites. The subsequent release of Ca^{2+} from the SR further increases $[Ca^{2+}]_C$ by feeding back to cause further RyR2 activation. This process, known as Ca^{2+}-induced Ca^{2+} release (CICR), provides a strong positive feedback to open more RyR2. In this way, Ca^{2+} release from the SR contributes up to 95% of the Ca^{2+} entering the cytoplasm during E–C coupling (Bers, 2002). Shortly after this, the positive feedback cycle of CICR is broken due to the large resultant decrease of Ca^{2+} in the SR lumen ($[Ca^{2+}]_L$) which causes RyR2 to close and hence SR Ca^{2+} release ceases. The excess Ca^{2+} in the cytoplasm is either pumped back into the SR by ATP-driven Ca^{2+} pumps in the SR membrane (SERCA2a) or extruded from the cell by the Na/Ca exchanger (NCX) in the sarcolemma during diastole. The major Ca^{2+} fluxes are Ca^{2+} uptake and release from the SR by SERCA2a and RyR2 and uptake and release from the cell by DHPRs and the NCX (Dibb, Graham, Venetucci, Eisner, & Trafford, 2007). These mechanisms serve a fundamental role in the large changes in free $[Ca^{2+}]$ in the cytoplasm ($[Ca^{2+}]_C \sim 0.1$–$1\ \mu M$) and SR lumen ($[Ca^{2+}]_L \sim 1$–0.3 mM) between diastole and systole, respectively (Bers, 2002).

It was first shown that oscillations in Ca^{2+} uptake and release across the SR underlie pacemaking in lymphatic smooth muscle (Van Helden, 1993) and that these provide the pacemaker mechanism by interacting as coupled oscillators within and between cells (Van Helden & Imtiaz, 2003). The SR in the heart has this same capability. The oscillating Ca^{2+} uptake/release occurs in two phases: First, SERCA2a causes loading of the SR to a point where Ca^{2+} refill causes spontaneous opening of RyR2 due to elevated $[Ca^{2+}]_L$. Second, during Ca^{2+} release, CICR provides positive reinforcement of RyR2 activity which continues until the stores sufficiently deplete to cause closure of RyR2 channels. Ca^{2+} release in turn activates the NCX to extrude Ca^{2+} out of the cell causing a net depolarization of the sarcolemma (i.e., 3 Na^+ enter for every Ca^{2+} extruded) and triggers an action potential. This prospect has generated considerable interest and there is now substantive evidence that the SR has a role in heart pacemaking (Ju & Allen, 1998, 2007; Rigg & Terrar, 1996; Vinogradova, Maltsev, Bogdanov, Lyashkov, & Lakatta, 2005).

IV. FOUR Ca^{2+} SENSING MECHANISMS FOR RYR2

One of the most striking and earliest discovered features of RyR regulation is cytoplasmic Ca^{2+}-activation (Hymel, Inui, Fleischer, & Schindler, 1988; Smith, Coronado, & Meissner, 1986). In the absence of luminal Ca^{2+}, or any other channel activator, cytoplasmic Ca^{2+} can increase channel open probability (P_o) from virtually zero up to $P_o \sim 0.6$ with a half-maximal activation concentration (K_a) of 5 μM (Sitsapesan & Williams, 1994a; Xu, Mann, & Meissner, 1996). The RyR has a high-order dependence of P_o on cytoplasmic $[Ca^{2+}]$ ($[Ca^{2+}]_C$) (i.e., a Hill coefficients of 2–4; Sitsapesan and Williams,

1994a), indicating that RyR2 activation resulted from cooperative involvement of multiple Ca^{2+} binding sites. Given the homotetrameric structure of the channel, it was suspected that one high-affinity (~1 μM) Ca^{2+} binding site on each subunit contributed to channel activation (Zahradnik, Gyorke, & Zahradnikova, 2005). This Ca^{2+}-activation site is referred to here as the *A*-site, a name originally coined by Balog, Fruen, Shomer, & Louis (2001) (*A* for activation).

High $[Ca^{2+}]_C$ is long known to inhibit RyRs (Meissner, 1986) and this phenomenon was first reported in single channel recordings of RyR2 by Laver et al. (1995). In cardiac RyRs, P_o markedly declines when $[Ca^{2+}]_C$ rises above 1 mM indicating the presence of low-affinity inhibitory sites on the RyR. In keeping with the nomenclature of Balog et al. (2001), this site is referred to here as the *I1*-site (*I* for inhibition and *I1* to distinguish it from a recently identified high-affinity inhibition site, *I2*-site, see below). Together, the *A*- and *I1*-sites generate the well-known, bell-shaped $[Ca^{2+}]_C$-dependence which characterizes RyR2 activation at μM $[Ca^{2+}]_C$ and inhibition at mM $[Ca^{2+}]_C$.

The first clear demonstration of activation of RyRs by luminal Ca^{2+} was made by Sitsapesan and Williams (1994b). Their interpretation of this finding was that there exists a luminal facing Ca^{2+} binding site that causes channel activation. In support of this hypothesis, they demonstrated that luminal Ca^{2+}-activation was abolished by tryptic digestion of the luminal domains of the RyR which presumably destroyed the Ca^{2+} sensing site (Ching, Williams, & Sitsapesan, 2000). However, considerable evidence has accrued to indicate that $[Ca^{2+}]_L$-activation of RyR2 also relies on the flow of Ca^{2+} through the pore (Ca^{2+} feed-through) to the cytoplasm where it can activate the RyR via the *A*-site (Laver, 2007b). Only recently have precise roles been assigned to cytoplasmic and luminal domains for Ca^{2+} regulation of RyR2 (Laver, 2007a; Laver & Honen, 2008). An important conceptual advance that lead to the assignment of these roles and functional characterization of a luminal Ca^{2+} site (*L*-site, L for luminal) was to consider RyR activity in terms of open and closed durations separately rather than open probability which is a combination of both. By considering only the properties of channel closed events, one could study the RyR gating mechanisms when Ca^{2+} was not flowing through the channel.

Finally, a cytoplasmic Ca^{2+}-inhibition phenomenon was only recently identified in single channel recordings of RyR2 and this has been attributed to a high-affinity cytoplasmic facing site on each RyR subunit (*I2*-site) (Laver, 2007a). This Ca^{2+}-inhibition is manifest as a $[Ca^{2+}]_C$-dependent reduction in channel open time at sub-μM range that can be observed in the presence of ATP (shown below). It is easily overlooked because it is not manifest as a reduction in P_o because the effect of reduced open time on P_o is swamped by a $[Ca^{2+}]_C$-dependent increase in P_o via a decrease in the closed times associated with Ca^{2+}-activation by the *A*-site.

Thus, four Ca^{2+} regulation sites have been identified on the RyR and the approximate locations of these sites and their binding affinities are shown on a schematic of the RyR (Fig. 1). Although there is good evidence to indicate the side of the membrane to which these sites belong, the precise locations of these sites on the RyR or its accessory proteins have not yet been determined. A clue to the relative proximities of the *A*- and *I2*-sites to the pore mouth has been gleaned from their response to Ca^{2+} feed-through (Laver, 2007a; Tripathy & Meissner, 1996). Although the *A*- and *I2*-sites have similar affinities for cytoplasmic Ca^{2+}, the apparent affinity of the *I2*-site for luminal Ca^{2+} is more than 10-fold lower than the *A*-site. The dilution and sequestering of Ca^{2+} emanating from the pore will create a concentration gradient such that the $[Ca^{2+}]$ at each site will depend on its distance from the pore. The lower sensitivity of *I2*-site is consistent with a location further from the pore.

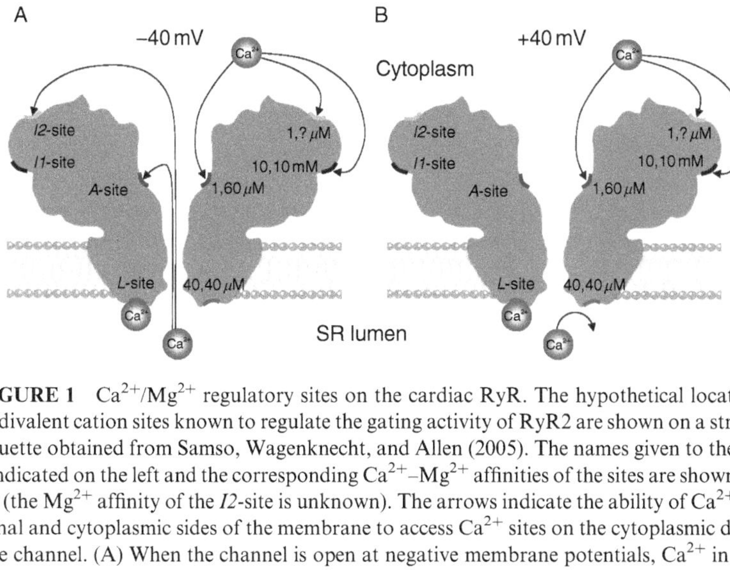

FIGURE 1 Ca^{2+}/Mg^{2+} regulatory sites on the cardiac RyR. The hypothetical locations of four divalent cation sites known to regulate the gating activity of RyR2 are shown on a structural silhouette obtained from Samso, Wagenknecht, and Allen (2005). The names given to these sites are indicated on the left and the corresponding Ca^{2+}–Mg^{2+} affinities of the sites are shown on the right (the Mg^{2+} affinity of the *I2*-site is unknown). The arrows indicate the ability of Ca^{2+} on the luminal and cytoplasmic sides of the membrane to access Ca^{2+} sites on the cytoplasmic domains of the channel. (A) When the channel is open at negative membrane potentials, Ca^{2+} in the SR lumen can flow through the channel and bind to cytoplasmic facing sites. (B) When the channel is closed, Ca^{2+} is unable to flow through the channel and positive membrane potentials oppose the flow through open channels. In these situations luminal Ca^{2+} only has access to luminal facing sites (after Laver, 2009). (See Color Insert.)

V. SYNERGISTIC Ca^{2+}-ACTIVATION VIA CYTOPLASMIC AND LUMINAL FACING BINDING SITES

Recent measurements of RyR2 gating have shown that activation of RyR2 by luminal and cytoplasmic Ca^{2+} cannot be thought of as independent processes (Laver & Honen, 2008). Data illustrating this point is shown in Fig. 2. Experiments measured RyR open and closed times over a wide range of $[Ca^{2+}]_L$ and $[Ca^{2+}]_C$ in the presence of maximally activating concentrations of ATP (2 mM). The RyR2 gating kinetics are expressed in terms of channel opening rate (k_o, reciprocal of the channel mean closed times, Fig. 2C and E) and mean open time (τ_o, Fig. 2D and F). The use of opening rate was preferred over mean closed time because τ_o and k_o both have the same positive correlation with P_o ($P_o = \tau_o k_o/(\tau_o k_o + 1)$). We first consider the simpler case of cytoplasmic and luminal activation via k_o (this is a property of channel closed intervals where Ca^{2+} is not flowing across the membrane).

In the absence of both cytoplasmic and luminal Ca^{2+}, RyR2 is virtually inactive with 1–5 ms openings occurring in single channels less than once per minute ($P_o < 10^{-5}$, Fig. 2C, ●). Figure 2C shows the response of RyR2 opening rate to $[Ca^{2+}]_C$ in the presence of $[Ca^{2+}]_L$ at 0.1 mM (○) and 10 μM (●). In low $[Ca^{2+}]_L$, the opening rate rises approximately threefold as $[Ca^{2+}]_C$ increases from 1 to 30 nM. The steepness of the $[Ca^{2+}]_C$-dependence markedly increases above this to reach a third-order dependence on $[Ca^{2+}]_C$ at about 1 μM. The opening rate peaks at ~1000 s^{-1} between 10 μM and 1 mM and then declines at higher concentrations (○). The dramatic activation of the channel is primarily mediated by the *A*-site and the inhibition by the *I1*-site.

Increasing $[Ca^{2+}]_L$ from 10 to 100 μM markedly increases the opening rate of the RyR but only at low-range $[Ca^{2+}]_C$ up to 0.3 μM. At higher $[Ca^{2+}]_C$, luminal Ca^{2+} has no effect on RyR activity. Luminal Ca^{2+} activates the channel with a hyperbolic $[Ca^{2+}]_L$-dependence with a K_a ~ 20 μM and Hill coefficient ~2 (Fig. 2E). This indicates that there are several *L*-sites (probably one on each of the four subunits) and that these sites have a Ca^{2+} affinity in the tens of micromolar range. In the presence of 100 μM $[Ca^{2+}]_L$, cytoplasmic Ca^{2+} activates the channel from an opening rate of ~0.4 s^{-1} to a maximum of ~3 s^{-1} and even in the virtual absence of cytoplasmic Ca^{2+} (~1 nM), luminal Ca^{2+} can still activate RyR2 with a maximal k_o ~ 0.7 s^{-1}. In comparison to the *A*-site, the *L*-site has much lower affinity and is only capable of activating the channel with relatively slow opening rate. The data also demonstrate a marked synergy that exists between *A*- and *L*-site-mediated activation of the channel. An increase in $[Ca^{2+}]_C$ from 1 to 100 nM amplifies the luminal Ca^{2+}-activation effect by fourfold (Fig. 2E, cf. ○, ●). Therefore, at low-range $[Ca^{2+}]_C$ (diastolic $[Ca^{2+}]_C$),

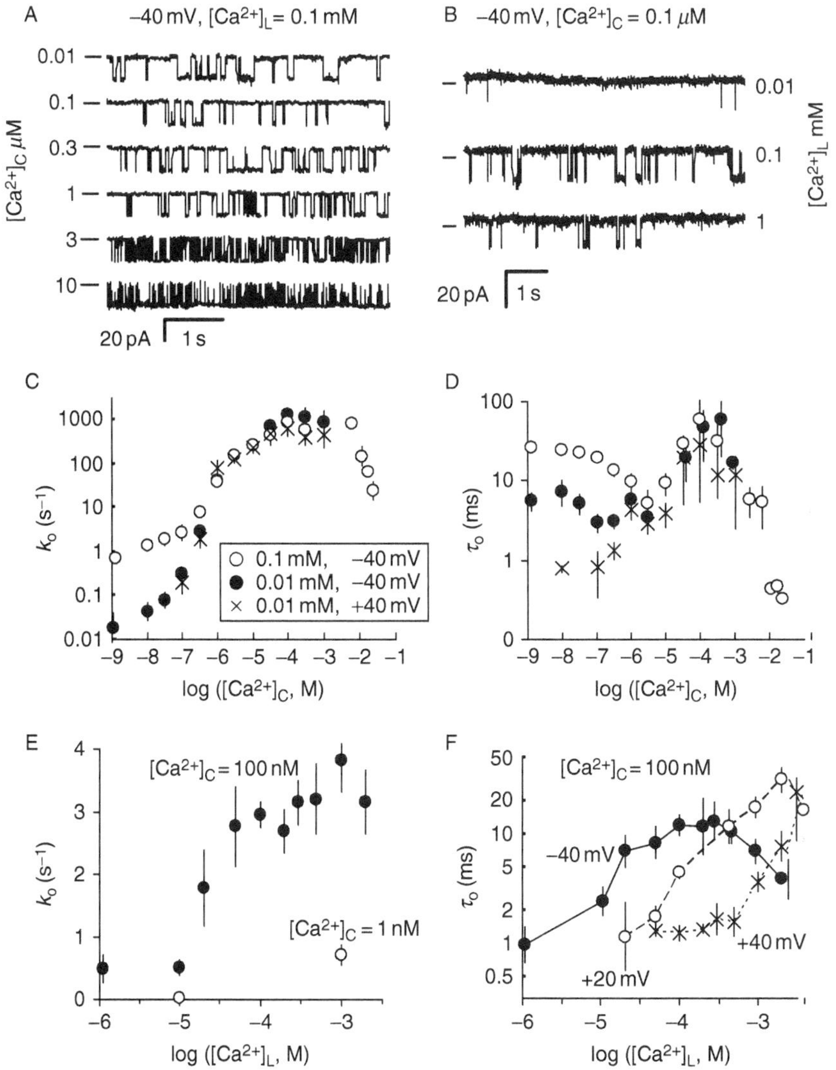

FIGURE 2 Activation of RyR2 by luminal and cytoplasmic Ca^{2+}. (A) The effect of $[Ca^{2+}]_C$ on the activity of a RyR in the presence of 0.1 mM $[Ca^{2+}]_L$ and (B) the effect of $[Ca^{2+}]_L$ on RyR activity in the presence of 0.1 μM $[Ca^{2+}]_C$ and 2 mM ATP. The bilayer voltage is −40 mV (cis–trans) which favors the flow of Ca^{2+} from luminal (trans) to cytoplasmic (cis) baths. Channel openings are downward current jumps from the baseline (indicated with a dash). $[Ca^{2+}]_C$-dependence of RyR2 opening rate (C) and mean open time (D) in the presence of 0.01 and 0.1 mM $[Ca^{2+}]_L$. The legend refers to panels (C) and (D). $[Ca^{2+}]_L$-dependence of RyR2 opening rate (E) and mean open time (F) in the presence of 0.1 μM $[Ca^{2+}]_C$. Mean channel open times are shown at three membrane voltages (data from Laver and Honen, 2008; Laver, 2007a).

Ca^{2+}-activation is not uniquely attributable to either the *A*- or *L*-sites but rather is a combination of both. At higher (systolic) $[Ca^{2+}]_C$ the RyR opening rate is insensitive to $[Ca^{2+}]_L$ and activation is primarily due to the *A*-site. The mechanism for this synergy, and the limited range of $[Ca^{2+}]_C$ over which this occurs, has not yet been confirmed experimentally but the data has been accurately modeled by a tetrameric RyR structure with *A*- and *L*-sites on each subunit where Ca^{2+} binding to three or more subunits will cause channel opening (see below).

VI. CHANNEL OPEN TIMES AND THE ROLE OF Ca^{2+} FEED-THROUGH

So far, the properties of RyR2 have been presented in terms of channel opening rate. However, this is only half the story because channel activity also depends on mean open duration, τ_o. In Fig. 2D and F, it can be seen that τ_o has a complex dependence on cytoplasmic and luminal $[Ca^{2+}]$. Moreover, the interpretation of these data is more complex than for the corresponding k_o data because one must consider the possibility that luminal Ca^{2+} can pass through the channel and act via the cytoplasmic facing sites (Fig. 1A, τ_o is a property of open channels). By employing a strong electrochemical gradient opposing a luminal to cytoplasmic Ca^{2+} flow, it has been possible to minimize Ca^{2+} feed-through in open channels and so be able to distinguish the roles of cytoplasmic and luminal domains of the RyR (Fig. 1B).

Figure 2D shows the $[Ca^{2+}]_C$-dependencies of τ_o corresponding to the k_o data in Fig. 2C in the presence of 10 and 100 μM $[Ca^{2+}]_L$. In addition, data is also presented from experiments where the membrane potential (+40 mV) opposes Ca^{2+} feed-through ($\times$) and the open times reflect only the properties of the cytoplasmic facing sites. At $[Ca^{2+}]_C$ above 1 μM, τ_o does not depend on bilayer voltage or luminal Ca^{2+} but rather depends on $[Ca^{2+}]_C$ (this was also the case for k_o). Under these conditions, τ_o increases from 5 to 50 ms when $[Ca^{2+}]_C$ is increased from 10 to 100 μM and then declines when $[Ca^{2+}]_C$ exceeds 1 mM. The activation and inhibition phases of this bell-shaped $[Ca^{2+}]_C$-dependence in τ_o is thought to be due to the action of the *A*- and *I1*-sites, respectively. At $[Ca^{2+}]_C$ below 1 μM, τ_o depends on both cytoplasmic and luminal $[Ca^{2+}]$. When RyRs are activated solely by luminal Ca^{2+} (i.e., $[Ca^{2+}]_L = 100$ μM, $[Ca^{2+}]_C = 1$ nM, Fig. 2D, ○), RyRs have $\tau_o \sim 20$ ms. The marked decline in τ_o that occurs when $[Ca^{2+}]_C$ is increased to 1 μM is an inactivation phenomenon which exhibits a hyperbolic $[Ca^{2+}]_C$-dependence with a half-inhibitory concentration (K_i) ~1 μM and a Hill coefficient ~2. These properties indicate the presence of several high-affinity Ca^{2+} sensing sites (*I2*-sites) which are quite distinct from the low-affinity *I1*-sites that produce the τ_o decline at mM $[Ca^{2+}]_C$.

The dependence of τ_o on voltage and $[Ca^{2+}]_L$ is shown in more detail in Fig. 2F. At −40 mV (●), τ_o can be clearly seen to have a bell-shaped $[Ca^{2+}]_L$-dependence with activation between 1 and 100 μM and inactivation at higher concentrations. The $[Ca^{2+}]_L$-dependence shifts to higher concentrations as the bilayer voltage is biased against Ca^{2+} feed-through (i.e., changed from −40 mV to positive voltages) which is one of the key indicators that activation and inhibition mechanisms involve Ca^{2+} feed-through. More evidence for the role of Ca^{2+} feed-through in these activation and inhibition mechanisms comes from the observation that the effects of $[Ca^{2+}]_L$ and $[Ca^{2+}]_C$ on open durations are not additive (Laver, 2007a). For example, luminal Ca^{2+} does not cause further lengthening of channel openings when $[Ca^{2+}]_C$ is high (Fig. 2D). This competitive action of cytoplasmic and luminal Ca^{2+} for RyR2 activation and inhibition indicates that cytoplasmic and luminal Ca^{2+} have an action at common sites which is could be easily explained by feed-through of luminal Ca^{2+} to the *A*-site and *I2*-sites, respectively.

VII. THREE MECHANISMS FOR Mg^{2+}-INHIBITION OF RYR2

Mg^{2+} is a strong inhibitor of RyRs which has an important role of shaping the cytoplasmic and luminal Ca^{2+}-dependencies of RyR activity in the cell (Laver & Honen, 2008; Meissner & Henderson, 1987). Three forms of Mg^{2+}-inhibition have been identified which have been linked to the binding of Mg^{2+} to the *A*-, *L*-, and *I1*-Ca^{2+} sensing sites on the RyR. The Mg^{2+} affinities for these sites have been measured and they are given in Fig. 1. At the Ca^{2+}-activation sites (*A*- and *L*-sites), Mg^{2+}-inhibition occurs because Mg^{2+} binding occludes the sites and prevents Ca^{2+} from binding and activating the RyR, and unlike Ca^{2+}, Mg^{2+} does not cause channel opening (Laver et al., 1997). At the Ca^{2+}-inhibition site (*I1*-site), Mg^{2+} is a surrogate for Ca^{2+} and both Ca^{2+} and Mg^{2+} cause channel closures (Laver et al., 1997).

Inhibition of RyRs by cytoplasmic Mg^{2+} was first observed in the 1980s, very soon after RyRs channels were identified and it was recognized even then that Mg^{2+}-inhibition occurred by its binding and competing with Ca^{2+} for the *A*-site (Smith et al., 1986). It was later found that Mg^{2+} binding did not only prevent Ca^{2+} from being an activator but Mg^{2+} was also an antagonist that could close the channel even in the presence of other activators such as luminal Ca^{2+} or ryanodine (Laver, O'Neill, & Lamb, 2004). The effect of Mg^{2+} was to shift K_a for Ca^{2+}-activation to higher $[Ca^{2+}]$ and examples of this are shown in Fig. 3A, C, and D. Mg^{2+}-inhibition was measured at +40 mV so that it would not be influenced by Ca^{2+} feed-through and simplify interpretation of the data. From first order enzyme kinetics, one

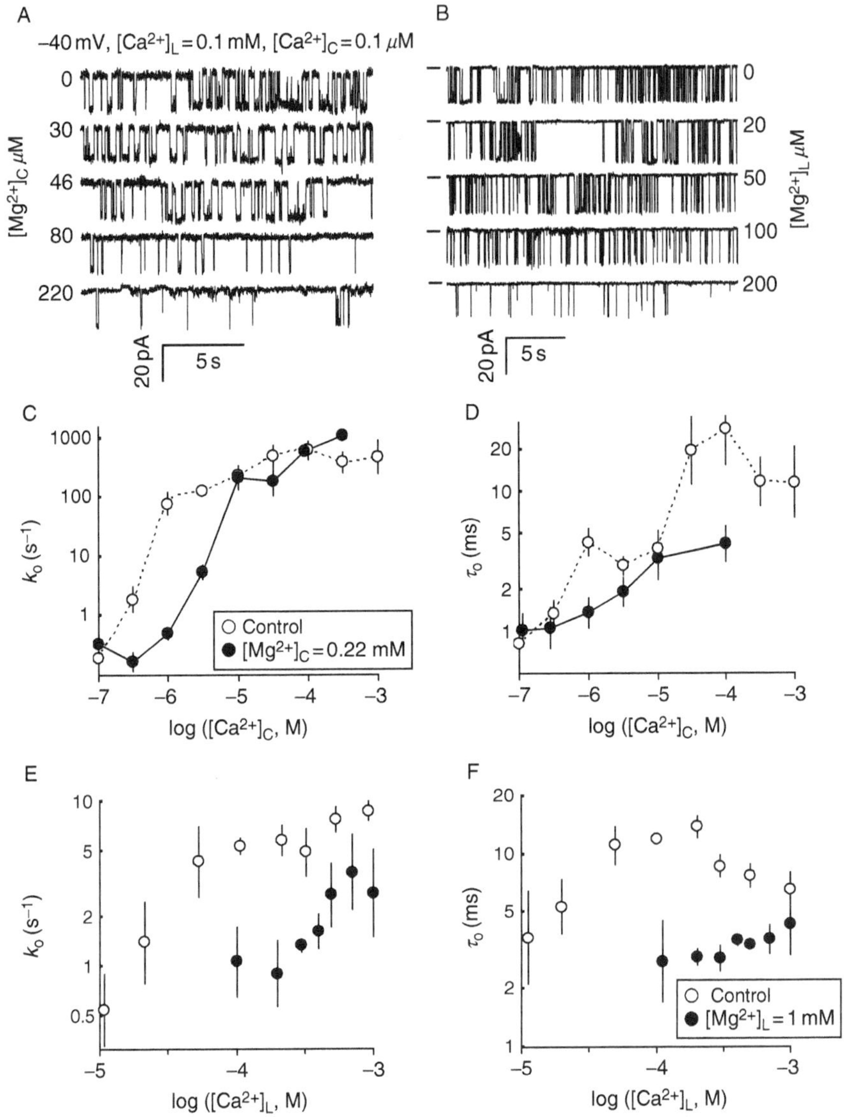

FIGURE 3 Inhibition of RyR2 by luminal and cytoplasmic Mg^{2+}. The effect of $[Mg^{2+}]_C$ (A) and $[Mg^{2+}]_L$ (B) on the activity of RyR2 in the presence of 0.1 mM $[Ca^{2+}]_L$, 0.1 μM $[Ca^{2+}]_C$, and 2 mM ATP. The bilayer voltage is −40 mV which favors the flow of Ca^{2+} from luminal to cytoplasmic baths. Channel openings are downward current jumps from the baseline (indicated with a dash). $[Ca^{2+}]_C$-dependence of RyR2 opening rate (C) and mean open time (D) in the presence of 0.01 mM $[Ca^{2+}]_L$ and the presence and absence of cytoplasmic Mg^{2+}. The bilayer voltage is +40 mV which opposes the flow of Ca^{2+} from luminal to cytoplasmic baths. The legend refers to panels (C) and (D). $[Ca^{2+}]_L$-dependence of RyR2 opening rate (E) and mean open time (F) in the presence of 0.1 μM $[Ca^{2+}]_C$ (−40 mV) and in the presence and absence of 1 mM luminal Mg^{2+} (data from Laver and Honen, 2008).

can approximate the increase in K_a for Ca^{2+}-activation to be a factor of $[Mg^{2+}]/K_{Mg}$ where K_{Mg} is the binding affinity for Mg^{2+}. The 10-fold shift in K_a in response to 0.22 mM cytoplasmic Mg^{2+} (Fig. 3C and D) would suggest a Mg^{2+} affinity for the *A*-site of ~22 μM (the more accurate value is 60 μM). Extrapolation of the data in Fig. 3 to physiological intracellular $[Mg^{2+}]$ (1 mM) gives a $K_a \sim 20$ μM for channel activation in the cell by the *A*-site.

The discovery of luminal Mg^{2+}-inhibition was quite recent (Laver & Honen, 2008) and it was only observed when $[Ca^{2+}]_C$ is less than 1 μM, the same $[Ca^{2+}]_C$ range over which luminal Ca^{2+}-activation can be detected (Fig. 3B). In bilayer experiments, luminal Mg^{2+} was found to have two inhibitory actions on RyR2. Firstly, it competes with Ca^{2+} for the *L*-site and second it can flow through the channel, bind to the *A*-site and terminate the channel opening (RyRs have the same permeability for both Mg^{2+} and Ca^{2+}). The former leads to a reduction in k_o and the latter to a reduction in τ_o (Fig. 3E and F). The effect of luminal Mg^{2+} is to increase K_a for luminal Ca^{2+}-activation (Fig. 3E). This is likely to be a very important factor in shaping the $[Ca^{2+}]_L$-response of RyRs in the cell. The free concentration of Ca^{2+} in the SR of cardiomyocytes cycles between 0.3 and 1 mM over the course of the heart beat and modulation of RyR activity by this change is important for cardiac pacemaking and contraction (see above). In the absence of Mg^{2+}, the luminal Ca^{2+}-activation of RyR2 reaches its maximum below 0.1 mM so that modulation of RyR2 activity over the physiological range would not be possible. However, in the presence of 1 mM Mg^{2+}, the K_a for $[Ca^{2+}]_L$-activation shifts from 20 μM to 1 mM where physiological changes in $[Ca^{2+}]_L$ will have a substantial effect of RyR2 activity.

Although Mg^{2+} competes with Ca^{2+} to cause inhibition at the *A*- and *L*-sites, there are some important differences in how they act at these two sites. Firstly, the *A*-site is selective for Ca^{2+} over Mg^{2+} by 50-fold whereas the *L*-site has the same affinity for Ca^{2+} and Mg^{2+}. Second, Mg^{2+} has different inhibitory actions at the *A*- and *L*-sites. Mg^{2+} at the *L*-site inhibits simply by preventing $[Ca^{2+}]_L$-activation whereas Mg^{2+} binding to the *A*-site will close the channel even if Ca^{2+} is bound to the *L*-site. While this difference seems to be a fine point, it does have important implications for RyR2 function. The fact that Mg^{2+} binding to the *A*-site will close the channel means that this form of Mg^{2+}-inhibition is very robust and will overpower the action of other channel activators. This form of Mg^{2+}-inhibition provides an effective break on Ca^{2+} release during diastolic and systolic conditions in the heart. On the other hand, Mg^{2+} binding to the *L*-site does not overpower channel activation by cytoplasmic Ca^{2+} so that this form of inhibition should only effective under diastolic conditions.

The inhibitory action of Mg^{2+} at the *I1*-site was discovered in the 1990s (Laver et al., 1997) where it was shown that the *I1*-site had no specificity between Ca^{2+} and Mg^{2+}. In cardiac muscle, Mg^{2+} is unlikely to have a sufficient inhibitory action via this site to make it a significant regulator in the heart.

VIII. A MODEL FOR Ca^{2+} AND Mg^{2+} REGULATION OF RYR2

Once the mechanisms for Ca^{2+}- and Mg^{2+}-dependent gating of the RyR had been identified and Ca^{2+}/Mg^{2+} binding characterized, it then became possible to develop a quantitative model to accurately fit the Ca^{2+} and Mg^{2+}-dependencies of k_o and τ_o (Laver, 2007a; Laver & Honen, 2008). The model serves two important functions. First, by quantitatively predicting the combined actions of the four Ca^{2+}/Mg^{2+} sensing sites it is possible to determine if these mechanisms fully account for the data or whether there exist other mechanisms which have not yet been identified. Second, by extrapolating the model to physiological ion concentrations, membrane potentials, and Ca^{2+} buffering, the model allows us to predict RyR2 activity in the cell. This is important because, even though bilayer experiments have proved extremely effective at demonstrating ligand regulation mechanisms, they are, by necessity, studied under nonphysiological conditions. This is because in the presence of diastolic concentrations of Ca^{2+} and Mg^{2+}, the P_o of RyRs is perishingly small. So low, in fact, that one would expect to observe a channel opening once every hour, far too infrequent to be reliably measured in bilayer experiments. Moreover, interpretation of the physiological significance of *in vitro* bilayer experiments requires careful consideration because processes that play an important role in RyR gating in bilayer experiments are not necessarily important for regulation of Ca^{2+} release in cells. For example, bilayer experiments show that $[Mg^{2+}]_L$-inhibition of RyR2 is mediated by a reduction of k_o via the Mg^{2+} binding at the *L*-site and a reduction in τ_o via Mg^{2+} feed-through to the *A*-site. In the cell, the latter will not occur because $[Mg^{2+}]$ is at mM levels on both sides of the SR membrane and Mg^{2+} feed-through will not significantly alter $[Mg^{2+}]$ near the pore. Therefore, $[Mg^{2+}]_L$-inhibition will occur via a different combination mechanisms in the cell than seen in bilayer experiments.

An overview of the model is shown in Fig. 4 (detailed explanations and equations are given by Laver and Honen, 2008). It is envisaged that channel openings are triggered by the combined action of the *A*- and *L*-sites. Once the channel is open, Ca^{2+} feed-through may further reinforce channel openings via the *A*-site or cause channel inactivation via the *I2*-site.

In the model of *A*/*L*-site activation (Fig. 4A and B), it is proposed that channel opening requires stimulation of at least three of the four RyR subunits by either the *A*- or *L*-sites. Each subunit can take on three stimulation levels according to Ca^{2+} and Mg^{2+} binding at these sites (Fig. 4A). For example, stimulation of three subunits by the *L*-sites (●) puts the channel in a conformation that has an opening rate, α, of 1 s^{-1} (this in the presence of ATP) whereas stimulation of three subunits by the *A*-sites (●) produces an opening rate of 60 s^{-1}. Combinations of *A*-site and *L*-site stimulated subunits leads to an opening rate of 7 s^{-1} and this accounts for the observed

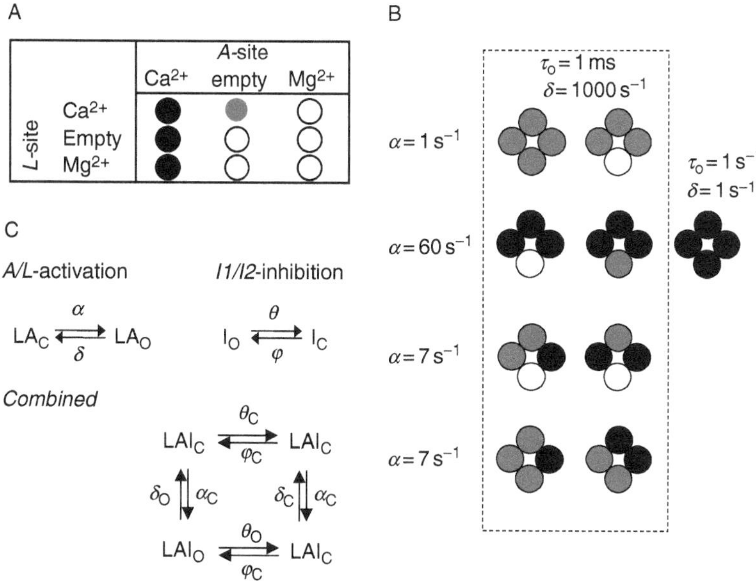

FIGURE 4 The schemes for regulation of RyR gating by the four Ca^{2+} sensing sites. (A) The scheme for how luminal and cytoplasmic activation sites (*A*- and *L*-sites) regulate the gating of RyR2. The key for the three different functional states of each subunit induced by combinations of Ca^{2+} and Mg^{2+} binding to the *A*- and *L*-sites. (B) The mean open times (τ_o), closing rates (δ), and opening rates (α) resulting from the various subunit stochiometries. A high opening rate (60 s^{-1}) will occur when either three or four subunits are (●), an intermediate opening rate (7 s^{-1}) when three subunits are not (○) *and* at least one subunit is (●), and a low opening rate (1 s^{-1}) when no subunits are (●) but at least three are (●). The mean open time of the channel is 1 ms when three or less subunits are (●) (the states enclosed by the dashed line). When all four are (●) the channel open configuration is stabilized such that the mean open time is extended 1000-fold. (C) Simplified schemes for channel activation by the *A*/*L*-sites, inhibition by the *I1*- and *I2*-sites and the combined action of activation and inhibition on channel gating. Uppercase subscripts, C and O, identify open and closed states of the RyR. Lowercase subscripts, identify rate constants that are affected by Ca^{2+} feed-through (after Laver, 2009).

synergy between cytoplasmic and luminal activation. The corresponding channel closing rate, δ, is 1000 s^{-1} for all conformations except where all four subunits are stimulated at the *A*-site. In that case the closing rate is very slow (<1 s^{-1} in the presence of ATP) and this leads to the long openings observed at 10–100 μM $[Ca^{2+}]_C$. The total rate of *A*/*L*-site mediated opening and closing depends on the proportional representation of all the binding stochiometries shown in Fig. 4B which can be calculated at each $[Ca^{2+}]$ and $[Mg^{2+}]$ using first order kinetic theory.

The opening and closing rates of the *I1*- and *I2*-sites are calculated in a similar fashion (not shown). Channel inactivation occurs when at least three subunits have Ca^{2+} bound to their inactivation sites. In the case of the *I2*-site, the model best accounts for the increase in τ_o observed at 10–100 μM $[Ca^{2+}]_C$ when inactivation *only* occurs when three of the four *I2*-sites bound to Ca^{2+}. The overall k_o and τ_o due to the combined effects of activation (*A*- and *L*-sites) and inhibition (*I1*- and *I2*-sites) are calculated using a reduced kinetic scheme for RyR gating (Fig. 4C). By combining all the *A*/*L*-site activation processes into a single step (also doing this for the inhibition mechanisms) the kinetic scheme is substantially simplified, but at a price. The price of this is that the lumped rate constants of the simplified scheme have complex dependencies on Ca^{2+} and Mg^{2+}.

The effects of Ca^{2+} feed-through appear in the model through the rate constants associated with closure of the channel (i.e., reaction paths leading away from the open state). The Ca^{2+} flux for each experimental condition is calculated using a model for ion permeation developed by Tinker, Lindsay, and Williams (1992). The Ca^{2+} feed-through effect is calculated by assuming that $[Ca^{2+}]$ at the cytoplasmic sites are increased above that of the bulk solution by an amount proportional to the Ca^{2+} flux through the channel. The proportionality constants for the *A*- and *I2*-sites are 12 and 0.35 μM/pA which were determined by fitting the model to the data.

The model was able to account for the increase in τ_o between 100 nM and 100 μM $[Ca^{2+}]_C$ (cytoplasmic activation, Fig. 2D, ×) and that observed between 1 and 100 μM $[Ca^{2+}]_L$ (luminal activation, Fig. 2F, ●) by Ca^{2+} binding to the *A*-site. Similarly, the model accounted for the decrease in τ_o between 1 nM and 3 μM $[Ca^{2+}]_C$ (cytoplasmic inactivation, Fig. 2D, ○) and that observed above 100 μM $[Ca^{2+}]_L$ (luminal inactivation, Fig. 2D, ●) by Ca^{2+} binding to the *I2*-site. The model also demonstrated that the lower apparent sensitivities of luminal activation and inactivation compared to their cytoplasmic counterparts arise from the dilution of Ca^{2+} as it emanates from the pore mouth and gets sequestered by Ca^{2+} buffers (i.e., the *A*- and *I2*-sites are exposed to lower Ca^{2+} concentrations than are present in the luminal bath).

IX. ADENINE NEUCLEOTIDES

Single channel recordings of RyRs and studies of ryanodine binding and Ca^{2+} release from SR vesicles have shown that millimolar levels of ATP activate RyRs (Meissner & Henderson, 1987). The by-products of ATP hydrolysis such as ADP (adenosine diphosphate) and AMP (adenosine monophosphate) are less effective activators of the RyR (Kermode, Williams, & Sitsapesan, 1998). In cardiac RyRs, ATP potentiates activation by Ca^{2+} but it cannot open the channel in the absence of Ca^{2+}. ATP activates RyRs with a K_a of 0.22 mM (Kermode et al., 1998) so that at concentrations above 1 mM, ATP exerts its near-maximal action on the RyR (intracellular [ATP] = 8 mM).

The overall effect of ATP on RyRs is to stabilize channel open conformations associated with the *A*-, *L*-, and *I2*-sites without significantly affecting the low-affinity Ca^{2+}/Mg^{2+}-inhibition mediated by the *I1*-site (Laver, 2007a). Stabilization is manifest as an increase in both the opening rates and open durations. As stated above, the $[Ca^{2+}]_C$-dependence of k_o above 1 μM is a unique property of the *A*-site. ATP (2 mM) causes an 11-fold increase in k_o under these conditions indicating a substantial increase in the *A*-site mediated opening rate (Fig. 5A). At $[Ca^{2+}]_C$ lower than 1 μM, k_o is a product of both *L*- and *A*-sites and k_o at 100 nM $[Ca^{2+}]_C$ is increased by 3.5-fold. This increase is seen again in the $[Ca^{2+}]_L$-dependencies of k_o in the presence and absence of ATP shown in Fig. 5C. ATP does not alter the K_a for $[Ca^{2+}]_L$-activation, indicating that ATP does not cause activation by altering the Ca^{2+} binding affinity of the *L*-site.

To get a measure of ATP stabilization of channel openings generated by the *A*-sites the $[Ca^{2+}]_C$-dependence of τ_o was measured when the membrane voltage (+40 mV) opposed Ca^{2+} feed-through, obviating the problem of local increases in $[Ca^{2+}]_C$ (Ca^{2+} microdomains). In the absence of ATP, τ_o was not dependent on $[Ca^{2+}]_C$ up to 10 μM (Fig. 5B, ●). However, as noted previously, in the presence of ATP, τ_o increases with increasing $[Ca^{2+}]_C$ (a sixfold increase at 3 μM Ca^{2+}, Fig. 5B, ○). This does not seem to be the case for the *L*-site, where in the presence of ATP and the absence of Ca^{2+} feed-through, τ_o does not increase with luminal Ca^{2+} (Fig. 2D, ×; $[Ca^{2+}]_L$ < 1 mM). Thus, ATP increases τ_o of openings triggered by the *A*-site but not those triggered by the *L*-site. However, when the membrane voltage favors Ca^{2+} feed-through, ATP causes a large increase in τ_o of openings triggered by luminal Ca^{2+} (Fig. 5D) which is consistent with Ca^{2+} flowing through the channel and binding to ATP-modified *A*-sites.

The key features of ATP-activation are (1) an increase in k_o at 100 nM $[Ca^{2+}]_C$ which reflects ATP modulation of the *A*/*L*-sites, (2) an increase in k_o at 1–3 μM $[Ca^{2+}]_C$ which reflects ATP modulation of the *A*-site only, (3) an increase in τ_o at 3 μM $[Ca^{2+}]_C$ and at +40 mV (negligible Ca^{2+} feed-through) which reflects combined action of ATP on the *A*-site and *I2*-site, and (4) an increase in τ_o at 100 nM $[Ca^{2+}]_C$ and at −40 mV where there is sufficient Ca^{2+} feed-through stimulate the *A*-site but not enough to cause inhibition via the *I2*-site.

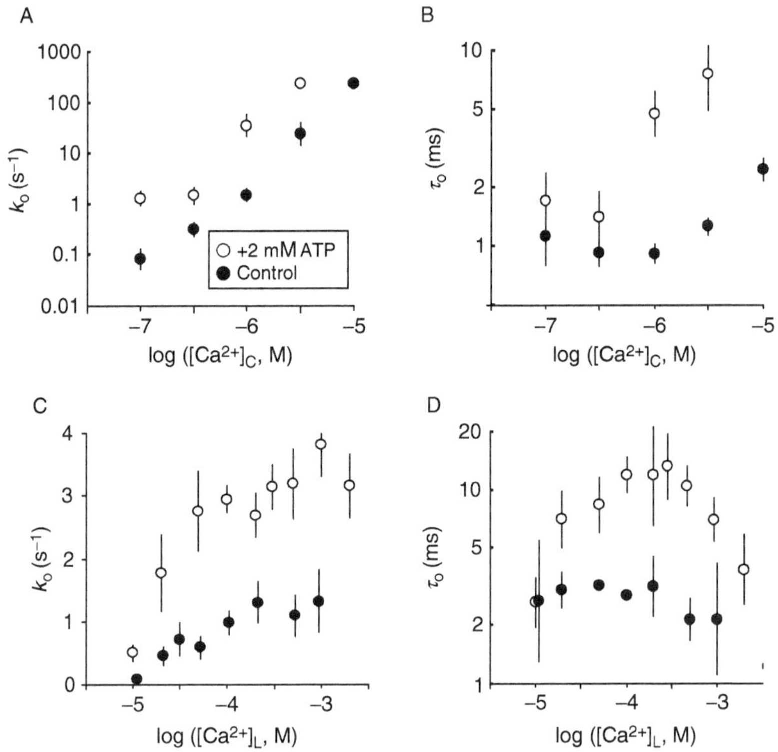

FIGURE 5 The effect of cytoplasmic ATP (2 mM) on Ca^{2+} regulation of RyR2. $[Ca^{2+}]_C$-dependence of RyR2 opening rate (A) and mean open time (B) in the presence of 0.1 mM $[Ca^{2+}]_L$ and in the presence and absence of 2 mM ATP. The bilayer voltage is +40 mV which opposes the flow of Ca^{2+} from luminal to cytoplasmic baths. $[Ca^{2+}]_L$-dependence of RyR2 opening rate (C) and mean open time (D) in the presence of 0.1 μM $[Ca^{2+}]_C$ and in the presence and absence of 2 mM ATP. The bilayer voltage is −40 mV (data from Laver, 2007a).

X. REGULATION OF RYR2 IN CARDIAC E–C COUPLING

Stimulation of Ca^{2+} release in cardiac cells can be induced by Ca^{2+} binding to either the luminal (*L*-site) or cytoplasmic (*A*-site) of RyR2. These different modes of RyR2-activation are likely to serve quite different functions in the heart. At the beginning of diastole, the RyRs in the heart are inactive because the SR is depleted of Ca^{2+} and the cytoplasmic Ca^{2+} has returned to resting levels. The model predicts that under these conditions, the *A*- and *L*-sites of the RyRs are mostly occupied by $[Mg^{2+}]$. During the course of diastole, the SERCA2a Ca^{2+}-pumps sequester Ca^{2+} into the SR, thus increasing $[Ca^{2+}]_L$

and RyR2-activation via the *L*-site. Also, the cardiac action potential causes Ca^{2+} influx through DHPRs and stimulation of RyR2 via their cytoplasmic (*A*-site). In the pacemaker cells of the sinoatrial node, luminal (*L*-site) stimulation of RyR2 is the driver of Ca^{2+} release and the timing of pacemaker potentials in the surface membrane (Vinogradova et al., 2005). However, in ventricular cells where the action potential initiates Ca^{2+} release, the *A*-site will trigger the release. Within this framework, one can see that excessive luminal stimulation of RyR2 in ventricular cells will cause the stores to drive the action potential, leading to action potentials that are not driven by the sinoatrial node.

Once open, Ca^{2+} flow through RyRs increases $[Ca^{2+}]_C$ in the region near the mouth of the pore thus generating Ca^{2+} microdomains. These Ca^{2+} microdomains cause a substantial prolongation of RyR open duration and they can trigger opening of neighboring RyRs in the triad junction. Bilayer experiments have shown Ca^{2+} driven coupling between adjacent RyRs via their *A*-sites (Laver, 2006). The coupled openings of RyRs have been visualized in the cell by fluorescent indicators and are referred to as Ca^{2+} sparks (Cheng, Lederer, & Cannell, 1993). Although there has been much debate about the role of Ca^{2+} feed-through in regard to luminal activation of RyRs in bilayer experiments (Gyorke et al., 2002; Sitsapesan & Williams, 1997), model calculations indicate that Ca^{2+} feed-through is *not* likely to be the main contributor to the $[Ca^{2+}]_L$-dependence of Ca^{2+} release *in vivo* (Laver & Honen, 2008). The model predicts that luminal Mg^{2+} is the main contributor to store $[Ca^{2+}]_L$-dependence of RyR activity in diastole and that removal of Mg^{2+} from the SR would remove 90% of the dependence of RyR2 activity on $[Ca^{2+}]_L$. In the absence of Mg^{2+}, the *L*-site would saturate at subphysiological $[Ca^{2+}]$ causing a loss of luminal regulation of RyR activity. Even though the Mg^{2+} concentrations within the cell are relatively constant, luminal triggering of RyR2 in the cell is an interplay between Ca^{2+} and Mg^{2+} where increased $[Ca^{2+}]_L$ acts by displacement of Mg^{2+} from the *L*-site.

When SR Ca^{2+} release increases $[Ca^{2+}]_C$ sufficiently, the Ca^{2+} sparks (localized Ca^{2+} release events) are replaced by a Ca^{2+} waves (global Ca^{2+} release) (Berridge, Lipp, & Bootman, 2000) that initial muscle contraction (i.e., systole). The model predicts that ligand regulation of RyR2 is very different between diastole and systole. Firstly, the $[Ca^{2+}]_L$-dependencies of RyR2 activity are different in diastole and systole. During diastole ($[Ca^{2+}]_C \sim 100$ nM), open probability of RyR2 increased eightfold over the physiological $[Ca^{2+}]_L$ range between 0.3 and 1 mM, but during systole ($[Ca^{2+}]_C \sim 1$ μM) activity increased by only 1.8-fold over this range. Secondly, the inhibitory effects of Mg^{2+} are quite different between diastole and systole. During diastole, Mg^{2+} is an important inhibitor at the *L*- and *A*-sites.

However, during systole, $[Ca^{2+}]_C$ is high enough to outcompete Mg^{2+} and the *A*-site. Moreover, during systole, the *L*-site is not a significant regulator of RyR2 activity which effectively sidelines Mg^{2+}-inhibition via the *L*-site.

XI. CONCLUDING REMARKS

Single channel recording has revealed a multiplicity of mechanisms by which the intracellular ligands Ca^{2+}, Mg^{2+}, and ATP can regulate RyR2 *in vitro*. The relative importance of these mechanisms *in vivo* will be different to how it is perceived from *in vitro* experiments. There now exists a kinetic model for RyR2 gating that permits prediction of the role of Mg^{2+}, ATP, and Ca^{2+} microdomains in stimulating RyR activity and SR Ca^{2+} release during cardiac E–C coupling. According to this model, the degree of Ca^{2+} loading of the SR is an important regulator of the SR Ca^{2+} leak during diastole but it has no significant effect on RyR2 gating during systole. (The release of Ca^{2+} during systole may terminate as a result of Ca^{2+} depletion in the SR even though RyRs are still active.) A corollary to this is that the properties of the *L*-sites can only be measured at low $[Ca^{2+}]_C$ (less than 1 μM) because the *L*-site does not contribute to RyR activity at higher concentrations. Therefore, the choice of correct experimental conditions is very important in assessing luminal regulation of RyRs. A major contributor to the shape of the luminal Ca^{2+} dependence of RyR activity is Mg^{2+} in the SR which competes with Ca^{2+} for the luminal activation site. Even though the Mg^{2+} concentrations within the cell are relatively constant, luminal triggering of RyR2 is an interplay (yin–yang) between Ca^{2+} and Mg^{2+} where increased luminal $[Ca^{2+}]$ leads to displacement of inhibitory Mg^{2+}.

A close link between luminal and cytoplasmic regulation of RyRs has been reported in many studies (Blayney & Lai, 2009; Gyorke & Gyorke, 1998; Lukyanenko, Gyorke, & Gyorke, 1996; Sitsapesan & Williams, 1994b). A common feature of channel activators such as ATP and caffeine is that they increase the potency of Ca^{2+} channel activation via the cytoplasmic *A*-site and increase the ability of luminal Ca^{2+} to activate the channel. The model which explains this, also predicts that regulation of RyR2 activity by store load, can result from changes in gating associated with either luminal or cytoplasmic domains on the channel protein. Firstly, Ca^{2+}-activation of subunits via the *A*-site contributes to luminal Ca^{2+}-activation via channel stoichiometries involving *A*- and *L*-sites. Secondly, the flow of luminal Ca^{2+} through the channel can regulate channel opening via the *A*-site. Therefore, compounds that increase the Ca^{2+}-activation by the *A*-site will render the RyR more sensitive to the level of Ca^{2+} loading inside the SR. The synergy observed between cytoplasmic and luminal Ca^{2+}-activation can be understood in terms of molecular processes that may also apply to transmembrane synergies reported for a wide range of regulators and disease causing mutations in the RyR.

Acknowledgments

Thanks to Ms. Meegan Jones for critically reading the manuscript. This work was supported by a Senior Brawn Fellowship from the University of Newcastle and by infrastructure grant from NSW Health through Hunter Medical Research Institute.

References

Balog, E. M., Fruen, B. R., Shomer, N. H., & Louis, C. F. (2001). Divergent effects of the malignant hyperthermia-susceptible arg(615)cys mutation on the Ca^{2+} and Mg^{2+} dependence of the RyR1. *Biophysical Journal, 81*, 2050–2058.

Berridge, M. J., Lipp, P., & Bootman, M. D. (2000). The versatility and universality of calcium signalling. *Nature Reviews. Molecular Cell Biology, 1*, 11–21.

Bers, D. M. (2002). Cardiac excitation-contraction coupling. *Nature, 415*, 198–205.

Blayney, L. M., & Lai, F. A. (2009). Ryanodine receptor-mediated arrhythmias and sudden cardiac death. *Pharmacology and Therapeutics, 123*, 151–177.

Cheng, H., Lederer, W. J., & Cannell, M. B. (1993). Calcium sparks: Elementary events underlying excitation-contraction coupling in heart muscle. *Science, 262*, 740–744.

Ching, L. L., Williams, A. J., & Sitsapesan, R. (2000). Evidence for Ca^{2+} activation and inactivation sites on the luminal side of the cardiac ryanodine receptor complex. *Circulation Research, 87*, 201–206.

Coronado, R., Morrissette, J., Sukhareva, M., & Vaughan, D. M. (1994). Structure and function of ryanodine receptors. *The American Journal of Physiology, 266*, C1485–C1504.

Dibb, K. M., Graham, H. K., Venetucci, L. A., Eisner, D. A., & Trafford, A. W. (2007). Analysis of cellular calcium fluxes in cardiac muscle to understand calcium homeostasis in the heart. *Cell Calcium, 42*, 503–512.

Gyorke, I., & Gyorke, S. (1998). Regulation of the cardiac ryanodine receptor channel by luminal Ca^{2+} involves luminal Ca^{2+} sensing sites. *Biophysical Journal, 75*, 2801–2810.

Gyorke, S., Gyorke, I., Lukyanenko, V., Terentyev, D., Viatchenko-Karpinski, S., & Wiesner, T. F. (2002). Regulation of sarcoplasmic reticulum calcium release by luminal calcium in cardiac muscle. *Frontiers in Bioscience, 7*, d1454–d1463.

Hymel, L., Inui, M., Fleischer, S., & Schindler, H. (1988). Purified ryanodine receptor of skeletal muscle sarcoplasmic reticulum forms Ca^{2+}-activated oligomeric Ca^{2+} channels in planar bilayers. *Proceedings of the National Academy of Sciences of the United States of America, 85*, 441–445.

Ju, Y. K., & Allen, D. G. (1998). Intracellular calcium and Na^{+}-Ca^{2+} exchange current in isolated toad pacemaker cells. *The Journal of Physiology, 508*, 153–166.

Ju, Y. K., & Allen, D. G. (2007). Store-operated Ca^{2+} entry and TRPC expression; possible roles in cardiac pacemaker tissue. *Heart, Lung and Circulation, 16*, 349–355.

Kermode, H., Williams, A. J., & Sitsapesan, R. (1998). The interactions of ATP, ADP, and inorganic phosphate with the sheep cardiac ryanodine receptor. *Biophysical Journal, 74*, 1296–1304.

Lamb, G. D. (2000). Excitation-contraction coupling in skeletal muscle: Comparisons with cardiac muscle. *Clinical and Experimental Pharmacology and Physiology, 27*, 216–224.

Lamb, G. D., & Stephenson, D. G. (1991). Effect of Mg^{2+} on the control of Ca^{2+} release in skeletal muscle fibres of the toad. *The Journal of Physiology, 434*, 507–528.

Lamb, G. D., & Stephenson, D. G. (1994). Effects of intracellular pH and [Mg^{2+}] on excitation-contraction coupling in skeletal muscle fibres of the rat. *The Journal of Physiology, 478*, 331–339.

Laver, D. R. (2006). Regulation of ryanodine receptors from skeletal and cardiac muscle during rest and excitation. *Clinical and Experimental Pharmacology and Physiology, 33*, 1107–1113.

Laver, D. R. (2007a). Ca^{2+} stores regulate ryanodine receptor Ca^{2+} release channels via luminal and cytosolic Ca^{2+} sites. *Biophysical Journal, 92*, 3541–3555.

Laver, D. R. (2007b). Ca^{2+} stores regulate ryanodine receptor Ca^{2+} release channels via luminal and cytosolic Ca^{2+} sites. *Clinical and Experimental Pharmacology and Physiology, 34*, 889–896.

Laver, D. R. (2009). Luminal Ca^{2+} activation of cardiac ryanodine receptors by luminal and cytoplasmic domains. *European Biophysics Journal, 39*, 19–26.

Laver, D. R., Baynes, T. M., & Dulhunty, A. F. (1997). Magnesium inhibition of ryanodine-receptor calcium channels: Evidence for two independent mechanisms. *The Journal of Membrane Biology, 156*, 213–229.

Laver, D. R., & Honen, B. N. (2008). Luminal Mg^{2+}, a key factor controlling RYR2-mediated Ca^{2+} release: Cytoplasmic and luminal regulation modeled in a tetrameric channel. *The Journal of General Physiology, 132*, 429–446.

Laver, D. R., Lenz, G. K., & Lamb, G. D. (2001). Regulation of the calcium release channel from rabbit skeletal muscle by the nucleotides ATP, AMP, IMP and adenosine. *The Journal of Physiology, 537*, 763–778.

Laver, D. R., O'Neill, E. R., & Lamb, G. D. (2004). Luminal Ca^{2+}-regulated Mg^{2+} inhibition of skeletal RyRs reconstituted as isolated channels or coupled clusters. *The Journal of General Physiology, 124*, 741–758.

Laver, D. R., Roden, L. D., Ahern, G. P., Eager, K. R., Junankar, P. R., & Dulhunty, A. F. (1995). Cytoplasmic Ca^{2+} inhibits the ryanodine receptor from cardiac muscle. *The Journal of Membrane Biology, 147*, 7–22.

Lukyanenko, V., Gyorke, I., & Gyorke, S. (1996). Regulation of calcium release by calcium inside the sarcoplasmic reticulum in ventricular myocytes. *Pflügers Archiv, 432*, 1047–1054.

Meissner, G. (1986). Ryanodine activation and inhibition of the Ca^{2+} release channel of sarcoplasmic reticulum. *The Journal of Biological Chemistry, 261*, 6300–6306.

Meissner, G. (1994). Ryanodine receptor/Ca^{2+} release channels and their regulation by endogenous effectors. *Annual Reviews in Physiology, 56*, 485–508.

Meissner, G., & Henderson, J. S. (1987). Rapid calcium release from cardiac sarcoplasmic reticulum vesicles is dependent on Ca^{2+} and is modulated by Mg^{2+}, adenine nucleotide, and calmodulin. *The Journal of Biological Chemistry, 262*, 3065–3073.

Rigg, L., & Terrar, D. A. (1996). Possible role of calcium release from the sarcoplasmic reticulum in pacemaking in guinea-pig sino-atrial node. *Experimental Physiology, 81*, 877–880.

Samso, M., Wagenknecht, T., & Allen, P. D. (2005). Internal structure and visualization of transmembrane domains of the RyR1 calcium release channel by cryo-EM. *Nature Structural and Molecular Biology, 12*, 539–544.

Sitsapesan, R., & Williams, A. J. (1994a). Gating of the native and purified cardiac SR Ca^{2+}-release channels with monovalent cations as permeant species. *Biophysical Journal, 67*, 1484–1494.

Sitsapesan, R., & Williams, A. J. (1994b). Regulation of the gating of the sheep cardiac sarcoplasmic reticulum Ca^{2+}-release channel by luminal Ca^{2+}. *The Journal of Membrane Biology, 137*, 215–226.

Sitsapesan, R., & Williams, A. J. (1997). Regulation of current flow through ryanodine receptors by luminal Ca^{2+}. *The Journal of Membrane Biology, 159*, 179–185.

Smith, J. S., Coronado, R., & Meissner, G. (1986). Single channel measurements of the calcium release channel from skeletal muscle sarcoplasmic reticulum. Activation by Ca^{2+} and ATP and modulation by Mg^{2+}. *The Journal of General Physiology, 88*, 573–588.

Tinker, A., Lindsay, A. R., & Williams, A. J. (1992). A model for ionic conduction in the ryanodine receptor channel of sheep cardiac muscle sarcoplasmic reticulum. *The Journal of General Physiology, 100*, 495–517.

Tripathy, A., & Meissner, G. (1996). Sarcoplasmic reticulum lumenal Ca^{2+} has access to cytosolic activation and inactivation sites of skeletal muscle Ca^{2+} release channel. *Biophysical Journal, 70*, 2600–2615.

Van Helden, D. F. (1993). Pacemaker potentials in lymphatic smooth muscle of the guinea-pig mesentery. *Journal de Physiologie, 471*, 465–479.

Van Helden, D. F., & Imtiaz, M. S. (2003). Ca^{2+} phase waves: A basis for cellular pacemaking and long-range synchronicity in the guinea-pig gastric pylorus. *The Journal of Physiology, 548*, 271–296.

Vinogradova, T. M., Maltsev, V. A., Bogdanov, K. Y., Lyashkov, A. E., & Lakatta, E. G. (2005). Rhythmic Ca^{2+} oscillations drive sinoatrial nodal cell pacemaker function to make the heart tick. *Annals of the New York Academy of Sciences, 1047*, 138–156.

Xu, L., Mann, G., & Meissner, G. (1996). Regulation of cardiac Ca^{2+} release channel (ryanodine receptor) by Ca^{2+}, H^+, Mg^{2+}, and adenine nucleotides under normal and simulated ischemic conditions. *Circulation Research, 79*, 1100–1109.

Zahradnik, I., Gyorke, S., & Zahradnikova, A. (2005). Calcium activation of ryanodine receptor channels–reconciling RyR gating models with tetrameric channel structure. *The Journal of General Physiology, 126*, 515–527.

CHAPTER 5

Regulation of Ryanodine Receptor Ion Channels Through Posttranslational Modifications

Gerhard Meissner

Department of Biochemistry and Biophysics, University of North Carolina, Chapel Hill, North Carolina, USA

I. OVERVIEW

The ryanodine receptors (RyRs) are cation-selective channels that release Ca^{2+} from an intracellular Ca^{2+} storing compartment, the endo/sarcoplasmic reticulum, during an action potential in a process known as excitation–contraction coupling. The released Ca^{2+} ions regulate a wide variety of biological functions. In striated muscle, the release of Ca^{2+} from the

Current Topics in Membranes, Volume 66 1063-5823/10 $35.00
Copyright 2010, Elsevier Inc. All right reserved.
DOI: 10.1016/S1063-5823(10)66005-X

sarcoplasmic reticulum into the cytoplasm leads to muscle contraction. This chapter focuses on the regulation of the skeletal muscle and cardiac muscle RyRs by protein kinases and redox active species. The RyRs are 2200 kDa Ca^{2+}-gated ion channels that are regulated, in addition to Ca^{2+} and endogenous effectors such as Mg^{2+} and ATP, by cAMP-dependent protein kinase A, calmodulin-dependent kinase II, protein kinase C, and protein phosphatases 1 and 2A. Thiols that serve as targets for reactive oxygen and nitrogen molecules determine the redox state and modulate the activity of the RyRs.

II. INTRODUCTION

The RyRs are large multiprotein complexes composed of four 560-kDa RyR subunits, four small FK506 binding protein (FKBP, also known as calstabin) subunits, and multiple associated proteins that include triadin, junctin, junctophilin, protein kinases and phosphatases, and calmodulin (Fill & Copello, 2002; Franzini-Armstrong & Protasi, 1997; Meissner, 1994; Zalk, Lehnart, & Marks, 2007). There are three mammalian RyR isoforms. RyR1 is found in the sarcoplasmic reticulum (SR) membrane of skeletal muscle, while RyR2 is present in heart muscle. In mammalian cells, RyR3 is coexpressed with one or two of the other RyR isoforms at low levels. All three isoforms are present in smooth muscle (Neylon, Richards, Larsen, Agrotis, & Bobik, 1995) and brain (Furuichi et al., 1994).

The mechanisms of RyR ion channel regulation have been most extensively studied in striated muscle. RyR1 and RyR2 reside in the junctional SR membrane near plasmalemmal L-type Ca^{2+} channels. An action potential in skeletal muscle initiates L-type Ca^{2+} channel protein conformational changes that alter the conformation of RyR1 by a direct physical interaction. In cardiac muscle, an action potential results in the influx of extracellular Ca^{2+}. Both mechanisms lead to the release of Ca^{2+} from the SR and subsequent muscle contraction (Fabiato, 1983; Rios & Pizarro, 1991). Sequestration of released Ca^{2+} by the SR Ca^{2+}-transporting ATPase (SERCA) and extrusion by the Na^{+}–Ca^{2+} exchanger restore the myofibrillar Ca^{2+} concentration from 10^{-6}–10^{-5} to $\sim 10^{-7}$ M, causing muscle to relax.

The RyRs share $\sim$70% sequence homology, with the greatest homology in the carboxyl-terminal region. In all isoforms, the C-terminal portion of the protein contains the transmembrane domain. Hydropathy analysis suggests between 4 and 12 transmembrane segments per RyR subunit (Takeshima et al., 1989; Zorzato et al., 1990). More recent studies using green fluorescence protein inserts and protease digestion indicate that each RyR1 subunit contains six to eight transmembrane helices (Du et al., 2004; Du, Sandhu, Khanna, Guo, & MacLennan, 2002). That six membrane spanning segments are sufficient to form a channel is supported by single channel recordings of tryptic fragments (Callaway et al., 1994), the presence of six transmembrane segments in the related inositol 1,4,5-trisphosphate receptor (Michikawa et al., 1994), and RyR cryo-electron microscopy data (Samso, Feng, Pessah, & Allen, 2009). The remaining RyR amino acids form the large catalytic cytoplasmic "foot" structure (Franzini-Armstrong & Protasi, 1997). As shall be discussed later, experimental evidence for several phosphorylation sites and redox-sensitive sites in the cytoplasmic structure has been described.

III. RYR1 AND RYR2 PHOSPHORYLATION

A. *RyR Phosphorylation Sites*

Primary sequence analysis suggests the presence of many phosphorylation sites in the large cytoplasmic foot region of the RyRs (Takeshima et al., 1989; Zorzato et al., 1990). Kinases and phosphatases that are part of the RyR1 and RyR2 multiprotein complexes include protein kinase A (PKA), Ca^{2+}/calmodulin-dependent protein kinase II (CaMKII), and protein phosphatase 1 (PP1) (Currie, Loughrey, Craig, & Smith, 2004; Dulhunty et al., 2001; Hohenegger & Suko, 1993; Marx et al., 2000; Marx, Reiken, et al., 2001; Ruehr et al., 2003). The RyR2 complex also contains protein phosphatase 2A (PP2A) (Marx et al., 2000). Protein phosphatase 2B (calcineurin) associates with RyR1 (Shin et al., 2002). Anchoring proteins mediate the interaction between RyRs and associated kinases and phosphatases through conserved leucine/isoleucine motifs. A-kinase anchoring protein targets PKA and phosphodiesterase 4D3, spinophilin targets PP1, and PR130 mediates the interaction of PP2A with RyR2 (Zalk et al., 2007).

Initial studies indicated that the RyRs have one phosphorylation site, RyR2-Ser2809 (Witcher, Kovacs, Schulman, Cefali, & Jones, 1991). Different functional effects following CaMK-, PKA-, and cGMP-dependent protein kinase phosphorylation indicated the presence of additional phosphorylation sites (Hain, Onoue, Mayrleitner, Fleischer, & Schindler, 1995; Takasago, Imagawa, Furukawa, Ogurusu, & Shigekawa, 1991). Xiao et al. (2005) observed that substitution of RyR2-Ser2809 with an alanine did not abolish PKA phosphorylation of RyR2 and by peptide mapping identified Ser2030 as a second phosphorylation site of RyR2. Wehrens, Lehnart, Reiken, and Marks (2004) showed that CaMKII uniquely phosphorylates Ser2815 near Ser2809 in recombinant RyR2 expressed in human embryonic kidney 293 cells. However, incorporation of more than one ^{32}P per monomer into the native, immunoprecipitated receptor indicated the presence of an additional CaMKII site in RyR2, in partial agreement with Rodriguez, Bhogal, and Colyer (2003) that there are four CaMKII phosphorylation sites relative to one PKA site, or eight sites based on two PKA sites per RyR2 monomer. The ryanodine receptor-calcium release channel in skeletal muscle was phosphorylated at Ser2843 (corresponding to Ser2809 in RyR2) by cAMP-, cGMP-, and CaM-dependent protein kinases (Suko et al., 1993). However, the presence of additional phosphorylation sites seems likely, since CaMKII also phosphorylated threonine residue(s).

B. RyR Modulation by PKA-Mediated Phosphorylation and the Role of FK506 Binding Proteins

During exercise and stress, β-adrenergic receptor stimulation results in increased PKA activity and changes in the activity of Ca^{2+} handling proteins involved in the release of Ca^{2+} (L-type Ca^{2+} channel, RyR2) and removal of Ca^{2+} (SERCA2a and its regulatory protein phospholamban) (Bers, 2002; Meissner, 2002). On the other hand, chronic stimulation of the sympathetic system can lead to maladaptive changes of β-adrenergic receptor signaling, cardiomyocyte death, and heart failure (Koch, 2004). *In vitro* phosphorylation of Ser2809 by CaMK, and to a lesser extent by PKA, activated the RyR ion channel isolated from cardiac muscle (Witcher et al., 1991). PKA regulated the ATP and Mg^{2+} sensitivity of RyR2 (Hain et al., 1995; Uehara, Yasukochi, Mejia-Alvarez, Fill, & Imanaga, 2002), changed channel kinetics, and increased the sensitivty to Ca^{2+} (Valdivia, Kaplan, Ellis-Davies, & Lederer, 1995). Single channel activity was lowest when RyR2-Ser2809 was phosphorylated to about 75% (Carter, Coyler, & Sitsapesan, 2006). Either a decreased or enhanced phosphorylation increased single channel activity. Marx et al. (2000) showed that PKA phosphorylation releases the small subunit FKBP12.6 (calstabin 2) from RyR2, which increased channel sensitivity to Ca^{2+} and induced the formation of channel substates. In failing hearts, RyR2 was PKA hyperphosphorylated, depleted of the FKBP12.6 subunit, and exhibited increased channel activity. Similarly, in skeletal muscle of animal models with heart failure and patients with heart diseases, exercise was linked to hyperphosphorylation and FKBP12 depletion of the skeletal muscle RyR1, increased RyR1 channel activity, and decreased exercise capacity (Reiken, Lacampagne, et al., 2003; Ward et al., 2003). These findings led Marks and coworkers to propose that PKA hyperphosphorylation and dissociation of the small FKBP subunit from the RyRs result in increased RyR sensitivity to cytosolic Ca^{2+}, referred to as leaky SR Ca^{2+} channels, and in the case of SR, Ca^{2+} handling referred to as SR Ca^{2+} leak.

The small 12 and 12.6 kDa FK506 binding proteins are predominantly associated with RyR1 and RyR2, respectively (Lam et al., 1995). They belong to the family of immunophilins, exhibit *cis/trans* isomerase activity, and their removal is reported to functionally uncouple groups of channels and increase channel activity (Brillantes et al., 1994; Marx, Gaburjakova, et al., 2001). Mechanisms implicated in RyR2 hyperphosphorylation and FKBP depletion include increased PKA activity and loss of PP1, PPA2, and phosphodiesterase 4D from the RyR2 macromolecular complex in failing hearts (Lehnart et al., 2005; Reiken, Gaburjakova, et al., 2003; Reiken, Lacampagne, et al., 2003). PKA-mediated hyperphosphorylation of RyR1 differed from that of RyR2 in that it failed to dissociate PP1 (Reiken, Lacampagne, et al., 2003).

Chen and colleagues (Jones et al., 2008; Meng et al., 2007) showed that the Ser2030 and Ser2809 phosphorylation sites in RyR2 are not located close to the FKBP12.6 binding site, as determined by cryo-electron microscopy. Therefore, it is unlikely that PKA-mediated phosphorylation of serines directly inhibits binding of FKBP12.6 to RyR2.

In attempts to define the specificity and functional consequences of RyR1 and RyR2 phosphorylation, mutants were prepared to mimic the dephosphorylated (alanine) or phosphorylated (aspartic acid) forms. A Ca^{2+}-dependent increase in open channel probability (P_o) was observed with recombinant RyR2-S2809D, whereas RyR2-S2815A open channel probability was comparable to WT-RyR2 (Wehrens, Lehnart, et al., 2004). The recombinant RyR1-S2843A mutant exhibited a single channel behavior similar to WT-RyR1, whereas the RyR1-S2843D mutation increased channel activity (Reiken, Lacampagne, et al., 2003). There was an increase in number of channel events, and current histograms indicated the appearance of substates similar to those observed in native, PKA-phosphorylated RyRs. In binding assays, RyR2-S2809A bound FKBP12.6 with an affinity comparable to nonphosphorylated WT-RyR2, whereas Ser2809D displayed a reduced binding affinity comparable to PKA-phosphorylated WT-RyR2 (Wehrens et al., 2004). These findings suggested that RyR phosphorylation and release of the accessory FKBPs lead to dysfunctional SR Ca^{2+} handling in striated muscle.

Several laboratories have challenged the hyperphosphorylation/FKBP depletion hypothesis of Marks and colleagues, which postulates that PKA hyperphosphorylation leads to FKBP depletion and leaky RyRs. Stange, Xu, Balshaw, Yamaguchi, and Meissner (2003) expressed RyR1-Ser2843A, RyR1-Ser2843D, RyR2-Ser2809A, and RyR2-Ser2809D phosphorylation mutants in HEK293 cells. In single channel and [^{3}H]-ryanodine binding assays, neither the skeletal nor cardiac mutants showed differences from wild type (WT) with regard to regulation by Ca^{2+}, Mg^{2+}, or ATP. Furthermore, no significant alterations in the FKBP/RyR binding ratios were observed. The PKA hyperphosphorylation/FKBP depletion hypothesis has also been disputed by Chen and colleagues (Xiao et al., 2006). FKBP12.6 bound to both the phosphorylated and nonphosphorylated forms of Ser2809. Ser-2030, but not Ser-2809, was the major RyR2 phosphorylation site responding to protein kinase A activation upon β-adrenergic stimulation in normal and failing hearts. RyR2-S2830D mutation sensitized the recombinant channel to luminal Ca^{2+}, referred to as store-overload-induced Ca^{2+} release (SOICR) (Xiao, Tian, Xie, et al., 2007).

Yano et al. (2000) assessed the functional interaction of FKBP12.6 with RyR2 in a canine model of pacing-induced heart failure, using the FKBP-interacting drug [^{3}H]-dihydro-FK506. The stoichiometry of FKBP12.6 to RyR2 was lowered from 3.6 in normal hearts to 1.6 in failing hearts. Xin et al. (2002) made the interesting observation that knockout of the FKBP12.6 gene resulted in cardiac hypertrophy in male but not female mice. Treatment with tamoxifen, an estrogen receptor α antagonist, resulted in cardiac hypertrophy, which suggested that estrogen protects the female FKBP12.6 null mice from cardiac hypertrophy. Additional studies indicated that FKBP12.6$^{-/-}$ mice exhibited exercise-induced arrhythmias that were similar to those observed in patients carrying RyR2 mutations associated with catecholaminergic polymorphic ventricular tachycardia (CPVT) (Wehrens et al., 2003). CPVT mutations decreased the binding affinity of FKBP12.6 and showed an increased single channel activity compared to WT following PKA phosphorylation. It was also shown that substitution of an aspartic residue with serine (FKBP12.6-D37S) increased the binding affinity of the FKBP12.6 mutant to PKA-phosphorylated WT-RyR2, RyR2-S2809D mutant, and CPVT-associated RyR2 mutants. Binding of the FKBP12.6 mutant restored normal channel function. Consistent with these findings, conditional overexpression of FKBP12.6 abrogated triggered ventricular tachycardia (Gellen et al., 2008).

Other studies have failed to establish that PKA phosphorylation of RyR2-S2809 and dissociation of FKBP12.6 from RyR2 are associated with a leaky SR Ca^{2+} channel. Li, Kranias, Mignery, and Bers (2002) measured resting Ca^{2+} sparks in permeabilized ventricular myocytes isolated from WT mice and mice lacking SERCA-regulatory subunit, phospholamban. The finding that PKA-mediated phosphorylation increased Ca^{2+} spark amplitude and duration in WT but not phospholamban knockout myocytes suggested that the primary mechanism of PKA action causing alterations in Ca^{2+}-homeostasis during heart failure involves phospholamban phosphorylation, and thus SERCA activity. Jiang et al. (2002) investigated the SR Ca^{2+} handling properties of human failing hearts and in a tachycardia-induced canine model of heart failure. The amplitude and kinetics of the Ca^{2+} transients were reduced. However, RyR2s isolated from canine and human heart failure tissues displayed no major structural or functional differences compared with controls. Furthermore, phosphorylation of RyRs by PKA did not appear to dissociate FKBP12.6 from the RyR2s. Xiao, Tian, Jones, et al. (2007) showed that removal of FKBP12.6 neither altered the functional properties of RyR2 nor increased the susceptibility of FKBP12.6$^{-/-}$ mice to stress-induced ventricular arrhythmias. Moreover, deletion of the Ser2809 phosphorylation site did not alter the β-adrenergic response in whole hearts and isolated cardiomyocytes. Aortic-banded RyR2-S2809A and WT mice both developed cardiac hypertrophy in absence of PKA-mediated phosphorylation of RyR2 (Benkusky et al., 2007; MacDonnell et al., 2008).

In skeletal muscle, during exercise, hyperphosphorylation of RyR1-Ser2843 by PKA, *S*-nitrosylation (see below), loss of phosphodiesterase PDE4D3, and the RyR1 subunit FKBP12 have been linked to decreased muscle function (Bellinger et al., 2008). A small molecule, S107 (whose action is discussed later in more detail), resulted in improved muscle function by stabilizing the RyR1–FKBP12 complex.

C. *RyR Modulation by CaMKII-Mediated Phosphorylation*

Phosphorylation of RyR2 by CaMKII has also been implicated in muscle diseases. Single channel recordings indicated that phosphorylation by CaMKII increases WT-RyR2 activity and Ca^{2+} sensitivity, but not of RyR2-S2815A that lacks the RyR2 CaMKII phosphorylation site (Wehrens, Lehnart, et al., 2004). Hain et al. (1995) proposed that phosphorylation of one subunit of the tetrameric RyR2 by endogenous CaMKII results in channel blockade by Mg^{2+}, whereas phosphorylation of all four subunits by exogenous CaMKII opens the channel. Transgenic mice that overexpressed the major cytosolic form of CaMKII (CaMKIIδ_c) had an increased phosphorylation of RyR2 and phospholamban, and died prematurely (Zhang et al., 2003). Ca^{2+} spark activities were enhanced in transgenic CaMKIIδ_c overexpressing hearts despite reduced SR Ca^{2+} content (Maier et al., 2003). This implied that CaMKII-mediated RyR2 phosphorylation results in the formation of leaky SR Ca^{2+} channels. In support of this view, creation of a genetic mouse model of CaMKII kinase inhibition protected the heart against excessive β adrenergic stimulation and myocardial infarction (Zhang et al., 2005). An opposing view is that acute overexpression of constitutively active CaMKII phosphorylates RyR2 and decreases the rate of local Ca^{2+} release events (Ca^{2+} sparks) and Ca^{2+} waves in cultured rat cardiomyocytes (Yang et al., 2007). In more recent studies using transgenic and KO mouse models, increased CaMKII activity contributed to increased SR leak, cardiac arrhythmogenesis and heart failure, whereas CaMKII deficiency reduced isoproterenol-induced arrhythmias and progression to heart failure (Ling et al., 2009; Sag et al., 2009). Chelu et al. (2009) reported that rapid atrial pacing unmasked increased vulnerability to atrial fibrillation in a genetic RyR2 gain-of-function mouse model compared to WT. This was ascribed to increased SR Ca^{2+} leak due to enhanced CaMKII-mediated phosphorylation of RyR2.

D. Protective Effects of 1,4-Benzothiazepine Derivatives JTV519 (K201) and S107

JTV519 (also known as K201) is a 1,4-benzothiazepine derivative that has a strong cardioprotective effect. This results from presumably inhibiting a broad spectrum of ion channel function (Hasumi, Matsuda, Shimamoto, Hata, & Kaneko, 2007). Yano et al. (2003) found that JTV519 reduced SR Ca^{2+} leak and improved cardiac function in dogs subjected to 4 weeks of chronic right ventricular pacing by minimizing RyR2 hyperphosphorylation and stabilizing the RyR2/FKBP12.6 complex. Single channel recordings showed that JTV519 stabilizes the closed state of RyR2 by promoting FKBP12.6 binding (Wehrens et al., 2004). The ineffectiveness of JTV519 to prevent exercise-induced ventricular tachyarrhythmias and death in FKBP12.6$^{-/-}$ mice suggested that binding of FkBP12.6 to RyR2 is a critical step for the drug having a beneficial effect. In two recent studies, the RyR2-specific derivative of JTV519, S107, enhanced the binding of FKBP12.6 to CPVT-associated RyR2-R2474S mutant (Lehnart et al., 2008), and in a dystrophic mouse model binding of FKBP12 to the hypernitrosylated RyR1 (Bellinger et al., 2009). In both cases, inhibition of channel leak was observed. At variance, JTV519 suppressed spontaneous Ca^{2+} release in FK506 treated HEK 293 cells and inhibited [^{3}H]-ryanodine binding to RyR2-N4104K linked to ventricular tachycardia, in the absence of FKPB12.6 (Hunt et al., 2007). Blayney, Jones, Griffiths, and Lai (2009) recently reported that both PKA-mediated phosphorylation and JTV519 change the affinity of FKBP12.6 binding to RyR2. Both reduced binding affinity to the closed RyR2 at 10^{-8} M Ca^{2+}, whereas a moderate increase in binding affinity was observed to the partially open channel at 10^{-3} M Ca^{2+}. Matsuzaki and colleagues (Oda et al., 2005; Tateishi et al., 2009) provided an explanation for the observation that JTV519 and its derivative S107 restore normal function for both phosphorylated and nitrosylated channels. These authors reported that disrupted (unzipped) domain interactions between the N-terminal and central domains of RyR2 are responsible for diastolic Ca^{2+} leak in failing hearts. Stabilization of these interactions by JTV519 improved cardiac function.

IV. RYR MODULATION BY REACTIVE OXYGEN AND NITROGEN SPECIES

A. RyRs and Reactive Oxygen Species

Active muscle produces reactive nitrogen and oxygen species that modulate contraction and relaxation (Ji, 2000; Reid, 1996; Stamler & Meissner, 2001). During strenuous exercise or short episodes of ischemia followed by reoxygenation, increased levels of reactive oxygen species impose oxidative stress. Reactive oxygen species can be formed by several mechanisms. These include the mitochondrial electron transport chain, xanthine oxidase, and NAD(P)H oxidases. Cells respond to excessive reactive oxygen species via the action of superoxide dismutase which converts superoxide ($O_2^{\bullet -}$) to O_2 and H_2O_2, and catalase and glutathione peroxidase which decompose H_2O_2.

Early studies showed that redox active reagents such as heavy metals, *N*-ethylmaleimide, diamide, $O_2^{\bullet -}$, and H_2O_2 modulate RyR activity (Abramson & Salama, 1989; Aghdasi, Zhang, Wu, Reid, & Hamilton, 1997; Anzai et al., 1998; Kawakami & Okabe, 1998; Oba, Ishikawa, & Yamaguchi, 1998). RyRs are good targets for redox active species because they contain a large number of thiols. The tetrameric mammalian skeletal and cardiac RyRs have 404 and 364 cysteines, with 100 and 89 cysteines per 560 kDa subunit, and 1 and 2 per FK506 binding protein, respectively.

The number of free thiols in purified RyR1 was determined using the lipophilic, thiol-specific probe, monobromobimane (Eu, Sun, Xu, Stamler, & Meissner, 2000). Nearly half of the 404 cysteines in the tetrameric RyR1 channel complex were labeled, that is, free in the presence of 5 mM reduced glutathione (GSH) at $pO_2 \sim 10$ mmHg. An increase in oxygen tension from ~10 mmHg (simulating tissue pO_2) to ambient air ($pO_2 \sim 150$ mmHg) in the presence of 5 mM GSH, resulted in loss of six to eight free thiols/RyR1 subunit without appreciably changing RyR1 activity. Substitution of GSH with oxidized glutathione (GSSG) at $pO_2 \sim 10$ mmHg or ambient air reduced the number of free thiols/RyR1 subunit and resulted in a large increase in RyR1 activity. The number of free thiols in RyR2-enriched fractions in the presence of 5 mM GSH was 50 per mg protein at $pO_2 \sim 150$ mmHg and 58 per mg protein ~10 mmHg (Sun et al., 2008). In the presence of 5 mM GSSG, a lower number of free cysteines were measured. There existed a good correlation between RyR2 activity and free RyR2 thiol content; however, because RyR2 was only partially purified, the exact number of glutathione and pO_2-sensitive thiols in native RyR2 remains to be determined. The results suggest that RyR1 and RyR2 have functional thiols that respond to the cells' pO_2 and GSH/GSSG redox state.

Additional cellular parameters influence the redox state of RyRs. Single channel measurements have shown that RyR1 responds to changes in cytosolic and SR luminal glutathione redox potential (Feng, Liu, Allen, &

Pessah, 2000). Micromolar activating Ca^{2+} concentrations lowered the redox state of RyR1 and favored channel opening, whereas inhibitory concentrations of Ca^{2+} and Mg^{2+} had opposite effects (Xia, Stangler, & Abramson, 2000). The observation that channel activators such as caffeine may act as electron acceptors, whereas inhibitors such as tetracaine may act as electron donors, suggested that nonthiol channel modulators regulate channel activity by shifting the thiol–disulfide ratio in the RyRs (Marinov, Olojo, Xia, & Abramson, 2007). RyR1-associated proteins have been shown to also affect receptor redox state. Channel closing unmasked the presence of hyperreactive cysteine residues in both RyR1 and triadin (Liu, Abramson, Zable, & Pessah, 1994), and SepN1, a selenoprotein, maintained the receptor's normal sensitivity to redox active species (Jurynec et al., 2008). Varied regulation of RyRs by plasmalemmal- and SR-associated NAD(P)H oxidases has been described (Espinosa et al., 2006; Hidalgo, Sanchez, Barrientos, & Aracena-Parks, 2006; Sanchez et al., 2008; Xia, Webb, Gnall, Cutler, & Abramson, 2003; Zima & Blatter, 2006).

A number of redox reactive cysteines among the 100 cysteines per RyR1 subunit have been identified, although the functional significance of specific cysteines remains elusive. Voss, Lango, Ernst-Russell, Morin, and Pessah (2004) determined by mass spectrometric analysis seven hyperreactive RyR1 cysteines (Cys-1040, 1303, 2436, 2565, 2606, 2611, and 3635) that were selectively labeled by 7-diethylamino-3-(4′-maleimidylphenyl)-4-methylcoumarin(CPM) under conditions that favored channel closing. Petrotchenko et al. (2006) identified by mass spectrometry one cysteine in RyR1 (Cys2327) that responded to a change in glutathione redox state. Aracena-Parks et al. (2006) showed that nine RyR1 cysteines (Cys-36, 315, 811, 906, 1591, 2326, 2363, 3193, and 3635) are endogenously modified and another three (Cys-253, 1040, and 1303) are modified by exogenous reactive oxygen and nitrogen molecules.

B. Regulation of RyRs by Nitric Oxide and Related Molecules

Nitric oxide (NO) is a ubiquitous regulator of cellular function and physiological modulator of cardiac and skeletal muscle excitation–contraction coupling. *S*-Nitrosylation of cardiac L-type Ca^{2+} channel (Sun et al., 2006) and RyRs (see below) have been described. Mammalian tissues express four isoforms of nitric oxide synthase: endothelial (eNOS), neuronal (nNOS), inducible (iNOS), and mitochondrial (mNOS) forms. In normal skeletal and cardiac muscle, the predominant isoforms, nNOS and eNOS, are targeted to the sarcolemma via interaction with the dystrophin-associated glycoprotein complex (Brenman, Chao, Xia, Aldape, & Bredt, 1995) and the caveolae structural protein caveolin-3 (Feron et al., 1996), respectively. Immunoelectron microscopy showed nNOS labeling in isolated cardiac SR vesicles, whereas no labeling was detected in skeletal muscle SR vesicles (Xu, Huso, Dawson, Bredt, & Becker, 1999). This suggested that NO regulates cardiac function by spatial confinement of nNOS and eNOS isoforms, *S*-nitrosylating RyR2 and plasmalemmal L-type Ca^{2+} channel, respectively. Studies with $eNOS^{-/-}$ and $nNOS^{-/-}$ mice support the idea that nNOS has a specific role in regulating SR Ca^{2+} release (Barouch et al., 2002). Although both $NOS^{-/-}$ mice developed age-related hypertrophy, only $eNOS^{-/-}$ mice were hypertensive. $nNOS^{-/-}$ mice exhibited a reduced ionotropic response, whereas $eNOS^{-/-}$ mice had enhanced contractility due to increased SR Ca^{2+} release. iNOS is absent or very low in normal heart but may increase in concentration, depending on the disease state (Arstall, Sawyer, Fukazawa, & Kelly, 1999).

NO exerts its cellular effects via cGMP-dependent and -independent pathways (Stamler & Hausladen, 1998). In the cGMP-dependent pathway, binding of NO to the heme group of guanylate cyclase increases the production of intracellular cGMP and activation of cGMP-dependent kinase, which was reported to phosphorylate RyR2 (Takasago et al., 1991). NO operates independently of cGMP through *S*-nitrosylation of proteins, most often at a single cysteine in acid–base or hydrophobic motifs (Hess, Matsumoto, Nudelman, & Stamler, 2001). Modification of RyR may involve other species, such as peroxynitrite ($OONO^-$), which is formed via a reaction of NO with superoxide. Peroxynitrite extensively oxidizes the RyRs (Sun, Xu, Eu, Stamler, & Meissner, 2001; Xu, Eu, Meissner, & Stamler, 1998) and has been implicated in myocardial injury and loss of RyR2 activity in the postischemic heart (Tang et al., 2009; Wang & Zweier, 1996).

Both RyR1 (Eu et al., 2000) and RyR2 (Xu et al., 1998) are endogenously *S*-nitrosylated, suggesting that NO is a physiological modulator of skeletal and cardiac muscle excitation–contraction coupling. In *in vitro* studies, NO or NO-related species activated or inhibited RyRs depending on donor concentration, membrane potential, the presence of channel agonists, and other sulfhydryl modifying reagents (Aghdasi, Reid, & Hamilton, 1997; Hart & Dulhunty, 2000; Meszaros, Minarovic, & Zahradnikova, 1996; Suko, Drobny, & Hellmann, 1999; Zahradnikova, Minarovic, Venema, & Meszaros, 1997). Low physiological concentrations of NO *S*-nitrosylated and activated RyR1 at tissue pO_2 ($\sim$10 mmHg) but not in ambient air ($pO_2 \sim$ 150 mmHg) (Eu et al., 2000). Changes in oxygen tension oxidized/reduced as many as six to eight thiols in each RyR1 subunit, which may explain the responsiveness of RyR1 to NO at tissue pO_2 but not ambient air. In intact muscle, NO modulated the O_2 tension dependence of SR Ca^{2+} release and contractility (Eu et al., 2003). The results suggest that RyR1 is an O_2 and NO sensing molecule, an idea questioned by Cheong, Tumbev,

Stoyanovsky, and Salama (2005). These investigators found that NO did not activate RyR1 under a range of pO_2. Like RyR1, RyR2 activity is dependent on pO_2 (Sun et al., 2008). However, unlike RyR1, RyR2 was not effectively activated and *S*-nitrosylated by NO. RyR2 was nonetheless modified and activated by *S*-nitrosoglutathione (GSNO) and $ONOO^-$.

3-Morpholinosydnonimine (SIN-1) (which generates peroxynitrite, $ONOO^-$), NOC12 (which generates a variety of reactive nitrogen oxides), and GSNO activate RyR1 independently of oxygen tension. NOC-12 activated by *S*-nitrosylation (Sun, Xu, Eu, Stamler, & Meissner, 2003), SIN-1 by oxidation of thiols (Sun, Xu, et al., 2001b), and GSNO by *S*-nitrosylation/oxidation (Sun et al., 2003) and *S*-glutathionylation (Aracena, Sanchez, Donoso, Hamilton, & Hidalgo, 2003). Peroxynitrite also modifies protein tyrosines (Souza, Peluffo, & Radi, 2008). The isolation of 3-nitrotyrosine containing fragments from type 2a Ca^{2+} ATPase (SERCA2a) and type 3 RyR (RyR3) isoforms in aging skeletal muscle was reported (Kanski, Hong, & Schoneich, 2005).

Site-directed mutagenesis studies demonstrated that at physiological O_2 concentrations, NO specifically *S*-nitrosylated Cys3635 out of ~50 free cysteines/RyR1 subunit (Sun, Xin, Eu, Stamler, & Meissner, 2001). Cys3635 is located in the CaM binding domain of RyR1 (Porter Moore, Zhang, & Hamilton, 1999; Yamaguchi, Xin, & Meissner, 2001), which provides an explanation that NO transduces its functional effect only in the presence of calmodulin. In contrast, activation of RyR1 by GSNO was independent of C3635 and calmodulin. Likewise, the corresponding RyR2 cysteine (C3602) was not required for RyR2 activation by GSNO (Sun et al., 2008).

C. RyR Oxidation and S-Nitrosylation in Normal and Diseased Muscle

Given the large number of free thiols, it is not surprising that abnormal modulation of RyRs by redox active molecules is implicated in muscle diseases. Durham et al. (2008) studied the role of RyR1-Y522S mutation in a knockin mouse model. In humans, the mutation is associated with malignant hyperthermia and a high incidence of central cores (Quane et al., 1994). Heterozygous expression of RyR1-Y522S increased SR Ca^{2+} leak, which led to the increased production of reactive nitrogen species. Increased *S*-nitrosylation of RyR1 further enhanced SR Ca^{2+} leak and resulted in increased susceptibility to heat-induced death. In fast-twitch fibers, the skeletal muscle-specific neuronal nitric oxide synthase (nNOSμ) is localized to the sarcolemma via interaction with the dystrophin-associated glycoprotein complex (Brenman et al., 1995). In mdx mice and in humans with Duchenne muscular dystrophy (the most common form of muscular dystrophy) disruption of the dystrophin-associated glycoprotein complex decreased the expression of nNOSμ. Despite a decreased nNOSμ expression, RyR1 was hypernitrosylated in mdx mice (Bellinger et al., 2009). This was due to increased expression and formation of a complex of inducible nitric oxide synthase (iNOS) with RyR1. Increased *S*-nitrosylation of RyR1 correlated with dissociation of FKBP12 from RyR1 and formation of leaky Ca^{2+} channels. S107, a compound previously shown by the authors to enhance the binding affinity of FKBP to hyperphosphorylated RyRs (Bellinger et al., 2008), enhanced muscle performance by increasing FKBP binding affinity to RyR1. SR Ca^{2+} leak and muscle damage were reduced. Interestingly, no increases in PKA-mediated phosphorylation of RyR1 at Ser2843 were observed in mdx mice. Thus, S107 may stabilize the FKBP–RyR1 complex under conditions where RyR1 is modified by very different mechanisms.

Aberrant oxidation and *S*-nitrosylation of RyR2 have been implicated in ischemic and failing hearts. Physical association of xanthine oxidoreductase (Khan et al., 2004) and nNOS (Khan et al., 2004; Xu et al., 1999) suggests that RyR2 is subjected to modifications by NO and superoxide ($O_2^{\bullet -}$) and their derivatives. NO can form HNO, an 1 electron reduction product of NO, reacts with $O_2^{\bullet -}$ to form peroxynitrite ($OONO^-$), and reacts with glutathione to generate GSNO. All three products activate RyR2 (Cheong, Tumbev, Abramson, Salama, & Stoyanovsky, 2005; Tocchetti et al., 2007; Xu et al., 1998). Adding to the complexity of the potential regulation of RyR2 by multiple oxygen and nitrogen reactive species, a translocation of nNOS from the SR to the sarcolemmal caveolae was observed in a rat model of heart failure (Bendall et al., 2004) and failing human hearts (Damy et al., 2004). Displacement of nNOS from RyR2 removes inhibition of xanthine oxidoreductase by NO, which can increase $O_2^{\bullet -}$ production and oxidative stress in failing hearts.

In the ischemic heart, increased NO production was associated with reduced myocardial contractility (Node et al., 1996; Zweier, Wang, & Kuppusamy, 1995). In the postischemic heart during the early period of reflow, NO production increased and reacted with superoxide to form peroxynitrite ($OONO^-$), which caused amino acid nitration and cellular injury (Wang & Zweier, 1996). Other studies have suggested that NO has a cardioprotective role. Physiologically relevant concentrations of peroxynitrite protected against myocardial reperfusion injury (Lefer et al., 1997). Cardiomyocyte-specific overexpression of eNOS limited left ventricular dysfunction after myocardial infarction (Janssens et al., 2004). Preconditioning and application of GSNO resulted in a similar pattern in protein *S*-nitrosylation and cardiac protection against ischemia/reperfusion (Sun, Morgan, Shen, Steenbergen, & Murphy, 2007). This suggested that *S*-nitrosylation of protein thiols protects cells from further oxidative damage. Although it is likely that RyR2 is modified in the ischemic/reperfused heart, the redox modifications that may occur are unclear.

Gonzalez, Beigi, Treuer, and Hare (2007) reported that diastolic Ca^{2+} levels increased in $nNOS^{-/-}$ but not $eNOS^{-/-}$ mice. nNOS elimination was associated with decreased *S*-nitrosylation, increased oxidation of RyR2 and SR Ca^{2+} leak, and arrhythmogenesis in cardiomyocytes. FKBP12.6 binding and phosphorylation of RyR2 were not altered in $nNOS^{-/-}$ mice. On the other hand, in hearts of mdx mice an increased *S*-nitrosylation and partial dissociation of FKBP12.6 from RyR2 was associated with a diastolic Ca^{2+} leak and arrhythmias (Fauconnier et al., 2010). S107 stabilized the RyR2/FKBP16.6 complex, reduced the SR Ca^{2+} leak, and prevented arrhythmias *in vivo*, without affecting the *S*-nitrosylation state of RyR2. The possibility of increased oxidation of RyR2 in mdx mice was not described by Fauconnier et al. (2010) but may have contributed to the formation of the SR Ca^{2+} leak, because treatment of cardiac and skeletal muscle SR membranes with the sulfhydryl oxidizing agents, H_2O_2 and diamide, diminished FKBP12.6 binding (Zissimopoulos, Docrat, & Lai, 2007). Consistent with this finding, Terentyev et al. (2008) observed that oxidizing agents increased SR Ca^{2+} leak in cardiomyocytes from normal hearts. In cardiomyocytes isolated from a chronic model of heart failure, reduced/oxidized glutathione ratio was decreased. The reduced level of free RyR2 thiols and enhanced SR Ca^{2+} leak were partially restored to normal levels by treating heart failure cardiomyocytes with sulfhydryl reducing agents.

V. CONCLUSIONS

Although recent work improved our understanding of the interaction of protein kinases and redox active molecules with the RyRs, their mechanisms in modulating Ca^{2+} handling in normal and dysfunctional muscle remain controversial. The third mammalian RyR isoform, RyR3, contains putative phosphorylation and redox active sites; however, coexpression with the other mammalian RyR isoforms has hindered establishment of a specific cellular role. Importantly, progress has been made in the development of drugs that stabilize RyR activity in dysfunctional muscle. However, their mechanism of action remains elusive. One major limitation in studying the RyRs is that their solution structure has not been solved. Cellular studies of protein kinases and redox active molecules are complicated by multiple interactions of the RyRs and potential changes in additional signaling mechanisms and transport proteins.

Acknowledgment

Support by National Institutes of Health grants AR018697 and HL073051 is gratefully acknowledged.

References

Abramson, J. J., & Salama, G. (1989). Critical sulfhydryls regulate calcium release from sarcoplasmic reticulum. *Journal of Bioenergetics and Biomembranes, 21*, 283–294.

Aghdasi, B., Reid, M. B., & Hamilton, S. L. (1997). Nitric oxide protects the skeletal muscle Ca^{2+} release channel from oxidation induced activation. *The Journal of Biological Chemistry, 272*, 25462–25467.

Aghdasi, B., Zhang, J. Z., Wu, Y., Reid, M. B., & Hamilton, S. L. (1997). Multiple classes of sulfhydryls modulate the skeletal muscle Ca^{2+} release channel. *The Journal of Biological Chemistry, 272*, 3739–3748.

Anzai, K., Ogawa, K., Kuniyasu, A., Ozawa, T., Yamamoto, H., & Nakayama, H. (1998). Effects of hydroxyl radical and sulfhydryl reagents on the open probability of the purified cardiac ryanodine receptor channel incorporated into planar lipid bilayers. *Biochemical and Biophysical Research Communications, 249*, 938–942.

Aracena, P., Sanchez, G., Donoso, P., Hamilton, S. L., & Hidalgo, C. (2003). S-glutathionylation decreases Mg^{2+} inhibition and S-nitrosylation enhances Ca^{2+} activation of RyR1 channels. *The Journal of Biological Chemistry, 278*, 42927–42935.

Aracena-Parks, P., Goonasekera, S. A., Gilman, C. P., Dirksen, R. T., Hidalgo, C., & Hamilton, S. L. (2006). Identification of cysteines involved in S-nitrosylation, S-glutathionylation, and oxidation to disulfides in ryanodine receptor type 1. *The Journal of Biological Chemistry, 281*, 40354–40368.

Arstall, M. A., Sawyer, D. B., Fukazawa, R., & Kelly, R. A. (1999). Cytokine-mediated apoptosis in cardiac myocytes. The role of inducible nitric oxide synthase induction and peroxynitrite generation. *Circulation Research, 85*, 829–840.

Barouch, L. A., Harrison, R. W., Skaf, M. W., Rosas, G. O., Cappola, T. P., Kobeissi, Z. A., et al. (2002). Nitric oxide regulates the heart by spatial confinement of nitric oxide synthase isoforms. *Nature, 416*, 337–339.

Bellinger, A. M., Reiken, S., Carlson, C., Mongillo, M., Liu, X., Rothman, L., et al. (2009). Hypernitrosylated ryanodine receptor calcium release channels are leaky in dystrophic muscle. *Natural Medicines, 15*, 325–330.

Bellinger, A. M., Reiken, S., Dura, M., Murphy, P. W., Deng, S. X., Landry, D. W., et al. (2008). Remodeling of ryanodine receptor complex causes "leaky" channels, a molecular mechanism for decreased exercise capacity. *Proceedings of the National Academy of Sciences of the United States of America, 105*, 2198–2202.

Bendall, J. K., Damy, T., Ratajczak, P., Loyer, X., Monceau, V., Marty, I., et al. (2004). Role of myocardial neuronal nitric oxide synthase-derived nitric oxide in beta-adrenergic hyporesponsiveness after myocardial infarction-induced heart failure in rat. *Circulation, 110*, 2368–2375.

Benkusky, N. A., Weber, C. S., Scherman, J. A., Farrell, E. F., Hacker, T. A., John, M. C., et al. (2007). Intact beta-adrenergic response and unmodified progression toward heart failure in mice with genetic ablation of a major protein kinase A phosphorylation site in the cardiac ryanodine receptor. *Circulation Research, 101*, 819–829.

Bers, D. M. (2002). Cardiac excitation-contraction coupling. *Nature, 415*, 198–205.

Blayney, L. M., Jones, J. L., Griffiths, J., & Lai, F. A. (2009). A mechanism of ryanodine receptor modulation by FKBP12/12.6, protein kinase A, and K201. *Cardiovascular Research, 85*, 68–78.

Brenman, J. E., Chao, D. S., Xia, H., Aldape, K., & Bredt, D. S. (1995). Nitric oxide synthase complexed with dystrophin and absent from skeletal muscle sarcolemma in Duchenne muscular dystrophy. *Cell, 82*, 743–752.

Brillantes, A. B., Ondrias, K., Scott, A., Kobrinsky, E., Ondriasova, E., Moschella, M. C., et al. (1994). Stabilization of calcium release channel (ryanodine receptor) function by FK506-binding protein. *Cell, 77*, 513–523.

Callaway, C., Seryshev, A., Wang, J. P., Slavik, K. J., Needleman, D. H., Cantu, C. I., et al. (1994). Localization of the high and low affinity [^{3}H]ryanodine binding sites on the skeletal muscle Ca^{2+} release channel. *The Journal of Biological Chemistry, 269*, 15876–15884.

Carter, S., Coyler, J., & Sitsapesan, R. (2006). Maximum phosphorylatioin of the cardiac ryanodine receptor at serine-2809 by protein kinase A produces unique modifications to channel gating and conductance not observed at lower levels to phosphorylation. *Circulation Research, 98*, 1506–1513.

Chelu, M. G., Sarma, S., Sood, S., Wang, S., van Oort, R. J., Skapura, D. G., et al. (2009). Calmodulin kinase II-mediated sarcoplasmic reticulum Ca^{2+} leak promotes atrial fibrillation in mice. *The Journal of Clinical Investigation, 119*, 1940–1951.

Cheong, E., Tumbev, V., Abramson, J., Salama, G., & Stoyanovsky, D. A. (2005). Nitroxyl triggers Ca^{2+} release from skeletal and cardiac sarcoplasmic reticulum by oxidizing ryanodine receptors. *Cell Calcium, 37*, 87–96.

Cheong, E., Tumbev, V., Stoyanovsky, D., & Salama, G. (2005). Effects of pO(2) on the activation of skeletal muscle ryanodine receptors by NO, A cautionary note. *Cell Calcium, 38*, 481–488.

Currie, S., Loughrey, C. M., Craig, M. A., & Smith, G. L. (2004). Calcium/calmodulin-dependent protein kinase IIdelta associates with the ryanodine receptor complex and regulates channel function in rabbit heart. *The Biochemical Journal, 277*, 357–366.

Damy, T., Ratajczak, P., Shah, A. M., Camors, E., Marty, I., Hasenfuss, G., et al. (2004). Increased neuronal nitric oxide synthase-derived NO production in the failing human heart. *Lancet, 363*, 1365–1367.

Du, G. G., Avila, G., Sharma, P., Khanna, V. K., Dirksen, R. T., & MacLennan, D. H. (2004). Role of the sequence surrounding predicted transmembrane helix M4 in membrane association and function of the Ca^{2+} release channel of skeletal muscle sarcoplasmic reticulum (ryanodine receptor isoform 1). *The Journal of Biological Chemistry, 279*, 37566–37574.

Du, G. G., Sandhu, B., Khanna, V. K., Guo, X. H., & MacLennan, D. H. (2002). Topology of the Ca^{2+} release channel of skeletal muscle sarcoplasmic reticulum (RyR1). *Proceedings of the National Academy of Sciences of the United States of America, 99*, 16725–16730.

Dulhunty, A. F., Laver, D., Curtis, S. M., Pace, S., Haarmann, C., & Gallant, E. M. (2001). Characteristics of irreversible ATP activation suggest that native skeletal ryanodine receptors can be phosphorylated via an endogenous CaMKII. *Biophysical Journal, 81*, 3240–3252.

Durham, W. J., Aracena-Parks, P., Long, C., Rossi, A. E., Goonasekera, S. A., Boncompagni, S., et al. (2008). RyR1 S-nitrosylation underlies environmental heat stroke and sudden death in Y522S RyR1 knockin mice. *Cell, 133*, 53–65.

Espinosa, A., Leiva, A., Pena, M., Muller, M., Debandi, A., Hidalgo, C., et al. (2006). Myotube depolarization generates reactive oxygen species through NAD(P)H oxidase; ROS-elicited Ca^{2+} stimulates ERK, CREB, early genes. *Journal of Cellular Physiology, 209*, 379–388.

Eu, J. P., Hare, J. M., Hess, D. T., Skaf, M., Sun, J., Cardenas-Navina, I., et al. (2003). Concerted regulation of skeletal muscle contractility by oxygen tension and endogenous nitric oxide. *Proceedings of the National Academy of Sciences of the United States of America, 100*, 15229–15234.

Eu, J. P., Sun, J., Xu, L., Stamler, J. S., & Meissner, G. (2000). The skeletal muscle calcium release channel, coupled O_2 sensor and NO signaling functions. *Cell, 102*, 499–509.

Fabiato, A. (1983). Calcium-induced release of calcium from the cardiac sarcoplasmic reticulum. *The American Journal of Physiology, 245*, C1–C14.

Fauconnier, J., Thireau, J., Reiken, S., Cassan, C., Richard, S., Matecki, S., et al. (2010). Leaky RyR2 trigger ventricular arrhythmias in Duchenne muscular dystrophy. *Proceedings of the National Academy of Sciences of the United States of America, 107*, 1559–1564.

Feng, W., Liu, G., Allen, P. D., & Pessah, I. N. (2000). Transmembrane redox sensor of ryanodine receptor complex. *The Journal of Biological Chemistry, 275*, 35902–35907.

Feron, O., Belhassen, L., Kobzik, L., Smith, T. W., Kelly, R. A., & Michel, T. (1996). Endothelial nitric oxide synthase targeting to caveolae. Specific interactions with caveolin isoforms in cardiac myocytes and endothelial cells. *The Journal of Biological Chemistry, 271*, 22810–22814.

Fill, M., & Copello, J. A. (2002). Ryanodine receptor calcium release channels. *Physiological Reviews, 82*, 893–922.

Franzini-Armstrong, C., & Protasi, F. (1997). Ryanodine receptors of striated muscles, a complex channel capable of multiple interactions. *Physiological Reviews, 77*, 699–729.

Furuichi, T., Furutama, D., Hakamata, Y., Nakai, J., Takeshima, H., & Mikoshiba, K. (1994). Multiple types of ryanodine receptor/Ca^{2+} release channels are differentially expressed in rabbit brain. *The Journal of Neuroscience, 14*, 4794–4805.

Gellen, B., Fernandez-Velasco, M., Briec, F., Vinet, L., LeQuang, K., Rouet-Benzineb, P., et al. (2008). Conditional FKBP12.6 overexpression in mouse cardiac myocytes prevents triggered ventricular tachycardia through specific alterations in excitation-contraction coupling. *Circulation, 117*, 1778–1786.

Gonzalez, D. R., Beigi, F., Treuer, A. V., & Hare, J. M. (2007). Deficient ryanodine receptor S-nitrosylation increases sarcoplasmic reticulum calcium leak and arrhythmogenesis in cardiomyocytes. *Proceedings of the National Academy of Sciences of the United States of America, 104*, 20612–20617.

Hain, J., Onoue, H., Mayrleitner, M., Fleischer, S., & Schindler, H. (1995). Phosphorylation modulates the function of the calcium release channel of sarcoplasmic reticulum from cardiac muscle. *The Journal of Biological Chemistry, 270*, 2074–2081.

Hart, J. D., & Dulhunty, A. F. (2000). Nitric oxide activates or inhibits skeletal muscle ryanodine receptors depending on its concentration, membrane potential and ligand binding. *The Journal of Membrane Biology, 173*, 227–336.

Hasumi, H., Matsuda, R., Shimamoto, K., Hata, Y., & Kaneko, N. (2007). K201, a multichannel blocker, inhibits clofilium-induced torsades de pointes and attenuates an increase in repolarization. *European Journal of Pharmacology, 555*, 54–60.

Hess, D. T., Matsumoto, A., Nudelman, R., & Stamler, J. S. (2001). S-nitrosylation, spectrum and specificity. *Nature Cell Biology, 3*, E46–E49.

Hidalgo, C., Sanchez, G., Barrientos, G., & Aracena-Parks, P. (2006). A transverse tubule NOX activity stimulates calcium release from isolated triads via RYR1 S-Glutathionylation. *The Journal of Biological Chemistry, 281*, 26473–26482.

Hohenegger, M., & Suko, J. (1993). Phosphorylation of the purified cardiac ryanodine receptor by exogenous and endogenous protein kinases. *The Biochemical Journal, 296*, 303–308.

Hunt, D. J., Jones, P. P., Wang, R., Chen, W., Bolstad, J., Chen, K., et al. (2007). K201 (JTV519) suppresses spontaneous Ca^{2+} release and [^{3}H]ryanodine binding to RyR2 irrespective of FKBP12.6 association. *The Biochemical Journal, 404*, 431–438.

Janssens, S., Pokreisz, P., Schoonjans, L., Pellens, M., Vermeersch, P., Tjwa, M., et al. (2004). Cardiomyocyte-specific overexpression of nitric oxide synthase 3 improves left ventricular performance and reduces compensatory hypertrophy after myocardial infarction. *Circulation Research, 94*, 1256–1262.

Ji, L. L. (2000). Exercise-induced oxidative stress in the heart. In C. K. Sen, L. Packer & O. Hanninen (Eds.), *Handbook of exercise and oxygen toxity* (pp. 689–712). Amsterdam, The Netherlands: Elsevier Science.

Jiang, M. T., Lokuta, A. J., Farrell, E. F., Wolff, M. R., Haworth, R. A., & Valdivia, H. H. (2002). Abnormal Ca^{2+} release, but normal ryanodine receptors, in canine and human heart failure. [comment]. *Circulation Research, 91*, 1015–1022.

Jones, P. P., Meng, X., Xiao, B., Cai, S., Bolstad, J., Wagenknecht, T., et al. (2008). Localization of PKA phosphorylation site, Ser(2030), in the three-dimensional structure of cardiac ryanodine receptor. *The Biochemical Journal, 410*, 261–270.

Jurynec, M. J., Xia, R., Mackrill, J. J., Gunther, D., Crawford, T., Flanigan, K. M., et al. (2008). Selenoprotein N is required for ryanodine receptor calcium release channel activity in human and zebrafish muscle. *Proceedings of the National Academy of Sciences of the United States of America, 105*, 12485–12490.

Kanski, J., Hong, S. J., & Schoneich, C. (2005). Proteomic analysis of protein nitration in aging skeletal muscle and identification of nitrotyrosine-containing sequences in vivo by nanoelectrospray ionization tandem mass spectrometry. *The Journal of Biological Chemistry, 280*, 24261–24266.

Kawakami, M., & Okabe, E. (1998). Superoxide anion radical-triggered Ca^{2+} release from cardiac sarcoplasmic reticulum through ryanodine receptor Ca^{2+} channel. *Molecular Pharmacology, 53*, 497–503.

Khan, S. A., Lee, K., Minhas, K. M., Gonzalez, D. R., Raju, S. V., Tejani, A. D., et al. (2004). Neuronal nitric oxide synthase negatively regulates xanthine oxidoreductase inhibition of cardiac excitation-contraction coupling. *Proceedings of the National Academy of Sciences of the United States of America, 101*, 15944–15948.

Koch, W. J. (2004). Genetic and phenotypic targeting of beta-adrenergic signaling in heart failure. *Molecular and Cellular Biochemistry, 263*, 5–9.

Lam, E., Martin, M. M., Timerman, A. P., Sabers, C., Fleischer, S., Lukas, T., et al. (1995). A novel FK506 binding protein can mediate the immunosuppressive effects of FK506 and is associated with the cardiac ryanodine receptor. *The Journal of Biological Chemistry, 270*, 26511–26522.

Lefer, D. J., Scalia, R., Campbell, B., Nossuli, T., Hayward, R., Salamon, M., et al. (1997). Peroxynitrite inhibits leukocyte-endothelial cell interactions and protects against ischemia-reperfusion injury in rats. *The Journal of Clinical Investigation, 99*, 684–691.

Lehnart, S. E., Mongillo, M., Bellinger, A., Lindegger, N., Chen, B. X., Hsueh, W., et al. (2008). Leaky Ca^{2+} release channel/ryanodine receptor 2 causes seizures and sudden cardiac death in mice. *The Journal of Clinical Investigation, 118*, 2230–2245.

Lehnart, S. E., Wehrens, X. H., Reiken, S., Warrier, S., Belevych, A. E., Harvey, R. D., et al. (2005). Phosphodiesterase 4D deficiency in the ryanodine-receptor complex promotes heart failure and arrhythmias. *Cell, 123*, 25–35.

Li, Y., Kranias, E. G., Mignery, G. A., & Bers, D. M. (2002). Protein kinase A phosphorylation of the ryanodine receptor does not affect calcium sparks in mouse ventricular myocytes. *Circulation Research, 90*, 309–316.

Ling, H., Zhang, T., Pereira, L., Means, C. K., Cheng, H., Gu, Y., et al. (2009). Requirement for Ca^{2+}/calmodulin-dependent kinase II in the transition from pressure overload-induced cardiac hypertrophy to heart failure in mice. *The Journal of Clinical Investigation, 119*, 1230–1240.

Liu, G., Abramson, J. J., Zable, A. C., & Pessah, I. N. (1994). Direct evidence for the existence and functional role of hyperreactive sulfhydryls on the ryanodine receptor-triadin complex selectively labeled by the coumarin maleimide 7-diethylamino-3-(4-maleimidylphenyl)-4-methylcoumarin. *Molecular Pharmaceutics, 45*, 189–200.

MacDonnell, S. M., Garcia-Rivas, G., Scherman, J. A., Kubo, H., Chen, X., Valdivia, H., et al. (2008). Adrenergic regulation of cardiac contractility does not involve phoshorylation of cardiac ryanodine receptor at serine 2808. *Circulation Research, 102*, e65–e72.

Maier, L. S., Zhang, T., Chen, L., DeSantiago, J., Brown, J. H., & Bers, D. M. (2003). Transgenic CaMKIIdeltaC overexpression uniquely alters cardiac myocyte Ca^{2+} handling, reduced SR Ca^{2+} load and activated SR Ca^{2+} release. *Circulation Research, 92*, 904–911.

Marinov, B. S., Olojo, R. O., Xia, R., & Abramson, J. J. (2007). Non-thiol reagents regulate ryanodine receptor function by redox interactions that modify reactive thiols. *Antioxidants Redox Signaling, 9*, 609–621.

Marx, S. O., Gaburjakova, J., Gaburjakova, M., Henrikson, C., Ondrias, K., & Marks, A. R. (2001). Coupled gating between cardiac calcium release channels (ryanodine receptors). *Circulation Research, 88*, 1151–1158.

Marx, S. O., Reiken, S., Hisamatsu, Y., Gaburjakova, M., Gaburjakova, J., Yang, Y. M., et al. (2001). Phosphorylation-dependent regulation of ryanodine receptors, a novel role for leucine/isoleucine zippers. *The Journal of Cell Biology, 153*, 699–708.

Marx, S. O., Reiken, S., Hisamatsu, Y., Jayaraman, T., Burkhoff, D., Rosemblit, N., et al. (2000). PKA phosphorylation dissociates FKBP12.6 from the calcium release channel (ryanodine receptor), defective regulation in failing hearts. *Cell, 101*, 365–376.

Meissner, G. (1994). Ryanodine receptor/Ca^{2+} release channels and their regulation by endogenous effectors. *Annual Review of Physiology, 56*, 485–508.

Meissner, G. (2002). Regulation of mammalian ryanodine receptors. *Frontiers in Bioscience, 7*, d2072–d2080.

Meng, X., Xiao, B., Cai, S., Huang, X., Li, F., Bolstad, J., et al. (2007). Three-dimensional localization of serine-2808, a phosphorylation site in cardiac ryanodine receptor. *The Journal of Biological Chemistry, 282*, 25929–25939.

Meszaros, L. G., Minarovic, I., & Zahradnikova, A. (1996). Inhibition of the skeletal muscle ryanodine receptor calcium release channel by nitric oxide. *FEBS Letters, 380*, 49–52.

Michikawa, T., Hamanaka, H., Otsu, H., Yamamoto, A., Miyawaki, A., Furuichi, T., et al. (1994). Transmembrane topology and sites of N-glycosylation of inositol 1,4,5-trisphosphate receptor. *The Journal of Biological Chemistry, 269*, 9184–9189.

Neylon, C. B., Richards, S. M., Larsen, M. A., Agrotis, A., & Bobik, A. (1995). Multiple types of ryanodine receptor/Ca^{2+} release channels are expressed in vascular smooth muscle. *Biochemical and Biophysical Research Communications, 215*, 814–821.

Node, K., Kitakaze, M., Kosaka, H., Komamura, K., Minamino, T., Inoue, M., et al. (1996). Increased release of NO during ischemia reduces myocardial contractility and improves metabolic dysfunction. *Circulation, 93*, 356–364.

Oba, T., Ishikawa, T., & Yamaguchi, M. (1998). Sulfhydryls associated with H_2O_2-induced channel activation are on luminal side of ryanodine receptors. *The American Journal of Physiology, 274*, C914–C921.

Oda, T., Yano, M., Yamamoto, T., Tokuhisa, T., Okuda, S., Doi, M., et al. (2005). Defective regulation of interdomain interactions within the ryanodine receptor plays a key role in the pathogenesis of heart failure. *Circulation, 111*, 3400–3410.

Petrotchenko, E. V., Pasek, D., Elms, P., Dokholyan, N. V., Meissner, G., & Borchers, C. H. (2006). Combining fluorescence detection and mass spectrometric analysis for comprehensive and quantitative analysis of redox-sensitive cysteines in native membrane proteins. *Analytical Chemistry, 78*, 7959–7966.

Porter Moore, C., Zhang, J. Z., & Hamilton, S. L. (1999). A role for cysteine 3635 of RYR1 in redox modulation and calmodulin binding. *The Journal of Biological Chemistry, 274*, 36831–36834.

Quane, K. A., Keating, K. E., Healy, J. M., Manning, B. M., Krivosic-Horber, R., Krivosic, I., et al. (1994). Mutation screening of the RYR1 gene in malignant hyperthermia, detection of a novel Tyr to Ser mutation in a pedigree with associated central cores. *Genomics, 23*, 236–239.

Reid, M. B. (1996). Reactive oxygen and nitric oxide in skeletal muscle. *News in Physiological Sciences, 11*, 114–121.

Reiken, S., Gaburjakova, M., Guatimosim, S., Gomez, A. M., D'Armiento, J., Burkhoff, D., et al. (2003). Protein kinase A phosphorylation of the cardiac calcium release channel (ryanodine receptor) in normal and failing hearts. Role of phosphatases and response to isoproterenol. *The Journal of Biological Chemistry, 278*, 444–453.

Reiken, S., Lacampagne, A., Zhou, H., Kherani, A., Lehnart, S. E., Ward, C., et al. (2003). PKA phosphorylation activates the calcium release channel (ryanodine receptor) in skeletal muscle, defective regulation in heart failure. *The Journal of Cell Biology, 160*, 919–928.

Rios, E., & Pizarro, G. (1991). Voltage sensor of excitation-contraction coupling in skeletal muscle. *Physiological Reviews, 71*, 849–908.

Rodriguez, P., Bhogal, M. S., & Colyer, J. (2003). Stoichiometric phosphorylation of cardiac ryanodine receptor on serine 2809 by calmodulin-dependent kinase II and protein kinase A. *The Journal of Biological Chemistry, 278*, 38593–38600.

Ruehr, M. L., Russell, M. A., Ferguson, D. G., Bhat, M., Ma, J., Damron, D. S., et al. (2003). Targeting of protein kinase A by muscle A kinase-anchoring protein (mAKAP) regulates phosphorylation and function of the skeletal muscle ryanodine receptor. *The Journal of Biological Chemistry, 278*, 24831–24836.

Sag, C. M., Wadsack, D. P., Khabbazzadeh, S., Abesser, M., Grefe, C., Neumann, K., et al. (2009). Calcium/calmodulin-dependent protein kinase II contributes to cardiac arrhythmogenesis in heart failure. *Circulation. Heart failure, 2*, 664–675.

Samso, M., Feng, W., Pessah, I. N., & Allen, P. D. (2009). Corodinated movement of cytoplasmic and transmembrane domains of RyR1 upon gating. *PLoS Biology, 7*, e85.

Sanchez, G., Escobar, M., Pedrozo, Z., Macho, P., Domenech, R., Hartel, S., et al. (2008). Exercise and tachycardia increase NADPH oxidase and ryanodine receptor-2 activity, possible role in cardioprotection. *Cardiovascular Research, 77*, 380–386.

Shin, D. W., Pan, Z., Bandyopadhyay, A., Bhat, M. B., Kim, D. H., & Ma, J. (2002). Ca^{2+}-dependent interaction between FKBP12 and calcineurin regulates activity of the $Ca(^{2+})$ release channel in skeletal muscle. *Biophysical Journal, 83*, 2539–2549.

Souza, J. M., Peluffo, G., & Radi, R. (2008). Protein tyrosine nitration–functional alteration or just a biomarker? *Free Radical Biology and Medicine, 45*, 357–366.

Stamler, J. S., & Hausladen, A. (1998). Oxidative modifications in nitrosative stress. *Nature Structural Biology, 5*, 247–249.

Stamler, J. S., & Meissner, G. (2001). Physiology of nitric oxide in skeletal muscle. *Physiological Reviews, 81*, 209–237.

Stange, M., Xu, L., Balshaw, D., Yamaguchi, N., & Meissner, G. (2003). Characterization of recombinant skeletal muscle (Ser-2843) and cardiac muscle (Ser-2809) ryanodine receptor phosphorylation mutants. *The Journal of Biological Chemistry, 278*, 51693–51702.

Suko, J., Drobny, H., & Hellmann, G. (1999). Activation and inhibition of purified skeletal muscle calcium release channel by NO donors in single channel current recordings. *Biochimica et Biophysica Acta, 1451*, 271–287.

Suko, J., Maurer-Fogy, I., Plank, B., Bertel, O., Wyskovsky, W., Hohenegger, M., et al. (1993). Phosphorylation of serine 2843 in ryanodine receptor-calcium release channel of skeletal muscle by cAMP-, cGMP- and CaM-dependent protein kinase. *Biochimica et Biophysica Acta, 1175*, 193–206.

Sun, J., Morgan, M., Shen, R. F., Steenbergen, C., & Murphy, E. (2007). Preconditioning results in S-nitrosylation of proteins involved in regulation of mitochondrial energetics and calcium transport. *Circulation Research, 101*, 1155–1163.

Sun, J., Picht, E., Ginsburg, K. S., Bers, D. M., Steenbergen, C., & Murphy, E. (2006). Hypercontractile female hearts exhibit increased S-nitrosylation of the L-type Ca^{2+} channel alpha1 subunit and reduced ischemia/reperfusion injury. *Circulation Research, 98*, 403–411.

Sun, J., Xin, C., Eu, J. P., Stamler, J. S., & Meissner, G. (2001). Cysteine-3635 is responsible for skeletal muscle ryanodine receptor modulation by NO. *Proceedings of the National Academy of Sciences of the United States of America, 98*, 11158–11162.

Sun, J., Xu, L., Eu, J. P., Stamler, J. S., & Meissner, G. (2001). Classes of thiols that influence the activity of the skeletal muscle calcium release channel. *The Journal of Biological Chemistry, 276*, 15625–15630.

Sun, J., Xu, L., Eu, J. P., Stamler, J. S., & Meissner, G. (2003). Nitric oxide, NOC-12 and S-nitrosoglutathione modulate the skeletal muscle calcium release channel/ryanodine receptor by different mechanisms, An allosteric function for O_2 in S-nitrosylation of the channel. *The Journal of Biological Chemistry, 278*, 8184–8189.

Sun, J., Yamaguchi, N., Xu, L., Eu, J. P., Stamler, J. S., & Meissner, G. (2008). Regulation of the cardiac muscle ryanodine receptor by O_2 tension and S-nitrosoglutathione. *Biochemistry, 47*, 13985–13990.

Takasago, T., Imagawa, T., Furukawa, K., Ogurusu, T., & Shigekawa, M. (1991). Regulation of the cardiac ryanodine receptor by protein kinase-dependent phosphorylation. *Journal of Biochemistry (Tokyo), 109*, 163–170.

Takeshima, H., Nishimura, S., Matsumoto, T., Ishida, H., Kangawa, K., Minamino, N., et al. (1989). Primary structure and expression from complementary DNA of skeletal muscle ryanodine receptor. *Nature, 339*, 439–445.

Tang, W. H., Kravtsov, G. M., Sauert, M., Tong, X. Y., Hou, X. Y., Wong, T. M., et al. (2010). Polyol pathway impairs the function of SERCA and RyR in ischemic-reperfused rat hearts by increasing oxidative modifications of these proteins. *Journal of Molecular and Cellular Cardiology, 49*, 58–69.

Tateishi, H., Yano, M., Mochizuki, M., Suetomi, T., Ono, M., Xu, X., et al. (2009). Defective domain-domain interactions within the ryanodine receptor as a critical cause of diastolic Ca^{2+} leak in failing hearts. *Cardiovascular Research, 81*, 536–545.

Terentyev, D., Gyorke, I., Belevych, A. E., Terentyeva, R., Sridhar, A., Nishijima, Y., et al. (2008). Redox modification of ryanodine receptors contributes to sarcoplasmic reticulum Ca^{2+} leak in chronic heart failure. *Circulation Research, 103*, 1466–1472.

Tocchetti, C. G., Wang, W., Froehlich, J. P., Huke, S., Aon, M. A., Wilson, G. M., et al. (2007). Nitroxyl improves cellular heart function by directly enhancing cardiac sarcoplasmic reticulum Ca^{2+} cycling. *Circulation Research, 100*, 96–104.

Uehara, A., Yasukochi, M., Mejia-Alvarez, R., Fill, M., & Imanaga, I. (2002). Gating kinetics and ligand sensitivity modified by phosphorylation of cardiac ryanodine receptors. *Pflugers Archiv, 444*, 202–212.

Valdivia, H. H., Kaplan, J. H., Ellis-Davies, G. C., & Lederer, W. J. (1995). Rapid adaptation of cardiac ryanodine receptors, modulation by Mg^{2+} and phosphorylation. *Science, 267*, 1997–2000.

Voss, A. A., Lango, J., Ernst-Russell, M., Morin, D., & Pessah, I. N. (2004). Identification of hyperreactive cysteines within ryanodine receptor type 1 by mass spectrometry. *The Journal of Biological Chemistry, 279*, 34514–34520.

Wang, P., & Zweier, J. L. (1996). Measurement of nitric oxide and peroxynitrite generation in the postischemic heart. Evidence for peroxynitrite-mediated reperfusion injury. *The Journal of Biological Chemistry, 271*, 29223–29230.

Ward, C. W., Reiken, S., Marks, A. R., Marty, I., Vassort, G., & Lacampagne, A. (2003). Defects in ryanodine receptor calcium release in skeletal muscle from post-myocardial infarct rats. *The FASEB Journal, 17*, 1517–1519.

Wehrens, X. H., Lehnart, S. E., Huang, F., Vest, J. A., Reiken, S. R., Mohler, P. J., et al. (2003). FKBP12.6 deficiency and defective calcium release channel (ryanodine receptor) function linked to exercise-induced sudden cardiac death. *Cell, 113*, 829–840.

Wehrens, X. H., Lehnart, S. E., Reiken, S. R., Deng, S. X., Vest, J. A., Cervantes, D., et al. (2004). Protection from cardiac arrhythmia through ryanodine receptor-stabilizing protein calstabin2. *Science, 304*, 292–296.

Wehrens, X. H., Lehnart, S. E., Reiken, S. R., & Marks, A. R. (2004). Ca^{2+}/calmodulin-dependent protein kinase II phosphorylation regulates the cardiac ryanodine receptor. *Circulation Research, 94*, e61–e70.

Witcher, D. R., Kovacs, R. J., Schulman, H., Cefali, D. C., & Jones, L. R. (1991). Unique phosphorylation site on the cardiac ryanodine receptor regulates calcium channel activity. *The Journal of Biological Chemistry, 266*, 11144–11152.

Xia, R., Stangler, T., & Abramson, J. J. (2000). Skeletal muscle ryanodine receptor is a redox sensor with a well defined redox potential that is sensitive to channel modulators. *The Journal of Biological Chemistry, 275*, 36556–36561.

Xia, R., Webb, J. A., Gnall, L. L., Cutler, K., & Abramson, J. J. (2003). Skeletal muscle sarcoplasmic reticulum contains a NADH-dependent oxidase that generates superoxide. *The American Journal of Physiology, 285*, C215–C221.

Xiao, B., Jiang, M. T., Zhao, M., Yang, D., Sutherland, C., Lai, F. A., et al. (2005). Characterization of a novel PKA phosphorylation site, serine-2030, reveals no PKA hyperphosphorylation of the cardiac ryanodine receptor in canine heart failure. *Circulation Research, 96*, 847–855.

Xiao, B., Tian, X., Xie, W., Jones, P. P., Cai, S., Wang, X., et al. (2007). Functional consequence of protein kinase A-dependent phosphorylation of the cardiac ryanodine receptor, sensitization of store overload-induced Ca^{2+} release. *The Journal of Biological Chemistry, 282*, 30256–30264.

Xiao, B., Zhong, G., Obayashi, M., Yang, D., Chen, K., Walsh, M. P., et al. (2006). Ser-2030, but not Ser-2808, is the major phosphorylation site in cardiac ryanodine receptors responding to protein kinase A activation upon beta-adrenergic stimulation in normal and failing hearts. *The Biochemical Journal, 396*, 7–16.

Xiao, J., Tian, X., Jones, P. P., Bolstad, J., Kong, H., Wang, R., et al. (2007). Removal of FKBP12.6 does not alter the conductance and activation of the cardiac ryanodine receptor or the susceptibility to stress-induced ventricular arrhythmias. *The Journal of Biological Chemistry, 282*, 34828–34838.

Xin, H. B., Senbonmatsu, T., Cheng, D. S., Wang, Y. X., Copello, J. A., Ji, G. J., et al. (2002). Oestrogen protects FKBP12.6 null mice from cardiac hypertrophy. *Nature, 416*, 334–338.

Xu, K. Y., Huso, D. L., Dawson, T. M., Bredt, D. S., & Becker, L. C. (1999). Nitric oxide synthase in cardiac sarcoplasmic reticulum. *Proceedings of the National Academy of Sciences of the United States of America, 96*, 657–662.

Xu, L., Eu, J. P., Meissner, G., & Stamler, J. S. (1998). Activation of the cardiac calcium release channel (ryanodine receptor) by poly-S-nitrosylation. *Science, 279*, 234–237.

Yamaguchi, N., Xin, C., & Meissner, G. (2001). Identification of apocalmodulin and Ca^{2+}-calmodulin regulatory domain in skeletal muscle Ca^{2+} release channel, ryanodine receptor. *The Journal of Biological Chemistry, 276*, 22579–22585.

Yang, D., Zhu, W. Z., Xiao, B., Brochet, D. X. P., Chen, S. R. W., Lakatta, E. G., et al. (2007). Ca^{2+}/calmodulin kinase II-dependent phopshorylation of ryanodine receptors suppresses Ca^{2+} sparks and Ca^{2+} waves in cardiac myocytes. *Circulation Research, 100*, 399–407.

Yano, M., Kobayashi, S., Kohno, M., Doi, M., Tokuhisa, T., Okuda, S., et al. (2003). FKBP12.6-mediated stabilizationof calcium-reelase channel (ryanodine receptor as a novel therapeutic strategey against heart failure. *Circulation, 107*, 477–484.

Yano, M., Ono, K., Ohkusa, T., Suetsugu, M., Kohno, M., Hisaoka, T., et al. (2000). Altered stoichiometry of FKBP12.6 versus ryanodine receptor as a cause of abnormal Ca^{2+} leak through ryanodine receptor in heart failure. *Circulation, 102*, 2131–2136.

Zahradnikova, A., Minarovic, I., Venema, R. C., & Meszaros, L. G. (1997). Inactivation of the cardiac ryanodine receptor calcium release channel by nitric oxide. *Cell Calcium, 22*, 447–454.

Zalk, R., Lehnart, S. E., & Marks, A. R. (2007). Modulation of the ryanodine receptor and intracellular calcium. *Ann. Rev. Biochem., 76*, 367–385.

Zhang, R., Khoo, M. S., Wu, Y., Yang, Y., Grueter, C. E., Ni, G., et al. (2005). Calmodulin kinase II inhibition protects against structural heart disease. *Natural Medicines, 11*, 409–417.

Zhang, T., Maier, L. S., Dalton, N. D., Miyamoto, S., Ross, J., Jr., Bers, D. M., et al. (2003). The deltaC isoform of CaMKII is activated in cardiac hypertrophy and induces dilated cardiomyopathy and heart failure. *Circulation Research, 92*, 912–919.

Zima, A. V., & Blatter, L. A. (2006). Redox regulation of cardiac calcium channels and transporters. *Cardiovascular Research, 71*, 310–321.

Zissimopoulos, S., Docrat, N., & Lai, F. A. (2007). Redox sensitivity of the ryanodine receptor interaction with FK506-binding protein. *The Journal of Biological Chemistry, 282*, 6976–6983.

Zorzato, F., Fujii, J., Otsu, K., Phillips, M., Green, N. M., Lai, F. A., et al. (1990). Molecular cloning of cDNA encoding human and rabbit forms of the Ca^{2+} release channel (ryanodine receptor) of skeletal muscle sarcoplasmic reticulum. *The Journal of Biological Chemistry, 265*, 2244–2256.

Zweier, J. L., Wang, P., & Kuppusamy, P. (1995). Direct measurement of nitric oxide generation in the ischemic heart using electron paramagnetic resonance spectroscopy. *The Journal of Biological Chemistry, 270*, 304–307.

CHAPTER 6

Crosstalk via the Sarcoplasmic Gap: The DHPR–RyR Interaction

Manfred Grabner and Anamika Dayal
Department of Medical Genetics, Molecular and Clinical Pharmacology, Innsbruck Medical University, Innsbruck, Austria

I. OVERVIEW

The release of Ca^{2+} from intracellular stores in striated muscle is the key link between electrical excitation of the sarcolemma and the contractile activation of the myofilaments. This process known as excitation–contraction

Copyright 2010, Elsevier Inc. All right reserved.
1063-5823/10 $35.00
DOI: 10.1016/S1063-5823(10)66006-1

(EC) coupling depends on the close interplay of two distinct Ca^{2+} channels located in close opposition to each other in junctional membrane domains. The two molecular partners are the voltage-gated L-type Ca^{2+} channel or 1,4-dihydropyridine receptor (DHPR) in the sarcolemma and the Ca^{2+} release channel or ryanodine receptor (RyR) in the sarcoplasmic reticulum (SR). There is fundamental difference in cardiac and skeletal muscle EC coupling which is defined by the mode how the two Ca^{2+} channels interact to bridge the sarcoplasmic gap. In cardiac muscle, a fast Ca^{2+} influx through the *c*DHPR triggers the release of Ca^{2+} from the SR by opening the Ca^{2+}-sensitive RyR2, whereas in skeletal muscle, the *sk*DHPR and RyR1 are in physical contact and depolarization-induced conformational changes in the *sk*DHPR are communicated to the RyR1, thereby inducing Ca^{2+} release from SR stores. Here, by subsuming results of physiological expression studies and biochemical peptide work, we try to give a brief overview about these structural–functional correlations of the DHPR–RyR interaction, that is, how cardiac- and skeletal-muscle-specific organization of the two channels in the membranes defines their functioning and which structural domains of the DHPR and of the RyR are involved in the channel crosstalk.

II. DHPR AND RYR ARRANGEMENT IN SKELETAL AND CARDIAC MUSCLE MEMBRANES—BASIS FOR DIFFERENCES IN THE EC COUPLING MECHANISM

The principal qualitative difference in the mode of EC coupling between skeletal and cardiac muscle is mirrored in skeletal- and cardiac-specific localization of DHPRs relative to the adjacent RyRs. While DHPRs and RyRs in both muscle types are coclustered into punctate foci as observed already under light microscopic resolution, either by immunolabeling (Flucher, Andrews, & Daniels, 1994; Jorgensen, Shen, Arnold, Leung, & Campbell, 1989; Protasi, Sun, & Franzini-Armstrong, 1996; Sun et al., 1995; Yuan, Arnold, & Jorgensen, 1991) or directly by GFP-tagging (Grabner, Dirksen, & Beam, 1998), the ultrastructural organization of these two Ca^{2+} channels is fundamentally distinct in both muscle types.

A. Skeletal Muscle

In skeletal muscle, the postulated physical skeletal *sk*DHPR–RyR1 interaction (direct or possibly via an intermediate, yet unidentified protein) requires stringent colocalization of these two channels. The structural basis for this tight interaction are the junctions between the SR terminal cisternae ("junctional SR") that are either closely apposed to the cell surface sarcolemma to form the "peripheral couplings" or to invaginations of the sarcolemma, the so-called transversal (t-) tubuli. These associations of (normally) two SR terminal cisternae with the cell surface membrane in the area of the t-tubules are called the *triads* (reviewed in Franzini-Armstrong, 1999). *sk*DHPRs and RyR1s are enriched in this triad region as colocalized molecular clusters that bridge the cytoplasmic gap.

As demonstrated by *freeze-fracture electron microscopy* of these clusters, *sk*DHPRs form groups of four—the so-called *tetrads*. These tetrads are positioned in strict correspondence to the homotetrameric RyR1 channels (Fig. 1), which are concentrated on the terminal cisternae to form highly ordered orthogonal arrays (Ferguson, Schwartz, & Franzini-Armstrong, 1984; Franzini-Armstrong & Nunzi, 1983). In such an array, every other

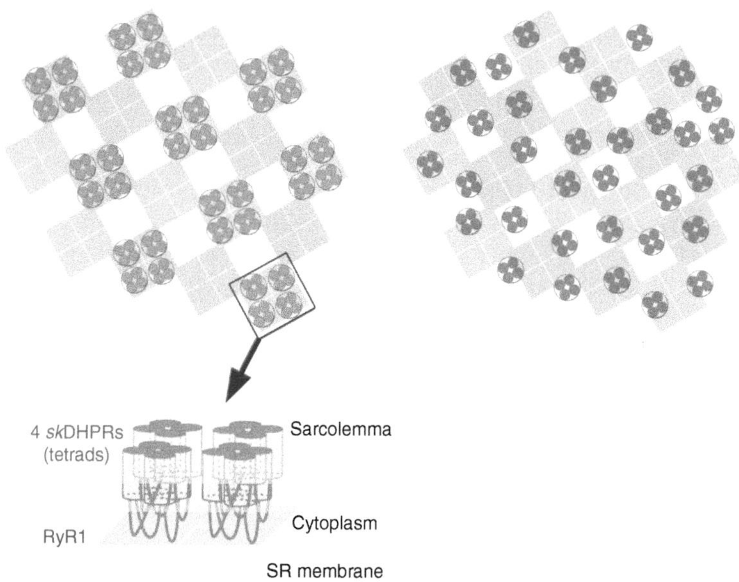

FIGURE 1 Scheme of typical DHPR–RyR clusters in skeletal (left panel) and cardiac (right panel) muscle cell membranes. *sk*DHPRs are marked in blue, *c*DHPRs in red, and RyR1 or RyR2 in gray. The black square marked in the left top view depicts one *sk*DHPR tetrad (four *sk*DHPRs) above one RyR1 and the arrow points to the side view scheme with *sk*DHPR and RyR1 in their respective membrane environment. *sk*DHPR tetrads are arranged into an orthogonal array. Alternatively, the *c*DHPRs are arbitrarily distributed with respect to the RyR2. (See Color Insert.)

RyR1 is associated with a sarcolemmal *sk*DHPR tetrad (Fig. 1), and consequently *sk*DHPR tetrads are also forced into an orthogonal array (Block, Imagawa, Campbell, & Franzini-Armstrong, 1988; Franzini-Armstrong & Kish, 1995; Takekura, Bennett, Tanabe, Beam, & Franzini-Armstrong, 1994).

The resulting close physical interchannel contact is prerequisite for *skeletal-type EC coupling*, a mechanism where membrane depolarizations are sensed by the *sk*DPHR, hereby inducing conformational changes which are transmitted to the RyR1 via protein–protein interaction, and as a consequence RyR1 responds by opening and releasing Ca^{2+} from the SR stores ("*mechanical hypothesis of skeletal muscle EC coupling*"; Rios & Brum, 1987; Schneider & Chandler, 1973).

Due to physical association of four *sk*DHPRs with only every other RyR1, 50% of RyR1s remain uncoupled in the array. These "orphan RyR1s" are believed to be either activated by the Ca^{2+} released from adjacent *sk*DHPR-coupled RyR1s (*stochastic gating*) or by direct physical signal transmission of *sk*DHPR-induced conformational changes from coupled to uncoupled RyR1s within the array (*coordinated or coupled gating*). According to the latter concept, FKBP12 plays a role in simultaneous gating of RyR1 clusters (Marx, Ondrias, & Marks, 1998).

1. *sk*DHPR–RyR1 Interaction Is Bidirectional

Studying biophysical properties of *sk*DHPRs in *dyspedic* (RyR1-null) myotubes (Takeshima et al., 1994) lead to the conclusion that in skeletal muscle in addition to the *orthograde sk*DHPR → RyR1 EC coupling signal, also a *retrograde* RyR1 → *sk*DHPR signal transmission exists whereby RyR1 somehow increases the amplitude of the L-type Ca^{2+} current mediated by the *sk*DHPR (Nakai et al., 1996). This fact aptly explained why earlier trials to express *sk*DHPR in nonmuscle cell systems lacking the *sk*DHPR–RyR1 interaction either failed or gave rather rudimentary expression results (Ren & Hall, 1997; Varadi, Lory, Schultz, Varadi, & Schwartz, 1991).

B. Cardiac Muscle

In cardiac muscle, the *c*DHPRs are clustered with RyR2s, but unlike *sk*DHPRs are not grouped in ordered arrays of tetrads (Protasi et al., 1996; Sun et al., 1995) but are randomly distributed with respect to the RyR2 arrays (Fig. 1). Thus, in contrast to *sk*DHPRs, the *c*DHPRs are apparently not anchored to the RyRs, providing no evidence for a physical *c*DHPR–RyR2 interaction. Instead, a fast Ca^{2+} influx through the *c*DHPR is the essential link between surface membrane depolarization and intracellular Ca^{2+} release (Nabauer, Callewaert, Cleemann, & Morad, 1989; Reuter &

Beeler, 1969; Tanabe, Mikami, Numa, & Beam, 1990). The rise in the local cytoplasmic Ca^{2+} concentration following *c*DHPR Ca^{2+} influx triggers the Ca^{2+}-sensitive cardiac RyR2 to open and release Ca^{2+} ions from the intracellular SR Ca^{2+} stores and this massive increase in cytoplasmic Ca^{2+} consequently leads to cardiac muscle contraction. In junctional microdomains, one *c*DHPR can induce Ca^{2+} release from 4 to 6 RyR2 channels (Wang et al., 2003). Thus, cardiac-type EC coupling includes the activation of RyR2 by Ca^{2+} influx as well as by SR-released Ca^{2+} that can further activate adjacent RyR2 channels. This process is known as *Ca^{2+}-induced Ca^{2+} release* (*CICR*) (Fabiato, 1983; Sham, Cleemann, & Morad, 1995). The close proximity of *c*DHPRs and RyR2s in colocalized clusters is indispensable for proper cardiac-type EC coupling, as heterologously expressed $Ca_V1.3$ channels (with intrinsic coclustering) were able to functionally substitute *c*DHPRs in cardiac-type EC coupling despite considerably lower Ca^{2+} influx through them. On the other hand, $Ca_V2.1$ channels which do not cluster, but have high Ca^{2+} conductance were unable to do so (Flucher, Kasielke, & Grabner, 2000; Grabner et al., 1998; Kasielke, Obermair, Kugler, Grabner, & Flucher, 2003).

During CICR no avalanche effect of Ca^{2+} release occurs, because CICR is rather graded and controlled by the amplitude and duration of the *c*DHPR Ca^{2+} current which is itself controlled by a negative feedback mechanism. It is understood that the Ca^{2+} released by CICR induces *Ca^{2+}-dependent inactivation* (*CDI*) of the *c*DHPR, a mechanism that plays an important role to control the amount of Ca^{2+} influx during action potentials in ventricular myocytes (Richard et al., 2006; Takamatsu, Nagao, Ichijo, & Adachi-Akahane, 2003). The termination of Ca^{2+} release implies a negative feedback on RyR2 itself to counter the inherent activating CICR effect—a mechanism which is still controversial (see reviews Bers, 2002; Fill & Copello, 2002; Stern & Cheng, 2004).

III. STRUCTURAL DOMAINS INVOLVED IN *sk*DHPR–RYR1 INTERACTION

A. RyR1–Interaction Domains Mapped on the sk*DHPR*

1. Domains of α_{1S} Involved in Orthograde Coupling

For identifying the orthograde interaction domain(s) on the DHPR α_{1S}, chimeras in which the intracellular loops as well as the N- and C-termini of the DHPR α_{1C} were substituted with the corresponding regions of the DHPR α_{1S} subunit, were expressed and analyzed in contraction studies to determine the type of (skeletal or cardiac) EC coupling. Only chimeras containing the DHPR α_{1S} II–III loop reconstituted Ca^{2+} independent, skeletal-type EC coupling, thereby identifying this molecular region to be essential for orthograde signal transmission (Tanabe, Beam, Adams, Niidome, & Numa, 1990). Later, a more refined study identified a 46-residue motif (residues 720–765; Fig. 2) positioned close to the center (referring to primary structure) of the α_{1S} II–III loop to be sufficient in transferring strong skeletal-type EC coupling properties to an otherwise pure α_{1C}, while an integral 18 residue motif (725–742; Fig. 2) could only establish a minor response (Nakai, Tanabe, Konno, Adams, & Beam, 1998).

2. Domains of α_{1S} Involved in Retrograde Coupling

Expression of a DHPR α_{1S}-based chimera with the II–III loop substituted with DHPR α_{1C} sequence disrupted the interaction of this chimeric DHPR with the RyR1 in both directions (Grabner et al., 1999). Due to this loss of interaction EC coupling was not restored and also the current amplitudes observed were comparably as small as RyR1-null myotubes (Nakai et al., 1996). By introducing the previously identified 46-residue motif (Fig. 2) responsible for strong orthograde interaction (Nakai, Tanabe, et al., 1998) into the cardiac II–III loop it was possible to restore *both*, EC coupling (*orthograde coupling*) and wild-type Ca^{2+} current densities (*retrograde*

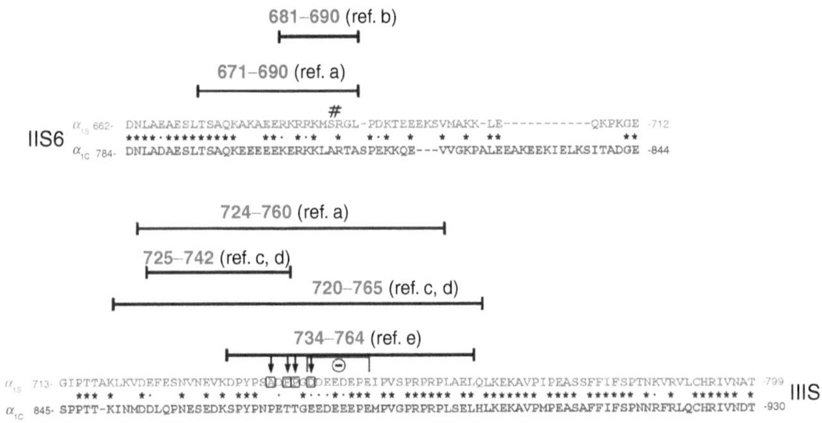

FIGURE 2 Putative RyR1-interaction domains on the DHPR α_{1S} II–III loop, analyzed by physiological and biochemical approaches. Alignment of the II–III loops of DHPR α_{1S} (blue) and DHPR α_{1C} (red). Asterisks (*) indicate identical residues, dots (·) indicate residues with identical charge, and # indicates phosphorylation site Ser-687. Regions used in DHPR chimeras or for peptide probes are indicated with bars with the corresponding references cited above, arrows indicate residues essential for orthograde coupling and bracket around the symbol ⊖ indicate the position of a negatively charged cluster: ref. a—El-Hayek et al. (1995), ref. b—El-Hayek and Ikemoto (1998), ref. c—Nakai, Tanabe, et al. (1998), ref. d—Grabner et al. (1999), and ref. e—Kugler et al. (2004). (See Color Insert.)

coupling). This *sk*DHPR ↔ RyR1 interaction via a joint *critical domain* (direct or mediated by other junctional proteins) was finally termed as "*bidirectional coupling*" (Grabner et al., 1999).

3. Minimal Sequence Requirements for Bidirectional Coupling

To elucidate the structural–functional basis of bidirectional *sk*DHPR ↔ RyR1 signaling, further fine mapping of the interaction domain was performed (Kugler et al., 2004). After reducing the α_{1S} II–III loop sequence required for full bidirectional coupling to the minimum (residues 734–748), eight residues heterologous between α_{1S} and α_{1C} were exchanged to the corresponding cardiac residues. Four of these skeletal-specific residues (A739, F741, P742, D744; see Figs. 2 and 3) turned out to be essential for orthograde, two of them (A739, F741) for retrograde coupling, indicating

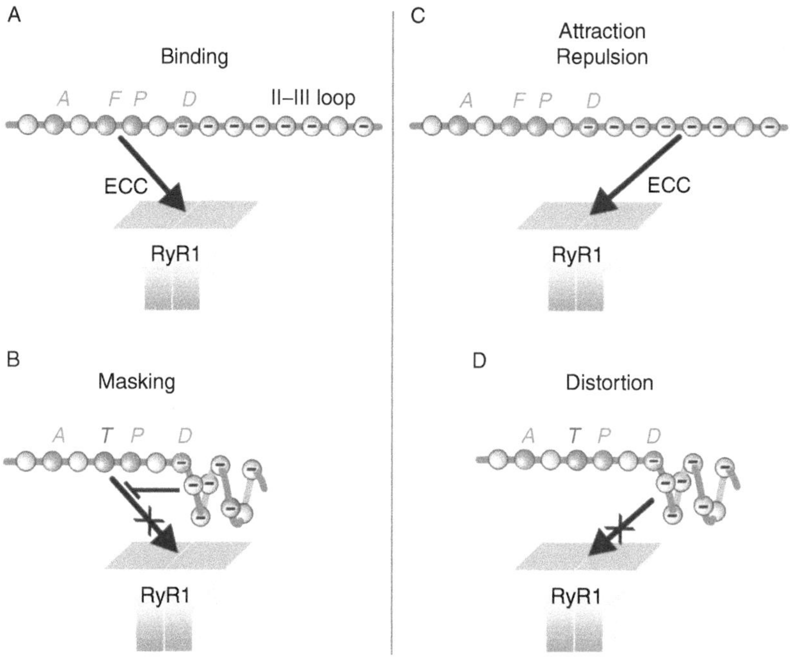

FIGURE 3 Two alternative models of the DHPR α_1 II–III loop–RyR1 interaction. Schematic representation of the domain of four critical residues (blue spheres, residues indicated in one letter code) within the DHPR α_{1S} II–III loop required for skeletal-type EC coupling (ECC). α_{1S} critical residues are indicated in *blue* and α_{1C} residues in *red*. The *minus sign* indicates the negatively charged residues of the cluster. (This research was originally published in *J. Biol. Chem.* (Kugler et al., 2004). © The American Society for Biochemistry and Molecular Biology). (See Color Insert.)

that on the single amino acid level orthograde coupling is not necessarily fully congruent with retrograde signaling. The results of Kugler et al. (2004) indicate a striking structure–function correlation between EC coupling properties of the individual chimeras/point mutants and the predicted secondary structure of the interaction domain. Whenever amino acid exchanges N-terminal to a negatively charged amino acid cluster in the center of the 734–748 minimum domain (Fig. 3) resulted in a conversion of a predicted random coil structure (skeletal sequence-type) to an α-helical structure (cardiac sequence-type), skeletal-type EC coupling was significantly reduced. Thus, a predicted random-coiled structure of the negatively charged cluster seems to be required for direct, skeletal-type EC coupling.

Together, analysis of the primary and secondary structure of the minimal essential EC coupling domain in the α_{1S} II–III loop allowed drafting two alternative models of II–III loop/RyR1 interactions (Fig. 3). First, the motif of critical residues may specifically interact with a corresponding sequence of RyR1 and the adjacent amino acids including the negatively charged cluster are important in enabling this interaction (Fig. 3A). Any changes in the four critical amino acids into their cardiac homologs (e.g., F $\rightarrow$ T) abolished the specific interaction with the RyR1 because these changes lead to an α-helical structure of the negative cluster that masked the specific interaction site (Fig. 3B). Alternatively, not the motif of critical residues but the negatively charged amino acid cluster may be the site of interaction with the RyR1. In this case, the interaction is likely to be an electrostatic attraction or repulsion and the adjacent critical residues determine the secondary structure and consequently the function of the interaction site (Fig. 3C). Changes of the four critical amino acids to their cardiac homologs and the subsequent formation of the α-helical structure of the negative cluster would in this model impair the interaction with RyR1 (Fig. 3D). In either case, primary and secondary structures of the participating sequences of the α_1 subunit and the RyR1 need to be considered in order to understand the protein–protein interactions involved in skeletal-type EC coupling.

4. The "*Peptide A*" versus the "*Critical Domain*" Hypothesis

Similar to the physiological expression studies also results of *in vitro* peptide binding experiments pointed to an essential role of the DHPR α_{1S} II–III loop in *sk*DHPR–RyR1 interaction. A synthetic skeletal II–III loop peptide (residues 666–791; Fig. 2) was able to activate RyR1 *in vitro*, by increasing open probability and ryanodine binding (Lu, Xu, & Meissner, 1994). But in contrast to the *in vivo* observations (Grabner et al., 1999; Nakai, Tanabe, et al., 1998; Tanabe, Mikami, et al., 1990) skeletal and cardiac II–III loop peptides were equally active on RyR1 (Lu et al., 1994). Subsequently, a shorter peptide close to the N-terminus of the skeletal II–III loop (residues 666–726) including a phosphorylation site was described to activate RyR1 by phosphorylation (Lu, Xu, & Meissner, 1995). A similar region (*peptide A*; residues 671–690) was described by El-Hayek et al. (1995), which was later confined to residues 681–690, containing a cluster of basic residues (*peptide A-10*; El-Hayek & Ikemoto, 1998) (Fig. 2), attracted a great deal of attention in the EC coupling field.

Several discrepancies between the *in vivo* and *in vitro* data were apparent. Neither peptide 666–726 nor peptides 671/681–690 nor the *sk*DHPR-specific phosphorylation site S687 (significantly) overlapped with the 46 residue long *critical domain* (residues 720–765; see Section III.A.2) identified in physiological experiments (Grabner et al., 1999; Nakai, Tanabe, et al., 1998). In addition, *in vivo* experiments indicated that a S687A mutant of *sk*DHPR was equally efficient in mediating skeletal-type EC coupling than the wild-type *sk*DHPR (Nakai, Tanabe, et al., 1998).

However, *peptide A-10* was considered by several groups as the key element for RyR-activation, after it was shown to activate RyR single channel properties in lipid bilayer studies as well as Ca^{2+} release and ryanodine binding in triad-enriched skeletal muscle microsomes (Casarotto et al., 2000; Dulhunty et al., 1999; El-Hayek & Ikemoto, 1998; El-Hayek et al., 1995).

Two *in vivo* studies, published later, shattered the *peptide A* hypothesis. Proenza, Wilkens, & Beam (2000) expressed in *dysgenic* myotubes an intact DHPR α_{1S} subunit with scrambled *peptide A-10* sequence. Interestingly, this construct restored all biophysical properties of the wild-type α_{1S} in terms of kinetics and voltage dependence of current and intracellular Ca^{2+} release. In an even more radical approach, Wilkens, Kasielke, Flucher, Beam, and Grabner (2001) tested the importance of the two distinct regions of the skeletal II–III loop (*peptide A-10;* residues 681–690 and the 46 critical residue motif; residues 720–765). A chimera (SkLM) comprising α_{1S} sequence except for the II–III loop which was from the α_1 subunit of housefly, *Musca domestica* (Grabner et al., 1994) was expressed in *dysgenic* myotubes. The dissimilar *Musca* α_1 II–III loop showed no resemblance to α_{1S} in the regions 681–690 and 720–765. Chimera SkLM was unable to restore EC coupling and displayed strongly reduced Ca^{2+} current densities despite normal surface expression levels and correct triad targeting. Introducing DHPR α_{1S} residues 720–764 into the *Musca* II–III loop completely restored bidirectional coupling, indicating its dependence on α_{1S} loop residues 720–764 but its independence from other regions of the loop like *peptide A-10*.

B. sk*DHPR–Interaction Domains Mapped on the RyR1*

To map regions on RyR1 critical for interaction with *sk*DHPR, RyR1/RyR2 chimeras were expressed in *dyspedic* myotubes (Nakai, Sekiguchi, Rando, Allen, & Beam, 1998). The cardiac RyR2 isoform provided the appropriate background for identifying essential *sk*DHPR-interacting domains on RyR1 because, unlike RyR1, the expressed RyR2 can neither increase the Ca^{2+} channel activity of *sk*DHPR nor RyR2 gating is controlled by *sk*DHPR (Nakai et al., 1997). On studying Ca^{2+} current amplitudes and depolarization-induced intracellular Ca^{2+} transients with a set of RyR1/RyR2 chimeras, two adjacent regions in RyR1 could be identified to be involved in coupling with the *sk*DHPR (see Fig. 4). While chimera R2 (containing RyR1 residues 1–1631), restored neither Ca^{2+} currents nor EC coupling, chimera R10 (containing RyR1 residues 1635–2636) was able to mediate both orthograde (skeletal-type EC coupling) and retrograde signaling (Ca^{2+} current amplitude enhancement). Chimera R9 (containing RyR1 residues 2659–3720) was only able to restore retrograde interaction with the *sk*DHPR. Consequently, chimera R9 enhanced the *sk*DHPR Ca^{2+} channel

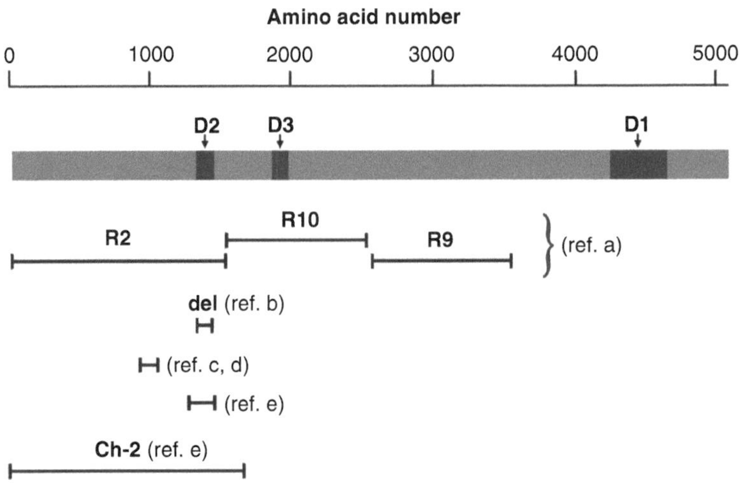

FIGURE 4 Schematic representation of the putative *sk*DPHR interaction domains mapped on RyR1 (symbolized with a light gray bar) by physiological and biochemical approaches. D1, D2, D3 indicate domains highly divergent between the different RyR isoforms (Sorrentino & Volpe, 1993; dark gray blocks). Regions mapped on RyR1 by chimeric studies, sequence deletions (del), or peptide probes are indicated with bars with the corresponding references: ref. a—Nakai, Sekiguchi, et al. (1998), ref. b—Yamazawa et al. (1997), ref. c—Leong and MacLennan (1998a), ref. d—Leong and MacLennan (1998b), and ref. e—Sheridan et al. (2006).

activity without restoring skeletal-type EC coupling indicates that the structures of RyR1 involved in orthograde signaling are not identical to those for retrograde interaction (Nakai, Sekiguchi, et al., 1998).

Deletion of the D2 region in RyR1 (1303–1406), representing one of three highly divergent regions (D1–D3) between RyR1 and RyR2 (see Fig. 4) resulted in the loss of EC coupling whereas the function as a Ca^{2+} release channel was preserved (Yamazawa et al., 1997). This pointed to the importance of the D2 region in the EC coupling mechanism, which seems incompatible with the results of Nakai, Sekiguchi, et al. (1998).

To more stringently test the role of the D2 domain, Sheridan et al. (2006) chimerized RyR1 with RyR3 that uniquely lacks the D2 region. Chimera Ch-2, bearing RyR1 residues 1–1681(Fig. 4) restored normal DHPR tetrad arrays and partial skeletal-type EC coupling (orthograde signaling) but failed to enhance DHPR Ca^{2+} currents (retrograde signaling). The D2 domain, even not effective if inserted alone (Perez, Mukherjee, & Allen, 2003), embedded within this region turned out to dramatically enhance the formation of tetrads and EC coupling rescue by constructs that otherwise were only partially effective. These findings support previous results (Perez, Mukherjee, et al., 2003) and altogether suggest that the D2 region of RyR1 is not only a critical element in stabilizing the RyR1–*sk*DHPR interaction, but also that the structural/functional link between RyR1 and *sk*DHPR requires the contributions of numerous interacting RyR1 regions, and that the regions responsible for tetrad formation do not correspond exactly to the ones required for functional coupling.

In binding experiments of *in vitro* translated RyR1 fragments to the immobilized *sk*DHPR II–III loop, one of the fragments (residues 922–1112; Fig. 4) was identified to bind to the skeletal (but not cardiac) II–III loop while the corresponding RyR2 fragment did not bind to either of the two (Leong and MacLennan, 1998a). In the same study using *in vitro* translated RyR1/RyR2 fusion peptides the binding motif was finally mapped down to 37 amino acids (residues 1076–1112). In a follow-up study (Leong and MacLennan, 1998b), it was shown that fragment 922–1112 also binds to the immobilized *sk*DHPR III–IV loop, again with no binding ability of the corresponding RyR2 sequence. The binding motif for the III–IV loop was found to exceed the 37 residue motif sufficient for II–III loop binding and was mapped to residues 954–1112.

Physiological RyR1/RyR2 chimera results are not entirely congruent with these *in vitro* results, because chimera R2 (RyR1 residues 1–1631 inserted into a RyR2 background; Fig. 4) restituted neither skeletal-type EC coupling nor Ca^{2+} current enhancement upon expression in *dyspedic* myotubes, while chimera R10 (RyR1 residues 1635–2636 in a RyR2 background) could do both (Nakai, Sekiguchi, et al., 1998). However, *in vivo* experiments with RyR1/RyR3 chimeras (Perez, Mukherjee, et al., 2003; Perez, Voss, Pessah, & Allen, 2003;

Sheridan et al., 2006) gave the N-terminal region (RyR1 residues 1–1681) an important role, regarding its physical interaction with the *sk*DHPR in promoting tetrad formation.

IV. THE ROLE OF INTRACELLULAR MOLECULAR REGIONS BESIDES THE α_{1S} II–III LOOP IN *sk*DHPR–RYR1 INTERACTION

Intracellular *sk*DHPR molecular elements other than the α_{1S} II–III loop are believed to be also involved in skeletal-type EC coupling and to directly or indirectly interact with RyR1. The role of the other intracellular loops, the N- or C-terminus, as well as other subunits of the *sk*DHPR complex than the α_{1S} subunit are briefly discussed here.

Functional expression studies in *dysgenic* myotubes with a modified α_{1S} construct lacking the critical domain as well as the peptide A region showed restoration of orthograde (but not retrograde) coupling to a small but significant extent (15% of wild type) (Ahern, Bhattacharya, Mortenson, & Coronado, 2001), suggesting that other α_{1S} regions outside the critical domain of the II–III loop are (partially) involved in coupling with RyR1. Similarly, orthograde coupling was not fully restored with a chimera comprising of α_{1C} with an α_{1S} II–III loop, in contrast to an α_{1C}-based chimera with all five intracellular domains substituted with α_{1S} sequence (Carbonneau, Bhattacharya, Sheridan, & Coronado, 2005). A model of multiple contacts of the *sk*DHPR with RyR1, beside the II–III loop, is consistent with findings that also multiple regions of RyR1 are involved in the crosstalk with the *sk*DHPR (Nakai, Sekiguchi, et al., 1998; Perez, Mukherjee, et al., 2003; Perez, Voss, et al., 2003; Proenza et al., 2002; Protasi et al., 2002; Sheridan et al., 2006).

A. The α_{1S} Amino-Terminus

A direct, significant role of the N-terminus of α_{1S} in bidirectional coupling can be excluded, considering that despite the strong sequence heterology chimeras with substitution of the α_{1S} N-terminus with corresponding α_{1C} sequence did not hamper induced contractions or Ca^{2+} release (Carbonneau et al., 2005; Tanabe, Beam, et al., 1990). In addition, clipping-off three-fourths of the distal N-terminus of α_{1S} had essentially no effect on the orthograde or retrograde *sk*DHPR $\leftrightarrow$ RyR1 signal transduction (Bannister & Beam, 2005). In addition more indirect effects, like an essential role in triad targeting could be excluded after the heterologous α_{1A} N-terminus was equally efficient when replaced in the α_{1S} subunit (Flucher et al., 2000).

B. The α_{1S} I-II Loop

In the classical mapping study by Tanabe, Beam, et al. (1990), the I–II loop was also proposed to support EC coupling, however, much weaker than the II–III loop. This might indicate a marginal direct effect on the interaction with the RyR1, but could also be due to a difference in interaction of α_{1S} or α_{1C} with the DHPR β_{1a} subunit that binds via the I–II loop (Pragnell et al., 1994) and–as discussed below–plays an essential role in enabling skeletal-type EC coupling.

C. The α_{1S} III-IV Loop

Because of the high-sequence homology between the α_{1S} and α_{1C} III–IV loops (46/53 residues conserved) the chimera approach of Tanabe, Beam, et al. (1990) can not fully exclude a contribution of the α_{1S} III–IV loop in skeletal-type EC coupling. A role as a potential site for direct *sk*DHPR $\leftrightarrow$ RyR1 interaction was implied by peptide interaction studies that showed that the recombinant III–IV loop competed with the II–III loop for binding to a RyR1 fragment (Leong & MacLennan, 1998b). Its identification as the first DHPR α_{1S} mutation locus (R1086H) linked to malignant hyperthermia (Monnier, Procaccio, Stieglitz, & Lunardi, 1997) additionally seemed to indicate it relevance to interact with RyR1. Functional analysis of the R1086H mutant channel suggested that the α_{1S} III–IV loop functions as a key negative allosteric modulator of RyR1 activation (Weiss et al., 2004). However, exchanging the α_{1S} III–IV loop with the heterologous P/Q-type channel α_{1A} subunit sequence indicated that the chimeric channels trafficked less well to the membrane but the ones that were in the membrane functioned as efficiently in EC coupling as wild-type *sk*DHPRs (Bannister, Grabner, & Beam, 2008). According to these data, the α_{1S} III–IV loop is unlikely to represent a DHPR–RyR1 interaction site that is essential for bidirectional coupling.

D. The α_{1S} Carboxyl-Terminus

The α_{1S} C-terminus apparently plays an important role in *sk*DHPR triad targeting (Flucher et al., 2000) and a PDZ domain within seems to have a crucial importance (Proenza et al., 2000). However, triad targeting seems to be rather a concerted action of several structures because singularly the α_{1S} C-terminus exchanged in the T-type channel α_{1H} subunit was insufficient to target this chimera to the triad junctions (Wilkens & Beam, 2003). Similarly, the α_{1S} C-terminus attached to an α_{1S} hemichannel consisting of only the homologous repeats I–II was incapable to target this construct into triads, despite its intact sarcolemmal expression (Flucher, Weiss, & Grabner, 2002). A more direct role of the α_{1S} C-terminus to interact with RyR1 was suggested by peptide binding studies. Peptides corresponding to segments of the α_{1S} C-terminus were reported to inhibit ryanodine binding to skeletal (but also to cardiac) muscle membranes and also to inhibit the activity of the reconstituted RyR1 (Sencer et al., 2001; Slavik et al., 1997). In addition to these binding studies also FRET and biotin–streptavidin accessibility experiments from the Beam laboratory indicate a strong closeness of the α_{1S} C-terminus and RyR1 (Lorenzon, Haarmann, Norris, Papadopoulos, & Beam, 2004; Papadopoulos, Leuranguer, Bannister, & Beam, 2004), supporting the role of the α_{1S} C-terminus in *sk*DPHR–RyR1 interaction.

V. INTRACELLULAR MOLECULAR REGIONS OF α_{1S} INVOLVED IN TETRAD FORMATION

As discussed above (Section II.A) tetrad formation is a prerequisite for skeletal-type EC coupling. Considering the dominant role of the α_{1S} II–III loop in bidirectional coupling (Section III.A) the question arose if this loop, and more precisely the *critical domain* in it, might determine the skeletal-muscle-specific arrangement of DHPRs into tetrads opposite the RyR1 arrays. Combined immunofluorescence and freeze-fracture analysis showed that wild-type α_{1S} or constructs containing the *critical domain* constantly were targeted into tetrads and restored skeletal-type EC coupling upon expression in *dysgenic* myotubes (Takekura et al., 2004). Conversely, wild-type α_{1C} and chimeras containing the α_{1C} II–III loop were targeted into the junctional membranes but never in tetradic arrays. Interestingly, an α_{1S}-based chimera with the heterologous II–III loop of the *Musca* α_1 (SkLM; see Fig. 5) produced perfect tetrads but was unable to restore skeletal-type EC coupling. Thus, it seems that the α_{1C} II–III loop plays an inhibitory role in tetrad formation and that the organization of DHPRs in tetrads *vis-à-vis* RyR1 is necessary but not sufficient for skeletal-type EC coupling. In addition, the results of Takekura et al. (2004) indicate that other α_{1S}-specific intracellular regions, in addition to the II–III loop, are involved in tethering *sk*DHPR to RyR1. However, the results with the *Musca* II–III loop chimeras indicate that although these additional interaction sites are sufficient for tetrad formation, they do not significantly support skeletal-type EC coupling without the critical II–III loop sequence.

VI. THE ROLE OF THE ACCESSORY *sk*DHPR SUBUNITS IN INTERACTION WITH RYR1

Beside the main pore-forming and voltage-sensing α_{1S} subunit the accessory subunits ($\alpha_2\delta$-1, β_{1a}, and γ_1) of the *sk*DHPR channel complex have also been investigated over the years for their role in skeletal-type EC coupling.

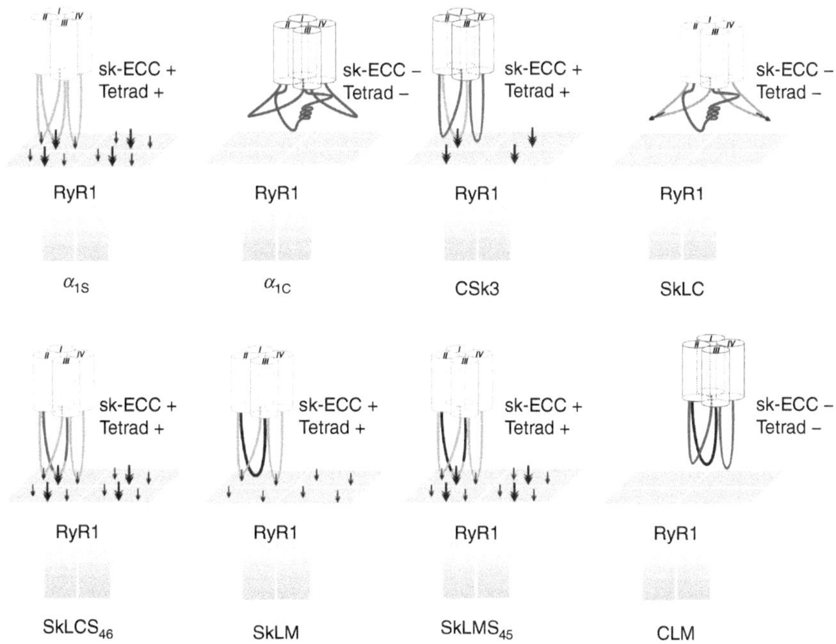

FIGURE 5 Model of *sk*DHPR DHPR–RyR1 interactions involved in tetrad formation. For clarity, only one of four DHPRs is shown opposite the RyR1, with the arrows indicating putative DHPR–RyR1 interactions. Bold arrows represent the primary interaction sites in the DHPR II–III loop, and small arrows represent additional unidentified α_{1S}-specific interaction sites; α_{1S} sequence is indicated in *blue*, α_{1C} in *red*, and *Musca* α_1 in *black*. Multiple interaction sites of α_{1S} cooperate in tetrad formation. Although, the α_{1S} II–III loop was sufficient to confer tetrad formation competence onto α_{1C} (CSk3), it is not absolutely necessary for tetrad formation, as evident when replaced with the heterologous II–III loop from the *Musca* α_1 subunit (SkLM). Nevertheless, the *Musca* II–III loop by itself was not able to coordinate α_{1S} into tetrads (CLM). The presence of a dominant inhibitory property in the α_{1C} II–III loop, as indicated by a putative change in secondary structure (see Kugler et al., 2004), can explain the lack of tetrad formation in α_{1C} and in the chimera SkLC. Upon transferring the 46-amino acid *critical domain* of the α_{1S} II–III loop into the α_{1C} II–III loop of chimera SkLC (resulting in chimera SkLCS46) lead to tetrad formation in parallel to skeletal-type EC coupling restoration (sk-ECC). In addition, strengthening of tetrad formation and restitution of skeletal-type EC coupling properties with SkLMS45 demonstrated that both, the primary interaction site in the α_{1S} II–III loop and the inhibitory signal in the α_{1C} II–III loop reside within this short *critical domain* sequence—an indispensible element for structural and functional interaction with RyR1. (This research was originally published in *Mol. Biol. Cell.* (Takekura et al., 2004). © The American Society for Cell Biology (ASCB)). (See Color Insert.)

The most direct way of testing the relevant function of channel subunits is via respective knockout animal models, as a wealth of fundamental insights was compellingly proven by expression studies in the *dysgenic* mouse muscle system over the last 20 years.

A. The $\alpha_2\delta$-1 Subunit

Recently, a conventional $\alpha_2\delta$-1 knockout mouse was created which is viable and motile—pointing to no gross effect on skeletal-type EC coupling (Fuller-Bicer et al., 2009). Absence of the $\alpha_2\delta$-1 subunit in cardiomyocytes did not show any effect on the expression level of the other DHPR channel subunits, but had a profound effect on the regulation of the α_{1C} subunit in biophysical studies. It results in decreased Ca^{2+} currents with slower current activation and inactivation kinetics. These observations confirmed earlier *in silico* models based on siRNA knockdown experiments, indicating an essential role of the $\alpha_2\delta$-1 in cardiac-type EC coupling (Tuluc, Kern, Obermair, & Flucher, 2007). Congruent with these results, *ex vivo* working heart experiments showed a significant decrease in basal contractility and relaxation, indicating a decrease in Ca^{2+} load (Fuller-Bicer et al., 2009).

Significant effects on the *sk*DHPR channel kinetics were also demonstrated in $\alpha_2\delta$-1 siRNA knockdown experiments in skeletal muscle cells (Obermair et al., 2005). $\alpha_2\delta$-1 depletion significantly accelerated current kinetics but current amplitudes and voltage dependence of the L-type currents and of the depolarization-induced Ca^{2+} transients were indistinguishable from wild-type muscle cells. Thus, in skeletal muscle the $\alpha_2\delta$-1 subunit seems, in contrast to conclusions drawn from heterologous coexpression studies (Arikkath & Campbell, 2003), neither essential for *sk*DHPRs membrane targeting nor for skeletal-type EC coupling.

B. The γ_1 Subunit

The γ_1 subunit is an integral part of *sk*DHPR but not of the *c*DHPR complex. γ_1 knockout mice are viable and do not show an abnormal phenotype (Freise et al., 2000). Functional analysis of γ_1-null myotubes and muscle fibers showed that amplitudes, voltage-dependence of Ca^{2+} currents and depolarization-induced Ca^{2+} release from the SR were not altered (Ursu, Schuhmeier, Freichel, Flockerzi, & Melzer, 2004). Thus, γ_1 seems not to be essential for EC coupling. However, the voltage-dependence of steady state inactivation of the currents and of Ca^{2+} release shifted to more depolarized potentials in γ_1 knockout muscle. Thus, the role of the γ_1 subunit in skeletal-type EC coupling was proposed (Andronache et al., 2007) to be like an endogenous Ca^{2+} antagonist to increase the voltage-sensitivity of inactivation and consequently to limit both Ca^{2+} influx and, more importantly, Ca^{2+} release under stress-induced conditions that increase plasmalemmal depolarization.

C. The β_{1a} Subunit

The *sk*DHPR-specific, intracellular β_{1a} subunit has multiple roles in targeting and modulating the central α_{1S} subunit. The lack of β_{1a} is incompatible with skeletal muscle EC coupling and leads to perinatal lethality in β_1-null mice due to asphyxia (Gregg et al., 1996) or to a paralyzed phenotype in β_1-null (*relaxed*) zebrafish which then die some days after hatching (Granato et al., 1996). As shown in the zebrafish *relaxed* system, lack of β_{1a} leads to a (i) decrease in α_{1S} membrane targeting, (ii) severe reduction in charge movement, and (iii) complete absence of the ultrastructural arrangement of *sk*DHPRs into tetrads in orthogonal arrays, which is essential for skeletal-type EC coupling (Schredelseker et al., 2005). Reconstitution studies in the *relaxed* expression system with β_{1a} and two heterologous β subunits (the cardiac/neuronal β_{2a} and the ancestral, neuronal *Musca* β_M) indicated that triad expression and thus facilitation of *sk*DHPR charge movement are common features of all tested β subunits, while tetrad formation and thus intact DHPR–RyR1 coupling is only promoted by the β_{1a} isoform (Schredelseker, Dayal, Schwerte, Franzini-Armstrong, & Grabner, 2009). Consequently, a model was postulated (see Fig. 6) that presents β_{1a} as an allosteric modifier of α_{1S} conformation and thereby enabling skeletal-type EC coupling. This model does not support hypotheses that understand β_{1a} as a direct signal transducer in EC coupling as was postulated to be preformed via the C-terminal heptad repeat (Coronado et al., 2004; see also Dayal, et al., 2010) or as a simple scaffold for α_{1S} tetrad targeting (see discussion in Schredelseker et al., 2009).

VII. CONCLUSION

It is now two decades ago that the fruitful cooperation between the Beam and Numa laboratories gave us the first deeper insight into the character of the fascinating physical interaction between DHPRs and RyRs in skeletal muscle (Tanabe, Beam, et al., 1990). Since then the II–III loop of the DHPR α_{1S} subunit is still the main candidate and till date the best characterized link between the two molecular partners that act together in the EC coupling crosstalk. Meanwhile more and more evidences arose that several other molecular elements (viz. sequence regions of the main subunit, accessory subunits, or associated proteins) of the DHPR as well as RyR are included in this bidirectional signal transduction, but a more refined picture of this interplay is still in the dark. We will see if it needs another two decades to have all open questions answered to fully understand this unique Ca^{2+} channel partnership that is the basis of our motility.

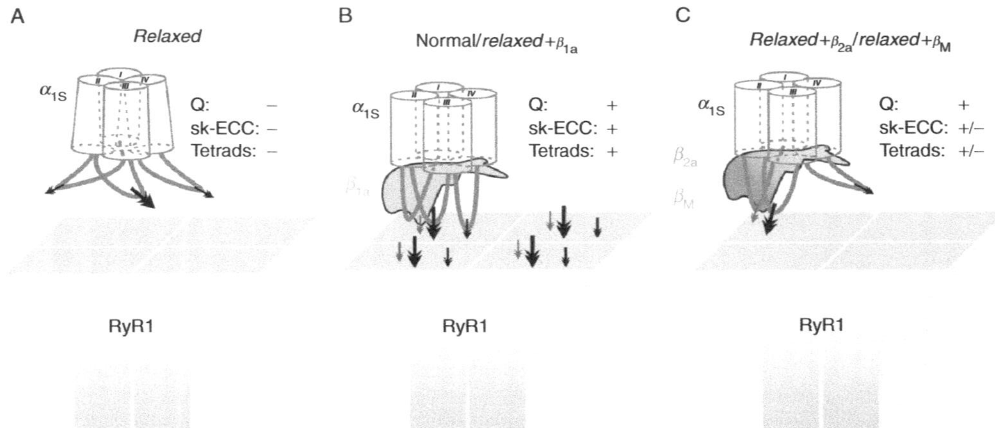

FIGURE 6 Model of β-induced DHPR α_{1S}–RyR1 interactions, incorporating results of Schredelseker et al. (2009) and of previous studies. (A) hypothetical situation of the lack of any DHPR α_{1S}–RyR1 interaction due to the lack of the β_{1a} subunit in β_1-null myotube triads. Without β the conformation of the α_{1S} subunit is severely distorted, which hampers α_{1S} charge movement and there is no activation of presumptive α_{1S}–RyR1 interaction sites (indicated by arrows). Bold arrow represents the primary interaction site in the α_{1S} II–III loop and small arrows represent additional unidentified α_{1S}-specific interaction sites (Takekura et al., 2004; Wilkens et al., 2001). The deficiency of direct α_{1S}–RyR1 interaction correlates with the lack of tetrad formation (Schredelseker et al., 2005). Consequently, by the lack of charge movement (Q –) and tetrad formation (Tetrads –), skeletal-type EC coupling is completely hampered (sk-ECC –). B, the situation in normal muscle or muscle from β_1-null muscle reconstituted with β_{1a} is depicted. The β_{1a} subunit leads to full and correct restoration of α_{1S} conformation, allowing charge movement (Q +) and appropriate targeting of the α_{1S} into tetrads opposite the RyR1 (Tetrads +). If the β subunit has additional interaction sites with RyR1 (red arrow) (Cheng et al., 2005), or not, is irrelevant for this model according to which β_{1a} acts as an allosteric modifier of α_{1S} conformation and thus function. C, if heterologous β subunits like the cardiac/neuronal β_{2a} or the ancestral, neuronal *Musca* β_M are expressed in β_1-null muscle cells, a partial restoration of α_{1S} conformation takes place. Charge movement is now fully restored, but tetrad formation and thus proper α_{1S}–RyR1 protein–protein interaction is still impaired. This "fuzzy targeting" of α_{1S} opposite the RyR1 leads to unstable Ca^{2+} release and thus to very weak (in the case of β_{2a}) or even no (with β_M) muscle motility. (This research was originally published in *J. Biol. Chem.* (Schredelseker et al., 2009). © The American Society for Biochemistry and Molecular Biology). (See Color Insert.)

References

Ahern, C. A., Bhattacharya, D., Mortenson, L., & Coronado, R. (2001). A component of excitation-contraction coupling triggered in the absence of the T671-L690 and L720-Q765 regions of the II-III loop of the dihydropyridine receptor α_{1S} pore subunit. *Biophysical Journal, 81*, 3294–3307.

Andronache, Z., Ursu, D., Lehnert, S., Freichel, M., Flockerzi, V., & Melzer, W. (2007). The auxiliary subunit γ_1 of the skeletal muscle L-type Ca^{2+} channel is an endogenous Ca^{2+} antagonist. *Proceedings of the National Academy of Sciences of the United States of America, 104*, 17885–17890.

Arikkath, J., & Campbell, K. P. (2003). Auxiliary subunits: Essential components of the voltage-gated calcium channel complex. *Current Opinion in Neurobiology, 13*, 298–307.

Bannister, R. A., & Beam, K. G. (2005). The α_{1S} N-terminus is not essential for bi-directional coupling with RyR1. *Biochemical and Biophysical Research Communications, 336*, 134–141.

Bannister, R. A., Grabner, M., & Beam, K. G. (2008). The α1S III-IV loop influences 1,4-dihydropyridine receptor gating but is not directly involved in excitation-contraction coupling interactions with the type 1 ryanodine receptor. *The Journal of Biological Chemistry, 283*, 23217–23223.

Bers, D. M. (2002). Cardiac excitation-contraction coupling. *Nature, 415*, 198–205.

Block, B. A., Imagawa, T., Campbell, K. P., & Franzini-Armstrong, C. (1988). Structural evidence for direct interaction between the molecular components of the transverse tubule/sarcoplasmic reticulum junction in skeletal muscle. *The Journal of Cell Biology, 107*, 2587–2600.

Carbonneau, L., Bhattacharya, D., Sheridan, D. C., & Coronado, R. (2005). Multiple loops of the dihydropyridine receptor pore subunit are required for full-scale excitation-contraction coupling in skeletal muscle. *Biophysical Journal, 89*, 243–255.

Casarotto, M. G., Gibson, F., Pace, S. M., Curtis, S. M., Mulcair, M., & Dulhunty, A. F. (2000). A structural requirement for activation of skeletal ryanodine receptors by peptides of the dihydropyridine receptor II-III loop. *The Journal of Biological Chemistry, 275*, 11631–11637.

Cheng, W., Altafaj, X., Ronjat, M., & Coronado, R. (2005). Interaction between the dihydropyridine receptor Ca^{2+} channel β-subunit and ryanodine receptor type 1 strengthens excitation-contraction coupling. *Proceedings of the National Academy of Sciences of the United States of America, 102*, 19225–19230.

Coronado, R., Ahern, C. A., Sheridan, D. C., Cheng, W., Carbonneau, L., & Bhattacharya, D. (2004). Functional equivalence of dihydropyridine receptor α_{1S} and β_{1a} subunits in triggering excitation-contraction coupling in skeletal muscle. *Biological Research, 37*, 565–575.

Dayal, A., Schredelseker, J., Franzini-Armstrong, C., & Grabner, M. (2010). Skeletal muscle excitation-contraction coupling is independent of a conserved heptad repeat motif in the C-terminus of the DHPR1a subunit. *Cell Calcium, doi:10.1016/j.ceca.2010.04.003.*

Dulhunty, A. F., Laver, D. R., Gallant, E. M., Casarotto, M. G., Pace, S. M., & Curtis, S. (1999). Activation and inhibition of skeletal RyR channels by a part of the skeletal DHPR II-III loop: Effects of DHPR Ser687 and FKBP12. *Biophysical Journal, 77*, 189–203.

El-Hayek, R., Antoniu, B., Wang, J., Hamilton, S. L., & Ikemoto, N. (1995). Identification of calcium release-triggering and blocking regions of the II-III loop of the skeletal muscle dihydropyridine receptor. *The Journal of Biological Chemistry, 270*, 22116–22118.

El-Hayek, R., & Ikemoto, N. (1998). Identification of the minimum essential region in the II-III loop of the dihydropyridine receptor α_1 subunit required for activation of skeletal muscle-type excitation-contraction coupling. *Biochemistry, 37*, 7015–7020.

Fabiato, A. (1983). Ca^{2+}-induced release of Ca^{2+} from the cardiac SR. *The American Journal of Physiology, 245*, 1–14.

Ferguson, D. G., Schwartz, H. W., & Franzini-Armstrong, C. (1984). Subunit structure of junctional feet in triads of skeletal muscle: A freeze-drying, rotary-shadowing study. *The Journal of Cell Biology, 99*, 1735–1742.

Fill, M., & Copello, J. A. (2002). Ryanodine receptor calcium release channels. *Physiological Reviews, 82*, 893–922.

Flucher, B. E., Andrews, S. B., & Daniels, M. P. (1994). Molecular organization of transverse tubule/SR junctions during development of EC coupling in skeletal muscle. *Molecular Biology of the Cell, 5*, 1105–1118.

Flucher, B. E., Kasielke, N., & Grabner, M. (2000). The triad targeting signal of the skeletal muscle calcium channel is localized in the COOH terminus of the α_{1S} subunit. *The Journal of Cell Biology, 151*, 467–477.

Flucher, B. E., Weiss, R. G., & Grabner, M. (2002). Cooperation of two-domain Ca^{2+} channel fragments in triad targeting and restoration of excitation-contraction coupling in skeletal muscle. *Proceedings of the National Academy of Sciences of the United States of America, 99*, 10167–10172.

Franzini-Armstrong, C. (1999). The SR and the control of muscle contraction. *The FASEB Journal, 13*, 266–270.

Franzini-Armstrong, C., & Kish, J. W. (1995). Alternate disposition of tetrads in peripheral couplings of skeletal muscle. *Journal of Muscle Research and Cell Motility, 16*, 319–324.

Franzini-Armstrong, C., & Nunzi, G. (1983). Junctional feet and particles in the triads of a fast-twitch muscle fibre. *Journal of Muscle Research and Cell Motility, 4*, 233–252.

Freise, D., Held, B., Wissenbach, U., Pfeifer, A., Trost, C., et al. (2000). Absence of the subunit of the skeletal muscle dihydropyridine receptor increases L-type Ca^{2+} currents and alters channel inactivation properties. *The Journal of Biological Chemistry, 275*, 14476–14481.

Fuller-Bicer, G. A., Varadi, G., Koch, S. E., Ishii, M., Bodi, I., et al. (2009). Targeted disruption of the voltage-dependent calcium channel $\alpha_2\delta$-1-subunit. *American Journal of Physiology. Heart and Circulatory Physiology, 297*, 117–124.

Grabner, M., Bachmann, A., Rosenthal, F., Striessnig, J., Schulz, C., Tautz, D., et al. (1994). Insect Ca^{2+}channels: Molecular cloning of an α_1-subunit from housefly (*Musca domestica*) muscle. *FEBS Letters, 339*(189), 194.

Grabner, M., Dirksen, R. T., & Beam, K. G. (1998). Tagging with green fluorescent protein reveals a distinct subcellular distribution of L-type and non-L-type Ca^{2+} channels expressed in dysgenic myotubes. *Proceedings of the National Academy of Sciences of the United States of America, 95*, 1903–1908.

Grabner, M., Dirksen, R. T., Suda, N., & Beam, K. G. (1999). The II-III loop of the skeletal muscle dihydropyridine receptor is responsible for the bi-directional coupling with the ryanodine receptor. *The Journal of Biological Chemistry, 274*, 21913–21919.

Granato, M., van Eeden, F. J., Schach, U., Trowe, T., Brand, M., et al. (1996). Genes controlling and mediating locomotion behavior of the zebrafish embryo and larva. *Development, 123*, 399–413.

Gregg, R. G., Messing, A., Strube, C., Beurg, M., Moss, R., et al. (1996). Absence of the β subunit (cchb1) of the skeletal muscle dihydropyridine receptor alters expression of the α_1 subunit and eliminates excitation-contraction coupling. *Proceedings of the National Academy of Sciences of the United States of America, 93*, 13961–13966.

Jorgensen, A. O., Shen, A. C., Arnold, W., Leung, A. T., & Campbell, K. P. (1989). Subcellular distribution of the 1, 4-dihydropyridine receptor in rabbit skeletal muscle in situ: An immunofluorescence and immunocolloidal gold-labeling study. *The Journal of Cell Biology, 109*, 135–147.

Kasielke, N., Obermair, G. J., Kugler, G., Grabner, M., & Flucher, B. E. (2003). Cardiac-type EC-coupling in dysgenic myotubes restored with Ca^{2+} channel subunit isoforms α_{1C} and α_{1D} does not correlate with current density. *Biophysical Journal, 84*, 3816–3828.

Kugler, G., Weiss, R. G., Flucher, B. E., & Grabner, M. (2004). Structural requirements of the dihydropyridine receptor α_{1S} II-III loop for skeletal-type excitation-contraction coupling. *The Journal of Biological Chemistry, 279*, 4721–4728.

Leong, P., & MacLennan, D. H. (1998a). A 37-amino acid sequence in the skeletal muscle ryanodine receptor interacts with the cytoplasmic loop between domains II and III in the skeletal muscle dihydropyridine receptor. *The Journal of Biological Chemistry, 273*, 7791–7794.

Leong, P., & MacLennan, D. H. (1998b). The cytoplasmic loops between domains II and III and domains III and IV in the skeletal muscle dihydropyridine receptor bind to a contiguous site in the skeletal muscle ryanodine receptor. *The Journal of Biological Chemistry, 273*, 29958–29964.

Lorenzon, N. M., Haarmann, C. S., Norris, E. E., Papadopoulos, S., & Beam, K. G. (2004). Metabolic biotinylation as a probe of supramolecular structure of the triad junction in skeletal muscle. *The Journal of Biological Chemistry, 279*, 44057–44064.

Lu, X., Xu, L., & Meissner, G. (1994). Activation of the skeletal muscle calcium release channel by a cytoplasmic loop of the dihydropyridine receptor. *The Journal of Biological Chemistry, 269*, 6511–6516.

Lu, X., Xu, L., & Meissner, G. (1995). Phosphorylation of dihydropyridine receptor II-III loop peptide regulates skeletal muscle calcium release channel function. *The Journal of Biological Chemistry, 270*, 18459–18464.

Marx, S. O., Ondrias, K., & Marks, A. R. (1998). Coupled gating between individual skeletal muscle Ca^{2+} release channels. *Science, 281*, 818–821.

Monnier, N., Procaccio, V., Stieglitz, P., & Lunardi, J. (1997). Malignant-hyperthermia susceptibility is associated with a mutation of the α_1-subunit of the human dihydropyridine-sensitive L-type voltage-dependent calcium-channel receptor in skeletal muscle. *American Journal of Human Genetics, 60*, 1316–1325.

Nabauer, M., Callewaert, G., Cleemann, L., & Morad, M. (1989). Regulation of calcium release is gated by calcium current, not gating charge in cardiac myocytes. *Science, 244*, 800–803.

Nakai, J., Dirksen, R. T., Nguyen, H. T., Pessah, I. N., Beam, K. G., & Allen, P. D. (1996). Enhanced dihydropyridine receptor channel activity in the presence of ryanodine receptor. *Nature, 380*, 72–75.

Nakai, J., Ogura, T., Protasi, F., Franzini-Armstrong, C., Allen, P. D., & Beam, K. G. (1997). Functional nonequality of the cardiac and skeletal ryanodine receptors. *Proceedings of the National Academy of Sciences of the United States of America, 94*, 1019–1022.

Nakai, J., Sekiguchi, N., Rando, T. A., Allen, P. D., & Beam, K. G. (1998). Two regions of the ryanodine receptor involved in coupling with L-type Ca^{2+} channels. *The Journal of Biological Chemistry, 273*, 13403–13406.

Nakai, J., Tanabe, T., Konno, T., Adams, B., & Beam, K. G. (1998). Localization in the II-III loop of the dihydropyridine receptor of a sequence critical for excitation-contraction coupling. *The Journal of Biological Chemistry, 273*, 24983–24986.

Obermair, G. J., Kugler, G., Baumgartner, S., Tuluc, P., Grabner, M., & Flucher, B. E. (2005). The Ca^{2+} channel $\alpha_2\delta$-1 subunit determines Ca^{2+} current kinetics in skeletal muscle but not targeting of α_{1S} or excitation-contraction coupling. *The Journal of Biological Chemistry, 280*, 2229–2237.

Papadopoulos, S., Leuranguer, V., Bannister, R. A., & Beam, K. G. (2004). Mapping sites of potential proximity between the dihydropyridine receptor and RyR1 in muscle using a cyan fluorescent protein-yellow fluorescent protein tandem as a fluorescence resonance energy transfer probe. *The Journal of Biological Chemistry, 279*, 44046–44056.

Perez, C. F., Mukherjee, S., & Allen, P. D. (2003). Amino acids 1–1,680 of ryanodine receptor type 1 hold critical determinants of skeletal type for excitation-contraction coupling. Role of divergence domain D2. *The Journal of Biological Chemistry, 278*, 39644–39652.

Perez, C. F., Voss, A., Pessah, I. N., & Allen, P. D. (2003). RyR/RyR3 chimeras reveal that multiple domains of RyR1 are involved in skeletal-type E-C coupling. *Biophysical Journal, 84*, 2655–2663.

Pragnell, M., De Waard, M., Mori, Y., Tanabe, T., Snutch, T. P., & Campbell, K. P. (1994). Calcium channel β-subunit binds to a conserved motif in the I-II cytoplasmic linker of the α_1-subunit. *Nature, 368*, 67–70.

Proenza, C., O'Brien, J., Nakai, J., Mukherjee, S., Allen, P. D., & Beam, K. G. (2002). Identification of a region of RyR1 that participates in allosteric coupling with the α_{1S} ($Ca_V1.1$) II-III loop. *The Journal of Biological Chemistry, 277*, 6530–6535.

Proenza, C., Wilkens, C. M., & Beam, K. G. (2000). Excitation-contraction coupling is not affected by scrambled sequence in residues 681-690 of the dihydropyridine receptor II-III loop. *The Journal of Biological Chemistry, 275*, 29935–29937.

Proenza, C., Wilkens, C., Lorenzon, N. M., & Beam, K. G. (2000). A carboxyl-terminal region important for the expression and targeting of the skeletal muscle dihydropyridine receptor. *The Journal of Biological Chemistry, 275*, 23169–23174.

Protasi, F., Paolini, C., Nakai, J., Beam, K. G., Franzini-Armstrong, C., & Allen, P. D. (2002). Multiple regions of RyR1 mediate functional and structural interactions with α_{1S}-dihydropyridine receptors in skeletal muscle. *Biophysical Journal, 83*, 3230–3244.

Protasi, F., Sun, X. H., & Franzini-Armstrong, C. (1996). Formation and maturation of the calcium release apparatus in developing and adult avian myocardium. *Developmental Biology, 173*, 265–278.

Ren, D., & Hall, L. M. (1997). Functional expression and characterization of skeletal muscle dihydropyridine receptors in *Xenopus oocytes*. *The Journal of Biological Chemistry, 272*, 22393–22396.

Reuter, H., & Beeler, G. W. Jr., (1969). Calcium current and activation of contraction in ventricular myocardial fibers. *Science, 163*, 399–401.

Richard, S., Perrier, E., Fauconnier, J., Perrier, R., Pereira, L., Gõmez, A. M., et al. (2006). 'Ca^{2+}-induced Ca^{2+} entry' or how the L-type Ca^{2+} channel remodels its own signalling pathway in cardiac cells. *Progress in Biophysics and Molecular Biology, 90*, 118–135.

Rios, E., & Brum, G. (1987). Involvement of dihydropyridine receptors in excitation-contraction coupling in skeletal muscle. *Nature, 325*, 717–720.

Schneider, M. F., & Chandler, W. K. (1973). Voltage dependent charge movement of skeletal muscle: A possible step in excitation-contraction coupling. *Nature, 242*, 244–246.

Schredelseker, J., Dayal, A., Schwerte, T., Franzini-Armstrong, C., & Grabner, M. (2009). Proper restoration of excitation-contraction coupling in the dihydropyridine receptor β_1-null zebrafish *relaxed* is an exclusive function of the β_{1a} subunit. *The Journal of Biological Chemistry, 284*, 1242–1251.

Schredelseker, J., Di Biase, V., Obermair, G. J., Felder, E. T., Flucher, B. E., Franzini-Armstrong, C., et al. (2005). The β_{1a} subunit is essential for the assembly of dihydropyridine-receptor arrays in skeletal muscle. *Proceedings of the National Academy of Sciences of the United States of America, 102*, 17219–17224.

Sencer, S., Papineni, R. V., Halling, D. B., Pate, P., Krol, J., Zhang, J. Z., et al. (2001). Coupling of RYR1 and L-type calcium channels via calmodulin binding domains. *The Journal of Biological Chemistry, 276*, 38237–38241.

Sham, J. S., Cleemann, L., & Morad, M. (1995). Functional coupling of Ca^{2+} channels and ryanodine receptors in cardiac myocytes. *Proceedings of the National Academy of Sciences of the United States of America, 92*, 121–125.

Sheridan, D. C., Takekura, H., Franzini-Armstrong, C., Beam, K. G., Allen, P. D., & Perez, C. F. (2006). Bidirectional signaling between calcium channels of skeletal muscle requires multiple direct and indirect interactions. *Proceedings of the National Academy of Sciences of the United States of America, 103*, 19760–19765.

Slavik, K. J., Wang, J. P., Aghdasi, B., Zhang, J. Z., Mandel, F., Malouf, N., et al. (1997). A carboxy-terminal peptide of the α_1-subunit of the dihydropyridine receptor inhibits Ca^{2+}-release channels. *The American Journal of Physiology, 272*, 1475–1481.

Sorrentino, V., & Volpe, P. (1993). Ryanodine receptors: How many, where and why? *Trends in Pharmacological Sciences, 14*, 98–103.

Stern, M. D., & Cheng, H. (2004). Putting out the fire: What terminates calcium-induced calcium release in cardiac muscle? *Cell Calcium, 35*, 591–601.

Sun, X. H., Protasi, F., Takahashi, M., Takeshima, H., Ferguson, D. G., & Franzini-Armstrong, C. (1995). Molecular architecture of membranes involved in excitation-contraction coupling of cardiac muscle. *Journal of Cell Biology, 129*, 659–671.

Takamatsu, H., Nagao, T., Ichijo, H., & Adachi-Akahane, S. (2003). L-type Ca^{2+} channels serve as a sensor of the SR Ca^{2+} for tuning the efficacy of Ca^{2+}-induced Ca^{2+} release in rat ventricular myocytes. *Journal de Physiologie, 552*, 415–424.

Takekura, H., Bennett, L., Tanabe, T., Beam, K. G., & Franzini-Armstrong, C. (1994). Restoration of junctional tetrads in dysgenic myotubes by dihydropyridine receptor cDNA. *Biophysical Journal, 67*, 793–803.

Takekura, H., Paolini, C., Franzini-Armstrong, C., Kugler, G., Grabner, M., & Flucher, B. E. (2004). Differential contribution of skeletal and cardiac II-III loop sequences to the assembly of dihydropyridine-receptor arrays in skeletal muscle. *Molecular Biology of the Cell, 15*, 5408–5419.

Takeshima, H., Iino, M., Takekura, H., Nishi, M., Kuno, J., Minowa, O., et al. (1994). Excitation-contraction uncoupling and muscular degeneration in mice lacking functional skeletal muscle ryanodine-receptor gene. *Nature, 369*, 556–559.

Tanabe, T., Beam, K. G., Adams, B. A., Niidome, T., & Numa, S. (1990a). Regions of the skeletal muscle dihydropyridine receptor critical for excitation-contraction coupling. *Nature, 346*, 567–569.

Tanabe, T., Mikami, A., Numa, S., & Beam, K. G. (1990b). Cardiac-type excitation-contraction coupling in *dysgenic* skeletal muscle injected with cardiac dihydropyridine receptor cDNA. *Nature, 344*, 451–453.

Tuluc, P., Kern, G., Obermair, G. J., & Flucher, B. E. (2007). Computer modeling of siRNA knockdown effects indicates an essential role of the Ca^{2+} channel $\alpha_2\delta$-1 subunit in cardiac excitation-contraction coupling. *Proceedings of the National Academy of Sciences of the United States of America, 104*, 11091–11096.

Ursu, D., Schuhmeier, R. P., Freichel, M., Flockerzi, V., & Melzer, W. (2004). Altered inactivation of Ca^{2+} current and Ca^{2+} release in mouse muscle fibers deficient in the DHP receptor γ_1 subunit. *The Journal of General Physiology, 124*, 605–618.

Varadi, G., Lory, P., Schultz, D., Varadi, M., & Schwartz, A. (1991). Acceleration of activation and inactivation by the β-subunit of the skeletal muscle calcium channel. *Nature, 352*, 159–162.

Wang, R., Zhang, L., Bolstad, J., Diao, N., Brown, C., Ruest, L., et al. (2003). Residue Gln4863 within a predicted transmembrane sequence of the Ca^{2+} release channel (ryanodine receptor) is critical for ryanodine interaction. *The Journal of Biological Chemistry, 278*, 51557–51565.

Weiss, R. G., O'Connell, K. M., Flucher, B. E., Allen, P. D., Grabner, M., & Dirksen, R. T. (2004). Functional analysis of the R1086H malignant hyperthermia mutation in the DHPR reveals an unexpected influence of the III-IV loop on skeletal muscle EC coupling. *American Journal of Physiology. Cell Physiology, 287*, 1094–1102.

Wilkens, C. M., & Beam, K. G. (2003). Insertion of α_{1S} II-III loop and C terminal sequences into α_{1H} fails to restore excitation-contraction coupling in dysgenic myotubes. *Journal of Muscle Research and Cell Motility, 24*, 99–109.

Wilkens, C. M., Kasielke, N., Flucher, B. E., Beam, K. G., & Grabner, M. (2001). Excitation-contraction coupling is unaffected by drastic alteration of the sequence surrounding residues L720-L764 of the α_{1S} II-III loop. *Proceedings of the National Academy of Sciences of the United States of America, 98*, 5892–5897.

Yamazawa, T., Takeshima, H., Shimuta, M., & Iino, M. (1997). A region of the ryanodine receptor critical for excitation-contraction coupling in skeletal muscle. *The Journal of Biological Chemistry, 272*, 8161–8164.

Yuan, S. H., Arnold, W., & Jorgensen, A. O. (1991). Biogenesis of transverse tubules and triads: Immunolocalization of the 1, 4-dihydropyridine receptor, TS28, and the ryanodine receptor in rabbit skeletal muscle developing in situ. *The Journal of Cell Biology, 112*, 289–301.

CHAPTER 7

Ryanodinopathies: RyR-Linked Muscle Diseases

Lan Wei and Robert T. Dirksen
Department of Pharmacology and Physiology, University of Rochester Medical Center, Rochester, New York, USA

I. OVERVIEW

Ryanodine receptors (RyRs) are the primary intracellular Ca^{2+} release channels responsible for providing Ca^{2+} used to drive muscle contraction. Given this central role, it is not surprising that several debilitating and life-threatening muscle disorders, collectively termed "ryanodinopathies," result from mutations and functional alterations in the skeletal (RyR1) and cardiac (RyR2) muscle RyRs. RyR1 mutations are linked to malignant hyperthermia (MH), a hypermetabolic pharmacogenetic disorder of skeletal muscle

Current Topics in Membranes, Volume 66 1063-5823/10 $35.00
Copyright 2010, Elsevier Inc. All right reserved.
DOI: 10.1016/S1063-5823(10)66007-3

triggered by volatile anesthetics, as well as several congenital myopathies, including central core disease (CCD), multi-minicore disease, core-rod myopathy, and centronuclear myopathy. RyR2 mutations are associated with catecholaminergic polymorphic ventricular tachycardia (CPVT) and arrhythmogenic right ventricular dysplasia type 2 (ARVD2), arrhythmogenic disorders linked to unexpected sudden death during physical or emotional stress. MH and CPVT/ARVD2 mutations hypersensitize channels to activation and promote Ca^{2+} leak from the sarcoplasmic reticulum (SR). CCD-linked mutations either enhance SR Ca^{2+} leak in a manner that compromises SR Ca^{2+} content and release or reduce RyR1-mediated Ca^{2+} flux from a full-complement SR Ca^{2+} store. Precisely, how RyR disease mutations alter channel sensitivity, leak, and release remain elusive, but potential mechanisms include (1) RyR hyperphosphorylation and FK506 binding protein (FKBP) dissociation, (2) disruption of critical RyR interdomain regulatory interactions, (3) enhanced store overload-induced Ca^{2+} release, and (4) reduced RyR-mediated Ca^{2+} flux due to altered channel gating or Ca^{2+} permeation. Although a few treatment options are available, further progress in elucidating the pathophysiological mechanisms that underlie the ryanodinopathies will undoubtedly lead to the development of new and more effective therapeutic interventions.

II. INTRODUCTION

The pivotal role of tetrameric ryanodine receptor (RyR) channels in mediating Ca^{2+} release from the sarcoplasmic reticulum (SR) during excitation–contraction (EC) coupling in striated muscle is covered in detail in previous chapters (e.g., Chapter 5). Briefly, RyR Ca^{2+} release channels embedded in the terminal cisternae of the SR are functionally coupled to voltage sensor dihydropyridine receptors (DHPRs) in the transverse tubule and/or surface membrane. Upon sarcolemmal depolarization as a result of a propagating action potential, voltage sensors are activated and transmit a signal to nearby RyRs either via intimate chemical (Ca^{2+} influx-induced Ca^{2+} release in the heart) or mechanical (depolarization-induced Ca^{2+} release in skeletal muscle) coupling mechanisms (Dirksen, 2002). In both cases, RyR channels are opened and release Ca^{2+} stored in the SR, which then diffuses to the contractile machinery to initiate crossbridge cycling and force production. During relaxation, released Ca^{2+} is resequestered back into the SR by sarcoplasmic/endoplasmic reticulum Ca^{2+} ATPase pumps (SERCA) with a minor fraction being taken up by mitochondria. Influx Ca^{2+} (e.g., in the heart) is extruded from the cell by sarcolemmal Na^{+}/Ca^{2+} exchanger and Ca^{2+}ATPase pump proteins. Collectively, the entire process of depolarization-induced Ca^{2+} release and removal is referred to as EC coupling. Defects in any component of the EC coupling machinery can result in a disturbance of proper control of intracellular Ca^{2+} homeostasis and lead to muscle dysfunction. The RyR complex serves as a central hub of the EC coupling and Ca^{2+} homeostatic machinery, and thus, even minor RyR dysfunction leads to a diverse assortment of clinically distinct skeletal and cardiac muscle disorders, collectively referred to here as "ryanodinopathies." While RyRs also contribute to intracellular Ca^{2+} signaling in smooth muscle cells, neurons, and nonexcitable cells, they do so in conjunction with the inositol 1,4,5-trisphosphate receptor (IP_3R), ligand-gated intracellular Ca^{2+} release channels (discussed in detail in subsequent chapters), and thus play a more supportive and less autonomous role in these cell types.

Single point mutations and small in-frame deletions in the skeletal (RyR1) and cardiac (RyR2) RyR isoforms are sufficient to result in defective Ca^{2+} regulation and severe skeletal and cardiac disease, respectively (Fig. 1). Thus far, a large number of RyR1 mutations have been associated with malignant hyperthermia (MH; a pharmacogenetic disorder triggered by halogenated volatile anesthetics and depolarizing muscle relaxants, see Section III.A) and several congenital myopathies, including central core disease (CCD; Section III.B), multi-minicore disease (MmD; Section III.B), core-rod myopathy (CRM; Section III.B), and centronuclear myopathy (CNM; Section III.C). RyR2 mutations are associated with two inherited malignant cardiac arrhythmias, catecholaminergic polymorphic ventricular tachycardia (CPVT; Section IV.A) and arrhythmogenic right ventricular dysplasia 2 (ARVD2; Section IV.B). Altered RyR1 mRNA splicing and release channel function may contribute to

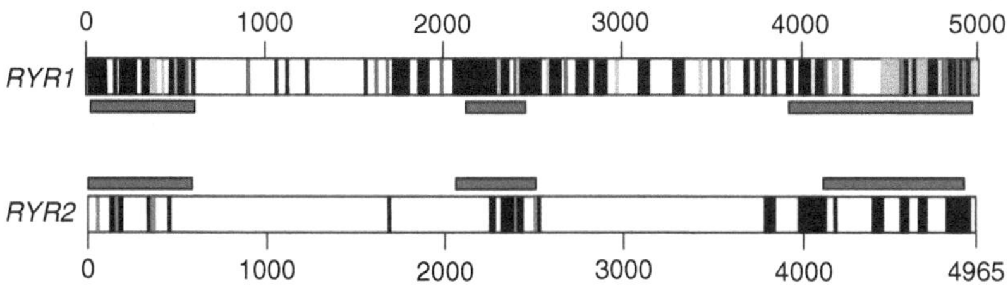

FIGURE 1 Spectrum of MH, CCD, MmD, CRM, and CNM related mutations in the *RYR1* gene and CPVT and ARVD related mutations in the *RYR2* gene. Gray horizontal bars indicate previously proposed MH/CCD "hot spot" regions. N-terminal and central regions are thought to form an interdomain regulatory module that may be disrupted, or "unzipped," by *RYR1* and *RYR2* disease mutations. Mutations are indicated by vertical colored bars: *RYR1*, MH, or MH/CCD mutations—black; CCD-only—green; CCD/MmD—blue; MH/CCD/MmD—red; MmD only—light blue; MH/CCD/CRM—pink; CNM—yellow; *RYR2*, CPVT—black; ARVD2—orange. (See Color Insert.)

myotonic dystrophy (Kimura et al., 2005) and RyR2 release channel dysfunction plays a role in heart failure (Section IV.C), although specific inherited RyR gene mutations have not been linked to either myotonic dystrophy or heart failure. Also, alterations in other participants in the Ca^{2+} cycling homeostatic machinery (e.g., DHPRs, SERCA, calsequestrin (CSQ)) can also lead to muscle dysfunction due to either direct or indirect effects on RyR activity. For example, in addition to RyR2 mutations, CSQ deficiency results in CPVT in mice (Knollmann et al., 2006) and humans (Lahat et al., 2001), as well as MH and heat susceptibility in mice (Dainese et al., 2009). The focus of this chapter is to examine the clinical characteristics, genetics, and pathophysiological mechanisms of the ryanodinopathies, an eclectic array of clinically distinct muscle disorders caused by inherited and acquired alterations in skeletal and cardiac muscle RyR function.

III. RYR1-LINKED DISEASES

A. Malignant Hyperthermia

MH is an autosomal dominant pharmacogenetic disorder characterized as a hypermetabolic state of skeletal muscle induced by potent volatile anesthetics (e.g., halothane, isoflurane, sevoflurane) and/or depolarizing muscle relaxants (e.g., succinylcholine). Triggering agents cause sustained Ca^{2+} release from the SR, which leads to contracture of skeletal muscle, glycogenolysis, and uncontrolled muscle metabolism that result in increased heat and lactate production. MH episodes are initially characterized by hypercapnia, tachycardia, masseter muscle rigidity, and metabolic acidosis. Later signs include pyrexia (Larach et al., 2010), myoglobinuria, and rhabdomyolysis (featuring hyperkalemia and a dramatic increase in serum creatine kinase levels), which may subsequently lead to fatal cardiac arrhythmias, disseminated intravascular coagulation, cerebral edema, and renal failure (Larach et al., 1994). Management of an acute MH crisis involves discontinuation of triggering agent, hyperventilation with 100% O_2, treatment of metabolic abnormalities, cooling/maintenance of body temperature, and timely intravenous administration of dantrolene sodium (2.5 mg/kg initial dose), the only known pharmacological antidote for the disorder.

MH-like crises triggered under nonanesthetic conditions have also been reported. Strenuous exercise during mild heat stress has induced sudden death in a few MHS patients (Tobin et al., 2001). A number of reports suggest a relationship between MH and exertional heatstroke, rhabdomyolosis, sudden infant death syndrome, and neuroleptic malignant syndrome (a MH-like syndrome that manifests after administration of neuroleptic agents including phenothiazines, haloperidol, and drugs used in the treatment of schizophrenia; Capacchione & Muldoon, 2009; Denborough, 1998; Downey et al., 1984). However, definitive links between MH and these diverse disorders have not been clearly established in controlled clinical studies.

MH was first reported by Denborough and Lovell in a family with more than 10 members that died during or after general anesthesia (Denborough & Lovell, 1960). The term "malignant" was used to provide a clear indication that the disorder was associated with a high mortality. However, with appropriate administration of i.v. dantrolene sodium (Krause et al., 2004), the mortality of MH attacks have now been reduced from more than 70% to ~5–10%. Estimates of MH incidence range widely from as low as 1:15,000 children and 1:50,000 adults undergoing anesthesia in North America and Europe to a prevalence based on genetic susceptibility that is as high as 1:2000 in some populations (Robinson et al., 2006; Rosenberg et al., 2010). However, actual frequency of MH susceptibility (MHS) is difficult to ascertain since <2% of the population receives triggering agents each year (Robinson et al., 2006). Surprisingly, males are more susceptible than females, with a male:female ratio of >2:1 (Larach et al., 2010).The reason for this gender inequity is not clear. Diagnosis of MHS is of great importance, especially for anesthesiologists, but is practically difficult, since patients with MHS lack an overt muscle phenotype in the absence of exposure to triggering agents. Following the discovery and recognition of MH as a life-threatening disorder during anesthesia, an *in vitro* contracture test (IVCT) was developed by the European Malignant Hyperthermia Group (EMHG) and a similar caffeine–halothane contracture test (CHCT) was developed by the North American MH Group (NAMHG), supported by the Malignant Hyperthermia Association of the United States (MHAUS). Both tests involve exposing fresh muscle biopsy tissue to sequential addition of increasing concentrations of caffeine and halothane while measuring muscle contracture. The tests determine the sensitivity of the muscle biopsy to contractures induced by normally subthreshold levels of caffeine and halothane (e.g., 2 mM caffeine and 3% halothane in the NAMHG CHCT). Individuals are designated as MHS if they exhibit a certain contracture threshold in the presence of caffeine and halothane. The tests are explicitly designed to exhibit high sensitivity (98–99%) and specificity (78–93%) in order to limit false negatives (i.e., to positively identify all MH susceptible individuals; Robinson et al., 2006; Rosenberg et al., 2010). For the past 30 years, the IVCT and CHCT have served as the gold standard in the diagnosis of MHS throughout the world. However, due to the painful, costly, and invasive nature of the procedure, IVCT/CHCT is used primarily for diagnosis of an index patient rather than for general screening purposes.

Molecular genetic testing was recently introduced as an alternative, non-invasive approach for diagnosis of MHS, particularly following positive IVCT/CHCT identification of an index patient or relative. MHAUS (http://www.mhaus.org/) and EMHG (http://www.emhg.org/) organizations have established guidelines for determining when molecular genetic screening for MHS is justified. A single missense mutation (R615C) in the gene encoding the porcine skeletal muscle ryanodine receptor (*RYR1*) results in a recessively inherited form of heat-, stress-, and anesthetic-induced MHS in pigs (Fujii et al., 1991). A homologous *RYR1* mutation (R614C) was identified in an MH susceptible family shortly thereafter (Gillard et al., 1991). Currently, more than 200 putative *RYR1* disease causing mutations have been identified (Fig. 1) and *RYR1* gene mutations account for up to 70% of all MH susceptible cases (MHS1; Robinson et al., 2006; Rosenberg et al., 2010). Of these mutations, over 160 are linked to MHS, though only a limited number of these have been confirmed as being causative using strict EMHG guidelines that require demonstration of a functional defect.

In addition to the *RYR1* gene, MH has also been linked to five other gene loci (MHS2–6), though specific non-*RYR1* causative mutations have so far only been identified in the gene encoding the α_{1S}-subunit of the skeletal muscle DHPR at locus 1q32 (*CACNA1S*; MHS5), which accounts for only ~1% of all MH cases (Robinson et al., 2006; Rosenberg et al., 2010). The identified DHPR mutations include substitution of a highly conserved arginine residue in the extracellular loop connecting transmembrane DHPR repeats III and IV to either a histidine (R1086H) or cysteine (R1086C; Monnier et al., 1997) residue. A tryptophan substitution for a conserved arginine residue (R174W) in the repeat I S4 voltage sensor region was also recently linked to MHS (Carpenter et al., 2009). In addition to 19q13.1 (MHS1, *RYR1*) and 1q32 (MHS5, *CACNA1S*), the other proposed MH loci include 17q11.2-q24 (MHS2), 7q21-q22 (MHS3), 3q13.1 (MHS4), and 5p (MHS6; Rosenberg et al., 2010). Interestingly, ablation of the type 1 calsequestrin (*CASQ1*) gene in mice was recently shown to result in a heat- and halothane-induced MH susceptible phenotype that is remarkably similar to that observed in both porcine and mouse models of MHS due to mutations in the *RYR1* gene (Dainese et al., 2009). However, genetic linkage of mutations in the *CASQ1* gene to MHS in humans has yet to be demonstrated.

B. Core Myopathies

CCD, MmD, and CRM belong to a family of related, yet histologically and clinically distinct, congenital myopathies. CCD is a rare, minimal, or nonprogressive myopathy characterized by variable degrees of proximal muscle weakness and hypotonia during infancy (floppy infant syndrome) that occurs in the absence of significant bulbar, respiratory, or extraocular muscle involvement (Magee & Shy, 1956). Proximal muscle weakness may persist throughout adolescence and be accompanied by a significant delay in the attainment of motor skill milestones (crawling, walking, etc.). Secondary involvement of skeletal abnormalities commonly includes hip dislocation, club foot, scoliosis, and joint contractures (Sewry et al., 2002; Shuaib et al., 1987). CCD is usually inherited as an autosomal dominant trait, with >90% of patients carrying RyR1 mutations (Fig. 1; Wu et al., 2006). However, cases of recessive inheritance of CCD that often overlap with characteristics observed in the other core myopathies (MmD, CRM) or MHS have been reported in a few families (Robinson et al., 2006).

The pathohistological hallmark of CCD is the presence of clearly demarcated, amorphous "core" lesions that run a substantial length along the longitudinal axis of type I muscle fibers. The cores may be central, eccentric, or peripheral and are devoid of mitochondria and oxidative enzyme activity. However, cores are often separated from adjacent normal tissue by a dark rim of intense oxidative staining indicative of a thin wall of mitochondria at the boundary (Monnier et al., 2001). Electron microscopic analyses of muscle biopsy samples from CCD patients reveal a wide variety of structural alterations, including damaged/disrupted sarcomere structure, heavily compacted myofibrils, Z-line streaming, or disruption, absence of mitochondria and glycogen granules, abnormal SR and T-tubule profiles, and in extreme cases, unstructured regions completely lacking contractile filaments (Isaacs et al., 1975), while surrounding noncore regions exhibit normal structure. However, the clinical presentation of CCD is highly variable and disease severity correlates poorly with the degree of structural alterations observed in muscle biopsy (Sewry et al., 2002). For example, approximately 40% of CCD patients exhibiting cores are clinically asymptomatic (Shuaib et al., 1987), while others may exhibit significant muscle weakness in daily life.

The underlying processes involved in the formation of mitochondrial-deficient and metabolically inert core regions in CCD muscle remain unclear. It has been hypothesized that altered RyR1 Ca^{2+} release channel activity in presumptive core regions results in local Ca^{2+} dysregulation, Ca^{2+}-mediated mitochondrial destruction, and altered ATP production that together conspire to alter muscle structure within the core regions (Boncompagni et al., 2009; Loke & MacLennan, 1998). Mitochondrial dysfunction and core formation may also result from altered local SR-mitochondrial coupling and Ca^{2+}-dependent ATP production due to impaired RyR1 Ca^{2+} release. The enrichment of mitochondria around the core circumference may reflect a "sealing-off" or protection of the rest of the cell from the defective core.

MH and CCD were initially thought to be strongly correlated with one another, as both disorders are linked to mutations in the *RYR1* gene and CCD patients were often found to be MH susceptible. However, while in some cases overlap between MH and CCD is clear, the correlation suggested in early studies may actually reflect a sampling bias introduced by evaluation of patients for CCD based on positive IVCT rather than possessing a demonstrated a clinical myopathy. Indeed, further analyses revealed that many CCD patients are MH negative with IVCT (Halsall et al., 1996; Monnier et al., 2001; Sewry et al., 2002). Thus, while some mutations result in both MH and CCD, others result in either only MH or only CCD (Robinson et al., 2006; Rosenberg et al., 2010; Treves et al., 2008). In order to improve MH detection and diagnosis, significant efforts have also been made to better define the relationship between genuine MHS and anesthetic hypersensitivity due to the presence of another unrelated underlying myopathy or muscular dystrophy (Davis & Brandom, 2009). However, except for CCD, only a weak association with MH is observed among the other congenital myopathies, including Duchenne/Becker muscular dystrophy, myotonic dystrophy, myotonia congenita, and Brody disease. However, even if only as a pragmatic approach, all individuals possessing MH, CCD, MmD, CRM, or CNM *RYR1* disease mutations should be considered as at risk for MHS until excluded by IVCT/CHCT.

MmD is a clinically and genetically heterogeneous, early-onset congenital myopathy that often affects bulbar, respiratory, and extraocular muscles. In contrast to CCD, MmD typically exhibits an autosomal recessive mode of inheritance (Robinson et al., 2006; Treves et al., 2008). The classical form of MmD ($\sim$75% cases) is associated with neonatal hypotonia, severe axial muscle weakness, scoliosis, and respiratory failure. A second early-onset form is associated with multiple persistent joint contractures (arthogryposis). A third form is more slowly progressive and presents with significant hand involvement (Jungbluth, 2007). Histopathologically, skeletal muscle from MmD patients is characterized by the presence of numerous poorly circumscribed and short unstructured core-like regions. Like classic central cores, minicores exhibit sarcomere disorganization and reduced mitochondrial and oxidative enzyme content, but differ in that minicores: (1) are multiple and randomly distributed throughout the fiber; (2) only extend a few sarcomeres along the longitudinal axis; and (3) are found in both type I and type II fibers. Thus, based on clinical presentation, mode of inheritance, and histological characteristics, CCD and MmD are regarded as distinct entities. However, since minicores and central cores are at time observed in the same biopsy and in different individuals of the same family, they may arise from a common pathogenic mechanism.

The most common type of MmD (classical form) results from recessive (and compound heterozygous) mutations in the selenoprotein N1 gene (*SEPN1*) on chromosome 1 (1p36; Jungbluth, 2007). The *SEPN1* gene encodes a 70 kDa ER/SR membrane glycoprotein that is involved in MmD, as well as other forms of muscular dystrophy (e.g., rigid spine muscular dystrophy 1; Moghadaszadeh et al., 2001). More than 20 recessive *SEPN1* mutations have been linked to MmD, the majority of which result in premature truncation and loss of function. SEPN1 was recently found to associate with muscle RyRs and be required for normal muscle development and Ca^{2+} flux in zebrafish embryos (Jurynec et al., 2008). Moreover, studies conducted in myotubes from MmD patients with SEPN1 null-mutations (Arbogast et al., 2009) and following SEPN1 knockdown in C2C12 myotubes (Moghadaszadeh et al., 2007) indicate that SEPN1 plays a critical role in protecting against cellular oxidative stress, as well as maintaining proper redox regulation and muscle Ca^{2+} homeostasis. MmD, particularly forms with extraocular muscle involvement, has also been linked to more than 15 recessive mutations in the *RYR1* gene (Fig. 1). While some of the recessive MmD mutations alter Ca^{2+} homeostasis, a significant fraction appears to lead to a marked reduction in the level of RyR protein expression (Monnier et al., 2003; Treves et al., 2008). Interestingly, recessive MmD can also result from compound heterozygosity (Zhou et al., 2006b) or when heterozygosity for a single causative mutation at the genomic level is coupled with wild-type allele silencing in muscle, thus resulting in monoallelic expression of the mutant (Zhou et al., 2006a).

Nemaline rod myopathy is a rare ($\sim$1:50,000 births) congenital neuromuscular disorder that ranges clinically from a severe neonatal form to a slowly progressive adult form (Sanoudou & Beggs, 2001). Clinical features include hypotonia in childhood, possibly attributed to altered Ca^{2+} regulation, small muscle bulk, multiple muscle weakness, and wild range of deformities. Nemaline rod myopathy is caused by mutations in genes that encode proteins of the sarcomere and is characterized histochemically by the presence of dense inclusions (called nemaline bodies or rods) composed of crystalline-like aggregates of electron dense material containing α-actinin and actin that emanate from the Z-line (Wallgren-Pettersson et al., 1995). The simultaneous occurrence of both central cores and nemaline rods have been reported in muscle biopsy samples from patients possessing *RYR1* gene mutations (Fig. 1) indicative of a distinct "CRM" that suggests either clinical overlap between CCD and nemaline rod myopathy or that rods reflect a secondary feature of CCD (Davis et al., 2003; Monnier et al., 2000). Also, patients with clinical features of CCD may exhibit rods or multiple minicores together or before the formation of central cores (Ferreiro et al., 2002; Sewry et al., 2002).

Thus, significant overlap in muscle pathology exists between CCD, MmD, and CRM, rendering definitive diagnostic distinction between these disorders challenging.

C. Centronuclear Myopathy

CNM is a genetically heterogeneous congenital myopathy exhibiting X-linked and both autosomal recessive and dominant variants. Mutations in the myotubularin (*MTM1*) and amphiphysin 2 (*BIN1*) genes are implicated in the X-linked and autosomal recessive forms, respectively, while mutations in the dynamin 2 (*DNM2*) and *RYR1* genes are associated with dominant inheritance of CNM (Jungbluth et al., 2008). Histopathological features of muscle biopsies from CNM patients reveal the presence of numerous centrally located nuclei and central zones either devoid or with enhanced oxidative enzyme activity. *DNM2*-linked forms often exhibit radial sarcoplasmic strands surrounding the central area. Significant necrosis and regeneration are not typically observed in CNM. A *de novo* RyR1 missense mutation (Fig. 1, S4112L) was recently linked to a young girl exhibiting clinical and histopathological features consistent with CNM (Jungbluth et al., 2007). Central nucleation affecting up to 50% of fibers, central accumulation of oxidative enzyme stains, and hypertrophy of type I fibers were observed upon muscle biopsy obtained at 1 year, while core-like regions devoid of oxidative enzyme staining were observed upon repeat biopsy 8 years later. Additional features including external opthalmoplegia, muscle MRI findings, and altered Ca^{2+} release channel function are all also suggestive of RyR1 involvement. Potentially, some RyR1 disease mutations may present early in life with a picture consistent with CNM, but then over time, develop features (e.g., cores, minicores, MHS) that are more commonly associated with RyR1-related disorders.

D. Functional Impact of RyR1-Linked Disease Mutations

More than 160 MH susceptible mutations and nearly 100 CCD mutations have been identified in the *RYR1* gene, with some mutations being associated with both disorders (Fig. 1; Rosenberg et al., 2010). Thus, three classes of mutations, referred to as MH-only, CCD-only, and MH/CCD have been proposed (Dirksen & Avila, 2004). Originally, the majority of MH and/or CCD mutations in RyR1 were thought to cluster in three "hot spot" regions including N-terminal (region 1), central (region 2), and C-terminal transmembrane (region 3) regions, corresponding to amino acid residues 35–614,

2163–2458, and 3916–4973, respectively (Fig. 1; Robinson et al., 2006; Treves et al., 2008; Yano et al., 2006). However, the recent implementation of whole *RYR1* gene sequencing has revealed more and more mutations that lie outside these regions and a few even occur within intronic regions of the gene. Thus, it has become apparent that RyR1 disease mutations are found throughout the entire *RYR1* gene and the proposed hot spot regions deduced from early candidate exon sequencing likely reflect a screening bias (Robinson et al., 2006; Wu et al., 2006). Nevertheless, in general MH-only and MH/CCD mutations are primarily found within the cytoplasmic region of the channel, while CCD-only mutations tend to be located in the transmembrane C-terminal region and recessive MmD mutations are scattered throughout the sequence (Fig. 1; Davis et al., 2003; Monnier et al., 2001; Treves et al., 2008).

Since identification of the first RyR1 mutation in a MH susceptible family (Gillard et al., 1991), elucidation of the pathophysiological mechanisms that underlie MHS has made tremendous progress. Due to the abundance and availability of muscle from MH susceptible pigs, initial studies involved characterizing the effects of the R615C mutation on RyR1 function. Muscle bundles, skinned muscle fibers, and SR vesicles isolated from MH susceptible pigs were found to exhibit enhanced sensitivity to activation by voltage, Ca^{2+}, halothane and caffeine, reduced Ca^{2+} and Mg^{2+} inhibition, and slowed relaxation following halothane-induced contractures (Laver et al., 1997; Mickelson & Louis, 1996; Owen et al., 1997). Enhanced RyR1 channel sensitivity to activation by triggers and reduced inhibition by antagonists was then extended to other identified MH susceptible mutants using patient-derived B-lymphocytes and myotubes (Ducreux et al., 2004; Girard et al., 2001; Wehner et al., 2002) and following both heterologous (e.g., HEK293, COS-7 cells; Lynch et al., 1999; Tong et al., 1999; Treves et al., 1994) and homologous (C2C12 and RyR1-null myotubes; Avila & Dirksen, 2001; Censier et al., 1998; Dirksen & Avila, 2004; Yang et al., 2003) expression. Additionally, expression of the DHPR α_{1S}-subunit containing the MHS R1086H mutation in myotubes derived from α_{1S}-null mice similarly enhances wild-type RyR1 sensitivity to activation by both voltage and caffeine (Weiss et al., 2004). Together, these results indicate that an increase in the sensitivity of the DHPR-RyR1 Ca^{2+} release mechanism to activation by a wide range of triggering agents as a result of MH-causing mutations to proteins of the EC coupling machinery represents a unifying principle that underlies increased muscle susceptibility to activation by anesthetics. Results obtained from patient-derived B-lymphocytes/myotubes and following heterologous/homologous expression have lead to several important advances in elucidating the functional effects of disease causing mutations in RyR1. Based on these studies, muscle weakness in CCD results, at least in part, from compromised RyR1-mediated Ca^{2+} release during EC coupling. However, the mechanism by which Ca^{2+} release is reduced differs between MH-only, MH/CCD, and CCD-only mutations. Two distinct pathogenic mechanisms have been proposed: "leaky" (Fig. 2B) and "EC uncoupled" (Fig. 2C) RyRs (Dirksen & Avila, 2002). MH-only and MH/CCD mutations are proposed to cause different levels of increased RyR1 sensitization and Ca^{2+} leak. While MH-only mutations (e.g., R614C) only modestly increase RyR1 sensitivity and Ca^{2+} leak is insufficient to alter net SR Ca^{2+} content (compensated leak), MH/CCD mutations (e.g., Y522S) enhance RyR1 release channel sensitivity and leak in a manner sufficient to promote RyR1 Ca^{2+} leak sufficient to cause SR store depletion, elevate resting intracellular Ca^{2+}, and reduce Ca^{2+} release during EC coupling ("decompensated leak"; Avila & Dirksen, 2001; Dirksen & Avila, 2004; Tong et al., 1999). On the other hand, CCD-only mutations (e.g., I4898T) result in a reduction in SR Ca^{2+} release during EC coupling without enhancing RyR1 Ca^{2+} leak or release channel sensitivity to activation by triggers (e.g., caffeine, halothane, DHPRs; Avila et al., 2001). For EC uncoupled channels, resting Ca^{2+} and SR Ca^{2+} content remain unaltered, while the release of Ca^{2+} from a full-complement SR store is still somehow attenuated. Several EC uncoupled mutants (Avila et al., 2003) have been localized to the putative pore lining/selectivity filter region of the channel located between the final two RyR1 transmembrane regions (Zhao et al., 1999). Mutations of residues within this critical region result in marked reductions in RyR1 single channel conductance, activation by Ca^{2+}, and Ca^{2+} permeation (Gao et al., 2000). Similar results are also observed for CCD mutations to pore-lining residues (Xu et al., 2008). In spite of these results, controversy regarding the validity of the EC uncoupling mechanism remains since Ca^{2+} measurements conducted in B-lymphocytes (Tilgen et al., 2001) and myotube cultures (Ducreux et al., 2004) derived from patients were interpreted to suggest that the I4898T mutation enhances Ryr1 Ca^{2+} leak. The discrepancies likely arise from differences between the preparations, approaches, and cellular systems used in these studies.

The precise molecular mechanisms by which MH and MH/CCD RyR1 disease mutations enhance RyR1 Ca^{2+} leak remain elusive. Ikemoto and colleagues have suggested that RyR1 mutations disrupt an important interdomain regulatory interaction that normally serves to stabilize the RyR channel closed state. Disruption of this RyR1 inhibitory interdomain interaction thus results increased basal RyR1 release channel activity or leak (Ikemoto & Yamamoto, 2000). An alternative mechanism for increased RyR1 Ca^{2+} leak proposed by Chen and colleagues is that the RyR1 disease mutations enhance release channel sensitivity to activation by luminal Ca^{2+}, resulting in an increase in store overload-induced Ca^{2+} release (SOICR) in

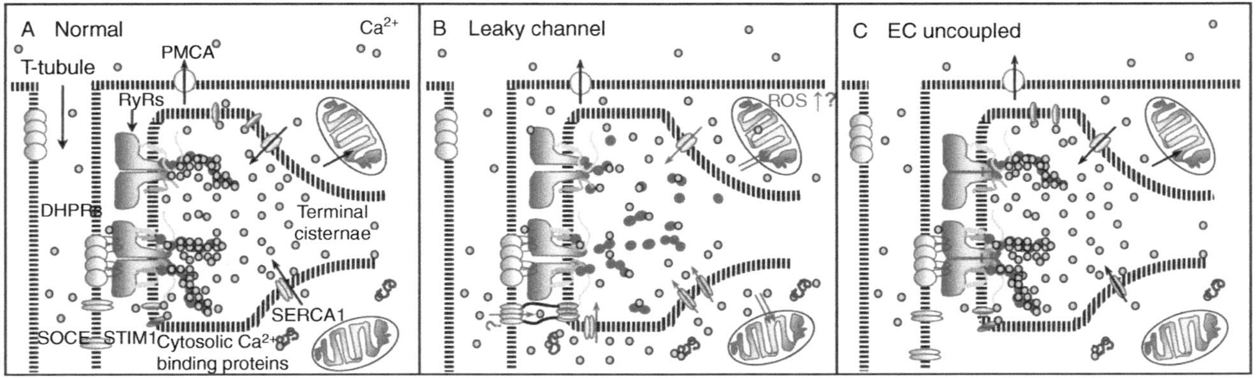

FIGURE 2 Proposed RyR1 pathogenic mechanisms: (A) RyR1 channel function in normal skeletal muscle, (B) MH/CCD RyR1 mutations enhance channel sensitivity to activation and promote SR Ca^{2+} leak and store depletion (leaky channels), (C) CCD RyR1 mutations reduce SR Ca^{2+} release without causing store depletion (EC uncoupled channels). SERCA, sarco/endoplasmic reticulum Ca^{2+} ATPase; STIM1, stromal interaction molecule 1; SOCE, store operated Ca^{2+} entry; PMCA, plasma membrane Ca^{2+} ATPase. (See Color Insert.)

MH/CCD (Jiang et al., 2008), an extension of a mechanism initially proposed to account for increased RyR2-mediated spontaneous Ca^{2+} release in CPVT (see Section IV.D).

Recently, several RyR1 knockin mice were generated in attempts to provide animal models of MH and CCD that could be used to systematically evaluate disease pathogenesis, progression, and to assess the validity of the proposed leaky and EC uncoupling mechanisms *in vivo*. Currently, three RyR1 knockin mouse model have been established and characterized: Y522S (MH/CCD, Chelu et al., 2006), R163C (MH/CCD, Yang et al., 2006), and I4898T (CCD; Zvaritch et al., 2007). For all three lines, homozygous knockin mice are either embryonic lethal or die immediately following birth, while heterozygous mice are viable and reproductive. Heterozygous Y522S/+ (Chelu et al., 2006) and R163C/+ (Yang et al., 2006) mice are MH susceptible as they experience lethal MH-like crises following either halothane/isofluorane exposure or heat stress (e.g., 41 °C for 15 min). Consistent with this MHS phenotype, muscle fibers and myotubes derived from both lines exhibit increased sensitivity to caffeine, 4-CMC, and voltage (Chelu et al., 2006; Yang et al., 2006). Interestingly, resting Ca^{2+} levels and SR store content were normal in Y522S/+ fibers and myotubes at room temperature, while resting Ca^{2+} is increased and store content reduced at temperatures above 30 °C (Chelu et al., 2006; Durham et al., 2008). This temperature-dependence of store depletion and elevation of resting Ca^{2+} was found to be due to increased ROS/RNS production that leads to RyR1 *S*-nitrosylation, which serves to further exacerbate heat-induced RyR1 Ca^{2+} leak (Durham et al., 2008).

Defining the mechanisms by which RyR1 disease mutations alter release channel function and lead to the development of central cores and minicores have been elusive. However, novel insights into this enigmatic process are now beginning to be revealed through systematic structural analyses of skeletal muscle from CCD RyR1 knockin mice. Proposed pathways for core formation due to enhanced RyR1 Ca^{2+} leak in Y522S/+ mice (Boncompagni et al., 2009) and EC uncoupled I4895T/+ mice (Zvaritch et al., 2009) were recently described. By 2 months of age, mitochondria within discrete regions (termed presumptive cores) of muscle fibers from Y522S/+ mice become swollen, misshapen, and disrupted (Boncompagni et al., 2009; Durham et al., 2008). Presumptive cores eventually progress to larger "early core" regions that lack mitochondria, SR, and oxidative enzyme activity. Early cores often exhibit contracted myofibrils, most likely due to the absence of SR and mitochondria required for Ca^{2+} removal. "Contraction cores" subsequently progress to form larger unstructured cores that also lack contractile elements (Boncompagni et al., 2009). The progression from early mitochondrial damage to contraction and then unstructured cores is proposed to result from increased local RyR1 Ca^{2+} leak and redox stress causing mitochondrial Ca^{2+} overload that triggers activation of the mitochondria permeability transition pore and subsequent mitochondria membrane depolarization, swelling, and damage. Regions of severe SR/mitochondrial damage result in hypercontractures due to loss of Ca^{2+} sequestration/extrusion and ATP production. Enhanced Ca^{2+} levels and oxidative damage within the contraction core may then activate proteolytic pathways that ultimately lead myofibrillar degradation and formation of metabolically inert unstructured cores. A different mechanical "shear and tear" theory was proposed to account for regions of Z-line streaming, compaction, minicores, and nemaline rods in muscle from I4895T/+ mice (Zvaritch et al., 2009). This theory proposes that the observed structural alterations result from the formation and expression of functionally heterogeneous WT:IT RyR1 tetramers that leads to inhomogeneous Ca^{2+} release that result in nonuniform contraction which, over time, cause focal myofibril tearing and shearing that underlie the progression from minicores to larger central cores and nemaline rods (Zvaritch et al., 2009). Alterations in both type I and type II fiber muscle structure are observed in Y522S/+ and I4895T/+ mice, which while inconsistent with the preferential type I fiber involvement of CCD in humans, is similar to that observed in MmD. Moreover, CCD in humans is typically nonprogressive while cores in both knockin mouse models exhibit marked progression between birth and 1 year of age. Thus, in spite of impressive advances in elucidating core pathogenesis in RyR1 knockin mouse models of CCD, the relevance of these findings to core formation and its importance in the phenotypic presentation of CCD in humans (see Section III.B) have yet to be determined.

IV. RYR2-LINKED DISEASES

A. *Catecholaminergic Polymorphic Ventricular Tachycardia*

CPVT is among the most malignant of the many cardiac arrhythmias caused by malfunction of cardiac ion channels. CPVT was first recognized in the early 1970s, and described in detail in 1995 (Leenhardt et al., 1995). The incidence of CPVT is estimated to be $\sim$1:10,000 (Yano et al., 2006) with about half of all patients exhibiting autosomal dominantly inherited mutations in the cardiac *RYR2* gene (locus 1q42-43, Laitinen et al., 2001; Fig. 1). Approximately 2% of CPVT patients are associated with autosomal recessive mutations in the cardiac calsequestrin (*CASQ2*) gene (locus 1p11-13.3, Lahat et al., 2001). CPVT arrhythmogenic events are triggered by increased sympathetic stimulation and catecholamine release (e.g., epinephrine) brought on by physical and/or emotional stress. Much like that observed during digitalis toxicity, aberrant spontaneous SR Ca^{2+} release, possibly mediated by defective RyR2 function, is thought to induce delayed afterdepolarizations (DADs) and triggered activity that create an arrhythmogenic substrate for initiation of premature ventricular complexes (PVCs) that can then degenerate into sustained polymorphic and bidirectional ventricular arrhythmias.

CPVT is typically observed in young individuals with a structurally normal heart who present after experiencing stress-induced syncope or sudden cardiac death. Electrocardiogram (ECG) recordings are characterized by bidirectional (180° rotation of QRS complex on a beat-to-beat basis) or polymorphic ventricular tachycardia in the absence of QT interval prolongation (Priori et al., 2001). The ventricular arrhythmia may self-terminate or deteriorate to ventricular fibrillation and result in loss of consciousness and eventually death if not promptly normalized by cardioversion. Under resting conditions, the ECG of affected CPVT patients is typically normal. In many cases, acute treatment with β-receptor blocker (e.g., nadolol, metoprolol, and propranolol) provides effective termination of tachycardia (Leenhardt et al., 1995; Swan et al., 1999), though patients may require long-term β-blocker treatment or an implanted cardioverter-defibrillator (ICD) in more severe cases ($\sim$30%). Early reports suggested the mortality rate of CPVT to be $\sim$30–50% by the age of 35 years (Leenhardt et al., 1995), but survival can be significantly extended by improved awareness, detection, and treatment. Differential diagnosis requires distinguishing CPVT from "long QT-syndrome" (LQTS; Monteforte & Priori, 2009; Priori et al., 2001), a distinct inherited arrhythmogenic disorder linked to mutations in genes encoding a broad range of other cardiac ion channels involved in repolarization of the ventricular action potential (Crotti et al., 2008).

B. Arrhythmogenic Right Ventricular Dysplasia Type 2

ARVD is a genetically heterogeneous, autosomal dominant atrophic cardiomyopathy characterized by progressive degeneration and fibro-fatty replacement of the myocardium of the right ventricle. ARVD type 2 (ARVD2) is a specific form associated with stress-induced polymorphic VT; essentially CPVT combined with cardiac structural abnormalities. The causative gene in ARVD2 is also mapped to locus 1q42-43 (Rampazzo et al., 1995; Tiso et al., 2001) and several RyR2 mutations have been identified (Fig. 1; d'Amati et al., 2005; Tiso et al., 2001). However, given the lack of severe classic ARVD presentation in patients with *RYR2* gene mutations coupled with the fact that the structural characteristics of ARVD are not recapitulated in knockin mouse models engineered with either AVRD2

(Kannankeril et al., 2006) or CPVT-linked RyR2 mutations (Cerrone et al., 2005; Lehnart et al., 2008), association of *RYR2* gene mutations with an ARVD phenotype is still controversial (Priori & Napolitano, 2005).

C. Heart Failure

In addition to being linked to CPVT and ARVD2, altered RyR2 function is also implicated in congestive heart failure and related arrhythmias that coincide with a chronic hyperadrenergic state (Lehnart et al., 2004a; Marx et al., 2000). Like CPVT and ARVD2, arrhythmias observed in heart failure are also primarily due to an increased incidence of spontaneous RyR2-mediated diastolic Ca^{2+} release events that trigger arrhythmogenic DADs, though in heart failure these events occur in the face of a significant reduction in SR Ca^{2+} content. The reduction in SR Ca^{2+} content in heart failure involves several factors including downregulation of SERCA2 expression, upregulation in phospholamban-mediated SERCA2 inhibition, increased NCX-mediated Ca^{2+} removal, and enhanced diastolic RyR2-mediated SR Ca^{2+} leak (Blayney & Lai, 2009). The reduction in SR Ca^{2+} available for release is an important reason for the decrease in systolic Ca^{2+} release and cardiac contractility in heart failure. The increase in RyR2-mediated SR Ca^{2+} leak is proposed to result from the hyperadrenergic state of heart failure resulting in RyR2 hyperphosphorylation, FKBP12.6 dissociation, and subsequent enhancement in basal RyR2 Ca^{2+} sensitivity and activity (Marx et al., 2000; Wehrens et al., 2005).

D. Functional Impact and Arrhythmogenic Mechanisms of RyR2-Linked Diseases

The underlying arrhythmogenic mechanism for CPVT results from ectopic ventricular trigger activity due to an increased incidence of DADs (Nakajima et al., 1997). DADs are abnormal membrane depolarizations that occur during diastole and arise from spontaneous SR Ca^{2+} release events that activate Ca^{2+} clearance via the electrogenic Na^+/Ca^{2+} exchanger (NCX), which transports three Na^+ ions into the cell in exchange for the extrusion of one Ca^{2+} ion. As a result, NCX-dependent Ca^{2+} clearance results in a net depolarizing inward current (termed "transient inward current" or I_{ti}). If the I_{ti}-induced diastolic membrane depolarization (or DAD) is large enough, then the membrane potential can reach threshold for opening sodium channels and initiating a premature arrhythmogenic beat. This DAD-mediated arrhythmogenic mechanism is supported by numerous studies in isolated cardiac myocytes as well as *in vivo* studies in RyR2 knockin and CASQ2$^{-/-}$ mouse models (Cerrone et al., 2005; Fernandez-Velasco et al., 2009; Goddard et al., 2008; Kannankeril et al., 2006; Knollmann et al., 2006; Lehnart et al., 2008; Liu et al., 2006).

The precise molecular mechanisms by which both RyR2 mutations and CASQ2 deficiency enhance both diastolic SR Ca^{2+} leak and the incidence arrhythmogenic spontaneous Ca^{2+} release events are still under investigation. Currently, over 70 RyR2 mutations and seven CSQ2 mutations are linked to CPVT (Fig. 1, see also http://www.fsm.it/cardmoc/). In general, CPVT mutations in RyR2 enhance release channel sensitivity to activation by triggers on both the cytosolic and luminal side of the channel. Specifically, several studies have demonstrated that RyR2 channels from CPVT patients are PKA hyperphosphorylated, which promotes FKBP12.6 dissociation and subsequent increased RyR2 Ca^{2+} leak due to enhanced sensitivity to activation by cytosolic Ca^{2+} (Blayney & Lai, 2009; Lehnart et al., 2004b; Wehrens et al., 2003; Fig. 3B). As discussed above, similar PKA-mediated RyR2 hyperphosphorylation, FKBP12.6 dissociation, and enhanced RyR2 Ca^{2+} sensitivity and leak have also been implicated in heart failure (Marx et al., 2000).

The central role of RyR2 hyperphosphorylation and FKBP12.6 dissociation in HF and CPVT/ARVD2 pathogenesis has been challenged by others (George et al., 2003; Jiang et al., 2005; Jiang et al., 2002b; Xiao et al., 2005). Thus, alternative mechanisms have also been proposed. For example, CPVT RyR2 mutants heterologously expressed in HEK293 cells exhibit increased basal activity as assessed from spontaneous Ca^{2+} release in intact cells, enhanced [^{3}H]-ryanodine binding, and single channel open probability (Jiang et al., 2002a). Subsequent studies suggested that the observed increase in spontaneous activity arises from an increased RyR2 channel sensitivity for activation by luminal Ca^{2+} via a mechanism termed SOICR (Fig. 3D; Jiang et al., 2004; Jiang et al., 2005). Consistent with the proposed SOICR mechanism, CPVT is also linked to loss-of-function mutations in the *CASQ2* gene, which reduce SR Ca^{2+} buffering, steepens the SR Ca^{2+} load–leak relationship, and increases the incidence of spontaneous SR Ca^{2+} release and DADs following adrenergic receptor stimulation (Chopra et al., 2007). Alternatively, RyR2 CPVT mutants expressed homologously in HL-1 cardiomyocytes exhibit normal spontaneous beating rate and intracellular Ca^{2+} levels, but exhibit increased responsiveness to caffeine and 4-CMC following interventions that increase cAMP (e.g., β-adrenergic stimulation, forskolin; George et al., 2003), consistent with the CPVT mutations inducing conformational instability in the channel during activation (Fig. 3C; George et al., 2006). A variation of this conformational instability is that CPVT mutations enhance RyR2 channel activity by altering a critical interdomain regulatory

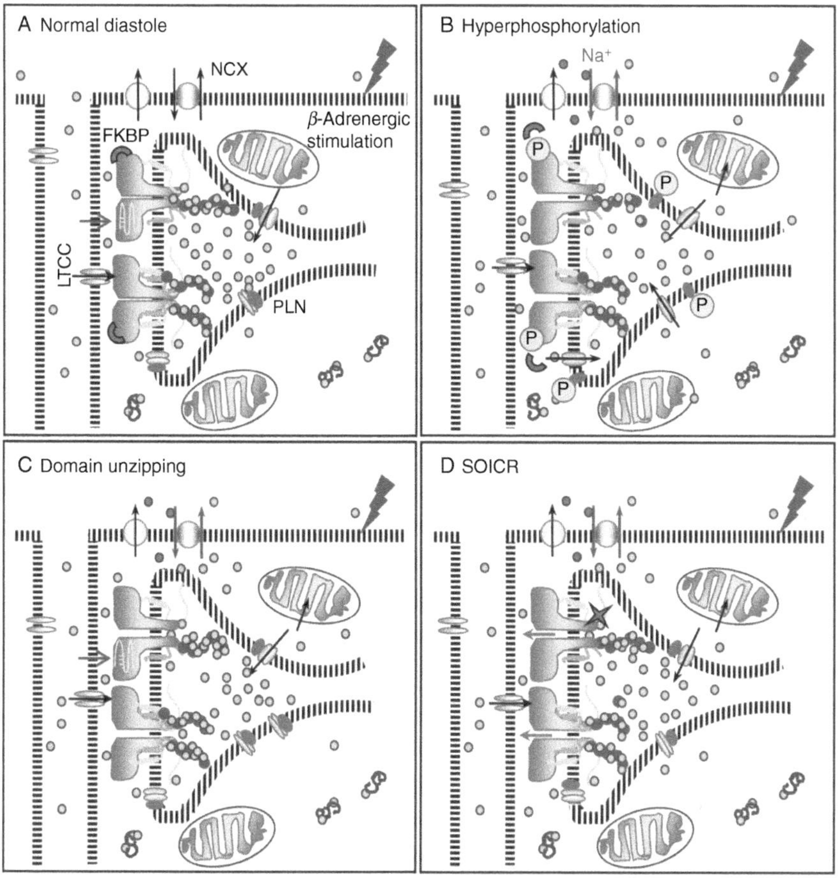

FIGURE 3 Proposed RyR2 pathogenic mechanisms. Schematic representation of the cardiac EC coupling machinery during diastole under β-adrenergic stimulation. (A) RyR2 channel function in normal cardiac muscle exhibiting FKBP12.6 association with RyR2 and an intact "zipped" interdomain interaction (red arrow). (B–D) Three proposed mechanisms for arrhythmogenic activity due to increased spontaneous RyR2 Ca^{2+} release and enhanced NCX activity. (B) Enhanced PKA-dependent RyR2 hyperphosphorylation causes FKBP12.6 dissociation and enhanced RyR2 channel activity. (C) RyR2 disease mutations destabilize or "unzip" (red arrow) an important interdomain regulatory interaction. (D) RyR2 disease mutations enhance luminal channel sensitivity to store overload-induced Ca^{2+} (SOICR) release. LTCC, L-type Ca^{2+} channel; NCX, Na^+/Ca^{2+} exchanger; PLN, phospholamban. (See Color Insert.)

interaction, a mechanism termed "interdomain unzipping" (Fig. 3C; Oda et al., 2005). Results obtained from the aforementioned *in vitro* and expression studies were subsequently validated *in vivo* using four different (R176Q, P2328S, R2474S, and R4496C) RyR2 knockin mouse models (Cerrone et al., 2005;

Fernandez-Velasco et al., 2009; Goddard et al., 2008; Kannankeril et al., 2006; Lehnart et al., 2008; Liu et al., 2006). Each of the mouse models reproduce several critical aspects of the CPVT phenotype including stress- and adrenoceptor-mediated bidirectional and polymorphic VT, DADs, and triggered activity, and enhanced spontaneous Ca^{2+} leak and release.

V. CONCLUSIONS AND PERSPECTIVES

This chapter reviews the presentation, genetics, pathophysiology, and treatment of the ryanodinopathies, an eclectic mix of clinically distinct muscle diseases that arise from genetically inherited and acquired alterations in skeletal and cardiac muscle RyR function. Given their central role in controlling dynamic and steady-state calcium homeostasis and striated muscle contraction, it is not surprising that alterations in RyR1 and RyR2 function result in several life-altering and -threatening skeletal and cardiac muscle disorders, respectively. While detailed understanding of the molecular mechanisms are still being worked out, both gain-of-function (increased Ca^{2+} leak) and loss-of-function (EC uncoupling, reduced RyR expression) alterations in RyR activity can be involved.

While the ryanodinopathies are classically associated with striated skeletal and cardiac muscle diseases, the widespread tissue distribution and functional roles of RyRs in neurons, glia, smooth muscle, and nonexcitable cells suggest that the ryanodinopathies might also extend to several non-muscle-related disorders. For example, since RyR2 is the most prominent isoform expressed in the brain, CPVT/ARVD2 mutations in RyR2 might be expected to contribute to altered neuronal activity. Indeed, children with CPVT due to mutations in the *RYR2* and *CASQ2* genes often present with seizures, convulsions, and involuntary incontinence (Lahat et al., 2001; Leenhardt et al., 1995) and heterozygous R2474S knockin mice exhibit exercise-induced generalized tonic–clonic seizures (Lehnart et al., 2008). Together, these results indicate that leaky RyR2 release channels promote neuronal hyperexcitability and epileptic seizure activity. Since low-level RyR1 expression is found in select brain regions (e.g., cerebellar Purkinje cells, dentate gyrus of the hippocampus, CA1 and CA3 cells of the Ammons' horn, and the olfactory bulb), a component of CNS involvement may also be observed for some of the RyR1-linked disorders. Consistent with this, mutations in the RyR1 gene were recently identified in psychiatric patients at autopsy suspected of having died of neuroleptic malignant syndrome, an MH-like syndrome observed following administration of neuroleptic agents (Sato et al., 2010). Altered RyR function in the ryanodinopathies may also impact proper blood pressure control. Quantal RyR-mediated Ca^{2+} release (termed Ca^{2+} sparks) in vascular smooth muscle cells leads to local activation of large conductance Ca^{2+}-activated potassium (BK) channels, which results in membrane hyperpolarization, closure of voltage-gated Ca^{2+} channels, and ultimately, arteriolar vasodilation (Nelson et al., 1995). Thus, enhanced RyR Ca^{2+} leak and store depletion in vascular smooth muscle cells could potentially contribute to essential hypertension by interfering with vasodilation via driven by Ca^{2+} spark-mediated BK channel activation. Similarly, since altered FKBP12.6 regulation of RyR2 function results in impaired insulin secretion in pancreatic beta cells (Noguchi et al., 2008), CPVT/ARVD2 mutations in RyR2 could also impact proper regulation of blood glucose levels. Finally, currently no diseases have yet to be linked to the type 3 RyR isoform (RyR3). However, given its wide tissue distribution, high conservation with RyR1 ($\sim$65% identity) and RyR2 ($\sim$68% identity), and potential role in synaptic plasticity, spatial learning (Balschun et al., 1999), and blood pressure regulation (Lohn et al., 2001), it would not be surprising if future studies were to identify disorders linked to mutations in the *RYR3* gene.

Continued progress in elucidating the molecular mechanisms that underlie muscle dysfunction in the ryanodinopathies will be critical for improving diagnosis and treatment. In addition to completing a comprehensive medical history, diagnosis of the ryanodinopathies involves IVCT/CHCT for MH, MRI and muscle biopsy pathology analyses for the core and CNMs, and EKG analysis during a controlled exercise stress test for CPVT. In addition, genetic testing is now being used for each disorder to confirm proband diagnosis and for subsequent evaluation of presymptomatic relatives. Effective therapeutic interventions for the RyR-linked disorders are limited. Since the landmark study by Harrison (1975) in MH pigs, management of MH crises has relied heavily on timely intravenous administration of the muscle relaxant dantrolene (2.5 mg/kg). Following its introduction, mortality of MH was reduced from 80% to currently $\sim$5–10% (Krause et al., 2004). The exact mechanism of action by which dantrolene is protective in MH is still under debate, but likely involves inhibition of aberrant RyR1-mediated Ca^{2+} release (Kobayashi et al., 2009; Krause et al., 2004) and/or Ca^{2+} entry (Cherednichenko et al., 2008; Zhao et al., 2006). Dantrolene is also used to treat neuroleptic malignant syndrome, spasticity, ecstasy intoxication and recent studies have suggested its potential utility in the management of neurodegenerative disorders including spinocerebellar ataxia types 2 and 3 (Chen et al., 2008; Liu et al., 2009). Although use of dantrolene has traditionally been significantly hampered by its relatively poor water solubility, a newer more highly soluble form (Revonto™) is now available. Currently, no effective therapeutic interventions are available for the treatment of the CRM and CNMs.

Beta blockers are currently the first line of defense in the effective management of CPVT. However, only ~70% of all affected patients are responsive to β-blockers and the efficiency of β-blocker termination of VT in responsive patients is not always complete. Thus, the efficacy of new "mechanism-based" interventions is currently being evaluated using *in vivo* animal models of heart failure and CPVT. Since RyR2 hyperphosphorylation and FKBP12.6 dissociation has been suggested to underlie increased SR Ca^{2+} leak in CPVT and heart failure, one approach has been to assess the effectiveness of small molecules purported to enhance FKBP12.6 association with RyR2. Initially, JTV519, a 1,4-benzothiazepine derivative, was found to stabilize FKBP12.6 association with RyR2 and improve cardiac function in a canine model of HF (Yano et al., 2003). More recently, S107, a more specific and orally bioavailable benzothiazepine, was shown to also stabilize FKBP12.6 binding to hyperphosphorylated RyR2 and protect against exercise- and catecholamine-induced arrhythmias in R2475S RyR2 knockin mice (Lehnart et al., 2008). However, since potential effects of JTV519 and S107 on inhibiting SOICR and stabilizing RyR interdomain regulatory interactions (Tateishi et al., 2009) have not been fully evaluated, the protective mechanisms of these agents remain to be determined. Also, the potential efficacy of these agents in management of the RyR1-related ryanodinopathies has also not yet been fully evaluated. As an alternate approach, Watanabe et al. (2009) recently reported that prior administration of flecainide, an FDA-approved class I antiarrhythmic agent, protects against arrhythmogenic events in both human CPVT subjects and CSQ2-deficient mice. The protective effect of flecainide was suggested to be due primarily to its ability to inhibit RyR2 SR Ca^{2+} leak rather than its sodium channel blocking activity since protection was not observed for lidocaine, a Na^+ channel blocker that does not inhibit RyR2 Ca^{2+} leak and lacks clinical efficacy in CPVT patients (Watanabe et al., 2009). Together, these studies indicate that interventions designed to limit RyR Ca^{2+} leak provide a powerful means for combating muscle dysfunction in at least some of the ryanodinopathies.

Acknowledgments

This work is supported by a research grant from the National Institutes of Health (AR044657 and AR053349 to R. T. D.) and the Academia Dei Lincei Fund (to L. W.).

References

Arbogast, S., Beuvin, M., et al. (2009). Oxidative stress in SEPN1-related myopathy: From pathophysiology to treatment. *Annals of Neurology, 65*, 677–686.

Avila, G., & Dirksen, R. T. (2001). Functional effects of central core disease mutations in the cytoplasmic region of the skeletal muscle ryanodine receptor. *The Journal of General Physiology, 118*, 277–290.

Avila, G., O'Brien, J. J., et al. (2001). Excitation–contraction uncoupling by a human central core disease mutation in the ryanodine receptor. *Proceedings of the National Academy of Sciences of the United States of America, 98*, 4215–4220.

Avila, G., O'Connell, K. M., et al. (2003). The pore region of the skeletal muscle ryanodine receptor is a primary locus for excitation-contraction uncoupling in central core disease. *The Journal of General Physiology, 121*, 277–286.

Balschun, D., Wolfer, D. P., et al. (1999). Deletion of the ryanodine receptor type 3 (RyR3) impairs forms of synaptic plasticity and spatial learning. *The EMBO Journal, 18*, 5264–5273.

Blayney, L. M., & Lai, F. A. (2009). Ryanodine receptor-mediated arrhythmias and sudden cardiac death. *Pharmacology and Therapeutics, 123*, 151–177.

Boncompagni, S., Rossi, A. E., et al. (2009). Characterization and temporal development of cores in a mouse model of malignant hyperthermia. *Proceedings of the National Academy of Sciences of the United States of America, 106*, 21996–22001.

Capacchione, J. F., & Muldoon, S. M. (2009). The relationship between exertional heat illness, exertional rhabdomyolysis, and malignant hyperthermia. *Anesthesia and Analgesia, 109*, 1065–1069.

Carpenter, D., Ringrose, C., et al. (2009). The role of CACNA1S in predisposition to malignant hyperthermia. *BMC Medical Genetics, 10*, 104.

Censier, K., Urwyler, A., et al. (1998). Intracellular calcium homeostasis in human primary muscle cells from malignant hyperthermia-susceptible and normal individuals. Effect Of overexpression of recombinant wild-type and Arg163Cys mutated ryanodine receptors. *Journal of Clinical Investigation, 101*, 1233–1242.

Cerrone, M., Colombi, B., et al. (2005). Bidirectional ventricular tachycardia and fibrillation elicited in a knock-in mouse model carrier of a mutation in the cardiacryanodine receptor. *Circulation Research, 96*, e77–e82.

Chelu, M. G., Goonasekera, S. A., et al. (2006). Heat- and anesthesia-induced malignant hyperthermia in an RyR1 knock-in mouse. *The FASEB Journal, 20*, 329–330.

Chen, X., Tang, T. S., et al. (2008). Deranged calcium signaling and neurodegeneration in spinocerebellar ataxia type 3. *The Journal of Neuroscience, 28*, 12713–12724.

Cherednichenko, G., Ward, C. W., et al. (2008). Enhanced excitation-coupled calcium entry in myotubes expressing malignant hyperthermia mutation R163C is attenuated by dantrolene. *Molecular Pharmacology, 73*, 1203–1212.

Chopra, N., Kannankeril, P. J., et al. (2007). Modest reductions of cardiac calsequestrin increase sarcoplasmic reticulum Ca^{2+} leak independent of luminal Ca^{2+} and trigger ventricular arrhythmias in mice. *Circulation Research, 101*, 617–626.

Crotti, L., Celano, G., et al. (2008). Congenital long QT syndrome. *Orphanet Journal of Rare Diseases, 3*, 18.

d'Amati, G., Bagattin, A., et al. (2005). Juvenile sudden death in a family with polymorphic ventricular arrhythmias caused by a novel RyR2 gene mutation: Evidence of specific morphological substrates. *Human Pathology, 36*, 761–767.

Dainese, M., Quarta, M., et al. (2009). Anesthetic- and heat-induced sudden death in calsequestrin-1-knockout mice. *The FASEB Journal, 23*, 1710–1720.

Davis, M. R., Haan, E., et al. (2003). Principal mutation hotspot for central core disease and related myopathies in the C-terminal transmembrane region of the RYR1 gene. *Neuromuscular Disorders, 13*, 151–157.

Davis, P. J., & Brandom, B. W. (2009). The association of malignant hyperthermia and unusual disease: When you're hot you're hot or maybe not. *Anesthesia and Analgesia, 109*, 1001–1003.

Denborough, M. (1998). Malignant hyperthermia. *Lancet, 352*, 1131–1136.

Denborough, M. A., & Lovell, R. R. H. (1960). Anaesthetic deaths in a family. *Lancet, 276*, 45.

Dirksen, R. T. (2002). Bi-directional coupling between dihydropyridine receptors and ryanodine receptors. *Frontiers in Bioscience, 7*, d659–d670.

Dirksen, R. T., & Avila, G. (2002). Altered ryanodine receptor function in central core disease: Leaky or uncoupled Ca^{2+} release channels? *Trends in Cardiovascular Medicine, 12*, 189–197.

Dirksen, R. T., & Avila, G. (2004). Distinct effects on Ca^{2+} handling caused by malignant hyperthermia and central core disease mutations in RyR1. *Biophysical Journal, 87*, 3193–3204.

Downey, G. P., Rosenberg, M., et al. (1984). Neuroleptic malignant syndrome. Patient with unique clinical and physiologic features. *American Journal of Medicine, 77*, 338–340.

Ducreux, S., Zorzato, F., et al. (2004). Effect of ryanodine receptor mutations on interleukin-6 release and intracellular calcium homeostasis in human myotubes from malignant hyperthermia-susceptible individuals and patients affected by central core disease. *The Journal of Biological Chemistry, 279*, 43838–43846.

Durham, W. J., Aracena-Parks, P., et al. (2008). RyR1 S-nitrosylation underlies environmental heat stroke and sudden death in Y522S RyR1 knockin mice. *Cell, 133*, 53–65.

Fernandez-Velasco, M., Rueda, A., et al. (2009). Increased Ca^{2+} sensitivity of the ryanodine receptor mutant RyR2R4496C underlies catecholaminergic polymorphic ventricular tachycardia. *Circulation Research, 104*, 201–209 (212p following 209).

Ferreiro, A., Monnier, N., et al. (2002). A recessive form of central core disease, transiently presenting as multi-minicore disease, is associated with a homozygous mutation in the ryanodine receptor type 1 gene. *Annals of Neurology, 51*, 750–759.

Fujii, J., Otsu, K., et al. (1991). Identification of a mutation in porcine ryanodine receptor associated with malignant hyperthermia. *Science, 253*, 448–451.

Gao, L., Balshaw, D., et al. (2000). Evidence for a role of the lumenal M3-M4 loop in skeletal muscle Ca^{2+} release channel (ryanodine receptor) activity and conductance. *Biophysical Journal, 79*, 828–840.

George, C. H., Higgs, G. V., et al. (2003). Ryanodine receptor mutations associated with stress-induced ventricular tachycardia mediate increased calcium release in stimulated cardiomyocytes. *Circulation Research, 93*, 531–540.

George, C. H., Jundi, H., et al. (2006). Arrhythmogenic mutation-linked defects in ryanodine receptor autoregulation reveal a novel mechanism of Ca^{2+} release channel dysfunction. *Circulation Research, 98*, 88–97.

Gillard, E. F., Otsu, K., et al. (1991). A substitution of cysteine for arginine 614 in the ryanodine receptor is potentially causative of human malignant hyperthermia. *Genomics, 11*, 751–755.

Girard, T., Cavagna, D., et al. (2001). B-lymphocytes from malignant hyperthermia-susceptible patients have an increased sensitivity to skeletal muscle ryanodine receptor activators. *The Journal of Biological Chemistry, 276*, 48077–48082.

Goddard, C. A., Ghais, N. S., et al. (2008). Physiological consequences of the P2328S mutation in the ryanodine receptor (RyR2) gene in genetically modified murine hearts. *Acta physiologica (Oxford, England), 194*, 123–140.

Halsall, P. J., Bridges, L. R., et al. (1996). Should patients with central core disease be screened for malignant hyperthermia? *Journal of Neurology, Neurosurgery & Psychiatry, 61*, 119–121.

Harrison, G. G. (1975). Control of the malignant hyperpyrexic syndrome in MHS swine by dantrolene sodium. *British Journal of Anaesthesia, 47*, 62–65.

Ikemoto, N., & Yamamoto, T. (2000). Postulated role of inter-domain interaction within the ryanodine receptor in Ca^{2+} channel regulation. *Trends in Cardiovascular Medicine, 10*, 310–316.

Isaacs, H., Heffron, J. J., et al. (1975). Central core disease. A correlated genetic, histochemical, ultramicroscopic, and biochemical study. *Journal of Neurology. Neurosurgery and Psychiatry, 38*, 1177–1186.

Jiang, D., Chen, W., et al. (2008). Reduced threshold for luminal Ca^{2+} activation of RyR1 underlies a causal mechanism of porcine malignant hyperthermia. *The Journal of Biological Chemistry, 283*, 20813–20820.

Jiang, D., Xiao, B., et al. (2002a). Enhanced basal activity of a cardiac Ca^{2+} release channel (ryanodine receptor) mutant associated with ventricular tachycardia and sudden death. *Circulation Research, 91*, 218–225.

Jiang, M. T., Lokuta, A. J., et al. (2002b). Abnormal Ca^{2+} release, but normal ryanodine receptors, in canine and human heart failure. *Circulation Research, 91*, 1015–1022.

Jiang, D., Xiao, B., et al. (2004). RyR2 mutations linked to ventricular tachycardia and sudden death reduce the threshold for store-overload-induced Ca^{2+} release (SOICR). *Proceedings of the National Academy of Sciences of the United States of America, 101*, 13062–13067.

Jiang, D., Wang, R., et al. (2005). Enhanced store overload-induced Ca2+ release and channel sensitivity to luminal Ca^{2+} activation are common defects of RyR2 mutations linked to ventricular tachycardia and sudden death. *Circulation Research, 97*, 1173–1181.

Jungbluth, H. (2007). Multi-minicore Disease. *Orphanet Journal of Rare Diseases, 2*, 31.

Jungbluth, H., Zhou, H., et al. (2007). Centronuclear myopathy due to a de novo dominant mutation in theskeletal muscle ryanodine receptor (RYR1)gene. *NeuromuscularDisorders, 17*, 338–345.

Jungbluth, H., Wallgren-Pettersson, C., et al. (2008). Centronuclear (myotubular) myopathy. *Orphanet Journal of Rare Diseases, 3*, 26.

Jurynec, M. J., Xia, R., et al. (2008). Selenoprotein N is required for ryanodine receptor calcium release channel activity in human and zebrafish muscle. *Proceedings of the National Academy of Sciences of the United States of America, 105*, 12485–12490.

Kannankeril, P. J., Mitchell, B. M., et al. (2006). Mice with the R176Q cardiac ryanodine receptor mutation exhibit catecholamine-induced ventricular tachycardia and cardiomyopathy. *Proceedings of the National Academy of Sciences of the United States of America, 103*, 12179–12184.

Kimura, T., Nakamori, M., et al. (2005). Altered mRNA splicing of the skeletal muscle ryanodine receptor and sarcoplasmic/endoplasmic reticulum Ca^{2+}-ATPase in myotonic dystrophy type 1. *Human Molecular Genetics, 14*, 2189–2200.

Knollmann, B. C., Chopra, N., et al. (2006). Casq2 deletion causes sarcoplasmic reticulum volume increase, premature Ca^{2+} release, and catecholaminergic polymorphic ventricular tachycardia. *Journal of Clinical Investigation, 116*, 2510–2520.

Kobayashi, S., Yano, M., et al. (2009). Dantrolene, a therapeutic agent for malignant hyperthermia, markedly improves the function of failing cardiomyocytes by stabilizing interdomain interactions within the ryanodine receptor. *Journal of the American College of Cardiology, 53*, 1993–2005.

Krause, T., Gerbershagen, M. U., et al. (2004). Dantrolene—a review of its pharmacology, therapeutic use and new developments. *Anaesthesia, 59*, 364–373.

Lahat, H., Pras, E., et al. (2001). A Missense Mutation in a Highly Conserved Region of CASQ2 Is Associated with Autosomal Recessive Catecholamine-Induced Polymorphic Ventricular Tachycardia in Bedouin Families from Israel. *American Journal of Human Genetics, 69*, 1378–1384.

Laitinen, P. J., Brown, K. M., et al. (2001). Mutations of the cardiac ryanodine receptor (*RYR2*) gene in familial polymorphic ventricular tachycardia. *Circulation, 103*, 485–490.

Larach, M. G., Gronert, G. A., et al. (2010). Clinical presentation, treatment, and complications of malignant hyperthermia in North America from 1987 to 2006. *Anesthesia and Analgesia, 110*, 498–507.

Larach, M. G., Localio, A. R., et al. (1994). A clinical grading scale to predict malignant hyperthermia susceptibility. *Anesthesiology, 80*, 771–779.

Laver, D. R., Owen, V. J., et al. (1997). Reduced inhibitory effect of Mg^{2+} on ryanodine receptor-Ca^{2+} release channels in malignant hyperthermia. *Biophysical Journal, 73*, 1913–1924.
Leenhardt, A., Lucet, V., et al. (1995). Catecholaminergic polymorphic ventricular tachycardia in children. A 7-year follow-up of 21 patients. *Circulation, 91*, 1512–1519.
Lehnart, S. E., Mongillo, M., et al. (2008). Leaky Ca^{2+} release channel/ryanodine receptor 2 causes seizures and sudden cardiac death in mice. *The Journal of Clinical Investigation, 118*, 2230–2245.
Lehnart, S. E., Wehrens, X. H., et al. (2004a). Cardiac ryanodine receptor function and regulation in heart disease. *Annals of the New York Academy of Sciences, 1015*, 144–159.
Lehnart, S. E., Wehrens, X. H. T., et al. (2004b). Sudden Death in Familial Polymorphic Ventricular Tachycardia Associated With Calcium Release Channel (Ryanodine Receptor) Leak. *Circulation 109*, 3208–3214.
Liu, J., Tang, T. S., et al. (2009). Deranged calcium signaling and neurodegeneration in spinocerebellar ataxia type 2. *The Journal of Neuroscience, 29*, 9148–9162.
Liu, N., Colombi, B., et al. (2006). Arrhythmogenesis in catecholaminergic polymorphic ventricular tachycardia: Insights from a RyR2 R4496C knock-in mouse model. *Circulation Research, 99*, 292–298.
Lohn, M., Jessner, W., et al. (2001). Regulation of calcium sparks and spontaneous transient outward currents by RyR3 in arterial vascular smooth muscle cells. *Circulation Research, 89*, 1051–1057.
Loke, J., & MacLennan, D. H. (1998). Malignant hyperthermia and central core disease: Disorders of Ca^{2+} release channels. *The American Journal of Medicine, 104*, 470–486.
Lynch, P. J., Tong, J., et al. (1999). A mutation in the transmembrane/luminal domain of the ryanodine receptor is associated with abnormal Ca^{2+} release channel function and severe central core disease. *Proceedings of the National Academy of Sciences of the United States of America, 96*, 4164–4169.
Magee, K. R., & Shy, G. M. (1956). A new congenital non-progressive myopathy. *Brain, 79*, 610–621.
Marx, S. O., Reiken, S., et al. (2000). PKA phosphorylation dissociates FKBP12.6 from the calcium release channel (ryanodine receptor): Defective regulation in failing hearts. *Cell, 101*, 365–376.
Mickelson, J. R., & Louis, C. F. (1996). Malignant hyperthermia: Excitation-contraction coupling, Ca^{2+} release channel, and cell Ca^{2+} regulation defects. *Physiological Reviews, 76*, 537–592.
Moghadaszadeh, B., Aracena–Parks, P., et al. (2007). SEPN1-related myopathy: A defect in redox regulation. In 12th International Congress of the World Muscle Society, Giardini Naxos-Taormina, Italy, 17–20 October 2007. *Neuromuscular Disorders, 17*, 899.
Moghadaszadeh, B., Petit, N., et al. (2001). Mutations in *SEPN1* cause congenital muscular dystrophy with spinal rigidity and restrictive respiratory syndrome. *Nature Genetics, 29*, 17–18.
Monnier, N., Procaccio, V., et al. (1997). Malignant-hyperthermia susceptibility is associated with a mutation of the alpha 1-subunit of the human dihydropyridine-sensitive L-type voltage-dependent calcium-channel receptor in skeletal muscle. *American Journal of Human Genetics, 60*, 1316–1325.
Monnier, N., Romero, N. B., et al. (2000). An autosomal dominant congenital myopathy with cores and rods is associated with a neomutation in the RYR1 gene encoding the skeletal muscle ryanodine receptor. *Human Molecular Genetics, 9*, 2599–2608.

Monnier, N., Romero, N. B., et al. (2001). Familial and sporadic forms of central core disease are associated with mutations in the C-terminal domain of the skeletal muscle ryanodine receptor. *Human Molecular Genetics, 10*, 2581–2592.

Monnier, N., Ferreiro, A., et al. (2003). A homozygous splicing mutation causing a depletion of skeletal muscle RYR1 is associated with multi-minicore disease congenital myopathy with ophthalmoplegia. *Human Molecular Genetics, 12*, 1171–1178.

Monteforte, N., & Priori, S. G. (2009). The long QT syndrome and catecholaminergic polymorphic ventricular tachycardia. *Pacing and Clinical Electrophysiology, 32*(Suppl 2), S52–S57.

Nakajima, T., Kaneko, Y., et al. (1997). The mechanism of catecholaminergic polymorphic ventricular tachycardia may be triggered activity due to delayed afterdepolarization. *European Heart Journal, 18*, 530–531.

Nelson, M. T., Cheng, H., et al. (1995). Relaxation of arterial smooth muscle by calcium sparks. *Science, 270*, 633–637.

Noguchi, N., Yoshikawa, T., et al. (2008). FKBP12.6 disruption impairs glucose-induced insulin secretion. *Biochemical and Biophysical Research Communications, 371*, 735–740.

Oda, T., Yano, M., et al. (2005). Defective regulation of interdomain interactions within the ryanodine receptor plays a key role in the pathogenesis of heart failure. *Circulation, 111*, 3400–3410.

Owen, V. J., Taske, N. L., et al. (1997). Reduced Mg^{2+} inhibition of Ca^{2+} release in muscle fibers of pigs susceptible to malignant hyperthermia. *The American Journal of Physiology, 272*, C203–C211.

Priori, S. G., & Napolitano, C. (2005). Cardiac and skeletal muscle disorders caused by mutations in the intracellular Ca^{2+} release channels. *The Journal of Clinical Investigation, 115*, 2033–2038.

Priori, S. G., Napolitano, C., et al. (2001). Mutations in the cardiac ryanodine receptor gene (hRyR2) underlie catecholaminergic polymorphic ventricular tachycardia. *Circulation, 103*, 196–200.

Rampazzo, A., Nava, A., et al. (1995). A new locus for arrhythmogenic right ventricular cardiomyopathy (ARVD2) maps to chromosome 1q42-q43. *Human Molecular Genetics, 4*, 2151–2154.

Robinson, R., Carpenter, D., et al. (2006). Mutations in RYR1 in malignant hyperthermia and central core disease. *Human Mutation, 27*, 977–989.

Rosenberg, H., Sambuughin, N., et al. (2010). *Malignant Hyperthermia Susceptibility In GeneReviews at GeneTests: Medical Genetics Information Resource [database online]. Copyright, University of Washington, Seattle, 1997-2010.* Available at: http://www.genetests.org.

Sanoudou, D., & Beggs, A. H. (2001). Clinical and genetic heterogeneity in nemaline myopathy—A disease of skeletal muscle thin filaments. *Trends in Molecular Medicine, 7*, 362–368.

Sato, T., Nishio, H., et al. (2010). Postmortem molecular screening for mutations in ryanodine receptor type 1 (RYR1) gene in psychiatric patients suspected of having died of neuroleptic malignant syndrome. *Forensic Science International, 194*, 77–79.

Sewry, C. A., Muller, C., et al. (2002). The spectrum of pathology in central core disease. *Neuromuscular Disorders, 12*, 930–938.

Shuaib, A., Paasuke, R. T., et al. (1987). Central core disease. Clinical features in 13 patients. *Medicine (Baltimore), 66*, 389–396.

Swan, H., Piippo, K., et al. (1999). Arrhythmic disorder mapped to chromosome 1q42-q43 causes malignant polymorphic ventricular tachycardia in structurally normal hearts. *Journal of the American College of Cardiology, 34*, 2035–2042.

Tateishi, H., Yano, M., et al. (2009). Defective domain-domain interactions within the ryanodine receptor as a critical cause of diastolic Ca^{2+} leak in failing hearts. *Cardiovascular Research, 81*, 536–545.

Tilgen, N., Zorzato, F., et al. (2001). Identification of four novel mutations in the C-terminal membrane spanning domain of the ryanodine receptor 1: Association with central core disease and alteration of calcium homeostasis. *Human Molecular Genetics, 10*, 2879–2887.

Tiso, N., Stephan, D. A., et al. (2001). Identification of mutations in the cardiac ryanodine receptor gene in families affected with arrhythmogenic right ventricular cardiomyopathy type 2 (ARVD2). *Human Molecular Genetics, 10*, 189–194.

Tobin, J. R., Jason, D. R., et al. (2001). Malignant hyperthermia and apparent heat stroke. *JAMA: Journal of the American Medical Association, 286*, 168–169.

Tong, J., McCarthy, T. V., et al. (1999). Measurement of resting cytosolic Ca^{2+} concentrations and Ca^{2+} store size in HEK-293 cells transfected with malignant hyperthermia or central core disease mutant Ca^{2+} release channels. *The Journal of Biological Chemistry, 274*, 693–702.

Treves, S., Larini, F., et al. (1994). Alteration of intracellular Ca^{2+} transients in COS-7 cells transfected with the cDNA encoding skeletal-muscle ryanodine receptor carrying a mutation associated with malignant hyperthermia. *The Biochemical Journal, 301*(Pt 3), 661–665.

Treves, S., Jungbluth, H., et al. (2008). Congenital muscle disorders with cores: The ryanodine receptor calcium channel paradigm. *Current Opinion in Pharmacology, 8*, 319–326.

Wallgren-Pettersson, C., Jasani, B., et al. (1995). Alpha-actinin in nemaline bodies in congenital nemaline myopathy: Immunological confirmation by light and electron microscopy. *Neuromuscular Disorders, 5*, 93–104.

Watanabe, H., Chopra, N., et al. (2009). Flecainide prevents catecholaminergic polymorphic ventricular tachycardia in mice and humans. *Natural Medicines, 15*, 380–383.

Wehner, M., Rueffert, H., et al. (2002). Increased sensitivity to 4-chloro-m-cresol and caffeine in primary myotubes from malignant hyperthermia susceptible individuals carrying the ryanodine receptor 1 Thr2206Met (C6617T) mutation. *Clinical Genetics, 62*, 135–146.

Wehrens, X. H., Lehnart, S. E., et al. (2003). FKBP12.6 deficiency and defective calcium release channel (ryanodine receptor) function linked to exercise-induced sudden cardiac death. *Cell, 113*, 829–840.

Wehrens, X. H., Lehnart, S. E., et al. (2005). Enhancing calstabin binding to ryanodine receptors improves cardiac and skeletal muscle function in heart failure. *Proceedings of the National Academy of Sciences of the United States of America, 102*, 9607–9612.

Weiss, R. G., O'Connell, K. M., et al. (2004). Functional analysis of the R1086H malignant hyperthermia mutation in the DHPR reveals an unexpected influence of the III-IV loop on skeletal muscle EC coupling. *American Journal of Physiology. Cell Physiology, 287*, C1094–C1102.

Wu, S., Ibarra, M. C., et al. (2006). Central core disease is due to RYR1 mutations in more than 90% of patients. *Brain, 129*, 1470–1480.

Xiao, B., Jiang, M. T., et al. (2005). Characterization of a novel PKA phosphorylation site, serine-2030, reveals no PKA hyperphosphorylation of the cardiac ryanodine receptor in canine heart failure. *Circulation Research, 96*, 847–855.

Xu, L., Wang, Y., et al. (2008). Single channel properties of heterotetrameric mutant RyR1 ion channels linked to core myopathies. *The Journal of Biological Chemistry, 283*, 6321–6329.

Yang, T., Ta, T. A., et al. (2003). Functional defects in six ryanodine receptor isoform-1 (RyR1) mutations associated with malignant hyperthermia and their impact on skeletal excitation-contraction coupling. *The Journal of Biological Chemistry, 278*, 25722–25730.

Yang, T., Riehl, J., et al. (2006). Pharmacologic and functional characterization of malignant hyperthermia in the R163C RyR1 knock-in mouse. *Anesthesiology, 105*, 1164–1175.

Yano, M., Kobayashi, S., et al. (2003). FKBP12.6-mediated stabilization of calcium-release channel (ryanodine receptor) as a novel therapeutic strategy against heart failure. *Circulation, 107*, 477–484.

Yano, M., Yamamoto, T., et al. (2006). Mechanisms of disease: Ryanodine receptor defects in heart failure and fatal arrhythmia. *Nature Clinical Practice. Cardiovascular Medicine, 3*, 43–52.

Zhao, M., Li, P., et al. (1999). Molecular identification of the ryanodine receptor pore-forming segment. *The Journal of Biological Chemistry, 274*, 25971–25974.

Zhao, X., Weisleder, N., et al. (2006). Azumolene inhibits a component of store-operated calcium entry coupled to the skeletal muscle ryanodine receptor. *The Journal of Biological Chemistry, 281*, 33477–33486.

Zhou, H., Brockington, M., et al. (2006a). Epigenetic allele silencing unveils recessive RYR1 mutations in core myopathies. *American Journal of Human Genetics, 79*, 859–868.

Zhou, H., Yamaguchi, N., et al. (2006b). Characterization of recessive RYR1 mutations in core myopathies. *Human Molecular Genetics, 15*, 2791–2803.

Zvaritch, E., Depreux, F., et al. (2007). An Ryr1I4895T mutation abolishes Ca^{2+} release channel function and delays development in homozygous offspring of a mutant mouse line. *Proceedings of the National Academy of Sciences of the United States of America, 104*, 18537–18542.

Zvaritch, E., Kraeva, N., et al. (2009). Ca^{2+} dysregulation in RyR1(I4895T/wt) mice causes congenital myopathy with progressive formation of minicores, cores, and nemaline rods. *Proceedings of the National Academy of Sciences of the United States of America, 106*, 21813–21818.

SECTION 2

IP_3R Ca^{2+} RELEASE CHANNELS

CHAPTER 8

3D Structure of IP_3 Receptor☆

Irina I. Serysheva* and Steven J. Ludtke†

*Department of Biochemistry and Molecular Biology, The University of Texas Medical School, Houston, Texas, USA
†National Center for Macromolecular Imaging, Verna and Marrs McLean Department of Biochemistry and Molecular Biology, Baylor College of Medicine, Houston, Texas, USA

I. OVERVIEW

Inositol 1,4,5-trisphosphate receptors (IP_3Rs) are members of the intracellular Ca^{2+} release channel family that are essential to a wide array of cellular processes, relying on transient changes in cytoplasmic (CY) Ca^{2+} concentration. Understanding how these channels function is an important frontier of structural biology. In this chapter, we highlight recent developments in the structure determination of IP_3R channel. To date, several three-dimensional

☆The amino acid sequence of mouse IP_3R1 (sequence accession number P1181) is cited as a reference in this review.

Current Topics in Membranes, Volume 66 1063-5823/10 $35.00
Copyright 2010, Elsevier Inc. All right reserved.
DOI: 10.1016/S1063-5823(10)66008-5

(3D) structures of the entire IP_3R channel at low resolution have been determined by single-particle reconstruction. However, the overall shape and dimensions of the tetrameric channel remain controversial. The only atomic structures available are for the IP_3-binding and ligand-binding suppressor regions comprising ~600 residues at the N-terminus of the receptor protein. These structures advanced our understanding of the IP_3R protein interaction with its primary native ligand, IP_3. Nevertheless, the molecular mechanism underlying IP_3R channel gating is poorly understood, since the 3D architecture of the entire IP_3R1 channel is yet to be determined at high resolution. We discuss future prospects for achieving high-resolution structure of the IP_3R channel.

II. INTRODUCTION

IP_3Rs are intracellular ligand-gated Ca^{2+} release channels occurring in the endoplasmic reticulum (ER) membrane of virtually all eukaryotic cells. IP_3 is the primary IP_3R ligand, which synergizes with Ca^{2+} to promote IP_3 channel opening. Numerous stimuli, such as hormones, growth factors, neurotransmitters, odorants, and light, lead to the generation of IP_3 as a result of phosphoinositide metabolism *in vivo*. Opening of IP_3Rs, in response to IP_3 binding, allows Ca^{2+} to be released into the cytoplasm of cells, an essential event in a wide range of cellular responses including proliferation, neurotransmitter release, fertilization, secretion, gene transcription, and apoptosis. Extensive information thus far available on the function of IP_3Rs and their interaction with the modulators (reviewed in Choe & Ehrlich, 2006; Foskett, White, Cheung, & Mak, 2007; Mikoshiba, 2007) establishes a complex regulation of IP_3R channels. However, the molecular mechanisms underlying their gating by IP_3 and Ca^{2+} are still poorly understood, largely due to the lack of high-resolution structure of the channel protein. The focus of this review is on the progress in the structure determination of IP_3Rs and on functional implications of structural information available to-date.

Three distinct types of IP_3R (type 1–3), encoded by three different genes, are expressed in mammals. These isoforms share ~70% sequence identity. Functional IP_3R channels are formed by these three types of protein monomers that can be assembled as homo- and heterotetrameric complexes, which differ in terms of their tissue distribution and channel modulation. Individual cell types can express more than one isoform; type-1 IP_3R (IP_3R1) is the predominant isoform in the ER of cerebellar Purkinje cells where it forms largely homotetramers. Consequently, the cerebellum is generally used as a primary source for purification of the IP_3R1. The mammalian IP_3R1 has been the subject of the most detailed analysis of structure–function. The structure of a whole channel has been characterized by electron microscopy, and two X-ray structures of its N-terminal region have been obtained. These studies are discussed in this review.

III. PREDICTED TOPOLOGY OF IP_3R MOLECULE

IP_3Rs are exceptionally large macromolecular complexes of ~1.3 MDa comprising four monomers of ~2700 residues each (Furuichi et al., 1989; Mignery, Sudhof, Takei, & De Camilli, 1989). Based on extensive biochemical, mutagenic and functional studies of structure–function relationships within IP_3Rs, three functional regions have been established in the primary structure of these proteins (Fig. 1): an N-terminal region that binds IP_3 (~600 residues); a C-terminal, channel-forming region (~400 residues); and a central modulatory region (~1700 residues; Mignery & Sudhof, 1990; Miyawaki et al., 1991).

Hydropathy analysis predicts six transmembrane (TM) helices in the C-terminal region of IP_3R (Mignery et al., 1990; Yoshikawa et al., 1992). The N- and C-termini of IP_3R are both intracellular and form a soluble CY portion of the channel protein that includes ~90% of the protein mass. The region responsible for binding of IP_3 is mapped to N-terminal residues 224–579 (IP_3-binding core region) in the protein sequence (Fig. 1; Mignery & Sudhof, 1990; Yoshikawa et al., 1996). The first 223 residues are identified as the IP_3-binding suppressor region, and deletion of these regions results in significant enhancement of IP_3-binding by the receptor protein (Yoshikawa et al., 1996). The activity of IP_3R Ca^{2+} release channel is regulated by an array of small molecules and proteins that bind to the central region of the channel protein that is named modulatory and transducing region (Uchida et al., 2003). The region comprising residues 2276–2589 forms the ion conduction pore in IP_3R (Mignery et al., 1990; Yoshikawa et al., 1992). Because the TM domain arrangement of Ca^{2+} release channels is homologous to that of K^+ channels (Welch, Rheault, West, & Williams, 2004; Williams, West, & Sitsapesan, 2001), similar structural elements in the ion conduction pore are proposed for IP_3R channels (reviewed in Chapter 9). There is some evidence suggesting that the last ~160 residues in the C-terminus might also be responsible for keeping the channel closed ("gatekeeper region"; Uchida et al., 2003). The C- and N-terminal regions also bind a number of intracellular molecules involved in the regulation of channel function (reviewed in Bosanac, Michikawa, Mikoshiba, & Ikura, 2004; Foskett et al., 2007; Mikoshiba, 2007).

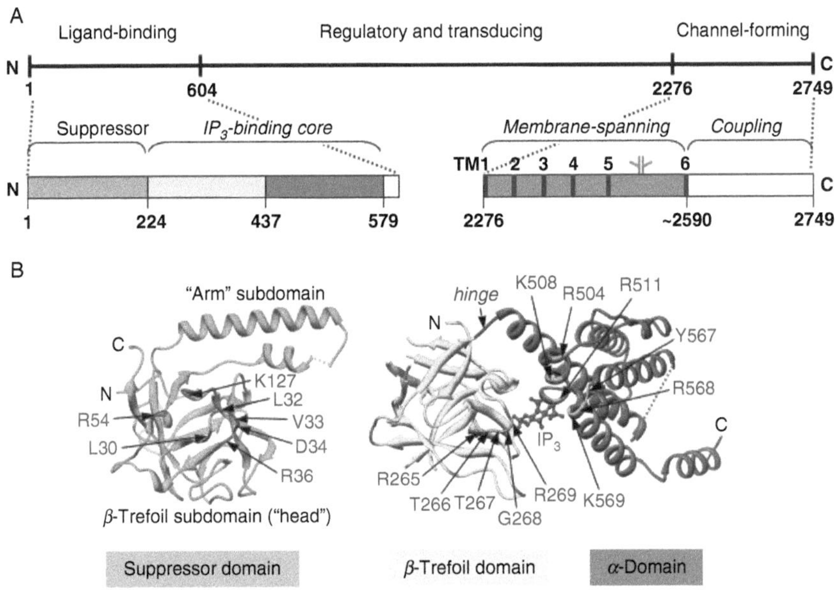

FIGURE 1 (A) Topology of functional regions in the primary sequence of IP_3R1. Upper panel shows a linear sequence of IP_3R1 with three putative functional regions: N-terminal ligand binding, central regulatory, and C-terminal channel forming (Mignery & Sudhof, 1990; Yoshikawa et al., 1996). Detailed arrangements of the N- and C-terminal regions are shown in the lower panel: the N-terminal region consists of the suppressor and the IP_3-binding core domains; the C-terminal portion of the receptor molecule comprises the membrane-spanning region with the six transmembrane (TM) segments depictured with numbered verticals bars (1 through 6) and the coupling domain, also known as the gatekeeper domain (Mignery et al., 1990; Uchida et al., 2003; Yoshikawa et al., 1992). The two *N*-glycosylation sites (residues N2474 and N2502) are identified in the TM region (green Y-shaped marks; Michikawa et al., 1994). (B) Crystal structures of the IP_3-binding suppressor domain (PDB ID: IXZZ; Bosanac et al., 2005) and the IP_3-binding core bound with IP_3 (PDB ID: 1N4K; Bosanac et al., 2002) are shown as ribbon diagrams. Conserved residues in the suppressor domain implicated in the suppression of IP_3 binding and in the interaction with the IP_3-binding core domains are shown with red and labeled. Residues highlighted in the IP_3-binding core domains highlighted with blue, form the positively charged IP_3-binding pocket. (See Color Insert.)

Based on the information from studies of structure–function relationships within IP_3Rs, it has been proposed that these channels are organized as scaffold membrane proteins, comprising multiple interaction and regulatory motifs with distinctive features that enable to convert multiple cellular signals into Ca^{2+} signals. However, the structural basis for such complex functionality of IP_3R channels remains to be established by direct structural analysis.

IV. ARRANGEMENT OF IP_3R IN THE NATIVE MEMBRANE

IP_3 receptor is widespread through the ER membrane of the cerebellar Purkinje cells, where it is expressed at very high level, making this tissue the best source for the purification of the receptor protein. The cerebellum IP_3R is present at particular high concentration in membranous stacks of ER cisternae (named "citernae stack"; Katayama et al., 1996; Takei, Mignery, Mugnaini, Sudhof, & De Camilli, 1994). These stacks have been noted in the cytoplasm of Purkinje cells since early studies of the ultrastructure of neurons (Herndon, 1963). Within cisternae stacks, IP_3Rs appear as electron-dense particles bridging space between adjacent cisternae membranes (Fig. 2). In some areas, these particles form 2D arrays similar to arrays of the foot structure on the junctional SR membrane in muscle (reviewed in Chapter 1). Freeze-fracture deep-etch visualization of Pukinje cells reveals that these arrays are formed by square-shaped compact particles with dimensions of 12 nm on a side (Katayama et al., 1996). This estimated size of IP_3R is consistent with results from earlier EM studies using negatively stained thin-sections of COS cells overexpressing IP_3R (Takei et al., 1994).

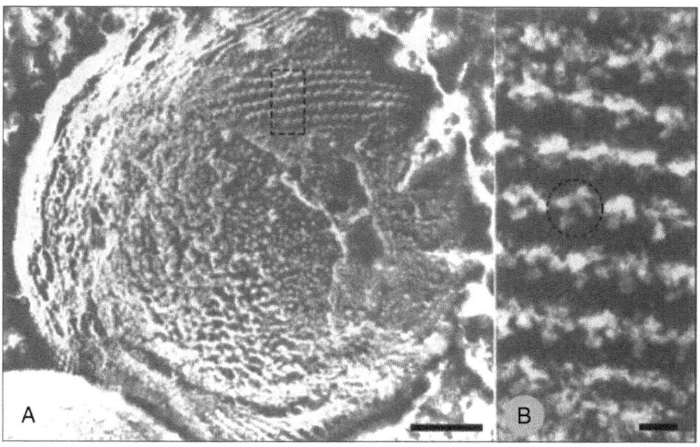

FIGURE 2 (A) Electron micrograph of a quick-freeze deep-etch replica of large globular vesicle visualized in Purkinje cell. Parallel striations formed by regularly spaced electron-dense particles of IP_3R are noticeable on the surface of the vesicle (Katayama et al., 1996; Takei et al., 1994). Scale bar is 1000 Å. (B) Closer view of the striation pattern similar to that indicated with a rectangle in (A). One putative IP_3R particle is indicated with a circle. Scale bar is 100 Å. Reproduced and adapted by permission from Macmillan Publishers Ltd. (Katayama et al., 1996).

V. 3D STRUCTURE OF IP_3R BY ELECTRON MICROSCOPY

Structural studies of Ca^{2+} release channels have been hampered by both their size and their interaction with membrane lipids in their native conformation. X-ray crystallography and nuclear magnetic resonance (NMR) spectroscopy are poorly suited methodologies to study massive integral membrane proteins due to this double complexity. Single-particle electron cryomicroscopy (cryo-EM) thus emerges as the most feasible way to study not only this kind of protein complexes, but also to choose conditions favoring a particular functional state of the subject protein.

A. Introduction to Single-Particle Reconstruction

Single-particle cryo-EM is a powerful technique for studying biological molecules/macromolecules under near-native conditions (Chiu et al., 1986; Dubochet et al., 1988; Serysheva, Chiu, & Ludtke, 2007). In this technique, the molecules to be studied are purified to homogeneity in an aqueous buffer. A few microliters of the molecule-containing buffer is then placed on the surface of a prepared cryo-EM grid, which is then blotted with filter paper, leaving behind a thin film of molecule bearing solution spanning the holes in the grid. This grid is then vitrified in liquid ethane, preserving the proteins in a snapshot of their solution conformation without introduction of stains or other perturbing elements. These particles are then in nearly random orientations in the cryo-EM images, each representing a 2D projection of the 3D structure. Through use of sophisticated image processing techniques, tens of thousands of such images are combined to produce a 3D structure from the ostensibly identical objects being imaged.

While this technique has used for decades, it is only over the last 3–4 years that it became possible to achieve the resolutions approaching those produced by X-ray crystallography. Unlike crystallography, this technique has no requirement for crystal growth, making it a favored method for studying fragile complexes and small quantities of purified proteins with no expression system. The last 2 years, have seen the publication of several structures at better than 5 Å resolution (Cheng et al., 2010; Cong et al., 2010; Jiang et al., 2008; Ludtke et al., 2008; Yu, Jin, & Zhou, 2008; Zhang et al., 2008, 2010), which demonstrates the veracity and capabilities of this method. In addition to high-resolution structures, variants of this technique are able to assess conformational flexibility of assemblies in solution and study mixtures where only a fraction of the molecules are in the desired state, such as partial ligand binding (Chen, Song, Chuang, Chiu, & Ludtke, 2006; Frank & Agrawal, 2000; Orlova, Serysheva, van Heel, Hamilton, & Chiu, 1996; Serysheva, Schatz, van Heel, Chiu, & Hamilton, 1999; Zhang et al., 2010).

B. 3D Structure of IP_3R1

Early EM studies of IP_3R1 channel using negative staining of purified receptors (Chadwick, Saito, & Fleischer, 1990; Hamada, Miyata, Mayanagi, Hirota, & Mikoshiba, 2002; Maeda, Niinobe, & Mikoshiba, 1990) revealed that, as viewed from the cytoplasm, the IP_3R channel exhibits a fourfold symmetry and has either a square-shaped or a pinwheel-like structure with a corresponding size of 190–250 Å. Subsequently, five 3D structures of IP_3R1 at 20–40 Å resolution have been reported based on single-particle reconstruction from specimens prepared by ice embedding (Jiang, Thrower, Chester, Ehrlich, & Sigworth, 2002; Sato et al., 2004; Serysheva et al., 2003) and negative staining (da Fonseca, Morris, Nerou, Taylor, & Morris, 2003; Hamada, Terauchi, & Mikoshiba, 2003). All of these 3D reconstructions are consistent in respect to the overall arrangement of the channel tetramer: it comprises two major regions that, based on the predicted topology of IP_3R molecule, represent the large CY and TM regions (Figs. 3 and 4). However, despite the fact that in these studies the channel protein has been imaged under similar buffer conditions and in the absence of primary channel agonists (i.e. IP_3 and Ca^{2+}) assuming the closed state of IP_3R, the resulting structures are strikingly different at the detailed level.

Jiang et al. (2002) reported the first 3D cryo-EM density map of IP_3R1 purified from mouse cerebellum (Fig. 3A). This structure appeared as an uneven dumbbell with a height of $\sim$170 Å and lateral dimensions of 155 × 155 Å for the bulky CY region and 100 × 100 Å for the TM region. The structure is strikingly compact compared with structures from other studies or with the structure of RyR (reviewed in Chapter 2). The majority of densities makes up a central core of the CY region with four laterally protruding densities ("arms"). The TM densities are splayed toward the lumen and form a pronounce indentation along the fourfold axis that contrasts to the TM structure of RyRs.

On other hand, the 3D structure of IP_3R1 from another single-particle cryo-EM study reveals almost opposite architecture of the receptor, which was immunoaffinity purified from bovine cerebellum (Serysheva et al., 2003). The CY region in this structure exhibits a "pinwheel" shape with dimensions of 250 × 250 Å, while the TM region is square-shaped with dimensions of 120 × 120 Å (Fig. 3B). The CY and TM regions are connected via column densities making up the entire channel structure of $\sim$170 Å in height. The CY region comprises four curved spokes connected through bridging densities with a central core region. While the overall shape of this cryo-EM map resembles that from studies performed with the negative staining EM (da Fonseca et al., 2003; Hamada et al., 2003), the dimensions of the CY region are significantly larger. In addition, the "pinwheel" conformation of the CY

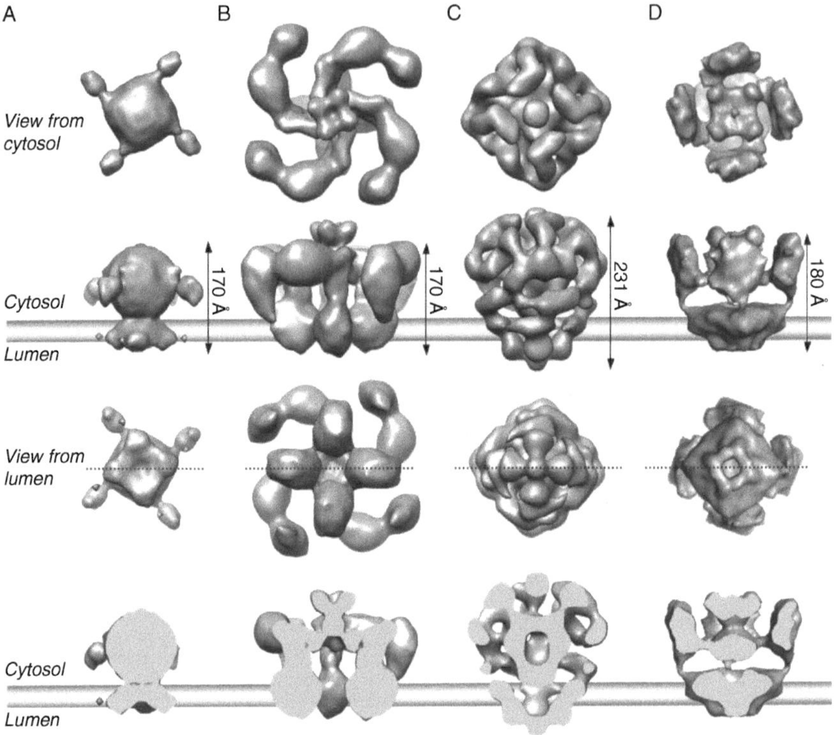

FIGURE 3 Surface representations of 3D structures of IP_3R1 as determined by single-particle reconstruction. (A–C) 3D structures determined in cryo-EM studies by Jiang et al. (2002), Serysheva et al. (2003), and Sato et al. (2004), respectively; (D) 3D structure determined by negative-staining EM by da Fonseca et al. (2003). Structure determination in all studies was performed in the absence of primary channel coagonists, Ca^{2+} and IP_3. Structures are shown in three characteristic views: the top (viewed from the cytoplasm), the side, and the bottom (viewed from the ER lumen). These maps represent true volumetric reconstructions, and each author independently determined the best contour level for optimal visualization of their map. We used these author-defined contours for display. The internal architecture of IP_3R1 is shown in the central cross-sections through 3D reconstructions (lower panel). (See Color Insert.)

region is not consistent with Ca^{2+} depleting conditions favoring a square-shape conformation of the CY region in other studies (da Fonseca et al., 2003; Hamada et al., 2002, 2003). Moreover, "bean-like" densities comprising the TM structure are separated by a large central opening, although in the absence of channel agonists the TM region is expected to be in a closed conformation, thus assuming more compact appearance of the TM region along its fourfold axis where the ion conduction pathway is presumably located.

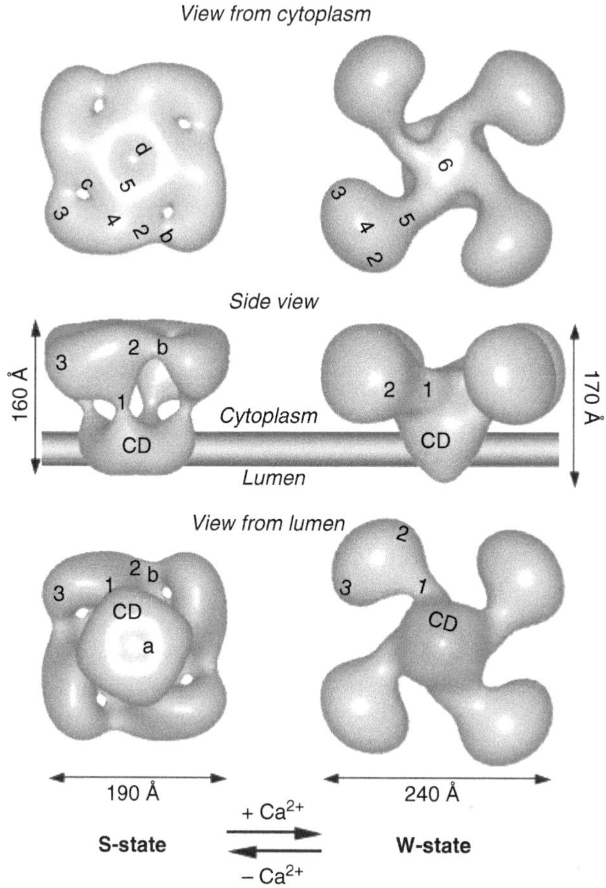

FIGURE 4 Surface representations of 3D structures of IP_3R1 determined in the absence (left column) and presence of Ca^{2+} (right column) by single-particle negative-staining EM at 34 and 43 Å resolutions, respectively (Hamada et al., 2003). Structures are viewed from the cytoplasm, side, and from the lumen. Subregions and hollows are arbitrarily labeled by numbers (1–6) and letters (a–d), respectively; CD is a putative TM region. Ca^{2+}-induced large-scale conformational changes might correlate with a functional channel transition from the S-state (square-shaped structure) to the W-state (windmill-like structure). Reproduced with some modifications from Hamada et al. (2003), by copyright permission of The American Society of Biochemistry and Molecular Biology.

The highest resolution (~20 Å based on the 0.5 cut-off FSC criterion) cryo-EM structure of IP_3R1 from mouse cerebellum was reported by Sato et al. (2004). This structure has a shape of a "hot air balloon" with the height of ~231 Å (Fig. 3C). It does not obviously exhibit the CY and TM regions connected via column densities like that in RyR (reviewed in Chapter 2) or via slender densities like in crystal structures of tetrameric K^+ channels (Jiang et al., 2002; Kuo et al., 2003; Long, Campbell, & Mackinnon, 2005; Tao, Avalos, Chen, & MacKinnon, 2009). Instead cryo-EM densities in this 3D map form a continuous network with multiple openings. The wider end of the structure is assigned to be the CY region (the "balloon" with a diameter of 175 Å), thus the opposite end represents the TM region that is described as a "basket" with a side length of 96 Å. As seen in the cross-section view of this structure (Fig. 3C, bottom panel), the TM region is quite hollow and exhibits no densities near the channel fourfold axis that would potentially make up an ion conduction pathway spanning a bilayer membrane. The overall shape and dimensions of this structure are noticeably different from the structures based on the negatively stained data from the same group (Hamada et al., 2003) as well as from the structure determined by da Fonseca et al. (2003).

It is noteworthy, that the 3D structures of the purified IP_3R, determined using single-particle negative staining EM, are quite consistent with each other and should likely be considered the most reliable, despite the limited resolution (da Fonseca et al., 2003; Hamada et al., 2003). The structure determined by da Fonseca et al. (2003) has a flower-like appearance and comprises essentially two square-shaped regions connected by slender densities (Fig. 3C): a small region with dimensions of $126 \times 126 \times 70$ Å, corresponding to the TM region, and a large region ($180 \times 180 \times 110$ Å), assigned to be the CY region. The TM region in this structure comprises ~35% of the total channel volume that would significantly exceed the mass predicted for the TM region (~11 % of the protein mass). Given that, authors suggested that the TM portion in their structure is composed of the membrane-spanning domains and also includes a part of the CY modulatory region and the C-terminal coupling domain. In addition, there is a possibility that an excessive mass in the TM region could be attributed to the residual lipid and/or detergent bound to the purified protein. The overall shape and dimensions of this structure are in agreement with the other IP_3R1 structure determined by the negative staining technique in the presence of low Ca^{2+} (Fig. 4; Hamada et al., 2003). The lower level of detail observed by Hamada et al. (2003) is consistent with the lower structure resolution achieved in this study as compared with the formerly mentioned structure (da Fonseca et al., 2003).

C. *Lessons from Cryo-EM Studies of Membrane Proteins*

While single-particle reconstruction technique would seem ideal for integral membrane proteins, and indeed, it is a much better technique than crystallography or NMR for such proteins, there are still difficulties related to the fundamental nature of these structures. Since objects studied using this technique must be monodisperse in solution, generally detergent-solubilized forms are used for imaging. This raises a number of issues with structure determination, and while membrane proteins have reached subnanometer resolution using this technique (Ludtke, Serysheva, Hamilton, & Chiu, 2005; Samso, Wagenknecht, & Allen, 2005), resolutions lag substantially behind soluble proteins.

The primary difficulties are image contrast and the presence of a detergent micelle around the hydrophobic structures of the protein. To maintain membrane proteins in a soluble state, detergent concentrations must typically be very close to CMC, raising the risk of the micelles themselves appearing in the images and obscuring or distorting the target. Even if sufficiently below CMC, the presence of detergent will act to lower the contrast in the images precipitously. The presence of detergent micelles on the surface of the target also causes difficulties, since they will also appear in any 3D reconstructions, and while such micelles are generally highly disordered, and thus do not produce high-resolution details, they can still easily impact the structure of the protein target itself.

Issues surrounding this low image contrast are the main impediment to producing a high-resolution structure of such integral membrane proteins. With typical soluble proteins, experimental requirements are far less stringent, and reasonable contrast can be obtained even when conditions are far from optimal. Ion channels are structurally dynamic complexes whose conformations are controlled by a range of ligands. In addition, integral membrane assemblies such as ion channels represent intricate physical–chemical systems, which properties depend on environmental conditions such as temperature, buffer composition, pH, ionic strength, and presence of contaminants. There are simply too many factors to control to permit the virtrification process to be fully controlled or repeatable, and optimizing conditions requires large numbers of trials. Although recent development of semiautomated computer-controlled vitrification apparatus significantly improved and facilitated the search of optimal conditions, producing images with good contrast remains the key to studies of such inherently difficult specimens.

In general, single-particle reconstruction is a highly reliable technique, a fact borne out as dozens of structures have gradually improved in resolution over the years, and in some cases been later verified via crystal structures, for example (Nakagawa et al., 2003; Zhou et al., 2001). Unfortunately, the various published cryo-EM structures of this channel clearly do not agree well, even as to the overall shape of the assembly. How is this possible? In the very few cases, where single-particle reconstructions have been found to be incorrect, it has been a problem with overinterpretation of marginal data. The issues discussed above combine to make images of membrane proteins exceptionally noisy, even in comparison to other single-particle studies, which are already considered very noisy. Single-particle analysis is susceptible to noise/model bias (Stewart & Grigorieff, 2004), which can be difficult to reliably detect and correct, even for domain experts. In the vast majority of specimens, this is not an issue at all, and image contrast is sufficient to clearly demonstrate that model bias is not an issue. As contrast is reduced, however, model bias cannot be disproved, but it cannot be readily proven either. There is no accepted standard for assessing a minimal acceptable contrast level in the raw data. In marginal cases, it is also quite possible for one portion of the structure to emerge from the reconstruction while other lower contrast regions appear to vanish.

We suspect that this is the current case with the various competing cryo-EM models of IP_3R1 (Jiang, Thrower, et al., 2002; Sato et al., 2004; Serysheva et al., 2003), since all of the images for the published cryo-EM structures have had extremely low contrast, our own included. It also remains possible that the difficulty lies in purification, and that the various specimens being imaged truly are different. Nonetheless, at this point in time, with no further facts, we must assume that the most reliable quaternary structure is the one obtained using negative stain, a process which enhances contrast at the expense of resolution. A truly believable structure at subnanometer resolution will clearly require a breakthrough in specimen imaging conditions, to produce images with contrast such that individual particles can clearly be observed.

VI. CRYSTAL STRUCTURES OF ISOLATED DOMAINS

While determining the high-resolution structure of the entire IP_3R channel remains a major challenge, two atomic resolution structures of the N-terminal ligand-binding region of the mouse IP_3R1 have been determined recently: the IP_3-binding suppressor domain (residues 2–223) and the IP_3-binding core (residues 226–576; Fig. 1). These domains have been expressed as soluble proteins and then crystallized. The IP_3-binding core structure was determined in an IP_3-bound form at 2.2-Å resolution (PDB ID: 1N4K; Bosanac et al., 2002). It includes two domains together forming an asymmetric L-shaped structure (Fig. 1B): the N-terminal domain (residues 225–436) comprises 12 β-sheets (a β-domain), and the C-terminal α-domain (residues 437–576) forms three armadillo repeats (Huber, Nelson, & Weis, 1997). The 11 residues were identified to coordinate IP_3 within a highly positively charged IP_3-binding pocket formed at the interface between these two domains (Bosanac et al., 2002). Two large conserved surfaces, identified on the IP_3-binding core structure, were proposed to create a potential Ca^{2+}-binding site that may be involved in modulation of channel gating (Bosanac et al., 2004; Taylor, da Fonseca, & Morris, 2004).

A 1.8-Å structure of the suppressor domain (PDB ID: 1XZZ) has a hammer-like shape that conforms to a globular β-trefoil head and an arm subdomain including a helix-turn-helix structure (Bosanac et al., 2005). Noteworthy that the β-trefoil suppressor subdomain and the N-terminal β-domain of the IP_3-binding core represent well overlapping structures that was somewhat expected based on earlier sequence-based predictions for this region (Ponting, 2000). Moreover, the N-terminal region of RyR1 (residues 12–565, sequence accession number P11716) is structurally homologous to the N-terminal domains of IP_3R channel (Ponting, 2000; Serysheva et al., 2008). Using comparative modeling and cryo-EM density fitting, three structural folds were generated for this region in RyR1, comprising two β-trefoil domains and the α-domain similar to that in IP_3R (Serysheva et al., 2008). Furthermore, crystal structures of the N-terminal $\sim$200 residues of RyR were determined recently and essentially confirmed its β-trefoil architecture (Amador et al., 2009; Lobo & Van Petegem, 2009). Despite the structural homology between the N-terminal domains of IP_3R and RyR channels, RyR lacks the IP_3-bidning ability (Serysheva et al., 2008), thus significance of this structural similarity is not clear.

Based on mutagenesis analysis combined with crystallographic studies of isolated domains of IP_3R1, it was proposed that the suppressor domain reduces IP_3-binding affinity of the channel protein through its direct interaction with the IP_3-binding core region (Bosanac et al., 2005; Iwai, Michikawa, Bosanac, Ikura, & Mikoshiba, 2007). Seven conserved residues clustered on the surface of the β-subdomain in the suppressor structure were found to be critical for the suppression of IP_3-binding and might form interface for the interaction with the IP_3-binding core region (Fig. 1B; Bosanac et al., 2005; Iwai et al., 2007). Since the binding sites for the suppressor in the IP_3-binding core region are not yet determined, the precise molecular mechanism of the interaction between these two regions remains ambiguous.

VII. CONFORMATIONAL TRANSITIONS IN IP_3R CHANNEL

Given the topology of IP_3R molecule, the critical sites for regulation of IP_3R gating are distant from the channel-forming region in the protein sequence (Fig. 1) suggesting an allosteric mechanism for channel modulation that relies on conformational coupling between the TM and the modulator-binding domains. Some data supporting this hypothesis have been obtained from EM studies of IP_3R1 (Hamada et al., 2002, 2003). Two distinct conformations of the CY region of the IP_3R1 have been found by these studies: one is square-shaped, abundant in samples depleted of Ca^{2+} by EGTA, and the other is pinwheel-shaped with the larger dimensions, predominant in samples obtained in the presence of Ca^{2+} (Fig. 4). Since Ca^{2+} is a coagonist for IP_3R1, it has been proposed that these two conformations may result from channel transitions between distinct functional states (i.e., open and closed) regulated by Ca^{2+}. Nevertheless, the precise nature of these conformational transitions is yet to be determined.

It has been suggested that the ligand-binding region rearranges upon binding IP_3 and Ca^{2+}. This hypothesis is supported by studies employing small-angle X-ray scattering in combination with circular dichroism and NMR (Chan et al., 2007). These studies were performed with the IP_3-binding and suppressor domains expressed as independent entities or linked. It has been shown that the N-terminal region can exist in multiple ligand-dependent conformations due to motions in the hinge regions between two β-trefoil domains in the N-terminus and the α- and β-domains of the IP_3-binding core. Addition of IP_3 promotes more compact conformation of the N-terminal region, while presence of Ca^{2+} favors more extended conformations of the N-terminus in both the IP_3-bound and IP_3-free forms. It remains to be established whether similar rearrangements take place in the full-length protein or whether they correlate with channel opening. Recently, mutagenesis scanning of cysteine residues in the N-terminal region of IP_3R1 showed significant changes in the protein expressed in COS cells as detected by changes in accessibility of surface exposed residues in the presence of Ca^{2+} (Anyatonwu & Joseph, 2009). Thus, this observation is in line with other studies suggesting ligand-induced global conformational changes in the channel structure.

A pivotal question in IP_3R gating is how ligand binding, which occurs in the N-terminal of the channel primary sequence, is transduced to the pore of the channel, located in the C-terminal portion of this sequence. The C-terminal region of IP_3R protein is implicated in the interaction with the N-terminus thus providing direct coupling between the TM- and IP_3-binding regions (Boehning & Joseph, 2000; Nakade, Maeda, & Mikoshiba, 1991; Schug & Joseph, 2006). The lack of the suppressor region (Uchida et al., 2003) or presence a mutation in its C-terminus affects IP_3R channel gating (Srikanth et al., 2004). It is conceivable that some domains in the central modulatory region are also be involved in transducing the signal from the ligand-binding region to the C-terminal channel-forming region (Nakayama et al., 2004; Uchida et al., 2003). It is obvious that understanding the molecular basis of the aforementioned architectural rearrangements in the IP_3R1 requires unambiguous mapping of the ligand-binding domains in the 3D structure of the channel protein. Location of the N-terminal

IP_3-binding region has been assigned to the peripheral domains of IP_3R by using negative-staining EM with gold-heparin and by fitting the crystal structure of the IP_3-binding core into the low-resolution EM-density map (Hamada et al., 2002; Serysheva et al., 2003). On other hand, Sato et al. (2004) mapped the IP_3-binding core closer to the central part of the CY region in their cryo-EM structure of IP_3R1 (Sato et al., 2004). In this study, the crystal structure of the IP_3-bidning core was bent at the linker between the α- and β-domains in order to achieve the best fit into cryo-EM densities. However, results on mapping of the IP_3-binding region in all these studies are compromised by controversy between the 3D structures of IP_3R1 generated in the same studies. Localization with gold-heparin was biased by the large size of heparin-gold, and was performed using 2D image analysis that cannot alone provide a precise location of the region of interest due to the lack of a reliable 3D structure of the whole channel protein (Hamada et al., 2002).

Noteworthy that the N-terminal region of RyR1 is localized into the clamp region of the 3D structure of RyR1 that based on cryo-EM studies undergoes significant conformational changes mediated by channel-specific ligands (Orlova et al., 1996; Samso, Feng, Pessah, & Allen, 2009; Serysheva et al., 1999; Serysheva et al., 2008). In addition, a number of mutations that affect RyR1 channel gating and are linked to malignant hyperthermia and central core disease were mapped to the N-terminal region of RyR1 (reviewed in Chapter 7).

VIII. FUTURE OUTLOOK

Despite the established importance of Ca^{2+}release channels in cell physiology and pathology, we currently lack insight into how any of these channels function at the molecular level, in either native or diseased states, since their 3D architecture has yet to be determined at high resolution. While we can gain some insights by comparison to the related class of K^+ channels, based on quaternary structure, any structural similarities are quite localized. In addition, even at low resolution, published structures of IP_3R1 clearly do not agree well. Prospects for a 3D crystal structure seem poor due to the inherent difficulties in crystallizing large flexible membrane proteins. The primary difficulty with cryo-EM has been and remains overall image contrast and contrast at high resolution. The solution to this problem will have to lie in development of better purification and vitrification protocols for these systems, as, once the contrast falls below some threshold, no existing algorithm can produce a reliable structure. It is also possible that completely different strategies, such as imaging channels reconstituted into membranes, will become the method of choice for these systems, as this is unarguably a more native environment than detergent solubilization. However, many research groups continue to work on this challenging problem and it is only a matter of time before someone devises a successful strategy for producing a high-resolution structure of these important proteins.

Acknowledgments

We thank Fred Sigworth, Qiu-Xing Jiang, Edward Morris, and Paula da Fonseca for their help in preparation of Fig. 3 in this chapter. This work was supported by NIH grants (P41RR02250, R01GM07284, and R01GM080139) and a grant from MDA.

References

Amador, F. J., Liu, S., Ishiyama, N., Plevin, M. J., Wilson, A., MacLennan, D. H., et al. (2009). Crystal structure of type I ryanodine receptor amino-terminal beta-trefoil domain reveals a disease-associated mutation "hot spot" loop. *Proceedings of the National Academy of Sciences of the United States of America, 106*, 11040–11044.

Anyatonwu, G., & Joseph, S. K. (2009). Surface accessibility and conformational changes in the N-terminal domain of type-I inositol trisphosphate receptors: Studies using cysteine substitution mutagenesis. *The Journal of Biological Chemistry, 284*, 8093–8102.

Boehning, D., & Joseph, S. K. (2000). Direct association of ligand-binding and pore domains in homo- and heterotetrameric inositol 1,4,5-trisphosphate receptors. *The EMBO Journal, 19*, 5450–5459.

Bosanac, I., Alattia, J. R., Mal, T. K., Chan, J., Talarico, S., Tong, F. K., et al. (2002). Structure of the inositol 1,4,5-trisphosphate receptor binding core in complex with its ligand. *Nature, 420*, 696–700.

Bosanac, I., Michikawa, T., Mikoshiba, K., & Ikura, M. (2004). Structural insights into the regulatory mechanism of IP3 receptor. *Biochimica et Biophysica Acta, 1742*, 89–102.

Bosanac, I., Yamazaki, H., Matsu-Ura, T., Michikawa, T., Mikoshiba, K., & Ikura, M. (2005). Crystal structure of the ligand binding suppressor domain of type 1 inositol 1,4,5-trisphosphate receptor. *Molecular Cell, 17*, 193–203.

Chadwick, C. C., Saito, A., & Fleischer, S. (1990). Isolation and characteriztion of the inositol triphosphate receptor from smooth muscle. *Proceedings of the National Academy of Sciences of the United States of America, 87*, 2132–2136.

Chan, J., Whitten, A. E., Jeffries, C. M., Bosanac, I., Mal, T. K., Ito, J., et al. (2007). Ligand-induced conformational changes via flexible linkers in the amino-terminal region of the inositol 1,4,5-trisphosphate receptor. *Journal of Molecular Biology, 373*, 1269–1280.

Chen, D. H., Song, J. L., Chuang, D. T., Chiu, W., & Ludtke, S. J. (2006). An expanded conformation of single-ring GroEL-GroES complex encapsulates an 86 kDa substrate. *Structure, 14*, 1711–1722.

Cheng, L., Zhu, J., Hui, W. H., Zhang, X., Honig, B., Fang, Q., et al. (2010). Backbone model of an aquareovirus virion by cryo-electron microscopy and bioinformatics. *Journal of Molecular Biology, 397*, 852–863.

Chiu, W., Downing, K. H., Dubochet, J., Glaeser, R. M., Heide, H. G., Knapek, E., et al. (1986). Cryoprotection in electron microscopy. *Journal de Microscopie, 141*, 385–391.

Choe, C. U., & Ehrlich, B. E. (2006). The inositol 1,4,5-trisphosphate receptor (IP3R) and its regulators: Sometimes good and sometimes bad teamwork. *Science STKE, 2006*, re15.

Cong, Y., Baker, M. L., Jakana, J., Woolford, D., Miller, E. J., Reissmann, S., et al. (2010). 4.0-Å resolution cryo-EM structure of the mammalian chaperonin TRiC/CCT reveals its unique subunit arrangement. *Proceedings of the National Academy of Sciences of the United States of America, 107*, 4967–4972.

da Fonseca, P. C., Morris, S. A., Nerou, E. P., Taylor, C. W., & Morris, E. P. (2003). Domain organization of the type 1 inositol 1,4,5-trisphosphate receptor as revealed by single-particle analysis. *Proceedings of the National Academy of Sciences of the United States of America, 100*, 3936–3941.

Dubochet, J., Adrian, M., Chang, J. J., Homo, J. C., Lepault, J., McDowall, A. W., et al. (1988). Cryo-electron microscopy of vitrified specimens. *Quarterly Reviews of Biophysics, 21*, 129–228.

Foskett, J. K., White, C., Cheung, K. H., & Mak, D. O. (2007). Inositol trisphosphate receptor Ca^{2+} release channels. *Physiological Reviews, 87*, 593–658.

Frank, J., & Agrawal, R. K. (2000). A ratchet-like inter-subunit reorganization of the ribosome during translocation. *Nature, 406*, 318–322.

Furuichi, T., Yoshikawa, S., Miyawaki, A., Wada, K., Maeda, N., & Mikoshiba, K. (1989). Primary structure and functional expression of the inositol 1,4,5-trisphosphate-binding protein P400. *Nature, 342*, 32–38.

Hamada, K., Miyata, T., Mayanagi, K., Hirota, J., & Mikoshiba, K. (2002). Two-state conformational changes in inositol 1,4,5-trisphosphate receptor regulated by calcium. *The Journal of Biological Chemistry, 277*, 21115–21118.

Hamada, K., Terauchi, A., & Mikoshiba, K. (2003). Three-dimensional rearrangements within inositol 1,4,5-trisphosphate receptor by calcium. *The Journal of Biological Chemistry, 278*, 52881–52889.

Herndon, R. M. (1963). The fine structure of the Purkinje cell. *The Journal of Cell Biology, 18*, 167–180.

Huber, A. H., Nelson, W. J., & Weis, W. I. (1997). Three-dimensional structure of the armadillo repeat region of beta-catenin. *Cell, 90*, 871–882.

Iwai, M., Michikawa, T., Bosanac, I., Ikura, M., & Mikoshiba, K. (2007). Molecular basis of the isoform-specific ligand-binding affinity of inositol 1,4,5-trisphosphate receptors. *The Journal of Biological Chemistry, 282*, 12755–12764.

Jiang, W., Baker, M. L., Jakana, J., Weigele, P. R., King, J., & Chiu, W. (2008). Backbone structure of the infectious epsilon15 virus capsid revealed by electron cryomicroscopy. *Nature, 451*, 1130–1134.

Jiang, Y., Lee, A., Chen, J., Cadene, M., Chait, B. T., & MacKinnon, R. (2002). Crystal structure and mechanism of a calcium-gated potassium channel. *Nature, 417*, 515–522.

Jiang, Q. X., Thrower, E. C., Chester, D. W., Ehrlich, B. E., & Sigworth, F. J. (2002). Three-dimensional structure of the type 1 inositol 1,4,5-trisphosphate receptor at 24 Å resolution. *The EMBO Journal, 21*, 3575–3581.

Katayama, E., Funahashi, H., Michikawa, T., SHiraishi, T., Ikemoto, T., Lino, M., et al. (1996). Native structure and arrangement of inositol-1,4,5-triphosphate receptor molecules in bovine cerebellar Purkinje cells as studied by quick-freeze deep-etch electron microscopy. *The EMBO Journal, 15*, 4844–4851.

Kuo, A., Gulbis, J. M., Antcliff, J. F., Rahman, T., Lowe, E. D., Zimmer, J., et al. (2003). Crystal structure of the potassium channel KirBac1.1 in the closed state. *Science, 300*, 1922–1926.

Lobo, P. A., & Van Petegem, F. (2009). Crystal structures of the N-terminal domains of cardiac and skeletal muscle ryanodine receptors: Insights into disease mutations. *Structure, 17*, 1505–1514.

Long, S. B., Campbell, E. B., & Mackinnon, R. (2005). Crystal structure of a mammalian voltage-dependent *Shaker* family K^+ channel. *Science, 309*, 897–903.

Ludtke, S. J., Baker, M. L., Chen, D. H., Song, J. L., Chuang, D. T., & Chiu, W. (2008). De novo backbone trace of GroEL from single particle electron cryomicroscopy. *Structure, 16*, 441–448.

Ludtke, S. J., Serysheva, I. I., Hamilton, S. L., & Chiu, W. (2005). The pore structure of the closed RyR1 channel. *Structure (Cambridge), 13*, 1203–1211.

Maeda, N., Niinobe, M., & Mikoshiba, K. (1990). A cerebellar Purkinje cell marker P400 protein is an inositol 1,4,5-triphosphate (insP3) receptor protein. Purification and characterization of InsP3 receptor complex. *The EMBO Journal, 9*, 61–67.

Michikawa, T., Hamanaka, H., Otsu, H., Yamamoto, A., Miyawaki, A., Furuichi, T., et al. (1994). Transmembrane toplogy and sites of glycosylation of the inositol triphosphate receptor. *The Journal of Biological Chemistry, 269*, 9184–9189.

Mignery, G. A., Newton, C. L., Archer, B. T., & Sudhof, T. C. (1990). Structure and expression of the rat inositol 1,4,5-trisphosphate receptor. *The Journal of Biological Chemistry, 265*, 12679–12685.

Mignery, G. A., & Sudhof, T. C. (1990). The ligand binding site and transduction mechanism in the inositol- 1,4,5-triphosphate receptor. *The EMBO Journal, 9*, 3893–3898.

Mignery, G. A., Sudhof, T. C., Takei, K., & De Camilli, P. (1989). Putative receptor for inositol 1,4,5-trisphosphate similar to ryanodine receptor. *Nature, 342*, 192–195.

Mikoshiba, K. (2007). IP_3 receptor/Ca^{2+} channel: From discovery to new signaling concepts. *Journal of Neurochemistry, 102*, 1426–1446.

Miyawaki, A., Furuichi, T., Ryou, Y., Yoshikawa, S., Nakagawa, T., Saitoh, T., et al. (1991). Structure-function relationships of the mouse inositol 1,4,5-trisphosphate receptor. *Proceedings of the National Academy of Sciences of the United States of America, 88*, 4911–4915.

Nakade, S., Maeda, N., & Mikoshiba, K. (1991). Involvement of the C-terminus of the inositol 1,4,5-trisphosphate receptor in Ca^{2+} release analysed using region-specific monoclonal antibodies. *The Biochemical Journal, 277*(Pt 1), 125–131.

Nakagawa, A., Miyazaki, N., Taka, J., Naitow, H., Ogawa, A., Fujimoto, Z., et al. (2003). The atomic structure of rice dwarf virus reveals the self-assembly mechanism of component proteins. *Structure, 11*, 1227–1238.

Nakayama, T., Hattori, M., Uchida, K., Nakamura, T., Tateishi, Y., Bannai, H., et al. (2004). The regulatory domain of the inositol 1,4,5-trisphosphate receptor is necessary to keep the channel domain closed: Possible physiological significance of specific cleavage by caspase 3. *The Biochemical Journal, 377*, 299–307.

Orlova, E. V., Serysheva, I. I., van Heel, M., Hamilton, S. L., & Chiu, W. (1996). Two structural configurations of the skeletal muscle calcium release channel. *Nature Structural Biology, 3*, 547–552.

Ponting, C. P. (2000). Novel repeats in ryanodine and IP3 receptors and protein O-mannosyltransferases. *Trends in Biochemical Sciences, 25*, 48–50.

Samso, M., Feng, W., Pessah, I. N., & Allen, P. D. (2009). Coordinated movement of cytoplasmic and transmembrane domains of RyR1 upon gating. *PLoS Biology, 7*, e85.

Samso, M., Wagenknecht, T., & Allen, P. D. (2005). Internal structure and visualization of transmembrane domains of the RyR1 calcium release channel by cryo-EM. *Nature Structural and Molecular Biology, 12*, 539–544.

Sato, C., Hamada, K., Ogura, T., Miyazawa, A., Iwasaki, K., Hiroaki, Y., et al. (2004). Inositol 1,4,5-trisphosphate receptor contains multiple cavities and L-shaped ligand-binding domains. *Journal of Molecular Biology, 336*, 155–164.

Schug, Z. T., & Joseph, S. K. (2006). The role of the S4–S5 linker and C-terminal tail in inositol 1,4,5-trisphosphate receptor function. *The Journal of Biological Chemistry, 281*, 24431–24440.

Serysheva, I. I., Bare, D. J., Ludtke, S. J., Kettlun, C. S., Chiu, W., & Mignery, G. A. (2003). Structure of the type 1 inositol 1,4,5-trisphosphate receptor revealed by electron cryomicroscopy. *The Journal of Biological Chemistry, 278*, 21319–21322.

Serysheva, I. I., Chiu, W., & Ludtke, S. J. (2007). Single-particle electron cryomicroscopy of the ion channels in the excitation-contraction coupling junction. *Methods in Cell Biology, 79*, 407–435.

Serysheva, I. I., Ludtke, S. J., Baker, M. L., Cong, Y., Topf, M., Eramian, D., et al. (2008). Subnanometer-resolution electron cryomicroscopy-based domain models for the cytoplasmic region of skeletal muscle RyR channel. *Proceedings of the National Academy of Sciences of the United States of America, 105*, 9610–9615.

Serysheva, I. I., Schatz, M., van Heel, M., Chiu, W., & Hamilton, S. L. (1999). Structure of the skeletal muscle calcium release channel activated with Ca^{2+} and AMP-PCP. *Biophysical Journal, 77*, 1936–1944.

Srikanth, S., Wang, Z., Tu, H., Nair, S., Mathew, M. K., Hasan, G., et al. (2004). Functional properties of the *Drosophila melanogaster* inositol 1,4,5-trisphosphate receptor mutants. *Biophysical Journal, 86*, 3634–3646.

Stewart, A., & Grigorieff, N. (2004). Noise bias in the refinement of structures derived from single particles. *Ultramicroscopy, 102*, 67–84.

Takei, K., Mignery, G. A., Mugnaini, E., Sudhof, T. C., & De Camilli, P. (1994). Inositol 1,4,5-trisphosphate receptor causes formation of ER cisternal stacks in transfected fibroblasts and in cerebellar Purkinje cells. *Neuron, 12*, 327–342.

Tao, X., Avalos, J. L., Chen, J., & MacKinnon, R. (2009). Crystal structure of the eukaryotic strong inward-rectifier K^+ channel Kir2.2 at 3.1 Å resolution. *Science, 326*, 1668–1674.

Taylor, C. W., da Fonseca, P. C., & Morris, E. P. (2004). IP_3 receptors: The search for structure. *Trends in Biochemical Sciences, 29*, 210–219.

Uchida, K., Miyauchi, H., Furuichi, T., Michikawa, T., & Mikoshiba, K. (2003). Critical regions for activation gating of the inositol 1,4,5-trisphosphate receptor. *The Journal of Biological Chemistry, 278*, 16551–16560.

Welch, W., Rheault, S., West, D. J., & Williams, A. J. (2004). A model of the putative pore region of the cardiac ryanodine receptor channel. *Biophysical Journal, 87*, 2335–2351.

Williams, A. J., West, D. J., & Sitsapesan, R. (2001). Light at the end of the Ca^{2+}-release channel tunnel: Structures and mechanisms involved in ion translocation in ryanodine receptor channels. *Quarterly Reviews of Biophysics, 34*, 61–104.

Yoshikawa, F., Morita, M., Monkawa, T., Michikawa, T., Furuichi, T., & Mikoshiba, K. (1996). Mutational analysis of the ligand binding site of the inositol 1,4,5- trisphosphate receptor. *The Journal of Biological Chemistry, 271*, 18277–18284.

Yoshikawa, S., Tanimura, T., Miyawaki, A., Nakamura, M., Yuzaki, M., Furuichi, T., et al. (1992). Molecular cloning and characterization of the inositol 1,4,5-trisphosphate receptor in *Drosophila melanogaster*. *The Journal of Biological Chemistry, 267*, 16613–16619.

Yu, X., Jin, L., & Zhou, Z. H. (2008). 3.88 Å structure of cytoplasmic polyhedrosis virus by cryo-electron microscopy. *Nature, 453*, 415–419.

Zhang, J., Baker, M. L., Schroder, G. F., Douglas, N. R., Reissmann, S., Jakana, J., et al. (2010). Mechanism of folding chamber closure in a group II chaperonin. *Nature, 463*, 379–383.

Zhang, X., Settembre, E., Xu, C., Dormitzer, P. R., Bellamy, R., Harrison, S. C., et al. (2008). Near-atomic resolution using electron cryomicroscopy and single-particle reconstruction. *Proceedings of the National Academy of Sciences of the United States of America, 105*, 1867–1872.

Zhou, Z. H., Baker, M. L., Jiang, W., Dougherty, M., Jakana, J., Dong, G., et al. (2001). Electron cryomicroscopy and bioinformatics suggest protein fold models for rice dwarf virus. *Nature Structural Biology, 8*, 868–873.

CHAPTER 9

Molecular Architecture of the Inositol 1,4,5-Trisphosphate Receptor Pore

Darren F. Boehning
Department of Neuroscience and Cell Biology, University of Texas Medical Branch, Galveston, Texas, USA

I. OVERVIEW

Inositol 1,4,5-trisphosphate receptors (IP_3Rs) are a family of ligand-gated channels which release calcium primarily from endoplasmic reticulum stores. Historically, structure/function studies of these channels were hampered by their intracellular localization and difficulties in measuring recombinant channel activity. These obstacles have recently been overcome, and a wealth of new

Current Topics in Membranes, Volume 66
1063-5823/10 $35.00
DOI: 10.1016/S1063-5823(10)66009-7
Copyright 2010, Elsevier Inc. All right reserved.

information regarding the structural details of the ion permeation pathway has revealed similarities to ancestral potassium channels. Here, I discuss molecular aspects of ion permeation through this important channel family.

II. INTRODUCTION

Inositol 1,4,5-trisphosphate receptors (IP_3Rs) are a family of ion channels which are activated by the second messenger inositol 1,4,5-trisphosphate (IP_3). IP_3 is generated by a wide variety of stimuli which are coupled to the phosphatidylinositol 4,5-bisphophate cleaving enzyme phospholipase C (PLC). Several PLC isoforms exists which can be activated by G-protein-coupled receptors or by tyrosine kinase receptors. Many extracellular agonists lead to the activation of PLC, such as hormones, growth factors, and neurotransmitters. This leads to the transient production of IP_3 and subsequent calcium release from endoplasmic reticulum (ER) calcium stores via the IP_3R. The resultant increase in cytoplasmic, mitochondrial, and nuclear calcium leads to varied cellular responses such as metabolic, transcriptional, proliferative, and apoptotic responses, depending upon the nature and strength of the stimulus. The importance of the IP_3R gene family in health and disease is exemplified by the autosomal dominant spinocerebellar ataxia SCA15/16 caused by deletions in the *ITPR1* gene (Hara et al., 2008; Iwaki et al., 2008; van de Leemput et al., 2007) and the lethal phenotype of IP_3R knockout mice (Matsumoto et al., 1996). Furthermore, much attention has focused on the role of the IP_3R gene family in regulating apoptotic cell death, with implications for neurodegeneration and cancer (Choe & Ehrlich, 2006; Joseph & Hajnoczky, 2007; Patterson, Boehning, & Snyder, 2004). Thus, elucidating the structure and regulation of IP_3R channels at the molecular level is of paramount importance to understanding disease progression and the development of targeted therapeutics. Here, I will discuss and critically examine what is known about the molecular architecture of the transmembrane domains and the channel pore.

A. The IP_3R Gene Family

There are three IP_3R paralogs in vertebrates termed IP_3R-1, -2, and -3. The gene duplication events likely occurred early in vertebrate evolution (up to 450 million years ago according to some estimates; Zhang, Boulware, Pendleton, Nogi, & Marchant, 2007). Urochordates and other deuterostomes appear to have a single IP_3R gene. The presence and function of IP_3R orthologs in nonmetazoans is much less well understood; however, there is evidence that several lower eukaryotes such as *Dictyostelium*

(Traynor, Milne, Insall, & Kay, 2000) and *Paramecium* (Ladenburger, Sehring, Korn, & Plattner, 2009) have IP_3R orthologs. Further increasing IP_3R protein diversity is the presence of multiple alternatively spliced variants in both vertebrates and nonvertebrates. For example, the *ITPR1* gene in humans has at least 17 spliceoforms which are regulated developmentally and in a tissue-specific manner (Regan, Lin, Emerick, & Agnew, 2005).

The IP_3R channel is a tetramer which forms a single ion conduction pore. It is generally accepted that all cells express at least one IP_3R isoform, and most tissues express a combination of isoforms (Wojcikiewicz, 1995) which can form homo- and heterotetramers (Joseph, Lin, Pierson, Thomas, & Maranto, 1995; Monkawa et al., 1995). Some tissues have differential expression of the three genes; for example, the cerebellum (and most of the central nervous system) expresses primarily IP_3R-1, while the pancreas expresses predominately IP_3R-2 and -3 (Wojcikiewicz, 1995). The presence of multiple IP_3R genes in vertebrates has stirred vigorous research into whether there exists isoform-specific functions. However, the modest phenotype of *itpr2* knockout mice (Li, Zima, Sheikh, Blatter, & Chen, 2005) and the lack of phenotype in *itpr3* knockout mice (Futatsugi et al., 2005) indicate significant functional redundancy. There is a severe neurological phenotype in the *itpr1* knockout mouse (Matsumoto et al., 1996); however, this is likely due to the restricted expression of IP_3R-1 in the central nervous system. Similarly, the cardiac phenotype of the *itpr2* knockout (Li et al., 2005) and the digestive phenotype of the *itpr2/3* double knockout (Futatsugi et al., 2005) closely mimic the restricted tissue expression pattern of these isoforms. Thus, *in vivo* it is evident that one IP_3R isoform can functionally replace another isoform when both are expressed in the same tissue. Interestingly, although mice heterozygous for loss of the *itpr1* gene have no phenotype (Matsumoto et al., 1996), in humans deletion of one copy results in spinocerebellar ataxia 15/16 (Hara et al., 2008; Iwaki et al., 2008; van de Leemput et al., 2007). Thus, the functional consequences and phenotypes of IP_3R gene loss in mice do not mimic the phenotype in humans.

B. The IP_3R is a Modular Protein

The IP_3R is a large protein, with one subunit being $\sim$300 kDa. Thus, a tetrameric receptor is over 1 million daltons. The N-terminus contains the ligand (IP_3) binding domain, and the C-terminus contains the channel domain. The intervening sequence between the ligand binding domain and the channel domain is loosely termed the regulatory or modulatory region, where there are multiple sites for allosteric regulation of channel activity by posttranslational modifications and binding proteins (Patterson et al., 2004).

Biochemical and functional evidence indicates that the N-terminal ligand binding domain is in close physical proximity to the C-terminal channel domain in the tetrameric receptor (Boehning & Joseph, 2000a; Riley et al., 2002). Additionally, the N-terminus of one subunit may associate with the C-terminus of an adjacent subunit (Boehning & Joseph, 2000a). Thus, although the ligand binding and channel domains are separated by almost 2000 amino acids in linear sequence, they are adjacent to each other in the native channel. The extreme N-terminus of the IP_3R (the first ~200 amino acids) has a suppressor region which is immediately upstream of the ligand binding domain. The suppressor region, as identified by mutagenesis, reduces the affinity of the ligand binding domain for IP_3 ~10-fold *in vitro* (Yoshikawa et al., 1996, 1999). Various regions on the channel have high homology to another closely related class of intracellular calcium channel, the ryanodine receptor. This homology is highest in the channel domain and parts of the modulatory region which have been termed "RIH" or ryanodine receptor and IP_3R homology domain. These domains are not present in any other class of protein (Ponting, 2000). In addition, there are a series of repeats near and overlapping the ligand binding domain which have similarity to fungal *O*-mannosyltransferases, and these are termed "MIR" domains for protein mannosyltransferase, IP_3R, and RyR. It is thought that these modules might be protein–protein interaction motifs (Ponting, 2000).

There are two atomic resolution structures of parts of the IP_3R protein: the ligand binding domain and the suppressor region. Thus, we have very little detailed information about the structure of most of the IP_3R protein (however, atomic resolution of the full-length protein is being approached using cryo-electron microscopy (cryo-EM); reviewed in Chapter 8). The ligand binding domain is formed by a pocket sandwiched between a beta-trefoil domain and an armadillo-like repeat (Bosanac et al., 2002). The suppressor domain also has a beta-trefoil like fold, and likely interacts with and stabilizes the IP_3 binding core (Bosanac et al., 2005; Chan et al., 2007). The field is anxiously awaiting more structures at atomic level resolution. In the meantime, significant advances have been made in understanding structure/function relationships in the channel domain using biochemistry and electrophysiology.

III. THE TRANSMEMBRANE DOMAINS

A. Biochemical Studies

Hydrophobicity profiles of the cloned mouse and rat IP_3R-1 genes suggested the presence of seven, possibly eight transmembrane segments (Furuichi et al., 1989; Mignery, Newton, Archer, & Sudhof, 1990). Detailed studies examining tagged mutant receptors in cells (Galvan, Borrego-Diaz, Perez, & Mignery, 1999), and *in vitro* translation and insertion into microsomal membranes (Joseph, Boehning, Pierson, & Nicchitta, 1997), have revealed that there are likely six transmembrane domains and a pore helix. It is now clear that the IP_3R and ryanodine receptor are family members of an ion channel superfamily which include voltage-gated Na^+, K^+, and Ca^{2+} channels, cyclic nucleotide-gated channels, and TRP channels. These channels share a similar transmembrane topology and have sequence homology within the pore forming segment (discussed in more detail below). The transmembrane domains of the IP_3R also contain the primary determinants for subunit oligomerization. Galvan et al. (1999) utilized deletion constructs expressed in COS-1 cells to demonstrate that transmembrane regions 5 and 6 contain the minimal determinants for subunit oligomerization. It was also noted that the same transmembrane segments had a coiled-coil or leucine zipper consensus motif which may mediate receptor oligomerization. We reached similar conclusions, and determined that transmembrane regions 5 and 6 also mediate heterooligomerization (Joseph et al., 1997). We further provided evidence that transmembrane regions 1–4 helped to stabilize heteroligomeric tetramers. In addition to providing the molecular determinants for oligomerization, transmembrane regions 5 and 6 encompass the ion conduction pore (Boehning, Mak, Foskett, & Joseph, 2001; Ramos-Franco, Galvan, Mignery, & Fill, 1999) and the determinants for association with the ligand binding domain (Boehning & Joseph, 2000a; Schug & Joseph, 2006).

B. Orientation of the TM Domains and the Packing of TM Helices

Significant insight into the structure of the transmembrane domains of this ion channel superfamily was revealed with the crystal structure of the *Streptomyces lividans* K^+ channel KcsA (Doyle et al., 1998). The structure indicated that the transmembrane domains equivalent to 5 and 6 in the IP_3R (and pore helix) formed an inverted cone. The interior of the channel had an aqueous cavity to accommodate a K^+ ion, which was thermodynamically stabilized within the cavity by a partial negative electrostatic potential provided by a pore helix dipole. The most remarkable finding was the mechanism of ion selectivity, which will be discussed in detail below. It is thus very tempting (and logical) to presume that transmembrane regions 5 and 6 of the IP_3R adopt a similar conformation (and indeed, has been modeled as such; Schug et al., 2008). This is further supported by mutational analysis of residues in the putative IP_3R ion conduction pore which are homologous to KcsA channels (Boehning & Joseph, 2000b; Boehning,

Mak, et al., 2001; Schug et al., 2008). Thus, it appears that transmembrane regions 5 and 6 of the IP_3R have at least some modest similarity to the KcsA structure. The next major question regarding the transmembrane domains is how are transmembrane domains 1–4 packed around the ion conduction pore. At first glance, it seems logical that the other four transmembrane domains would pack in behind the 5/6 helices in each monomer (as suggested by others; Foskett, White, Cheung, & Mak, 2007; Taylor, da Fonseca, & Morris, 2004). The first structure of a six-transmembrane domain voltage-gated K^+ channel (KvAP) fit this model, although with some surprising (and controversial) helix packing, with helices 3 and 4 parallel to the lipid bilayer (Jiang et al., 2003). A subsequent structure of KvAP determined in the presence of lipid revealed a more traditional (i.e., perpendicular) orientation of the transmembrane helices (Lee, Lee, Chen, & MacKinnon, 2005). The first structure of a *Shaker* K^+ channel similarly revealed a conventional transmembrane helix orientation; however, an additional surprise was that the voltage sensor of one subunit was cradling the pore forming 5/6 helices of an adjacent subunit (Long, Campbell, & Mackinnon, 2005). As shown in Fig. 1A, transmembrane (TM) regions 1–4 are separated from TM5–6 by the TM4/5 linker, thus allowing the voltage sensing TM4 helix to wrap around an adjacent pore forming subunit. In the IP_3R, the TM4/5 linker is quite large, which would permit a similar subunit topology. It is easier to appreciate the arrangement of helices when viewed from below (i.e., looking through the pore of the IP_3R channel from the ER lumen). As shown schematically in Fig. 1B, the TM5/6 and pore helix are offset from TM1–4 in one subunit, and wrap around the TM5/6 of an adjacent subunit. What evidence is there that this could be a plausible topology for the IP_3R? Recently, several high-resolution (~10 Å) cryo-EM ryanodine receptor structures have partially resolved the pore forming helices (Ludtke, Serysheva, Hamilton, & Chiu, 2005; Samso, Feng, Pessah, & Allen, 2009; Samso, Wagenknecht, & Allen, 2005). The densities of the pore forming helices overlap significantly with the atomic structure of the *Shaker* transmembrane domains (as depicted in Fig. 1A), but not KvAP structure (Samso et al., 2009), thus supporting the topology depicted in Fig. 1B. Ludtke et al. (2005) additionally provided evidence that pore lining helix (TM6) in the ryanodine receptor was kinked, likely at a highly conserved glycine residue (Schug et al., 2008). This model overlaps significantly with the crystal structure of the two-transmembrane K^+ channel MthK in the open state (Ludtke et al., 2005). In some potassium channels, a hinged TM6 helix functions as a mechanism for gating (Jiang et al., 2002). However, there is little evidence at present that a hinged helix functions as a mechanism for channel gating in ryanodine receptors and IP_3Rs (Ludtke et al., 2005; Schug et al., 2008). Due to the high degree of sequence homology of the ryanodine receptor with the IP_3R in the

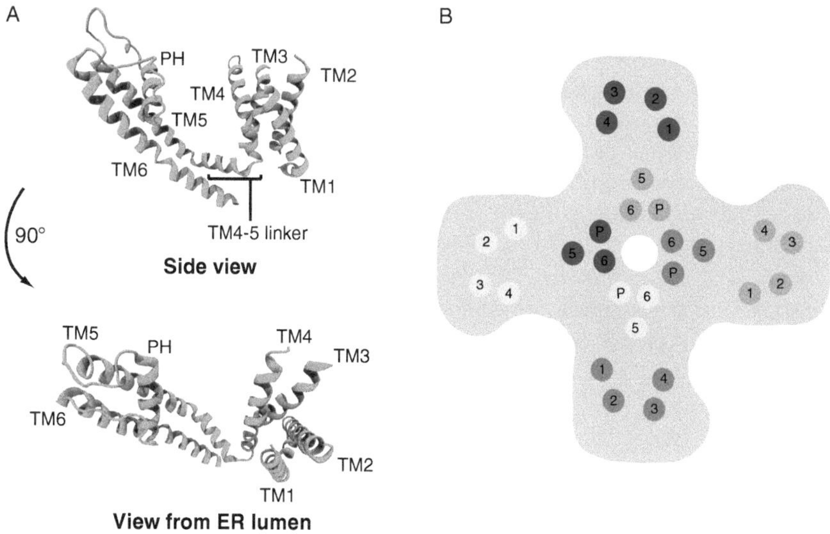

FIGURE 1 Helix packing in the *Shaker* potassium channel. (A) side view (along the membrane) and a 90° rotation looking through the ion conduction pathway of the transmembrane domains of a single *Shaker* subunit (PDB accession number 2A79). Transmembrane (TM) domains and the 4/5 linker are labeled. (B) Schematic view down the ion conduction pore of helix packing in a tetramer. All of the helices from one subunit are assigned the same color. See text for details. Structures presented here and elsewhere in this chapter were created with visual molecular dynamics (VMD; Humphrey, Dalke, & Schulten, 1996). (See Color Insert.)

transmembrane domains, it is likely the IP_3R transmembrane helices adopt a similar orientation as depicted in Fig. 1. Refinement of the cryo-EM structure of the IP_3R to 10 Å or better should provide significant insight into whether the transmembrane domains of the IP_3R fit the model of helix packing depicted in Fig. 1B.

IV. THE ION CONDUCTION PORE: ELECTROPHYSIOLOGIC STUDIES

A. Measuring IP_3R Activity

Ion channels are unique in the protein world in that it is relatively simple to quantitatively measure the activity of a single molecule in real time using electrophysiological techniques. Thus, much of what is known about structure/function of ion channels has been determined using mutagenesis combined with electrophysiological recording. As the IP_3R is primarily an intracellular channel (but see below), it has historically precluded traditional electrophysiological analysis. Much of the early work was done using planar lipid bilayers, which was invaluable for investigating multiple aspects of IP_3R physiology such as ion permeation and gating (reviewed in Foskett et al., 2007). However, the peculiarities of this system were not amiable to the analysis of large numbers of mutant recombinant channels, and in particular, there was always the concern that the ubiquitous endogenous channels may contaminate the recordings. Nonetheless, some groups were successful in using this technique to identify unambiguously that the ion channel is formed by transmembrane domains 5 and 6 (Ramos-Franco et al., 1999), and to characterize several mutant receptors (Srikanth, Wang, Tu, et al., 2004; Tu et al., 2002), including a pore mutant (Srikanth, Wang, Hasan, & Bezprozvanny, 2004). More recently, modified patch-clamp techniques which rely on patching isolated nuclei have been developed using *Xenopus* oocytes (Mak & Foskett, 1994), insect (White et al., 2005), and mammalian (Boehning, Joseph, Mak, & Foskett, 2001) cells. These techniques rely on cell types with relatively low densities of IP_3R channels such that the probability of detecting an endogenous channel is very low. For example, in COS-7 cells the detection rate of endogenous channels is approximately 1.5%, whereas expression of recombinant IP_3R channels increases the probability of detection to 45% (Boehning et al., 2001). A true "zero background" system has now been achieved with the successful recording of recombinant IP_3R channels on the nuclei of DT40 triple IP_3R KO cells (Li et al., 2007). However, from my personal experience (Boehning, Mak, et al., 2001; Boehning et al., 2001), recording from isolated mammalian nuclei can be a technically demanding endeavor.

Recently, it has been reported that a single or several recombinant IP_3R channels are trafficked to the plasma membrane in DT40 IP_3R triple KO cells, and endogenous IP_3R channels in mouse B cells (Dellis et al., 2006). As a result, single-channel currents through recombinant and endogenous IP_3Rs can be measured using whole-cell recording. Although extremely interesting with regards to the physiologic mechanisms of calcium entry, this finding is also significant with regards to structure/function studies of the IP_3R. It is now possible to use the zero background DT40 system with the relatively simple technique of whole-cell recording to analyze the effects of IP_3R mutations on single-channel currents. This system has already been exploited to systematically analyze mutations affecting ion permeation through IP_3Rs (discussed in more detail below; Schug et al., 2008). It is likely that this much more accessible technique will open the door for many more laboratories to perform single-channel studies on recombinant IP_3R channels.

B. Selectivity and Permeability

The structure of the KcsA channel was revolutionary in that it solved the apparent paradox of how a K^+ channel can discriminate with high fidelity over a Na^+ ion. The K^+ channel achieves this feat by passing only dehydrated K^+ ions through the selectivity filter. It accomplishes this task by using the carbonyl oxygen atoms of the polypeptide backbone of the selectivity filter as surrogate waters (Doyle et al., 1998). The selectivity filter is composed of the residues GYGD, which is also known as the "signature sequence" of K^+ channels. It is held in a rigid conformation such that it cannot precisely coordinate a Na^+ or other cation of different crystal radius. As shown in Fig. 2, the selectivity filter is close to the exterior of the cell (the exit point of the physiologic K^+ current). The analogous helices on the IP_3R are indicated. Other features of the ion conduction pathway also promote high-throughput rates such as an aqueous vestibule lined with hydrophobic residues, the electronegative dipole induced by the pore helix, and negative charges at the mouth of the channel to concentrate cations (Doyle et al., 1998).

The IP_3R has a sequence within the putative selectivity filter of GVGD, and the ryanodine receptor has the sequence GIGD. Thus, it is tempting to speculate that the pore of the IP_3R/RyR may be very similar in structure to

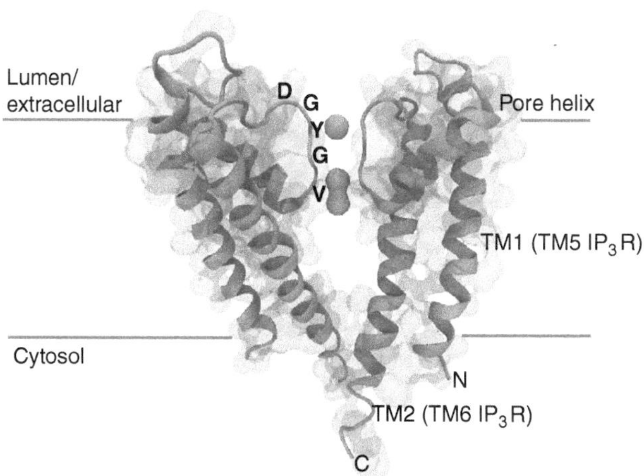

FIGURE 2 Structure of the KcsA channel from *Streptomyces lividans*. Two subunits of a KcsA tetramer are presented (PDB accession number 1BL8). The bilayer is represented by thin lines. The TM helices in KcsA and equivalent helices in IP_3R are indicated. The single amino acid code of the selectivity filter VGYGD is overlaid on the selectivity filter. Electron density for potassium ions within the filter are in magenta. (See Color Insert.)

the KcsA channel. It is first prudent to examine the ion permeation properties of the IP_3R channel. The channel is modestly selective for Ca^{2+} over monovalent cations (about 4- to 10-fold) and has a very high unitary conductance for most cations (reviewed in Foskett et al., 2007). The lack of selectivity is not surprising, since the only ionic gradient across the ER membrane is calcium (and thus, there has been no selective pressure for the channel to be more discriminatory against other ions). The selectivity sequence for cations is similar for ryanodine receptors. The ion permeation properties of the ryanodine receptor have been investigated in much more detail, albeit with conflicting conclusions. For example, one area of contention is whether the channel is a single- or multi-ion occupancy pore. Early work suggested that the channel was single occupancy (Tinker & Williams, 1995). However, more recent work suggests the channel is a multioccupancy pore (as is the KcsA channel) (Gillespie, Giri, & Fill, 2009). Much less is known about ion occupancy within the IP_3R pore, but it has been suggested to be single occupancy (reviewed in Foskett et al., 2007). The permeation properties of the ryanodine receptor and IP_3R are not consistent with dehydration of the permeant ion, and therefore there must be fundamental differences in the mechanism(s) of selectivity between K^+ channels and IP_3Rs.

C. Mutagenesis Studies

In order to determine the molecular basis for ion selectivity in IP_3R channels, we mutated residues within the putative selectivity filter GVGD. These mutations were first characterized using a $^{45}Ca^{2+}$ microsomal flux assay which selectively measures recombinant channel activity (Boehning & Joseph, 2000b), and then were characterized in more detail using the nuclear patch-clamp technique (Boehning, Mak, et al., 2001). We chose to focus primarily on valine 2548 (V2548) and aspartic acid 2550 (D2550). The rationale behind mutating V2548 is that this residue is an isoleucine in the ryanodine receptor, and a tyrosine in K^+ channels. We predicted that the mutation V2548I would make the IP_3R more ryanodine receptor-like. Indeed, mutating V2548 to I resulted in functional calcium channels that had a much higher K^+ conductance (like the RyR; Boehning, Mak, et al., 2001), whereas another group demonstrated that mutating the analogous isoleucine residue to valine in RyR decreased channel conductance (Gao et al., 2000). This result is consistent with this residue controlling pore size in IP_3Rs and RyRs, although it is not readily apparent why a larger side chain would increase conductance. Mutating V2548 to tyrosine (mimicking the K^+ channel sequence) resulted in channels which could no longer permeate Ca^{2+} (Boehning & Joseph, 2000b). It is currently not known if these channels can permeate K^+ ions.

The rationale behind mutating D2550 was that this residue may mediate Ca^{2+} selectivity in a manner analogous to voltage-gated Ca^{2+} channels, whereas a ring of acidic residues act as a high-affinity Ca^{2+} binding site excluding other cations from the ion permeation pore (discussed in more detail below). Mutating D2550 to asparagine or alanine eliminated $^{45}Ca^{2+}$ flux (Boehning & Joseph, 2000b); however, D2550A channels are still capable of permeating K^{+} ions (Dellis, Rossi, Dedos, & Taylor, 2008). These data are consistent with residue D2550 forming a high-affinity calcium binding site within the ion conduction pathway. In addition, as several lines of evidence suggest that permeant ions through the IP_3R are not dehydrated, this model is currently the most attractive hypothesis. However, there are clearly some steric constraints on ion permeation as discussed with V2548 earlier and flanking glycine residues as discussed later. The "pore-dead" D2550A mutant, in addition to providing insights into the pore architecture, has also been useful as a control for various studies examining IP_3R function (e.g., see van Rossum et al., 2004).

Using the aforementioned $^{45}Ca^{2+}$ flux assay combined with DT40 whole-cell recording, Schug et al. (2008) systematically mutated residues in the putative selectivity filter, the S6 helix, and the C-terminal tail. One area of focus was the glycine residues flanking (or part of) the selectivity filter **GGGVGD** (G2545, G2546, G2547, G2549). Of note, most substitutions at these positions eliminated $^{45}Ca^{2+}$ flux, with the exception of G2549A. In a separate study, substitution at glycine 2547 to serine in the *Drosophila* IP_3R was also found to inactivate the channel (Srikanth et al., 2004). Similar to the "pore-dead" D2550A mutant, G2546A was unable to flux $^{45}Ca^{2+}$, but was still able to permeate K^{+} ions. The underlying mechanisms are unclear, but it was hypothesized that the additional methyl group of the alanine side chain resulted in a steric constraint which allowed the K^{+} ion with a much smaller hydrated radius pass through, whereas Ca^{2+} was excluded because of a much larger hydrated radius. This hypothesis of course necessitates that the permeant ions are hydrated, since the crystal radii of Ca^{2+} is smaller than K^{+}. These results are also consistent with the effects of mutating V2548 as described earlier, demonstrating that the region of the selectivity filter encompassing G2546 to G2549 (GGVG) is likely the narrowest portion of the ion conduction pathway, similar to potassium channels. This study also identified phenylalanine 2592 as putative a gaiting residue in the TM6, although the F2592A channel unexpectedly retains some function, suggesting that other residues or possibly the concerted movement of several residues occludes the pore in the closed state (Schug et al., 2008).

V. THE ION CONDUCTION PORE: MODELING STUDIES

A. *Mechanisms of Ion Selectivity in Calcium Channels and Weakly Selective Cation Channels*

In contrast to potassium channels, the molecular mechanisms for calcium selectivity in voltage-gated Ca^{2+} channels is thought to be mediated by a high-affinity calcium/cation binding site within the ion conduction pathway (Sather, Yang, & Tsien, 1994). In the voltage-gated L-type calcium channel, this binding site is mediated by a ring of four glutamic acid residues (Yang, Ellinor, Sather, Zhang, & Tsien, 1993). This high-affinity site excludes other more numerous monovalent cations, and high-throughput ion permeation is facilitated by the large ionic gradient of Ca^{2+} ions and competition for binding at this site. Evidence for a single high-affinity Ca^{2+} binding site in several voltage-gated Ca^{2+} channels has been established using mutagenesis (Kim, Morii, Sun, Imoto, & Mori, 1993; Yang et al., 1993). Similar rings of charged glutamate residues control ion permeation through weakly Ca^{2+} selective cyclic nucleotide-gated channels (Hackos & Korenbrot, 1999). Of note, voltage-gated Ca^{2+} channels, like the IP_3R, are a clear evolutionary descendent of an ancestral K^+ channel based on both the transmembrane topology and clear sequence similarity surrounding the selectivity filter.

Let us now consider a weakly selective cation channel, which evolved separately from K^+ channels. The atomic structure of several bacterial homologs of the pentameric Cys-loop neurotransmitter receptors have been solved (Bocquet et al., 2009; Hilf & Dutzler, 2008, 2009). These channels can be either cationic channels (nicotinic acetylcholine and serotonin receptor) or anionic channels (GABA and glycine receptors). The cationic family members have variable selectivity for Ca^{2+} depending on subunit composition, however, all are relatively nonselective for monovalent cations. Selectivity for cations is mediated by a ring of either glutamic or aspartic acid residues in up to four separate places: one in the extracellular domain, two in the transmembrane domains (one of these sites also being the narrowest portion of the ion conduction pathway), and one in the cytosolic domain (Hansen, Wang, Taylor, & Sine, 2008). In the anionic family members, the analogous residue is predictably a positive charge (lysine or arginine). Thus, in both highly selective and weakly selective cation channels, electrostatic interactions of Ca^{2+} with acidic side chains are the dominant forces regulating selectivity. Although less well characterized, this is likely the mechanism for selectivity in IP_3R channels as well.

B. *Atomic Structure of K^+ Channels: Appropriate Models for IP_3R Channels?*

We have presented here circumstantial evidence that the IP_3R pore may look similar to the K^+ channel superfamily. This is most evident when looking at the transmembrane topology of the IP_3R and sequence homology within the selectivity filter. In addition, the few studies which have done mutational analysis of the selectivity filter indeed show that these residues are potent mediators of ion permeation through the channel (Boehning & Joseph, 2000b; Boehning, Mak, et al., 2001; Schug et al., 2008). Finally, high-resolution cryo-EM structures of the ryanodine receptor suggest that the orientation of the transmembrane helices in the *Shaker* and MthK potassium channels are likely very similar in the ryanodine receptor (and by homology, the IP_3R; Ludtke et al., 2005; Samso et al., 2005, 2009). This is where I suggest the similarity ends. The selectivity filter of the potassium channel is held in a precise diameter to allow only K^+ ions to be dehydrated and enter (Fig. 3A). The permeant ions are first held and stabilized in an aqueous vestibule which overcomes some of the thermodynamic barriers associated with ion transport across a lipid bilayer. This structure appears to be conserved in a variety of prokaryotic and eukaryotic channels (Doyle et al., 1998; Jiang et al., 2003; Lee et al., 2005; Long et al., 2005). By necessity, the

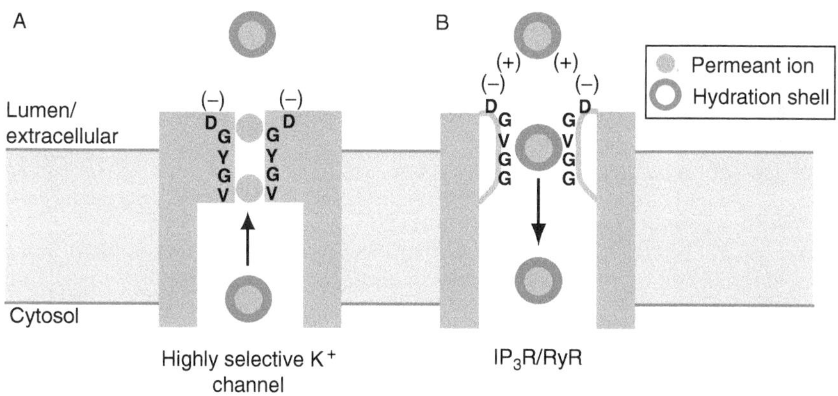

FIGURE 3 A model for ion permeation through K^+ and IP_3R channels. (A) Schematic of ion permeation through the KcsA channel. Potassium ion is in green; waters of hydration in blue. The direction of physiologic ion flux is indicated with an arrow. (B) Schematic of cation permeation through an IP_3R/ryanodine receptor (RyR). Structural flexibility within the selectivity filter is represented by a loop. Interaction of a hydrated Ca^{2+} ion with an aspartic acid residue at the mouth of the pore is also shown. The direction of physiologic ion flux is indicated with an arrow. (See Color Insert.)

nonselective nature of the IP_3R pore indicates that the selectivity filter is likely flexible to accommodate ions of different sizes, and furthermore the permeant ions are not dehydrated as they pass through the selectivity filter (Fig. 3B). The dominant force mediating the Ca^{2+} selectivity is likely a divalent cation binding site conferred by aspartic acid residue 2550. This is most simply illustrated by the observation that mutation of this single residue eliminates Ca^{2+} selectivity. There are other considerations as well. The internal vestibule in K^+ channels precedes the entry of ions into selectivity filter (Figs. 2 and 3A), and is thought to contribute significantly to high-throughput ion flux and selectivity (Doyle et al., 1998). This is not the case for the IP_3R, where the dominant movement of Ca^{2+} ions is in precisely the opposite direction. Thus, the Ca^{2+} ions would not encounter the internal vestibule until *after* passing through the selectivity filter (Fig. 3B). Thus, great caution should be exercised when modeling the ion permeation pathway of the IP_3R on the KcsA/KvAP/*Shaker* channels, as the functional evidence indicates there must be significant differences in the pore architecture.

References

Bocquet, N., Nury, H., Baaden, M., Le Poupon, C., Changeux, J. P., Delarue, M., et al. (2009). X-ray structure of a pentameric ligand-gated ion channel in an apparently open conformation. *Nature, 457*, 111–114.

Boehning, D., & Joseph, S. K. (2000a). Direct association of ligand-binding and pore domains in homo- and heterotetrameric inositol 1, 4, 5-trisphosphate receptors. *The EMBO Journal, 19*, 5450–5459.

Boehning, D., & Joseph, S. K. (2000b). Functional properties of recombinant type I and type III inositol 1, 4, 5-trisphosphate receptor isoforms expressed in COS-7 cells. *The Journal of Biological Chemistry, 275*, 21492–21499.

Boehning, D., Joseph, S. K., Mak, D. O., & Foskett, J. K. (2001). Single-channel recordings of recombinant inositol trisphosphate receptors in mammalian nuclear envelope. *Biophysical Journal, 81*, 117–124.

Boehning, D., Mak, D. O., Foskett, J. K., & Joseph, S. K. (2001). Molecular determinants of ion permeation and selectivity in inositol 1,4,5-trisphosphate receptor Ca^{2+} channels. *The Journal of Biological Chemistry, 276*, 13509–13512.

Bosanac, I., Alattia, J. R., Mal, T. K., Chan, J., Talarico, S., Tong, F. K., et al. (2002). Structure of the inositol 1,4,5-trisphosphate receptor binding core in complex with its ligand. *Nature, 420*, 696–700.

Bosanac, I., Yamazaki, H., Matsu-Ura, T., Michikawa, T., Mikoshiba, K., & Ikura, M. (2005). Crystal structure of the ligand binding suppressor domain of type 1 inositol 1,4,5-trisphosphate receptor. *Molecular Cell, 17*, 193–203.

Chan, J., Whitten, A. E., Jeffries, C. M., Bosanac, I., Mal, T. K., Ito, J., et al. (2007). Ligand-induced conformational changes via flexible linkers in the amino-terminal region of the inositol 1,4,5-trisphosphate receptor. *Journal of Molecular Biology, 373*, 1269–1280.

Choe, C. U., & Ehrlich, B. E. (2006). The inositol 1,4,5-trisphosphate receptor (IP3R) and its regulators: Sometimes good and sometimes bad teamwork. *Science's STKE: Signal Transduction Knowledge Environment, re15*.

Dellis, O., Dedos, S. G., Tovey, S. C., Taufiq Ur, R., Dubel, S. J., & Taylor, C. W. (2006). Ca^{2+} entry through plasma membrane IP3 receptors. *Science, 313*, 229–233.

Dellis, O., Rossi, A. M., Dedos, S. G., & Taylor, C. W. (2008). Counting functional inositol 1,4,5-trisphosphate receptors into the plasma membrane. *The Journal of Biological Chemistry, 283*, 751–755.

Doyle, D. A., Morais Cabral, J., Pfuetzner, R. A., Kuo, A., Gulbis, J. M., Cohen, S. L., et al. (1998). The structure of the potassium channel: molecular basis of K^+ conduction and selectivity. *Science, 280*, 69–77.

Foskett, J. K., White, C., Cheung, K. H., & Mak, D. O. (2007). Inositol trisphosphate receptor Ca^{2+} release channels. *Physiological Reviews, 87*, 593–658.

Furuichi, T., Yoshikawa, S., Miyawaki, A., Wada, K., Maeda, N., & Mikoshiba, K. (1989). Primary structure and functional expression of the inositol 1,4,5-trisphosphate-binding protein P400. *Nature, 342*, 32–38.

Futatsugi, A., Nakamura, T., Yamada, M. K., Ebisui, E., Nakamura, K., Uchida, K., et al. (2005). IP3 receptor types 2 and 3 mediate exocrine secretion underlying energy metabolism. *Science, 309*, 2232–2234.

Galvan, D. L., Borrego-Diaz, E., Perez, P. J., & Mignery, G. A. (1999). Subunit oligomerization, and topology of the inositol 1,4,5-trisphosphate receptor. *The Journal of Biological Chemistry, 274*, 29483–29492.

Gao, L., Balshaw, D., Xu, L., Tripathy, A., Xin, C., & Meissner, G. (2000). Evidence for a role of the lumenal M3-M4 loop in skeletal muscle Ca(2+) release channel (ryanodine receptor) activity and conductance. *Biophysical Journal, 79*, 828–840.

Gillespie, D., Giri, J., & Fill, M. (2009). Reinterpreting the anomalous mole fraction effect: The ryanodine receptor case study. *Biophysical Journal, 97*, 2212–2221.

Hackos, D. H., & Korenbrot, J. I. (1999). Divalent cation selectivity is a function of gating in native and recombinant cyclic nucleotide-gated ion channels from retinal photoreceptors. *The Journal of General Physiology, 113*, 799–818.

Hansen, S. B., Wang, H. L., Taylor, P., & Sine, S. M. (2008). An ion selectivity filter in the extracellular domain of Cys-loop receptors reveals determinants for ion conductance. *The Journal of Biological Chemistry, 283*, 36066–36070.

Hara, K., Shiga, A., Nozaki, H., Mitsui, J., Takahashi, Y., Ishiguro, H., et al. (2008). Total deletion and a missense mutation of ITPR1 in Japanese SCA15 families. *Neurology, 71*, 547–551.

Hilf, R. J., & Dutzler, R. (2008). X-ray structure of a prokaryotic pentameric ligand-gated ion channel. *Nature, 452*, 375–379.

Hilf, R. J., & Dutzler, R. (2009). Structure of a potentially open state of a proton-activated pentameric ligand-gated ion channel. *Nature, 457*, 115–118.

Humphrey, W., Dalke, A., & Schulten, K. (1996). VMD: Visual molecular dynamics. *Journal of Molecular Graphics, 14*(33–38), 27–38.

Iwaki, A., Kawano, Y., Miura, S., Shibata, H., Matsuse, D., Li, W., et al. (2008). Heterozygous deletion of ITPR1, but not SUMF1, in spinocerebellar ataxia type 16. *Journal of Medical Genetics, 45*, 32–35.

Jiang, Y., Lee, A., Chen, J., Cadene, M., Chait, B. T., & MacKinnon, R. (2002). The open pore conformation of potassium channels. *Nature, 417*, 523–526.

Jiang, Y., Lee, A., Chen, J., Ruta, V., Cadene, M., Chait, B. T., et al. (2003). X-ray structure of a voltage-dependent K^+ channel. *Nature, 423*, 33–41.

Joseph, S. K., Boehning, D., Pierson, S., & Nicchitta, C. V. (1997). Membrane insertion, glycosylation, and oligomerization of inositol trisphosphate receptors in a cell-free translation system. *The Journal of Biological Chemistry, 272*, 1579–1588.

Joseph, S. K., & Hajnoczky, G. (2007). IP3 receptors in cell survival and apoptosis: Ca^{2+} release and beyond. *Apoptosis, 12*, 951–968.

Joseph, S. K., Lin, C., Pierson, S., Thomas, A. P., & Maranto, A. R. (1995). Heteroligomers of type-I and type-III inositol trisphosphate receptors in WB rat liver epithelial cells. *The Journal of Biological Chemistry, 270*, 23310–23316.

Kim, M. S., Morii, T., Sun, L. X., Imoto, K., & Mori, Y. (1993). Structural determinants of ion selectivity in brain calcium channel. *FEBS Letters, 318*, 145–148.

Ladenburger, E. M., Sehring, I. M., Korn, I., & Plattner, H. (2009). Novel types of Ca^{2+} release channels participate in the secretory cycle of Paramecium cells. *Molecular and Cellular Biology, 29*, 3605–3622.

Lee, S. Y., Lee, A., Chen, J., & MacKinnon, R. (2005). Structure of the KvAP voltage-dependent K^+ channel and its dependence on the lipid membrane. *Proceedings of the National Academy of Sciences of the United States of America, 102*, 15441–15446.

Li, C., Wang, X., Vais, H., Thompson, C. B., Foskett, J. K., & White, C. (2007). Apoptosis regulation by Bcl-x(L) modulation of mammalian inositol 1,4,5-trisphosphate receptor channel isoform gating. *Proceedings of the National Academy of Sciences of the United States of America, 104*, 12565–12570.

Li, X., Zima, A. V., Sheikh, F., Blatter, L. A., & Chen, J. (2005). Endothelin-1-induced arrhythmogenic Ca^{2+} signaling is abolished in atrial myocytes of inositol-1,4,5-trisphosphate(IP3)-receptor type 2-deficient mice. *Circulation Research, 96*, 1274–1281.

Long, S. B., Campbell, E. B., & Mackinnon, R. (2005). Crystal structure of a mammalian voltage-dependent Shaker family K^+ channel. *Science, 309*, 897–903.

Ludtke, S. J., Serysheva, I. I., Hamilton, S. L., & Chiu, W. (2005). The pore structure of the closed RyR1 channel. *Structure, 13*, 1203–1211.

Mak, D. O., & Foskett, J. K. (1994). Single-channel inositol 1,4,5-trisphosphate receptor currents revealed by patch clamp of isolated Xenopus oocyte nuclei. *The Journal of Biological Chemistry, 269*, 29375–29378.

Matsumoto, M., Nakagawa, T., Inoue, T., Nagata, E., Tanaka, K., Takano, H., et al. (1996). Ataxia and epileptic seizures in mice lacking type 1 inositol 1,4,5-trisphosphate receptor. *Nature, 379*, 168–171.

Mignery, G. A., Newton, C. L., Archer, B. T., 3rd, Sudhof, T. C. (1990). Structure and expression of the rat inositol 1,4,5-trisphosphate receptor. *The Journal of Biological Chemistry, 265*, 12679–12685.

Monkawa, T., Miyawaki, A., Sugiyama, T., Yoneshima, H., Yamamoto-Hino, M., Furuichi, T., et al. (1995). Heterotetrameric complex formation of inositol 1,4,5-trisphosphate receptor subunits. *The Journal of Biological Chemistry, 270*, 14700–14704.

Patterson, R. L., Boehning, D., & Snyder, S. H. (2004). Inositol 1,4,5-trisphosphate receptors as signal integrators. *Annual Review of Biochemistry, 73*, 437–465.

Ponting, C. P. (2000). Novel repeats in ryanodine and IP3 receptors and protein O-mannosyltransferases. *Trends in Biochemical Sciences, 25*, 48–50.

Ramos-Franco, J., Galvan, D., Mignery, G. A., & Fill, M. (1999). Location of the permeation pathway in the recombinant type 1 inositol 1,4,5-trisphosphate receptor. *The Journal of General Physiology, 114*, 243–250.

Regan, M. R., Lin, D. D., Emerick, M. C., & Agnew, W. S. (2005). The effect of higher order RNA processes on changing patterns of protein domain selection: A developmentally regulated transcriptome of type 1 inositol 1,4,5-trisphosphate receptors. *Proteins, 59*, 312–331.

Riley, A. M., Morris, S. A., Nerou, E. P., Correa, V., Potter, B. V., & Taylor, C. W. (2002). Interactions of inositol 1,4,5-trisphosphate (IP(3)) receptors with synthetic poly(ethylene glycol)-linked dimers of IP(3) suggest close spacing of the IP(3)-binding sites. *The Journal of Biological Chemistry, 277*, 40290–40295.

Samso, M., Feng, W., Pessah, I. N., & Allen, P. D. (2009). Coordinated movement of cytoplasmic and transmembrane domains of RyR1 upon gating. *PLoS Biology, 7*, e85.

Samso, M., Wagenknecht, T., & Allen, P. D. (2005). Internal structure and visualization of transmembrane domains of the RyR1 calcium release channel by cryo-EM. *Nature Structural & Molecular Biology, 12*, 539–544.

Sather, W. A., Yang, J., & Tsien, R. W. (1994). Structural basis of ion channel permeation and selectivity. *Current Opinion in Neurobiology, 4*, 313–323.

Schug, Z. T., da Fonseca, P. C., Bhanumathy, C. D., Wagner, L., 2nd, Zhang, X., Bailey, B., et al. (2008). Molecular characterization of the inositol 1,4,5-trisphosphate receptor pore-forming segment. *The Journal of Biological Chemistry, 283*, 2939–2948.

Schug, Z. T., & Joseph, S. K. (2006). The role of the S4–S5 linker and C-terminal tail in inositol 1,4,5-trisphosphate receptor function. *The Journal of Biological Chemistry, 281*, 24431–24440.

Srikanth, S., Wang, Z., Hasan, G., & Bezprozvanny, I. (2004). Functional properties of a pore mutant in the *Drosophila melanogaster* inositol 1,4,5-trisphosphate receptor. *FEBS Letters, 575*, 95–98.

Srikanth, S., Wang, Z., Tu, H., Nair, S., Mathew, M. K., Hasan, G., et al. (2004). Functional properties of the Drosophila melanogaster inositol 1,4,5-trisphosphate receptor mutants. *Biophysical Journal, 86*, 3634–3646.

Taylor, C. W., da Fonseca, P. C., & Morris, E. P. (2004). IP(3) receptors: The search for structure. *Trends in Biochemical Sciences, 29*, 210–219.

Tinker, A., & Williams, A. J. (1995). Measuring the length of the pore of the sheep cardiac sarcoplasmic reticulum calcium-release channel using related trimethylammonium ions as molecular calipers. *Biophysical Journal, 68*, 111–120.

Traynor, D., Milne, J. L., Insall, R. H., & Kay, R. R. (2000). Ca(2+) signalling is not required for chemotaxis in *Dictyostelium*. *The EMBO Journal, 19*, 4846–4854.

Tu, H., Miyakawa, T., Wang, Z., Glouchankova, L., Iino, M., & Bezprozvanny, I. (2002). Functional characterization of the type 1 inositol 1,4,5-trisphosphate receptor coupling domain SII(+/−) splice variants and the Opisthotonos mutant form. *Biophysical Journal, 82*, 1995–2004.

van de Leemput, J., Chandran, J., Knight, M. A., Holtzclaw, L. A., Scholz, S., Cookson, M. R., et al. (2007). Deletion at ITPR1 underlies ataxia in mice and spinocerebellar ataxia 15 in humans. *PLoS Genetics, 3*, e108.

van Rossum, D. B., Patterson, R. L., Kiselyov, K., Boehning, D., Barrow, R. K., Gill, D. L., et al. (2004). Agonist-induced Ca^{2+} entry determined by inositol 1,4,5-trisphosphate recognition. *Proceedings of the National Academy of Sciences of the United States of America, 101*, 2323–2327.

White, C., Li, C., Yang, J., Petrenko, N. B., Madesh, M., Thompson, C. B., et al. (2005). The endoplasmic reticulum gateway to apoptosis by Bcl-X(L) modulation of the InsP3R. *Nature Cell Biology, 7*, 1021–1028.

Wojcikiewicz, R. J. (1995). Type I, II, and III inositol 1,4,5-trisphosphate receptors are unequally susceptible to down-regulation and are expressed in markedly different proportions in different cell types. *The Journal of Biological Chemistry, 270*, 11678–11683.

Yang, J., Ellinor, P. T., Sather, W. A., Zhang, J. F., & Tsien, R. W. (1993). Molecular determinants of Ca^{2+} selectivity and ion permeation in L-type Ca^{2+} channels. *Nature, 366*, 158–161.

Yoshikawa, F., Morita, M., Monkawa, T., Michikawa, T., Furuichi, T., & Mikoshiba, K. (1996). Mutational analysis of the ligand binding site of the inositol 1,4,5-trisphosphate receptor. *The Journal of Biological Chemistry, 271*, 18277–18284.

Yoshikawa, F., Uchiyama, T., Iwasaki, H., Tomomori-Satoh, C., Tanaka, T., Furuichi, T., et al. (1999). High efficient expression of the functional ligand binding site of the inositol 1,4,5-triphosphate receptor in Escherichia coli. *Biochemical and Biophysical Research Communications, 257*, 792–797.

Zhang, D., Boulware, M. J., Pendleton, M. R., Nogi, T., & Marchant, J. S. (2007). The inositol 1,4,5-trisphosphate receptor (Itpr) gene family in Xenopus: Identification of type 2 and type 3 inositol 1,4,5-trisphosphate receptor subtypes. *The Biochemical Journal, 404*, 383–391.

CHAPTER 10

Adenophostins: High-Affinity Agonists of IP_3 Receptors

Ana M. Rossi,* Andrew M. Riley,† Barry V. L. Potter,† and Colin W. Taylor*

*Department of Pharmacology, University of Cambridge, Cambridge, United Kingdom
†Department of Pharmacy and Pharmacology, Wolfson Laboratory of Medicinal Chemistry, University of Bath, Claverton Down, Bath, United Kingdom

I. OVERVIEW

Adenophostins are naturally occurring high-affinity agonists of inositol 1,4,5-trisphosphate receptors (IP_3R). Synthesis of adenophostin analogs, while perhaps more difficult than for simple analogs of IP_3, is feasible, allowing medicinal chemistry to be combined with biological evaluation of adenophostin analogs to address the basis of adenophostin activity. Extensive analyses suggest that IP_3R behave similarly whether activated by IP_3 or adenophostin, but the latter typically binds with ~10-fold greater affinity to all IP_3R subtypes. The equatorial 3″,4″-phosphates and 2″-hydroxyl of adenophostin mimic the essential 4,5-phosphates and 6-hydroxyl groups of IP_3, but the structural basis of the enhanced affinity of adenophostins is less clear. We review evidence showing that the IP_3-binding core (IBC) of the IP_3R (residues 224–604) is sufficient to mediate this high-affinity binding. The 2′-phosphate of adenophostin has been suggested to provide a supraoptimal mimic of the 1-phosphate of IP_3 and

Current Topics in Membranes, Volume 66 1063-5823/10 $35.00
Copyright 2010, Elsevier Inc. All right reserved.
DOI: 10.1016/S1063-5823(10)66010-3

thereby to enhance considerably the affinity of adenophostin for IP_3R. We suggest that a more likely explanation for high-affinity binding of adenophostin is that its adenine moiety forms a cation–π interaction with Arg-504 within the IBC. We conclude by assessing whether the high-affinity and metabolic stability of adenophostin are sufficient to account entirely for its biological actions.

II. DISCOVERY AND INITIAL CHARACTERIZATION OF ADENOPHOSTINS

Penicillium brevicompactum, a fungus cultured from Japanese soil samples, entered the limelight in late 1993 when adenophostins A and B (Fig. 1A), isolated from a culture broth, were shown to bind to cerebellar IP_3R with higher affinity than any other known ligand (Takahashi, Kagasaki, Hosoya, & Takahashi, 1994). Subsequent papers from the same laboratory first established the structures of adenophostins (Takahashi, Takeshi, & Takahashi, 1994) and soon afterward their absolute configurations when total syntheses of adenophostin A were reported (reviewed in Chrétien,

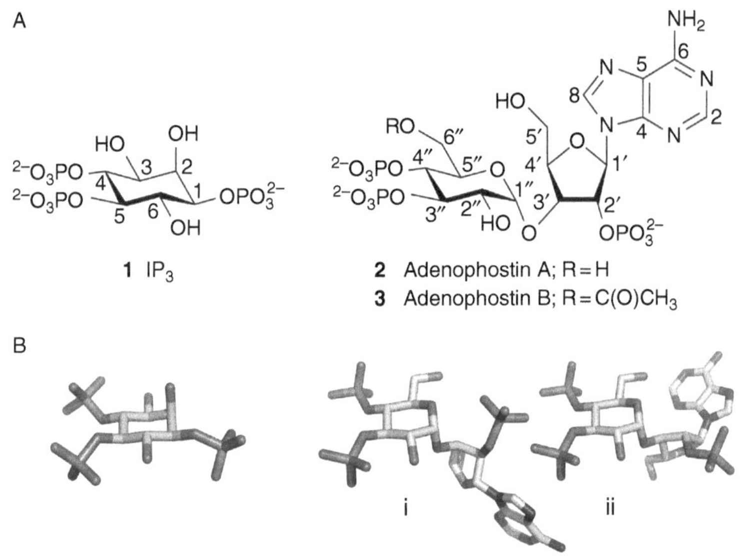

FIGURE 1 Structures of IP_3 and adenophostins. (A) Structures show the essential 4,5-bisphosphate and 6-hydroxyl groups of IP_3 (**1**) and the equivalent 3″,4″-bisphosphate and 2″-hydroxyl of adenophostins A and B (**2**, **3**). (B) Structure of IP_3 taken from the X-ray crystal structure of the IBC of IP_3R1 complexed with IP_3 (Bosanac et al., 2002) alongside two possible low-energy conformations of adenophostin A (i) and (ii).

Moitessier, Roussel, Mauger, & Chapleur, 2000; Section III). Takahashi, Kagasaki, et al. (1994) showed that adenophostins were neither antimicrobial nor acutely toxic to mice, there were no detectable interactions with cell-surface receptors or Ca^{2+} channels, and most importantly, adenophostins stimulated release of Ca^{2+} from intracellular stores via IP_3R. Adenophostins thus became the most potent known agonists of IP_3R1, and the first fungal metabolites known to interfere with mammalian Ca^{2+} signaling (Takahashi, Tanzawa, & Takahashi, 1994; Takahashi, Takeshi, et al., 1994).

In many papers that followed this original work, adenophostin has been said to bind to IP_3R with 100-fold greater affinity than IP_3. This claim derives largely from data in the original report (Takahashi, Tanzawa, et al., 1994), which was then reproduced in subsequent papers (Hotoda et al., 1999; Takahashi, Kagasaki, et al., 1994; Takahashi, Takeshi, et al., 1994), where an error in the calculation of the equilibrium dissociation constant (K_D) from the experimentally determined half-maximal inhibitory concentration (IC_{50}) caused the affinity of adenophostin relative to IP_3 to be exaggerated. The corrected value, in line with many subsequent analyses (Beecroft et al., 1999; Correa et al., 2001; de Kort, Regenbogen, et al., 2000), suggests that adenophostin usually binds with about 10-fold greater affinity than IP_3 to IP_3R. A similar $\sim$10-fold difference in potency is typical of most functional analyses of adenophostin, whether performed on permeabilized cells (Hirota et al., 1995; Rossi et al., 2009), microsomes (Hirota et al., 1995), purified IP_3R reconstituted into liposomes (Hirota et al., 1995), or by patch-clamp recording from the nuclear envelope (where adenophostin was estimated to be $\sim$60-fold more potent than IP_3; Mak, McBride, & Foskett, 2001).

Takahashi's group showed also that adenophostins do not interact with IP_4-binding sites, and nor are they metabolized by IP_3-5-phosphatase or IP_3-3-kinase (Takahashi, Tanzawa, et al., 1994). The metabolic stability of adenophostins accounts for their very long-lasting effects on Ca^{2+} signaling (hours in some cases; DeLisle et al., 1997; Marchant & Parker, 1998; Murphy, Riley, Lindley, Westwick, & Potter, 1997; Sato et al., 1998; Yoshida, Sensui, Inoue, Morisawa, & Mikoshiba, 1998). Finally, they suggested that despite the very different structures of IP_3 and adenophostins (Fig. 1A), the latter might mimic the key elements of IP_3 known to be important for its interaction with IP_3R, namely its vicinal 4- and 5-phosphates, and 6-hydroxyl (Nerou, Riley, Potter, & Taylor, 2001). Although adenophostins are based on a glucose ring, rather than the inositol ring of IP_3, the 3″- and 4″-phosphates and 2″-hydroxyl of adenophostin were proposed to mimic the essential 4,5-bisphosphate and 6-hydroxyl of IP_3 (Takahashi, Tanzawa, et al., 1994; Fig. 1A). Adenophostins lack a structure equivalent to the axial 2-hydroxyl of IP_3, but abundant evidence indicates that the 2-hydroxyl forms no important contacts with the IP_3-binding site of the IP_3R (Bosanac et al., 2002; Rossi et al., 2009). Furthermore, the 3-hydroxyl of IP_3 is replaced by a 5″-hydroxymethyl group in adenophostin, but the equivalent change in IP_3 analogs causes only a twofold decrease in affinity (Nerou et al., 2001). It is, however, noteworthy that adenophostins A and B, which differ in the size of the 5″-substituent (Fig. 1A), have the same affinity for IP_3R1 (Takahashi, Tanzawa, et al., 1994). Relatively bulky substitutions at the 5″-position of adenophostin thus appear to have lesser effects than equivalent modifications to the 3-position of IP_3. Presumably, as with IP_3 analogs (Nerou et al., 2001; Wilcox, Primrose, Nahorski, & Challiss, 1998), lack of a moiety equivalent to the axial 2-hydroxyl of IP_3 allows adenophostins to tolerate more bulky substitutions at the 5″-position. Finally, it was suggested that the 2′-phosphate of adenophostin may mimic the 1-phosphate of IP_3. The latter is not essential for activity but increases the affinity of inositol phosphates for IP_3R (Nerou et al., 2001; Section V).

By 1994, the scene had been set: adenophostins were the most potent known agonists of IP_3R1, they were selective, metabolically stable, there was a basic understanding of why they bound to IP_3R, and synthetic routes were starting to be explored for adenophostin analogs. Here, we review progress in understanding the structural determinants of the high-affinity binding of adenophostins. Adenophostins A and B differ only in that the 5″-hydroxymethyl of adenophostin B is acetylated (Fig. 1A). The two analogs have the same affinity and appear to have indistinguishable effects on IP_3R (Takahashi, Takeshi, et al., 1994) and most studies have used adenophostin A, we therefore refer to the latter as "adenophostin" throughout this review.

III. STRUCTURE AND SYNTHESIS OF ADENOPHOSTIN

The three components of adenophostin (bisphosphorylated glucose, phosphorylated ribose, and adenine; Fig. 2) are connected by potentially flexible linkages, and the central five-membered ribose ring is also intrinsically more flexible than a six-membered ring such as inositol or glucose. Many different conformations of adenophostin are therefore possible and the receptor-bound conformation is difficult to predict. Molecular modeling suggests that adenophostin can adopt extended conformations in which the adenosine holds the 2′-phosphate close to the glucose ring in a position similar to that occupied by the 1-phosphate of IP_3 (Fig. 1B; Hotoda et al., 1999; Marwood, Correa, Taylor, & Potter, 2000). NMR studies of adenophostin in solution also support some aspects of this type of model (Felemez, Marwood, Potter, & Spiess, 1999; Hotoda et al., 1999). Measurements of ^1H–^1H coupling constants

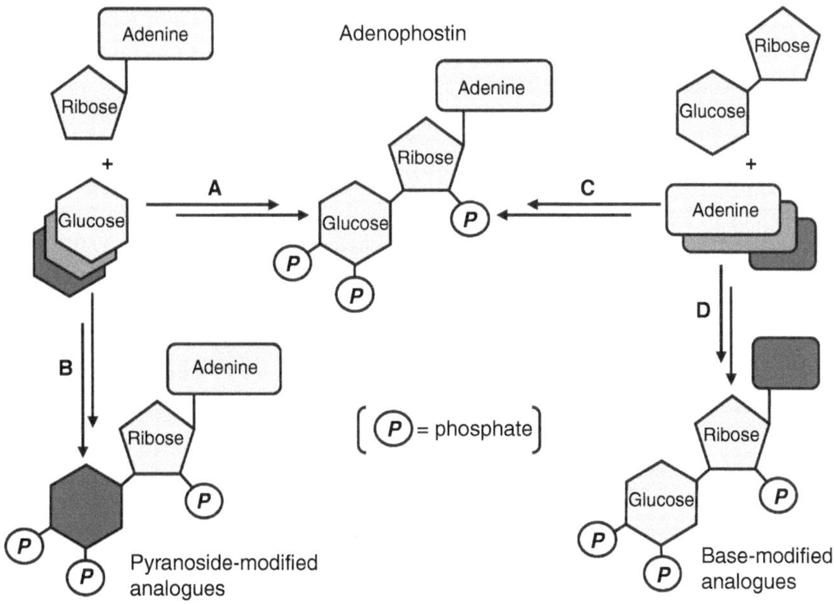

FIGURE 2 Strategies for the synthesis of adenophostin and its analogs. Adenophostin (center) can be considered to consist of three structural building blocks: a bisphosphorylated glucose, a phosphorylated ribose, and a base (adenine). The strategies used to synthesize adenophostin (routes A and C) can therefore be adapted to produce adenophostin analogs by using different six-membered ring sugars (pyranosides) in place of glucose (route B) or alternative bases and other aromatic structures in place of adenine (route D).

show that the glucose ring adopts a chair conformation with two equatorial phosphates (similar to the inositol ring of IP_3), and a Nuclear Overhauser Effect (NOE) between the H-3′ on ribose and H-1″ on glucose suggests they are relatively close to each other (Borissow et al., 2005; Hotoda et al., 1999). Further evidence comes from NMR titration experiments where protonation of phosphates alters their local electrostatic fields; these changes can then disturb the ^{1}H NMR shift of nearby protons. Such analyses suggest that in solution the 2′-phosphate of adenophostin is close to the H-1″ of glucose (Felemez et al., 1999). They further indicate that there is an interaction between the 2′- and 3″-phosphate groups of adenophostin, although it is less pronounced than the corresponding interaction between the 1- and 5-phosphates of IP_3. Finally, at physiological pH, the protonation states of the three phosphate groups in adenophostin are remarkably similar to those of the corresponding groups in IP_3 (Felemez et al., 1999). Molecular modeling and NMR data for adenophostin in solution are thus consistent with the notion that its 3″,4″-bisphosphate and 2″-hydroxyl group mimic the 4,5-bisphosphate and 6-hydroxyl of IP_3, while the 2′-phosphate of adenophostin is held in a region similar to that occupied by the 1-phosphate of IP_3 (Fig. 1).

Four total syntheses of adenophostin have been reported (Borissow et al., 2005; Hotoda, Takahashi, Tanzawa, Takahashi, & Kaneko, 1995; Marwood, Correa, et al., 2000; Marwood, Riley, Jenkins, & Potter, 2000; Van Straten, Van der Marel, & Van Boom, 1997a). Each involves linking structural building blocks (carbohydrates and nucleosides) in which most functional groups (hydroxyls and amines) are masked using temporary or semipermanent protecting groups (Fig. 2). Coupling of the protected building blocks, under conditions chosen to establish the required stereochemistry, is followed by removal of the protecting groups to expose key hydroxyls, which are then converted into protected phosphate esters in a two-step phosphitylation–oxidation sequence. Finally, all protecting groups are removed and the analog is purified using HPLC or ion-exchange chromatography. Although adenophostin is more complex than IP_3, its multicomponent structure means that convergent strategies can be used in its synthesis. These strategies can be generalized, allowing a range of adenophostin analogs to be synthesized by linking structural building blocks in a combinatorial manner (Fig. 2).

Two complementary synthetic strategies have been employed that differ in the building blocks chosen for coupling. In the first strategy, a protected glucose is coupled to a protected adenosine, which already contains the β-nucleosidic linkage (Fig. 2, route A; Hotoda et al., 1995; Marwood, Correa, et al., 2000; Marwood, Riley, et al., 2000). The method is readily adapted to give pyranoside-modified adenophostin analogs (Fig. 2, route B) by using different protected pyranoses, such as xylose or mannose, in place of glucose (e.g., as in analogs **4** and **5**, Fig. 3; Marwood, Riley, et al., 2000). In the second strategy, a protected disaccharide containing the glucose and ribose units is reacted with a protected adenine (Fig. 2, route C; Van Straten et al., 1997a). This method can be tailored to give base-modified analogs (Fig. 2, route D) in which the adenine is replaced by other nucleobases or by unrelated aromatic structures, for example, **6–8** (Marwood, Jenkins, Correa, Taylor, & Potter, 2000; Marwood, Shuto, Jenkins, & Potter, 2000; Shuto, Horne, Marwood, & Potter, 2001) and **9** (Sureshan, Trusselle, Tovey, Taylor, & Potter, 2006, 2008). Several elaborated adenophostins with added structures or groups have also been synthesized, for example, **10** and **11** (Borissow et al., 2005), **12** (Mochizuki et al., 2006), and **13** (Rosenberg, Riley, Laude, Taylor, & Potter, 2003). In order to determine which features of adenophostin are essential for activity, a range of simplified analogs has been developed, in which chemical groups or bonds were deleted, for example, **14–16** (Sureshan et al., 2009), **17** and **18** (Van Straten, Van der Marel, & Van Boom, 1997b), **19** (de Kort, Correa, et al., 2000), **20** (Jenkins, Marwood, &

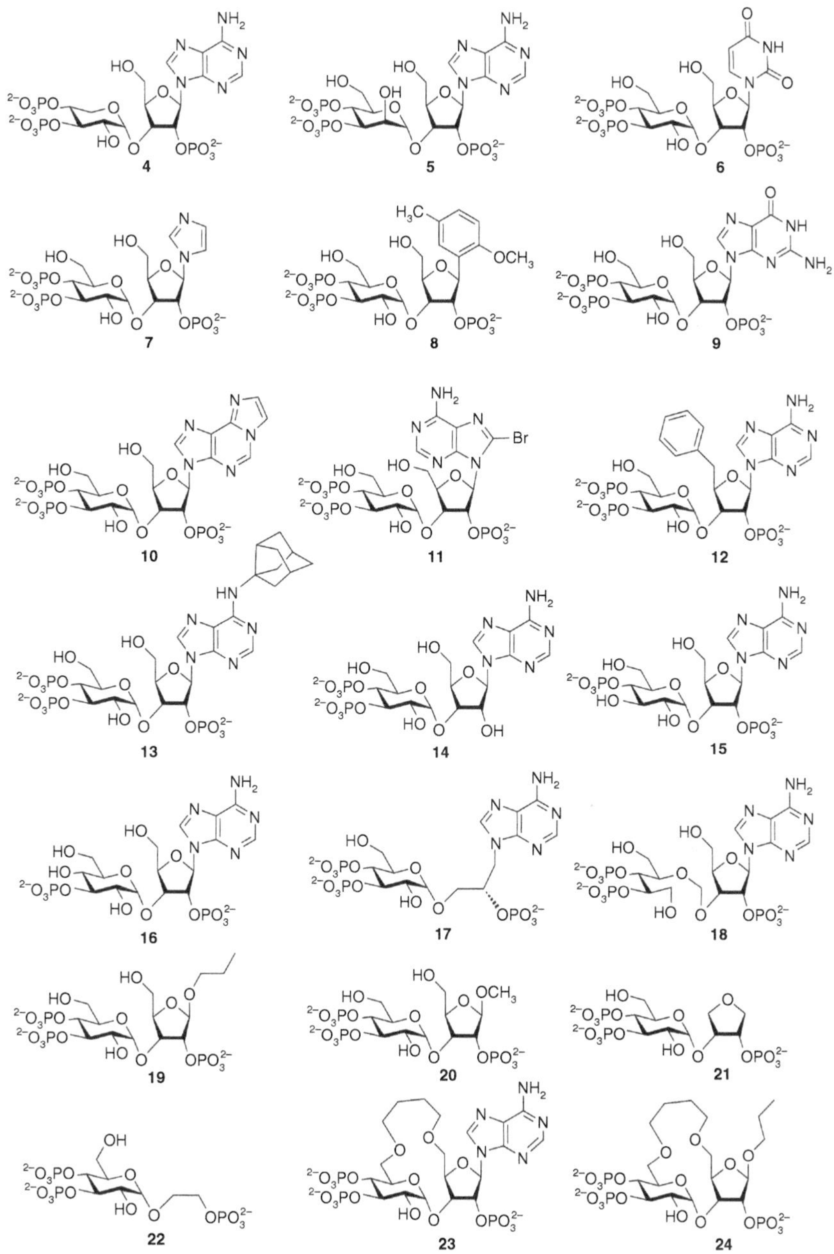

FIGURE 3 Structures of key adenophostin analogs.

Potter, 1997), **21** (Marwood, Riley, Correa, Taylor, & Potter, 1999; Shuto, Tatani, Ueno, & Matsuda, 1998) and related minimal analogs (Rosenberg, Riley, Marwood, Correa, Taylor & Potter, 2001), and **22** (Jenkins & Potter, 1995; Wilcox, Erneux, Primrose, Gigg, & Nahorski, 1995). Finally, conformationally restrained analogs such **23** and **24** (de Kort, Regenbogen, et al., 2000) have also been explored in attempts to gain insights into the receptor-bound conformation of adenophostin.

The synthetic advantages, together with the high-affinity of adenophostin for IP_3R, make adenophostin analogs attractive ligands with which to apply medicinal chemistry approaches to analysis of IP_3R activation and as a possible route to novel high-affinity ligands of IP_3R. The structures of the adenophostin analogs mentioned in this review, together with their numerical codes, are shown in Fig. 3.

IV. ACTIVATION OF IP_3R BY ADENOPHOSTIN

Adenophostin and IP_3 release Ca^{2+} from the same intracellular stores (Correa et al., 2001; DeLisle et al., 1997; Hirota et al., 1995; Marchant et al., 1997; Fig. 4A), the effect of each is competitively antagonized by heparin (Murphy et al., 1997; Takahashi, Tanzawa, et al., 1994), and in cells expressing largely IP_3R1, by an inhibitory anti-IP_3R1 antibody (Sato et al., 1998). In equilibrium binding experiments, IP_3 and adenophostin compete for the same binding site (Correa et al., 2001; Marchant et al., 1997; Murphy et al., 1997; Takahashi, Tanzawa, et al., 1994; Fig. 4B), and each only partially displaces an analog of phosphatidylinositol 4,5-bisphosphate from the IP_3R (Glouchankova, Krishna, Potter, Falck, & Bezprozvanny, 2000). Several reports suggest that binding of adenophostin and the Ca^{2+} release evoked by it (Borissow et al., 2005; de Kort, Regenbogen, et al., 2000; Ding et al., 2010; Hirota et al., 1995; Vanlingen et al., 2000) are more positively cooperative than for IP_3. The differences have not, however, been consistently observed (Marchant et al., 1997), and nor is it easy to explain cooperative binding of adenophostin to isolated IP_3-binding domains that appear to be monomeric (Ding et al., 2010; Vanlingen et al., 2000).

Submaximal concentrations of IP_3 or adenophostin cause quantal Ca^{2+} release (Hirota et al., 1995; Murphy et al., 1997). The biphasic kinetics of Ca^{2+} release from permeabilized cells are also similar for each (Adkins, Wissing, Potter, & Taylor, 2000; Hirota et al., 1995; Fig. 4C), consistent with IP_3 and adenophostin first activating IP_3R and then, at least in its native setting, causing it to switch to a partially inactivated state (Marchant & Taylor, 1998). The pattern of Ca^{2+} release from purified IP_3R is also similar after stimulation with half-maximally effective concentrations of IP_3 or

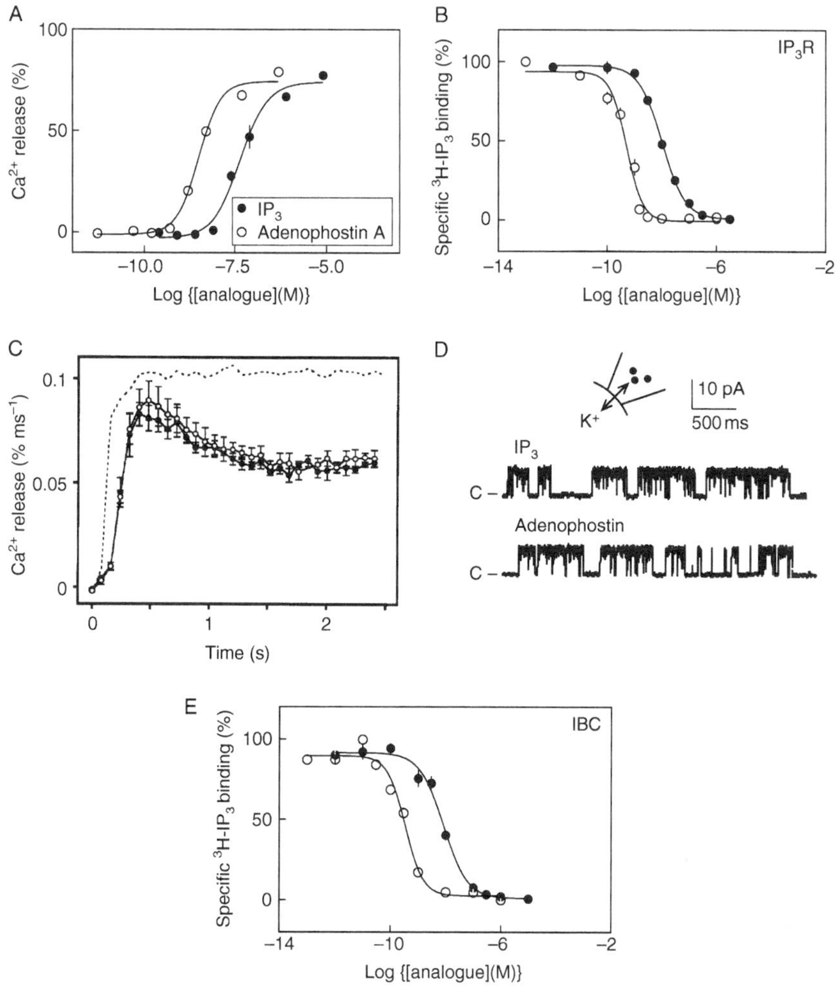

FIGURE 4 Adenophostin is a high-affinity full agonist of IP_3R. (A) Concentration-dependent release of Ca^{2+} from the intracellular stores evoked by IP_3 and adenophostin (the same code is used in all panels) from permeabilized DT40 cells expressing rat IP_3R1. (B) Specific binding of ^3H-IP_3 to full-length IP_3R1 purified from rat cerebellum is shown in the presence of the indicated concentrations of IP_3 or adenophostin. (C) Kinetics of $^{45}Ca^{2+}$ release from permeabilized hepatocytes evoked by maximally effective concentrations of IP_3 or adenophostin. The arrival of each ligand is shown by the dashed line. Results show the fractional release rates (i.e., the amount of $^{45}Ca^{2+}$ released in each 80-ms interval expressed as a fraction of the $^{45}Ca^{2+}$ remaining within the IP_3-sensitive Ca^{2+} stores at the beginning of that interval). Data are reproduced from Adkins, Wissing, et al. (2000) with permission from the Biochemical Society. (D) Typical recordings from excised nuclear patches taken from DT40 cells expressing rat IP_3R1 and stimulated with maximally effective concentrations of IP_3 or adenophostin in the presence of 0.5 mM ATP. The holding potential was +40 mV. C denotes the closed state. (E) Specific binding of ^3H- IP_3 to the IBC in the presence of the indicated concentrations of IP_3 or adenophostin.

adenophostin (Hirota et al., 1995). The effects of both IP_3 and adenophostin on Ca^{2+} release are similarly biphasically regulated by cytosolic Ca^{2+} (Mak et al., 2001), and inhibition of IP_3R1 by calmodulin is similar whether the receptors are occupied by IP_3 or adenophostin (Adkins, Morris, De Smedt, Török, & Taylor, 2000). Finally, in single-channel recordings from excised nuclear patches of DT40 cells expressing IP_3R1, maximally effective concentrations of IP_3 and adenophostin cause the IP_3R to open to the same single-channel conductance (γ_K) and open probability (P_o), and with similar mean channel open and closed times (Rossi et al., 2009; Fig. 4D). For IP_3R expressed in the plasma membrane (Dellis et al., 2006) and in the nuclear envelope of *Xenopus* oocytes (Mak et al., 2001), the single-channel properties are also similar whether they are activated by IP_3 or adenophostin (but see Section VI).

In intact cells too, the effects of adenophostin and IP_3 can be very similar: in ascidian (Yoshida et al., 1998) and mammalian (Sato et al., 1998) eggs, IP_3 and adenophostin similarly mimic egg activation, the post-fertilization Ca^{2+} oscillations (Sato et al., 1998), and degradation of IP_3R1 (Brind, Swann, & Carroll, 2000; He et al., 1999; Jellerette, He, Wu, Parys, & Fissore, 2000). The ability of adenophostin to cause the long-lasting activation of IP_3R required during early embryogenesis raises the possibility that it may be used in combination with injection of a spermatid for *in vitro* fertilization for men unable to produce functional spermatozoa (Sato et al., 1998).

There is presently no evidence to suggest that adenophostin and IP_3 differ significantly in their interactions with different IP_3R subtypes. Adenophostin binds with greater affinity than IP_3 to insect IP_3R (Swatton, Morris, & Taylor, 2001), to mammalian cells expressing predominantly IP_3R1 (Hirota et al., 1995; Murphy et al., 1997; Takahashi, Tanzawa, et al., 1994), IP_3R2 (Correa et al., 2001), or IP_3R3 (Missiaen et al., 1998), to Sf9 cells expressing each of the three mammalian IP_3R subtypes (Morris, Nerou, Riley, Potter, & Taylor, 2002), and to DT40 cells expressing rat IP_3R1 (Rossi et al., 2009; Fig. 4A).

Structure–activity analyses using a range of synthetic adenophostin analogs have systematically addressed the relative importance of each of the three components of adenophostin for its binding to IP_3R. Although adenophostin analogs differ considerably in their affinities for IP_3R, the active analogs, like adenophostin itself, appear to be full agonists. This conclusion must, however, be tempered with the qualification that single-channel recordings provide the best means of resolving the efficacy of IP_3R ligands, but these have been performed only for adenophostin itself (Mak et al., 2001; Rossi et al., 2009). For adenophostin analogs, the conclusion that they are full agonists rests on evidence that they release the same amount of Ca^{2+}as IP_3 from permeabilized cells, that the ratio of the concentrations required for half-maximal Ca^{2+} release (EC_{50}) and half-maximal occupancy of IP_3-binding sites (K_D) is similar for IP_3 and adenophostin analogs, and that the latter do not antagonize

Ca^{2+} release by IP_3 (Beecroft et al., 1999; Correa et al., 2001). An exception is acyclophostin (**17**, Fig. 3), in which the glucose and adenine are flexibly linked: it is a full agonist under physiological conditions, but a partial agonist at high pH (Beecroft et al., 1999). We tentatively suggest, although further work is required, that adenophostin is a high-affinity full agonist of all IP_3R subtypes, and that under physiological conditions its many analogs are also full agonists that differ only in their relative affinities for IP_3R.

Evaluation of pyranoside-modified analogs of adenophostin confirms that the glucopyranoside ring is analogous to the *myo*-inositol ring of IP_3. Thus, *xylo*-adenophostin (**4**, Fig. 3), which lacks the 5″-CH_2OH of adenophostin A, is only twofold less potent than adenophostin. Furthermore, a conformationally restricted analog, cyclophostin (**23**), in which the 5″-CH_2OH and 4′-CH_2OH are tethered to each other, is only fivefold less potent than adenophostin (de Kort, Regenbogen, et al., 2000). These results are consistent with the idea that the 5″-CH_2OH mimics the relatively unimportant 3-OH of IP_3, rather than the essential 6-OH. By contrast, *manno*-adenophostin (**5**), in which the orientation of the 2″-OH is switched from equatorial to axial, is 12-fold less potent than adenophostin (Correa et al., 2001), consistent with the notion that the 2″-OH of adenophostin mimics the essential 6-OH of IP_3. Opening of the glucose ring abolishes activity (**18**; Beecroft et al., 1999). As expected from the suggestion that the 3″,4″-bisphosphate of adenophostin mimics the 4,5-bisphosphate of IP_3, removal of either phosphate from adenophostin (**15** and **16**) greatly attenuates activity (Sureshan et al., 2009). We return in Section V to consider the relative roles of the 1-phosphate of IP_3 and 2′-phosphate of adenophostin.

Complete removal of the adenine moiety, as in ribophostin, for example (**20**), reduces the affinity to a value similar to that for IP_3 (Correa et al., 2001; Shuto et al., 1998; Rossi, Sureshan, Riley, Potter, & Taylor, 2010). However, even substantial modifications to the adenine moiety are well tolerated (Rosenberg et al., 2003). Its replacement with larger (e.g., etheno adenophostin, **10**, and N^6-noradamantyl adenophostin, **13**), brominated (**11**) or smaller (e.g., uridophostin, **6**) aromatic structures, even when entirely unrelated to adenine (e.g., **8**), leads to analog that are only two to threefold less potent than adenophostin (Borissow et al., 2005; Correa et al., 2001; Rosenberg et al., 2003). Even replacement of the 1′-OCH_3 of ribophostin with an imidazole ring (imidophostin, **7**) is sufficient to improve affinity by about threefold (Correa et al., 2001). It is noteworthy that all adenophostin analogs that are more potent than IP_3 have a ring system attached to the 1′-position of the ribose ring that would be capable of forming a cation–π interaction (Mecozzi, West, & Dougherty, 1996) with an appropriately placed cationic residue within the IP_3R (Rosenberg et al., 2003). We return to this point in Section V.

The third component of adenophostin is the ribose ring that links the glucose and adenine moieties. Acyclophostin (**17**), in which the rigid ribose ring is replaced by a flexible linker, is ~13-fold less potent than adenophostin (Beecroft et al., 1999; Correa et al., 2001), but still 9-fold more potent than the equivalent analog (Glc(2′,3,4)P_3, **22**) without an adenine moiety. Analyses of analogs that changed the stereochemistry of the *O*-glycosidic linkage between the glucose and ribose rings, and of the linkage between the ribose and adenine groups indicate that the former requires an α-linkage, while the latter requires β-stereochemistry (Correa et al., 2001).

The results aforementioned define the attributes of adenophostin that allow it to mimic IP_3 and so activate IP_3R (Fig. 5A), but we must now consider the features that allow it to bind to IP_3R with greater affinity than IP_3.

V. WHY DOES ADENOPHOSTIN BIND TO IP_3R WITH HIGH-AFFINITY?

Activation of an IP_3R begins with IP_3 binding to the IBC (residues 224–604), located toward the N-terminal of each subunit of the tetrameric IP_3R. A high-resolution structure of the IBC with IP_3 bound (Bosanac et al., 2002; Fig. 5B) revealed the structural basis for IP_3 recognition confirming many of the earlier predictions from structure–activity studies (Nerou et al., 2001; Wilcox et al., 1998). The two domains of the IBC (α and β) form a clam-like structure, the inside of which includes many basic residues that coordinate the phosphate groups of IP_3 (Bosanac et al., 2002; Fig. 5B). The essential 4,5-bisphosphate and 6-hydroxyl groups of IP_3 engage in an extensive network of interactions with the IBC. The 4-phosphate is hydrogen-bonded with several residues in the β-domain, the 5-phosphate forms hydrogen bonds with residues predominantly within the α-domain, and the 6-hydroxyl interacts via a water molecule with the backbone of K569 in the α-domain (Bosanac et al., 2002; Fig. 5B). We can envisage, although it remains speculative in the absence of structures of the IBC without IP_3 bound, that IP_3 binding might pull the α- and β-domains together, closing its clam-like structure in a manner reminiscent of glutamate binding to ionotropic glutamate receptors (Taylor, da Fonseca, & Morris, 2004), thereby causing the IBC to adopt a more constrained structure (Chan et al., 2007). In this simple scheme, the requirement for a vicinal 4,5-bisphosphate moiety in all agonists of IP_3R reflects the need to "cross-bridge" the α- and β-domains of the IBC via the 5- and 4-phosphates, respectively. It is not yet clear how these conformational changes in the IBC cause gating of the pore, formed by transmembrane regions close to the C-terminal of the IP_3R. N-terminal residues that comprise the so-called "suppressor domain" (SD; residues 1-223) are clearly

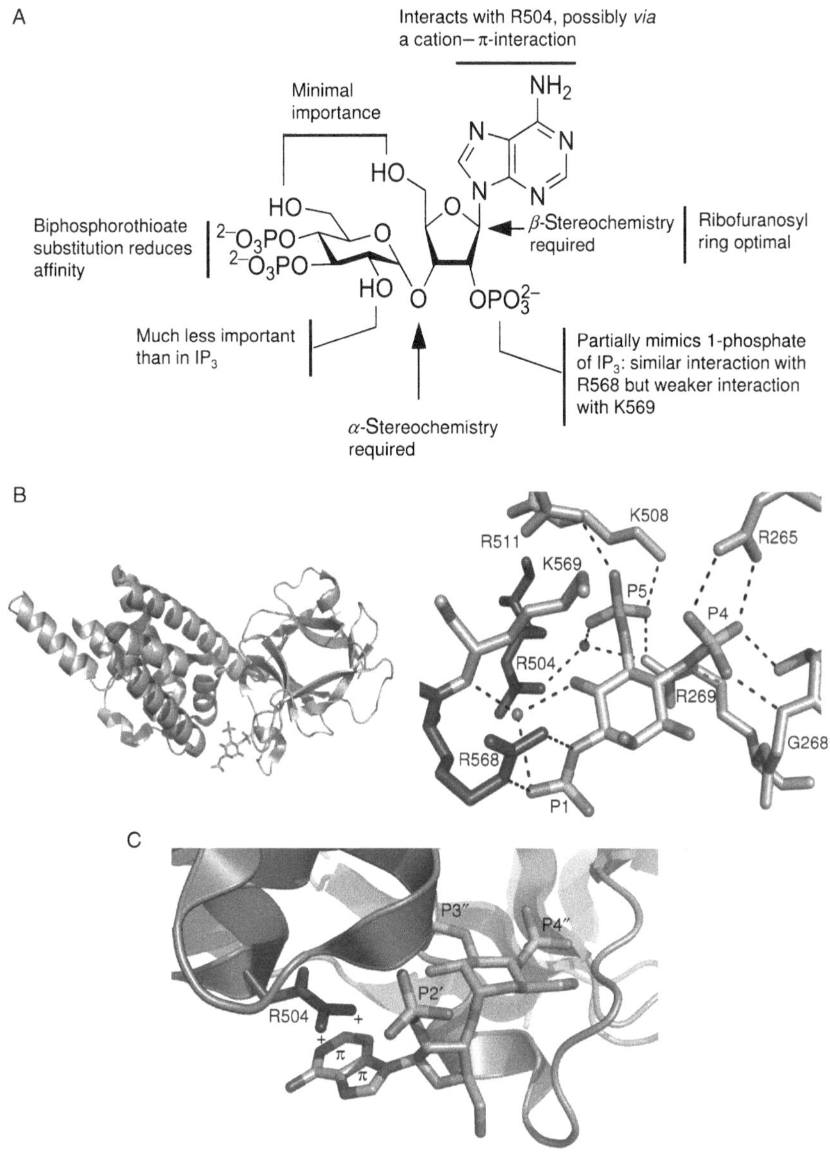

FIGURE 5 Key features contributing to recognition of adenophostin by IP_3R. (A) Summary of the structure–activity relationships for adenophostin binding to IP_3R. (B) Structure of the IBC (Bosanac et al., 2002; PDB 1N4K) with IP_3 bound. The right enlarged panel highlights some of the residues involved in IP_3-binding. Spheres represent water. (C) Model of adenophostin binding to the IBC (Rosenberg et al., 2003) showing the proposed cation–π-interaction between the adenine ring of adenophostin and R504 of the IBC. (See Color Insert.)

essential for activation, and we have suggested that binding of IP_3 evokes conformational changes in the IBC that then propagate entirely via the SD to the pore region (Rossi et al., 2009). The details, which have yet to be resolved, are less important in the present context than the observation that full agonists, like IP_3, divert more binding energy into rearranging the relationship between the IBC and SD than do partial agonists (Rossi et al., 2009). The result is that the presence of the SD reduces the affinity for full agonists more than it does for partial agonists (Rossi et al., 2009). Key points for the present discussion are that the IBC alone mediates the initial recognition of IP_3, and comparison of ligand binding to the IBC alone or with the SD attached (NT, residues 1–604) can provide insight into the initial conformational changes that culminate in opening of the pore.

Competitive binding of IP_3 and adenophostin shows that they bind to an overlapping site on IP_3R (Fig. 4B), but there have been suggestions that neither the IBC nor the NT is alone sufficient to mediate fully the 10-fold increase in adenophostin affinity relative to IP_3 (Morris et al., 2002; Vanlingen et al., 2000). We recently revisited this issue using untagged N-terminal fragments of IP_3R1 and in media that more faithfully replicate the cytosol-like conditions in which the functional assays are performed. These analyses showed that adenophostin binds to the IBC with $\sim$10-fold greater affinity than IP_3 (Ding et al., 2010; Fig. 4E). Furthermore, the effects of the SD are indistinguishable for adenophostin and IP_3 binding: both ligands bind to the NT with lesser affinity than to the IBC, consistent with binding of each ligand diverting similar amounts of binding energy ($\sim$6 kJ/mol) into rearranging the relationship between the IBC and SD (Ding et al., 2010). These results are consistent with the idea that IP_3 and adenophostin evoke similar conformational changes in the IP_3R leading to similar gating of the pore (Section IV, Fig. 4D). We conclude that, in seeking to explain the structural determinants of high-affinity binding of adenophostin to IP_3R, we need, at least initially, to consider only its interactions with the IBC.

Two possible explanations have been proposed to account for the greater affinity of adenophostin. The first recognizes that the 2′-phosphate of adenophostin is probably functionally equivalent to the 1-phosphate of IP_3. The latter is not essential for binding of IP_3, but increases its affinity (Nerou et al., 2001). The suggestion is that the 2′-phosphate of bound adenophostin may be forced into a position that allows it to bind more strongly than the 1-phosphate of IP_3 (Takahashi, Tanzawa, et al., 1994; Section III). This idea is consistent with the first molecular modeling studies, which suggested that in some conformations the 2′-phosphate of adenophostin is further from the 3″,4″-bisphosphate motif than is the 1-phosphate from the 4,5-bisphosphate motif of IP_3 (Hotoda et al., 1999; Wilcox et al., 1995). Alternatively, direct interactions between elements of the adenine moiety and the IBC may strengthen binding. Such interactions would need to be tolerant of even radical changes to this moiety (Section IV), although all high-affinity adenophostin analogs retain structures at this position that would be capable of forming a cation–π interaction (Mecozzi et al., 1996; Rosenberg et al., 2003). The two possibilities are not mutually exclusive.

The first suggestion predicts that the high affinity of adenophostin requires its 2′-phosphate, and that this should contribute more to adenophostin binding than does the 1-phosphate to IP_3 binding. An initial report suggested that enzymatic removal of the 2′-phosphate from adenophostin caused a much larger (1800-fold) decrease in affinity than did removal of the 1- phosphate from IP_3 (<300-fold; Takahashi, Kagasaki, et al., 1994). However, as noted earlier (Section II), it is difficult from these published data to calculate reliably the K_D for 2′-dephospho-adenophostin (**14**, Fig. 3). Nevertheless, the idea that the 2′-phosphate of adenophostin is a supraoptimal mimic of the 1-phosphate of IP_3 has dominated the field for 15 years. The first serious challenge to this idea was provided by studies of analogs of adenophostin in which the 2′-phosphate was either conformationally restricted (cyclophostin, **23** and cycloribophostin, **24**, Fig. 3; de Kort, Regenbogen, et al., 2000) or freed of the constraints imposed by the ribose ring (Glc(2′,3,4)P_3, **22** and acyclophostin, **17**; Correa et al., 2001). In each of these pairs of analogs, the affinity was increased by at least 10-fold when the adenine moiety was present, which is similar to the increase in affinity when adenine is added to ribophostin to give adenophostin (Correa et al., 2001; Shuto et al., 1998). These results suggest that the adenine moiety increases binding affinity similarly whether the orientation of the 2′-phosphate is partially restricted by the ribose ring (adenophostin, **2**), more severely restricted by an additional tethering of the glucose and ribose rings (cyclophostin, **23**) or less restricted by opening of the ribose ring (acyclophostin, **17**). It is difficult to reconcile these observations with the suggestion that the primary effect of the adenine moiety is to position optimally the 2′-phosphate of adenophostin. Our recent results, combining systematic site-directed mutagenesis of the IBC with structure–activity studies, further suggest that supraoptimal positioning of the 2′-phosphate is not the key determinant of the high-affinity binding of adenophostin (Rossi et al., 2010).

In binding analyses to both full-length IP_3R and its N-terminal fragments, removal of the 2′-phosphate from adenophostin has lesser effects on affinity than does removal of the 1-phosphate from IP_3. Functional analyses of Ca^{2+} release further support the conclusion that the 2′-phosphate of adenophostin contributes less to its activity than does the 1-phosphate of IP_3 (Rossi et al., 2010). The 1-phosphate of IP_3 interacts directly with only R568 of the IBC and indirectly, via water, with the backbone of K569 (Bosanac et al., 2002; Fig. 5B). Mutation of R568 (R568Q) similarly reduced (by ~40-fold) the affinity of the IP_3R for IP_3 and adenophostin, while minimally affecting the binding of 2′-dephospho-adenophostin and inositol 4,5-bisphosphate (4,5-IP_2). This suggests that the 1-phosphate and 2′-phosphate groups interact similarly with R568. The greater role of the 1-phosphate in IP_3 binding probably results from it interacting more strongly than the 2′-phosphate of adenophostin with the backbone of K569 (Bosanac et al., 2002). Collectively, these results demonstrate that the greater affinity of adenophostin for IP_3R does not result from its 2′-phosphate behaving as a supraoptimal mimic of the 1-phosphate of IP_3.

We have used molecular modeling to dock models of adenophostin into the IBC (Rosenberg et al., 2003). This kind of approach has its limitations; although the ligand is treated as flexible, the binding site is effectively rigid, and it is difficult to deal with water molecules that may play a role in ligand binding. Nevertheless, one of the predicted binding modes suggested a cation–π interaction between the guanidinium side-chain of R504 and the adenine group of adenophostin (Fig. 5C). Similar interactions have often been observed in protein structures with side chains bound to nucleobase ligands, and arginine–adenine is the most frequent cation–π pair (Biot, Buisine, Kwasigroch, Wintgens, & Rooman, 2002). R504 is one of several residues to form hydrogen bonds with the 5-phosphate of IP_3 (Bosanac et al., 2002) and presumably also with the equivalent 3″-phosphate of adenophostin. Unsurprisingly, therefore, mutating R504 to an uncharged residue (R504Q) reduced the affinity of the IBC for both IP_3 and adenophostin, but the decrease was more than 30-fold greater for adenophostin (Rossi et al., 2010). These selective effects of R504Q on adenophostin binding were confirmed by functional analyses of cells expressing only the mutant IP_3R, where again the sensitivity to adenophostin was more dramatically reduced than was the sensitivity to IP_3 (Rossi et al., 2010). Collectively, these results are consistent with our suggestion that a cation–π interaction between the adenine of adenophostin and the side-chain of R504 is an important determinant of the high-affinity binding of adenophostin. It is noteworthy that this proposed cation–π interaction is with a residue within the α-domain of the IBC. This led us to consider whether this interaction might be capable of partially substituting for that between the 3″-phosphate and the α-domain and so allow activation of the IBC in the absence of the otherwise essential vicinal pair of phosphate groups. Our results demonstrate that 3″-dephospho-adenophostin (**15**, Fig. 3) is a low-affinity agonist of IP_3R ($\sim$1500-fold less potent than IP_3) that requires interaction with R504, whereas 4″-dephospho-adenophostin (**16**) is effectively inactive (Sureshan et al., 2009). 3″-Dephospho-adenophostin is the first agonist of the IP_3R, albeit with low affinity, to lack a vicinal bisphosphate moiety. This is significant because it suggests the possibility of developing less polar ligands of

IP_3R in which an essential contact with the α-domain of the IBC is mediated by a cation–π interaction with R504 rather than the usual contact with the 3″-phosphate of adenophostin.

In summary, the IBC alone mediates high-affinity binding of adenophostin and it seems reasonable to suppose that the major ensuing conformational changes are similar whether the IP_3R has IP_3 or adenophostin bound. It is clear that the 3″,4″-phosphates and 2″-hydroxyl of adenophostin mimic the essential 4,5-phosphates and 6-hydroxyl of IP_3. The 2′-phosphate of adenophostin substantially mimics the 1-phosphate of IP_3 and thereby increases affinity, but it does not underlie the ability of adenophostin to bind with greater affinity than IP_3 to IP_3R. A more likely explanation for the high-affinity binding of adenophostin is that its adenine moiety forms cation–π-interaction with the side-chain of R504 in the IBC (Fig. 5C).

VI. IS ADENOPHOSTIN MORE THAN A STABLE, HIGH-AFFINITY AGONIST OF IP_3R?

Two sets of observations sit uncomfortably with the simple notion that IP_3 and adenophostin differ only in their metabolic stability and affinity for IP_3R. First, single-channel analyses have suggested that ATP differentially influences IP_3R activated by IP_3 and adenophostin (Mak et al., 2001). Second, Ca^{2+} signals evoked by microinjection of adenophostin into intact cells have been reported to differ from those evoked by IP_3. Clearly with IP_3 and adenophostin making different contacts with the IBC (Section V), it is entirely plausible that active IP_3R behave differently when activated by each agonist. In this final section, we briefly consider whether available evidence can be adequately explained by assuming that IP_3 and adenophostin differ only in their metabolic stability and affinities for IP_3R.

Patch-clamp recording from endogenous IP_3R of *Xenopus* nuclear envelope showed that in the presence of ATP, a coregulator of all IP_3R (Betzenhauser et al., 2008), the behavior of active IP_3R was indistinguishable when activated by IP_3 or adenophostin. The latter, however, bound with substantially greater affinity than IP_3 (Mak et al., 2001). But in the absence of ATP, the affinity of adenophostin appeared to be reduced to a level similar to that of IP_3, and the efficacy of adenophostin and its analogs (**20** and **21**) was substantially reduced. The latter was reflected in shortened mean channel open times. From these persuasive observations, a model was proposed in which ATP binding to an allosteric site is required for the IBC to adopt a conformation that allows its effective interaction with the 2′-phosphate of adenophostin. In the absence of ATP, therefore, adenophostin was proposed to bind more weakly to the IBC, and in this binding mode it was presumably also less effectively coupled to channel gating. By contrast, the 1-phosphate of IP_3 was proposed to bind effectively to the IBC irrespective of the presence of ATP. The details of this scheme are clearly wrong. 2′-Dephospho-adenophostin appears not to be a partial agonist (Sureshan et al., 2009). In the absence of ATP, IP_3 and adenophostin evoke the same peak rate of Ca^{2+} release (Adkins, Wissing, et al., 2000; Fig. 4C), and Ca^{2+} release via IP_3R reconstituted into liposomes is, in the absence of ATP, 10-times more sensitive to adenophostin than to IP_3 (Hirota et al., 1995). There is no evidence that ATP, ADP, or AMP selectively affect binding of adenophostin (Beecroft et al., 1999; Morris et al., 2002). Most analyses of IP_3 and adenophostin binding are performed in the absence of ATP, and all concur in showing that adenophostin binds with substantially greater affinity than IP_3 (Section II). Adenophostin binds to the isolated IBC, which lacks an ATP-binding site, with 10-fold greater affinity than IP_3 (Ding et al., 2010; Fig. 4E). Finally, our analysis of mutant IBC suggests that in the absence of ATP, the 1-phosphate of IP_3 and 2′-phosphate of adenophostin interact primarily with the same residue (R568; Rossi et al., 2010; Section V). The results from Mak et al. (2001) are intriguing and single-channel recording can provide insight that lower resolution methods fail to reveal, but the counterbalancing weight of evidence suggests that neither the high-affinity binding of adenophostin nor its ability to activate IP_3R fully requires ATP. Further comparisons of the activity of IP_3 and adenophostin at the single-channel level are clearly required to resolve the issue.

Two different observations in intact cells have also suggested that adenophostin and IP_3 may differ more profoundly in their interactions with IP_3R. First, the behavior of elementary Ca^{2+} release events is different when they are evoked by IP_3 and adenophostin. In *Xenopus* oocytes, Ca^{2+} waves propagate more slowly when initiated by adenophostin rather than by IP_3 (Bird, Takahashi, Tanzawa, & Putney, 1999; Machaca & Hartzell, 1999). This has been proposed to result from the high-affinity of the IP_3R for adenophostin allowing IP_3R substantially to deplete the cytosol of the low concentrations of adenophostin required for IP_3R activation. Slow diffusion of adenophostin as it is detained by its binding to IP_3R might then give rise to a slowly propagating Ca^{2+} wave. Clearly with this explanation, the high-affinity of adenophostin for IP_3R is sufficient to account for the different Ca^{2+} signals evoked by it and IP_3. More difficult to explain is the observation that IP_3, but not adenophostin, evoked repetitive Ca^{2+} waves, and the Ca^{2+} puffs evoked by adenophostin rose more rapidly and then terminated more quickly than those evoked by IP_3 (Marchant & Parker, 1998). These results suggest that the rate of agonist dissociation from the IP_3R is not immediately responsible for terminating a puff, and recent evidence suggests that local depletion of luminal Ca^{2+} is not the cause either (Smith & Parker, 2009).

The most probable cause of puff termination is that local Ca^{2+} feedback rapidly inactivates neighboring IP_3R preventing any further recruitment (Smith & Parker, 2009), but available evidence suggests that IP_3R activated by adenophostin and IP_3 are inhibited by cytosolic Ca^{2+} (Mak et al., 2001). In the absence of a better understanding of the processes that terminate puffs, which might include such processes as Ca^{2+} feedback, regulation by luminal Ca^{2+} and disassembly of IP_3R clusters (Rahman, Skupin, Falcke, & Taylor, 2009), it is difficult to account specifically for the ability of IP_3 and adenophostin to evoke puffs with different time courses. The possibility clearly remains that subtle, though yet unseen, differences in the gating of IP_3R by the two ligands may underlie the differences.

Second, under some conditions, adenophostin and its analogs appear more effective than IP_3 in causing activation of store-operated Ca^{2+} entry (SOCE). The latter requires activation of Orai channels in the plasma membrane by STIM1 in the ER when it sheds Ca^{2+} from its luminal Ca^{2+}-binding site (Park et al., 2009). The essential feature of SOCE is that it is active only when intracellular stores are substantially depleted of Ca^{2+} (Parekh & Putney, 2005): under physiological conditions, sustained activation of SOCE therefore requires sustained activation of IP_3R. A second feature of SOCE that is relevant to the present discussion is that Ca^{2+} both inhibits the Orai channels that mediate SOCE (Park et al., 2009) and it inhibits associated IP_3R.

Several studies have suggested that depletion of intracellular Ca^{2+} stores by adenophostin, acting via IP_3R, more effectively activates SOCE than does IP_3 (DeLisle et al., 1997; Gregory et al., 1999; Hartzell, Machaca, & Hirayama, 1997; Huang, Takahashi, Tanzawa, & Putney, 1998; Machaca & Hartzell, 1999). The difference is most pronounced when SOCE is measured with limited cytosolic Ca^{2+}-buffering so that Ca^{2+} entry is allowed, as occurs physiologically, to cause changes in the free cytosolic Ca^{2+} concentration (Broad, Armstrong, & Putney, 1999; Huang et al., 1998; Parekh, Riley, & Potter, 2002). Why should adenophostin be better able than IP_3 to sustain activation of SOCE in the face of an elevated cytosolic Ca^{2+} concentration? The inability of a stable analog of IP_3 (2,4,5-IP_3) to replicate the effect of adenophostin suggests that metabolic stability is probably not the key factor (Huang et al., 1998; Parekh et al., 2002). Nor is it likely that adenophostin targets an intracellular Ca^{2+} store that is more favorably coupled to activation of SOCE (Turner, Fleig, Stokes, Kinet, & Penner, 2003) because there is no evidence that adenophostin interacts differently with different IP_3R subtypes (Section II). Perhaps the most likely explanation is that adenophostin is better able than IP_3 to protect IP_3R from the inhibitory effect of Ca^{2+} entering via the SOCE pathway (Broad et al., 1999). We have demonstrated that Ca^{2+} inhibits IP_3R only when IP_3 (or presumably adenophostin) has dissociated (Adkins & Taylor, 1999). The high-affinity of adenophostin for IP_3R and its slower dissociation rate (Adkins, Wissing, et al., 2000) may therefore ensure that in the face of sustained Ca^{2+} entry, adenophostin-activated IP_3R remain active for longer than IP_3R activated by IP_3. A fly in this ointment is, however, provided by the observation that ribophostin (**20**, which is about twofold less potent than IP_3; Correa et al., 2001) effectively mimics adenophostin, while *manno*-adenophostin (**5**, which has a similar potency to IP_3) does not (Parekh et al., 2002). These results raise the possibility that adenophostin may have effects distinct from one of its closest structural analogs (**5**, which differs from adenophostin only in the orientation of its 2″-OH, Fig. 3). Further work is required to assess this possibility, but in the interim it seems reasonable to conclude that the different effects of IP_3 and adenophostin on SOCE may be substantially explained by the greater affinity of adenophostin for IP_3R.

In conclusion, we suggest that the key features of adenophostin are that it is a high-affinity full agonist of IP_3R. The 3″,4″-phosphates and 2″-hydroxyl of adenophostin effectively mimic the essential elements of all active ligands of IP_3R, and its 2′-phosphate group partially mimics the 1-phosphate of IP_3. The increased affinity of adenophostin seems, at least in substantial part, to derive from a cation–π interaction between its adenine moiety and the side-chain of R504 within the IBC.

Acknowledgments

We thank the Wellcome Trust and BBSRC for financial support. Ana M. Rossi is a Fellow of Queens' College, Cambridge.

References

Adkins, C. E., Morris, S. A., De Smedt, H., Török, K., & Taylor, C. W. (2000). Ca^{2+}-calmodulin inhibits Ca^{2+} release mediated by type-1, -2 and -3 inositol trisphosphate receptors. *The Biochemical Journal, 345*, 357–363.

Adkins, C. E., & Taylor, C. W. (1999). Lateral inhibition of inositol 1,4,5-trisphosphate receptors by cytosolic Ca^{2+}. *Current Biology, 9*, 1115–1118.

Adkins, C. E., Wissing, F., Potter, B. V. L., & Taylor, C. W. (2000). Rapid activation and partial inactivation of inositol trisphosphate receptors by adenophostin A. *The Biochemical Journal, 352*, 929–933.

Beecroft, M. D., Marchant, J. S., Riley, A. M., Van Straten, N. C. R., Van der Marel, G. A., Van Boom, J. H., et al. (1999). Acyclophostin: A ribose-modified analog of adenophostin A with high affinity for inositol 1,4,5-trisphosphate receptors and pH-dependent efficacy. *Molecular Pharmacology, 55*, 109–117.

Betzenhauser, M. J., 2nd, Wagner, L. E., Iwai, M., Michikawa, T., Mikoshiba, K., & Yule, D. I. (2008). ATP modulation of Ca^{2+} release by type-2 and type-3 $InsP_3R$: Differing ATP sensitivities and molecular determinants of action. *The Journal of Biological Chemistry, 283*, 21579–21587.

Biot, C., Buisine, E., Kwasigroch, J.-M., Wintgens, R., & Rooman, M. (2002). Probing the energetic and structural role of amino acid/nucleobase cation–π interactions in protein-ligand complexes. *The Journal of Biological Chemistry, 277*, 40816–40822.

Bird, G. S. J., Takahashi, M., Tanzawa, K., & Putney, J. W. J. (1999). Adenophostin A induces spatially restricted calcium signaling in *Xenopus laevis* occytes. *The Journal of Biological Chemistry, 274*, 20643–20649.

Borissow, C. N., Black, S. J., Paul, M., Tovey, S. C., Dedos, S. G., Taylor, C. W., et al. (2005). Adenophostin A and analogues modified at the adenine moiety: Synthesis, conformational analysis and biological activity. *Organic & Biomolecular Chemistry, 3*, 245–252.

Bosanac, I., Alattia, J.-R., Mal, T. K., Chan, J., Talarico, S., Tong, F. K., et al. (2002). Structure of the inositol 1,4,5-trisphosphate receptor binding core in complex with its ligand. *Nature, 420*, 696–700.

Brind, S., Swann, K., & Carroll, J. (2000). Inositol 1,4,5-trisphosphate receptors are down-regulated on mouse oocytes in response to sperm or adenophostin A but not to increases in intracellular Ca^{2+} or egg activation. *Developmental Biology, 223*, 251–265.

Broad, L. M., Armstrong, D. L., & Putney, J. W., Jr. (1999). Role of the inositol 1,4,5-trisphosphate receptor in Ca^{2+} feedback inhibition of calcium release-activated calcium current I_{CRAC}. *The Journal of Biological Chemistry, 274*, 32881–32888.

Chan, J., Whitten, A. E., Jeffries, C. M., Bosanac, I., Mal, T. K., Ito, J., et al. (2007). Ligand-induced conformational changes via flexible linkers in the amino-terminal region of the inositol 1,4,5-trisphosphate receptor. *Journal of Molecular Biology, 373*, 1269–1280.

Chrétien, F., Moitessier, N., Roussel, F., Mauger, J.-P., & Chapleur, Y. (2000). Carbohydrate-based mimics of D-*myo*-inositol 1,4,5-trisphosphate. *Current Organic Chemistry, 4*, 513–534.

Correa, V., Riley, A. M., Shuto, S., Horne, G., Nerou, E. P., Marwood, R. D., et al. (2001). Structural determinants of adenophostin A activity at inositol trisphosphate receptors. *Molecular Pharmacology, 59*, 1206–1215.

de Kort, M., Correa, V., Valentijn, A. R. P. M., van der Marel, G. A., Potter, B. V. L., Taylor, C. W., et al. (2000). Synthesis of potent agonists of the D-*myo*-inositol 1,4,5-trisphosphate receptor based on clustered disaccharide polyphosphate analogs of adenophostin A. *Journal of Medicinal Chemistry, 43*, 3295–3303.

de Kort, M., Regenbogen, A. D., Overleeft, H. S., Challis, R. A. J., Iwata, Y., Miyamoto, S., et al. (2000). Synthesis and biological evaluation of cyclophostin A 5′, 6″-tethered analog of adenophostin. *Tetrahedron, 56*, 5915–5928.

DeLisle, S., Marksberry, E. W., Bonnett, C., Jenkins, D. J., Potter, B. V. L., Takahashi, M., et al. (1997). Adenophostin A can stimulate Ca^{2+} influx without depleting the inositol 1,4,5-trisphosphate-sensitive Ca^{2+} stores in the *Xenopus* oocyte. *The Journal of Biological Chemistry, 272*, 9956–9961.

Dellis, O., Dedos, S., Tovey, S. C., Rahman, T.-U., Dubel, S. J., & Taylor, C. W. (2006). Ca^{2+} entry through plasma membrane IP_3 receptors. *Science, 313*, 229–233.

Ding, Z., Rossi, A. M., Riley, A. M., Rahman, T., Potter, B. V. L., & Taylor, C. W. (2010). Binding of IP_3 and adenophostin A to the N-terminal of IP_3 receptors: Thermodynamic analysis using fluorescence polarization with a novel IP_3 receptor ligand. *Molecular Pharmacology, 77*, 995–1004.

Felemez, M., Marwood, R. D., Potter, B. V., & Spiess, B. (1999). Inframolecular studies of the protonation of adenophostin A: Comparison with 1-D-*myo*-inositol 1,4,5-trisphosphate. *Biochemical and Biophysical Research Communications, 266*, 334–340.

Glouchankova, L., Krishna, U. M., Potter, B. V. L., Falck, J. R., & Bezprozvanny, I. (2000). Association of the inositol-(1,4,5) trisphosphate receptor ligand binding site with phosphatidylinositol (4, 5)-bisphosphate and adenophostin A. *Molecular Cell Biology Research Communications, 3*, 153–158.

Gregory, R. B., Wilcox, R. A., Berven, L. A., van Straten, N. C., van der Marel, G. A., van Boom, J. H., et al. (1999). Evidence for the involvement of a small subregion of the endoplasmic reticulum in the inositol trisphosphate receptor-induced activation of Ca^{2+} inflow in rat hepatocytes. *The Biochemical Journal, 341*, 401–408.

Hartzell, H. C., Machaca, K., & Hirayama, Y. (1997). Effects of adenophostin-A and inositol 1,4,5-trisphosphate on Cl^- currents in *Xenopus laevis* oocytes. *Molecular Pharmacology, 51*, 683–692.

He, C. L., Damiani, P., Ducibella, T., Takahashi, M., Tanzawa, K., Parys, J. B., et al. (1999). Isoforms of the inositol 1,4,5-trisphosphate receptor are expressed in bovine oocytes and ovaries: The type-1 isoform is down-regulated by fertilization and by injection of adenophostin A. *Biology of Reproduction, 61*, 935–943.

Hirota, J., Michikawa, T., Miyawaki, A., Takahashi, M., Tanzawa, K., Okura, I., et al. (1995). Adenophostin-mediated quantal Ca^{2+} release in the purified and reconstituted inositol 1,4,5-trisphosphate receptor type 1. *FEBS Letters, 368*, 248–252.

Hotoda, H., Murayama, K., Miyamoto, S., Iwata, Y., Takahashi, M., Kawase, Y., et al. (1999). Molecular recognition of adenophostin, a very potent Ca^{2+} inducer, at the D-*myo*-inositol 1,4,5-trisphosphate receptor. *Biochemistry, 38*, 9234–9241.

Hotoda, H., Takahashi, M., Tanzawa, K., Takahashi, S., & Kaneko, M. (1995). IP_3 receptor ligand. 1: Synthesis of adenophostin A. *Tetrahedon Letters, 36*, 5037–5040.

Huang, Y., Takahashi, M., Tanzawa, K., & Putney, J. W., Jr. (1998). Effect of adenophostin A on Ca^{2+} entry and calcium release-activated calcium current (I_{crac}) in rat basophilic leukemia cells. *The Journal of Biological Chemistry, 273*, 31815–31821.

Jellerette, T., He, C. L., Wu, H., Parys, J. B., & Fissore, R. A. (2000). Down-regulation of the inositol-trisphosphate receptor in mouse eggs following fertilization or parthenogenetic activation. *Developmental Biology, 223*, 238–250.

Jenkins, D. J., Marwood, R. D., & Potter, B. V. L. (1997). A disaccharide polyphosphate mimic of 1D-*myo*-inositol 1,4,5-trisphosphate. *Chemical Communications, 5*, 449–450.

Jenkins, D. J., & Potter, B. V. L. (1995). (2-Hydroxyethyl)-α-D-glucopyranoside-2′,3,4-trisphosphate – synthesis of a second messenger mimic related to adenophostin A. *Journal of Chemical Society, Chemical Communications, 1169–1170.*

Machaca, K., & Hartzell, H. C. (1999). Adenophostin A and inositol 1,4,5-trisphosphate differentially activate Cl^- currents in *Xenopus* oocytes because of disparate Ca^{2+} release kinetics. *The Journal of Biological Chemistry, 274*, 4824–4831.

Mak, D.-O., McBride, S., & Foskett, J. K. (2001). ATP-dependent adenophostin activation of inositol 1,4,5-trisphosphate receptor channel gating. Kinetic implications for the duration of calcium puffs in cells. *The Journal of General Physiology, 117*, 299–314.

Marchant, J. S., Beecroft, M. D., Riley, A. M., Jenkins, D. J., Marwood, R. D., Taylor, C. W., et al. (1997). Disaccharide polyphosphates based upon adenophostin A activate hepatic D-*myo*-inositol 1,4,5-trisphosphate receptors. *Biochemistry, 36*, 12780–12790.

Marchant, J. S., & Parker, I. (1998). Kinetics of elementary Ca^{2+} puffs evoked in *Xenopus* oocytes by different Ins(1,4,5)P_3 receptor agonists. *The Biochemical Journal, 334*, 505–509.

Marchant, J. S., & Taylor, C. W. (1998). Rapid activation and partial inactivation of inositol trisphosphate receptors by inositol trisphosphate. *Biochemistry, 37*, 11524–11533.

Marwood, R. D., Correa, V., Taylor, C. W., & Potter, B. V. L. (2000). Synthesis of adenophostin A. *Tetrahedron: Asymmetry, 11*, 397–403.

Marwood, R. D., Jenkins, D. J., Correa, V., Taylor, C. W., & Potter, B. V. L. (2000). Contribution of the adenine base to the activity of adenophostin A investigated using a base replacement strategy. *Journal of Medicinal Chemistry, 43*, 4278–4287.

Marwood, R. D., Riley, A. M., Correa, V., Taylor, C. W., & Potter, B. V. L. (1999). Simplification of adenophostin A defines a minimal structure for potent glucopyranoside-based mimics of D-*myo*-inositol 1,4,5-trisphosphate. *Bioorganic & Medicinal Chemistry Letters, 9*, 453–458.

Marwood, R. D., Riley, A. M., Jenkins, D. J., & Potter, B. V. L. (2000). Synthesis of adenophostin A and congeners modified at glucose. *Journal of Chemical Society Perkin Transactions, 1935–1947.*

Marwood, R. D., Shuto, S., Jenkins, D. J., & Potter, B. V. L. (2000). Convergent synthesis of adenophostin analogues *via* a base replacement strategy. *Chemical Communications, 219–220.*

Mecozzi, S., West, A. P., Jr., Dougherty, D. A. (1996). Cation–π interactions in aromatics of biological and medicinal interest: Electrostatic potential surfaces as a useful qualitative guide. *Proceedings of the National Academy of Sciences of the United States of America, 93*, 10566–10571.

Missiaen, L., Parys, J. B., Sienaert, I., Maes, K., Kunzelmann, K., Takahashi, M., et al. (1998). Functional properties of the type-3 $InsP_3$ receptor in 16HBE14o- bronchial mucosal cells. *The Journal of Biological Chemistry, 273*, 8983–8986.

Mochizuki, T., Kondo, Y., Abe, H., Tovey, S. C., Dedos, S. G., Taylor, C. W., et al. (2006). Synthesis of adenophostin A analogues conjugating an aromatic group at the 5′-position as potent IP_3 receptor ligands. *Journal of Medicinal Chemistry, 49*, 5750–5758.

Morris, S. A., Nerou, E. P., Riley, A. M., Potter, B. V. L., & Taylor, C. W. (2002). Determinants of adenophostin A binding to inositol trisphosphate receptors. *The Biochemical Journal, 367*, 113–120.

Murphy, C. T., Riley, A. M., Lindley, C. J., Westwick, J., & Potter, B. V. L. (1997). Structural analogues of D-*myo*-inositol-1,4,5-trisphosphate and adenophostin A: Recognition by cerebellar and platelet inositol-1,4,5-trisphosphate receptors. *Molecular Pharmacology, 52*, 741–748.

Nerou, E. P., Riley, A. M., Potter, B. V. L., & Taylor, C. W. (2001). Selective recognition of inositol phosphates by subtypes of inositol trisphosphate receptor. *The Biochemical Journal, 355*, 59–69.

Parekh, A. B., & Putney, J. W. (2005). Store-operated calcium channels. *Physiological Reviews, 85*, 757–810.

Parekh, A. B., Riley, A. M., & Potter, B. V. L. (2002). Adenophostin A and ribophostin, but not inositol 1,4,5-trisphosphate or *manno*-adenophostin, activate a Ca^{2+} release-activated Ca^{2+} current, I_{CRAC}, in weak intracellular Ca^{2+} buffer. *The Biochemical Journal, 361*, 133–141.

Park, C. Y., Hoover, P. J., Mullins, F. M., Bachhawat, P., Covington, E. D., Raunser, S., et al. (2009). STIM1 clusters and activates CRAC channels via direct binding of a cytosolic domain to Orai1. *Cell, 136*, 876–890.

Rahman, T.-U.-., Skupin, A., Falcke, M., & Taylor, C. W. (2009). Clustering of IP_3 receptors by IP_3 retunes their regulation by IP_3 and Ca^{2+}. *Nature, 458*, 655–659.

Rosenberg, H. J., Riley, A. M., Laude, A. J., Taylor, C. W., & Potter, B. V. L. (2003). Synthesis and Ca^{2+}-mobilizing activity of purine-modified mimics of adenophostin A: A model for the adenophostin-Ins(1,4,5)P_3 receptor interaction. *Journal of Medicinal Chemistry, 46*, 4860–4871.

Rosenberg, H. J., Riley, A. M., Marwood, R. D., Correa, V., Taylor, C. W., & Potter, B. V. L. (2001). Xylopyranoside-based agonists of D-*myo*-inositol 1,4,5-trisphosphate receptors: Synthesis and effect of stereochemistry on biological activity. *Carbohydrate Research, 332*, 53–66.

Rossi, A., Sureshan, K. M., Riley, A. M., Potter, B. V. L., Taylor, C. W. (2010) Selective determinants of IP_3 and adenophostin A interactions with type 1 IP_3 receptors. *British Journal of Pharmalogy,* in press.

Rossi, A. M., Riley, A. M., Tovey, S. C., Rahman, T., Dellis, O., Taylor, E. J. A., et al. (2009). Synthetic partial agonists reveal key steps in IP_3 receptor activation. *Nature Chemical Biology, 5*, 631–639.

Sato, Y., Miyazaki, S., Shikano, T., Mitsuhashi, N., Takeuchi, H., Mikoshiba, K., et al. (1998). Adenophostin, a potent agonist of the inositol 1,4,5-trisphosphate receptor, is useful for fertilization of mouse oocytes injected with round spermatids leading to normal offspring. *Biology of Reproduction, 58*, 867–873.

Shuto, S., Horne, G., Marwood, R. D., & Potter, B. V. L. (2001). Total synthesis of nucleobase-modified adenophostin A mimics. *Chemistry A. European Journal, 7*, 4937–4946.

Shuto, S., Tatani, K., Ueno, Y., & Matsuda, A. (1998). Synthesis of adenophostin analogues lacking the adenine moiety as novel potent IP_3 receptor ligands: Some structural requirements for the significant activity of adenophostin A. *The Journal of Organic Chemistry, 63*, 8815–8824.

Smith, I. F., & Parker, I. (2009). Imaging the quantal substructure of single IP_3R channel activity during Ca^{2+} puffs in intact mammalian cells. *Proceedings of the National Academy of Sciences of the United States of America, 106*, 6404–6409.

Sureshan, K. M., Riley, A. M., Rossi, A. M., Tovey, S. C., Dedos, S. G., Taylor, C. W., et al. (2009). Activation of IP_3 receptors by synthetic bisphosphate ligands. *Chemical Communications, 14*, 1204–1206.

Sureshan, K. M., Trusselle, M., Tovey, S. C., Taylor, C. W., & Potter, B. V. L. (2006). Guanophostin A: Synthesis and evaluation of a high affinity agonist of the D-*myo*-inositol 1,4,5-trisphosphate receptor. *Chemical Communications, 19*, 2015–2017.

Sureshan, K. M., Trusselle, M., Tovey, S. C., Taylor, C. W., & Potter, B. V. L. (2008). 2-Position base-modified analogues of adenophostin A as high-affinity agonists of the D-*myo*-inositol trisphosphate receptor: *In vitro* evaluation and molecular modeling. *The Journal of Organic Chemistry, 73*, 1682–1692.

Swatton, J. E., Morris, S. A., & Taylor, C. W. (2001). Functional properties of *Drosophila* inositol trisphosphate receptors. *The Biochemical Journal, 359*, 435–441.

Takahashi, M., Kagasaki, T., Hosoya, T., & Takahashi, S. (1994). Adenophostins A and B: Potent agonists of inositol-1,4,5-trisphosphate receptor produced by *Penicillium brevicompactum*. Taxonomy, fermentation, isolation, physico-chemical and biological properties. *The Journal of Antibiotics, 46*, 1643–1647.

Takahashi, M., Tanzawa, K., & Takahashi, S. (1994). Adenophostins, newly discovered metabolites of *Penicillium brevicompactum*, act as potent agonists of the inositol 1,4,5-trisphosphate receptor. *The Journal of Biological Chemistry, 269*, 369–372.

Takahashi, S., Takeshi, K., & Takahashi, M. (1994). Adenophostins A and B: Potent agonists of inositol-1,4,5-trisphosphate receptors produced by *Penicillium brevicompactum*. Structure elucidation. *The Journal of Antibiotics, 47*, 95–100.

Taylor, C. W., da Fonseca, P. C. A., & Morris, E. P. (2004). IP_3 receptors: The search for structure. *Trends in Biochemical Sciences, 29*, 210–219.

Turner, H., Fleig, A., Stokes, A., Kinet, J.-P., & Penner, R. (2003). Discrimination of intracellular calcium store subcompartments using TRPV1 (transient receptor potential channel, vanilloid subfamily member 1) release channel activity. *The Biochemical Journal, 371*, 341–350.

Vanlingen, S., Sipma, H., De Smet, P., Callewaert, G., Missiaen, L., De Smedt, H., et al. (2000). Ca^{2+} and calmodulin differentially modulate myo-inositol 1,4,5-trisphosphate (IP_3)-binding to the recombinant ligand-binding domains of the various IP_3 receptor isoforms. *The Biochemical Journal, 346*, 275–280.

Van Straten, N. C. R., Van der Marel, G. A., & Van Boom, J. H. (1997a). An expeditious route to the synthesis of adenophostin A. *Tetrahedron, 53*, 6509–6522.

Van Straten, N. C. R., Van der Marel, G. A., & Van Boom, J. H. (1997b). Synthesis of 2′,3″,4″-trisphosphate-containing analogues of adenophostin A. *Tetrahedron, 53*, 6523–6538.

Wilcox, R. A., Erneux, C., Primrose, W. U., Gigg, R., & Nahorski, S. R. (1995). 2-Hydroxy-α-D-glucopyranoside-2,3′,4′-trisphosphate, a novel, metabolically resistant, adenophostin A and myo-inositol-1,4,5-trisphosphate analogue, potently interacts with the *myo*-inositol-1,4,5-trisphosphate receptor. *Molecular Pharmacology, 47*, 1204–1211.

Wilcox, R. A., Primrose, W. U., Nahorski, S. R., & Challiss, R. A. J. (1998). New developments in the molecular pharmacology of the *myo*-inositol 1,4,5-trisphosphate receptor. *Trends in Pharmacological Sciences, 19*, 467–475.

Yoshida, M., Sensui, N., Inoue, T., Morisawa, M., & Mikoshiba, K. (1998). Role of two series of Ca^{2+} oscillations in activation of ascidian eggs. *Developmental Biology, 203*, 122–133.

CHAPTER 11

Regulation of IP_3R Channel Gating by Ca^{2+} and Ca^{2+} Binding Proteins

J. Kevin Foskett*,† and Don-On Daniel Mak*

*Department of Physiology, University of Pennsylvania, Philadelphia, Pennsylvania, USA
†Department of Cell and Developmental Biology, University of Pennsylvania, Philadelphia, Pennsylvania, USA

I. OVERVIEW

The activity of the inositol 1,4,5-trisphosphate (IP_3) receptor (IP_3R) Ca^{2+} release channel is intricately regulated by a multitude of ligands. The most important ligands regulating IP_3R channel activity are IP_3 and Ca^{2+}, its physiological permeant ion. Generally, cytoplasmic Ca^{2+} regulates steady-state IP_3R channel gating with a biphasic concentration dependence—Ca^{2+} at low concentrations activates the channel and increases its open probability P_o, whereas at higher concentrations, Ca^{2+} inhibits the channel. Ca^{2+} modulates channel activity by binding to several apparently distinct sites that regulate

Current Topics in Membranes, Volume 66 1063-5823/10 $35.00
Copyright 2010, Elsevier Inc. All right reserved.
DOI: 10.1016/S1063-5823(10)66011-5

Ca^{2+} activation, IP_3-dependent Ca^{2+} inhibition, IP_3-independent Ca^{2+} inhibition, channel inactivation, and channel recruitment. In addition, a separate Ca^{2+}-binding site appears to regulate the functionality of the IP_3-independent Ca^{2+} inhibition site. Insights into Ca^{2+} regulation of channel gating have been advanced by recent studies of the kinetic responses of single channels in endoplasmic reticulum membranes to rapid changes in the concentrations of Ca^{2+} and IP_3. Quantitative and qualitative molecular models can account for much of the steady-state and kinetic behaviors of IP_3R channel gating regulation by cytoplasmic Ca^{2+}. The identification of modal gating has revealed that Ca^{2+} modulates IP_3R channel activity primary by regulating its propensity to gate in particular modes. Ca^{2+} regulation of the channel is impinged upon by other channel regulators, including IP_3 and ATP, as allosteric modulators. Ca^{2+} can also impinge on the activity of the channel indirectly, by binding to other proteins that interact with it.

II. INTRODUCTION

The activity of the inositol 1,4,5-trisphosphate (IP_3) receptor (IP_3R) Ca^{2+} release channel is regulated by a wide range of ligands. The most important ligands regulating IP_3R channel activity are IP_3 and Ca^{2+}, its physiological permeant ion. Generally, Ca^{2+} regulates steady-state IP_3R channel gating with a biphasic concentration dependence—Ca^{2+} at low concentrations activates the channel and increases its open probability P_o, whereas at higher concentrations, Ca^{2+} inhibits the channel. IP_3 and Ca^{2+} are essential for IP_3R channel activation, but other physiological ligands, such as ATP, are not. ATP-free acid potentiates IP_3 channel activity allosterically by enhancing the sensitivity of the channel to Ca^{2+} activation. Regulation of channel activity by IP_3 is also by mainly modifying the sensitivity of the channels to Ca^{2+} regulation. The most systematic studies have revealed that IP_3, as well as adenophostin A and its analogs, regulates the IP_3R channel by allosterically modulating the sensitivity of the channels to Ca^{2+} inhibition, with little effect on the Ca^{2+} activation properties.

Besides steady-state channel gating kinetics, other aspects of IP_3R channel activity are also regulated by Ca^{2+} and IP_3. In the constant presence of IP_3, IP_3R channel activity inevitably terminates. This reversible, IP_3-induced inactivation of the channel progresses faster in high Ca^{2+} concentrations and in subsaturating concentrations of IP_3. In addition, channel recruitment is a distinct process in which suboptimal Ca^{2+} concentrations (too low or too high) and subsaturating IP_3 concentrations activate fewer IP_3R channels than optimal concentrations of either ligand.

Thus, ligand regulation of the IP_3R channel is complex, with Ca^{2+} as the major determinant of the channel properties. Ca^{2+} modulates channel activity by binding to several apparently distinct binding sites that regulate Ca^{2+} activation, IP_3-dependent Ca^{2+} inhibition, IP_3-independent Ca^{2+} inhibition, channel inactivation, and channel recruitment. In addition, a separate Ca^{2+}-binding site appears to regulate the properties of the IP_3-independent Ca^{2+} inhibition site. Ca^{2+} regulation of the channel is impinged on by other channel regulators, including IP_3 and ATP, as allosteric modulators. Furthermore, Ca^{2+} can impinge on the activity of the channel indirectly, by binding to other proteins that interact with the channel (Foskett, White, Cheung, & Mak, 2007; and Section III).

III. CYTOPLASMIC Ca^{2+} REGULATION OF IP_3R CHANNEL GATING

A. *Insights from Steady-State Measurements*

1. Biphasic Regulation of IP_3R Channel Gating by Ca^{2+}

The most widely studied aspect of IP_3R channel activity regulation is that by Ca^{2+}. Patch-clamp experiments on outer membranes of isolated nuclei or whole cell recordings of plasma membrane-localized IP_3R isoforms from different species (*Xenopus* type 1, rat types 1 and 3, and IP_3R from insect *Spodoptera*; Boehning, Joseph, Mak, & Foskett, 2001; Gin, Falcke, Wagner, Yule, & Sneyd, 2009; Ionescu et al., 2006; Mak, McBride, & Foskett, 1998, 2001b; Wagner, Joseph, & Yule, 2008) have revealed that in the presence of saturating concentrations of IP_3, IP_3R channels have very low activity in resting cytoplasmic Ca^{2+} concentrations ($[Ca^{2+}]_i \sim 50$ nM), but that as Ca^{2+} is raised through the sub-micromolar range ($< 1\ \mu M$), IP_3R channel activity increases to a maximum level with $P_o \sim 0.8$ (Fig. 1). In nuclear patch-clamp recordings, P_o remains at the maximal level over a wide range of $[Ca^{2+}]_i$ before further increases begin to inhibit it, at $[Ca^{2+}]_i > 10\ \mu M$. Thus, the Ca^{2+} dependence of channel activity for these IP_3R channels is biphasic, with a broad, plateau-shaped P_o versus $[Ca^{2+}]_i$ curve (Fig. 2).

Although the general shapes of the channel P_o versus $[Ca^{2+}]_i$ curves in nuclear patch-clamp studies are similar, the Ca^{2+} dependencies have minor differences among different IP_3R isoforms, which may be physiologically important. Ca^{2+} activation of type 1 IP_3R channels is positively cooperative, enabling P_o to increase sharply and reach the maximum value within a narrow range of $[Ca^{2+}]_i$. Such a behavior is ideal for Ca^{2+}-induced Ca^{2+} release (CICR). In contrast, P_o of the type 3 IP_3R channel increases over a broader range of $[Ca^{2+}]_i$ and with a higher Ca^{2+} affinity. Such a behavior is ideal as a trigger that responds to low IP_3 concentrations. Ca^{2+} inhibition of

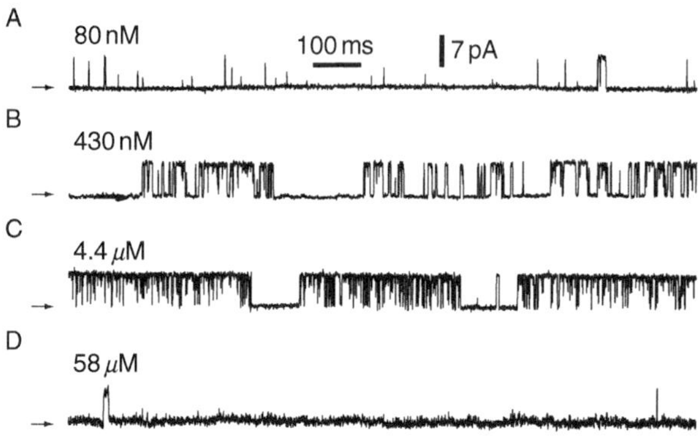

FIGURE 1 Typical single-channel current traces of *Xenopus* type 1 IP_3R in various cytoplasmic Ca^{2+} concentrations and saturating 10 μM IP_3. Current traces were recorded during nuclear patch-clamp experiments at cytoplasmic Ca^{2+} concentrations as tabulated in 0.5 mM free ATP (modified from Mak et al., 1998). All current traces in this and other graphs were recorded at 20 mV. Arrows indicate closed-channel current level in all current traces. Channel P_o evaluated for the single-channel patch-clamp experiments yielding the current traces shown in A, B, C, and D are 0.008, 0.50, 0.89, and 0.002, respectively. From Foskett et al. (2007).

the vertebrate IP_3R channels in these studies is highly cooperative, so channel P_o decreases rapidly over a narrow range of $[Ca^{2+}]_i$, whereas Ca^{2+} inhibition of the insect Sf9 cell channel exhibits no cooperativity, and P_o decreases gently over a broader range of $[Ca^{2+}]_i$. The physiological relevance of these differences is not clear.

In some other single-channel studies, different $[Ca^{2+}]_i$ dependencies have been observed. In some (Marchenko, Yarotskyy, Kovalenko, Kostyuk, & Thomas, 2005; Ramos-Franco, Caenepeel, Fill, & Mignery, 1998; Ramos-Franco, Fill, & Mignery, 1998; Striggow & Ehrlich, 1996; Tu et al., 2002, 2003; Tu, Wang, & Bezprozvanny, 2005), IP_3R channels in saturating IP_3 were very sensitive to Ca^{2+} inhibition, such that their biphasic P_o versus $[Ca^{2+}]_i$ curves are narrow and bell-shaped. In others, high Ca^{2+} inhibition was significantly reduced or completely absent (Hagar, Burgstahler, Nathanson, & Ehrlich, 1998; Mak, McBride, Petrenko, & Foskett, 2003; Ramos-Franco et al., 2000; Ramos-Franco, Fill, et al., 1998), so that the channel remained active in physiologically attainable supra-micromolar $[Ca^{2+}]_i$ (>500 μM).

2. Use of Biphasic Hill Equation to Describe Ca^{2+} Regulation of IP_3R Channel Activity

A useful way to quantify Ca^{2+} regulation of IP_3R channel activity is to fit the P_o versus $[Ca^{2+}]_i$ data with an empirical, model-independent Hill equation. The biphasic equation

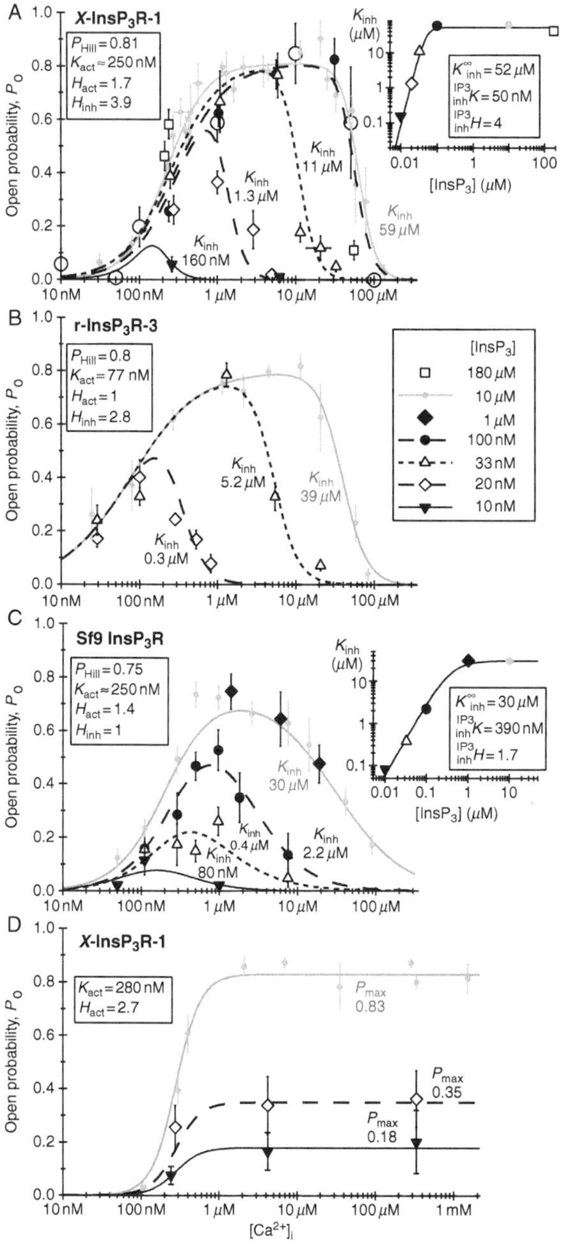

FIGURE 2 $[Ca^{2+}]_i$ and $[IP_3]$ regulation of IP_3R channel activity. (A) $[Ca^{2+}]_i$ dependence of mean P_o of endogenous X-IP_3R-1 channels (filled symbols) in various $[IP_3]$ as tabulated (modified from Mak et al., 1998). Each data point in this and subsequent P_o versus $[Ca^{2+}]_i$

$$P_o = P_{Hill}\left\{1 + \left(\frac{K_{act}}{[Ca^{2+}]_i}\right)^{H_{act}}\right\}^{-1}\left\{1 + \left(\frac{[Ca^{2+}]_i}{K_{inh}}\right)^{H_{inh}}\right\}^{-1} \quad (1)$$

has been used to describe the P_o of channels that are both activated and inhibited by Ca^{2+} (Foskett et al., 2007). The Ca^{2+} dependencies of the channel can be characterized in terms of five Hill equation parameters: P_{Hill}, K_{act}, H_{act}, K_{inh}, H_{inh}. For IP_3R channels that have very different sensitivities to Ca^{2+} activation and inhibition, with a plateau-shaped P_o versus $[Ca^{2+}]_i$ curve in saturating $[IP_3]$ (Ionescu et al., 2006; Mak et al., 1998, 2001b), the data can be fitted by the Hill equation with a unique set of parameters, with each parameter describing one aspect of the Ca^{2+} dependence. P_{Hill} is the maximum P_o that the channel could attain if there was no Ca^{2+} inhibition. Because of the low sensitivity of the channel to Ca^{2+} inhibition in saturating IP_3, the channel is fully activated before Ca^{2+} starts to inhibit it. Thus, the observed maximum open probability P_{max} under optimal ligand conditions is equal to P_{Hill}. The parameters K_{act} and K_{inh} are then EC_{50} and IC_{50} for Ca^{2+}, that is, the $[Ca^{2+}]_i$ at which $P_o = 0.5P_{Hill}$. They are inversely related to the functional sensitivity of the channel to Ca^{2+} activation and inhibition, respectively. H_{act} and H_{inh} describe the level of cooperativity of Ca^{2+} activation and inhibition, respectively.

IP_3R channels that have similar sensitivities to Ca^{2+} activation and inhibition display narrow bell-shaped P_o versus $[Ca^{2+}]_i$ curves, because the channel is not yet fully activated by Ca^{2+} when Ca^{2+} inhibition starts to reduce P_o. In such cases, the set of Hill equation parameters that provide a good fit to

---

plots is the average of channel P_o from at least four experiments using the same ligand concentrations. The curves are least-square fit of the data points using the biphasic Ca^{2+} regulation Hill equation (Eq. (1)) with parameters as tabulated. The large open circles represent P_o for recombinant rat IP_3R-1 channels in various $[Ca^{2+}]_i$ in saturating 10 μM IP_3 (modified from Boehning et al., 2001). Inset: Plot of K_{inh} derived from the biphasic Hill equation fit of P_o data versus $[IP_3]$ used. The curve is the least-square fit of the K_{inh} values using the activation Hill equation $\log(K_{inh}) = \log(K^{\infty}_{inh}) - {}^{IP_3}_{inh}H \log(1 + {}^{IP_3}_{inh}K/[IP_3R])$ with parameters as tabulated. (B) $[Ca^{2+}]_i$ dependence of mean P_o of recombinant rat type 3 IP_3R channels in various $[IP_3]$ as tabulated (modified from Mak et al., 2001b). Data points and fitted curves are obtained as described for A. (C) $[Ca^{2+}]_i$ dependence of mean P_o of endogenous IP_3R channels from Sf9 cells in various $[IP_3]$ as tabulated (modified from Ionescu et al., 2006). Data points, fitted curves, and inset graph are obtained as described for (A). (D) $[Ca^{2+}]_i$ dependence of mean P_o of X-IP_3R-1 IP_3R channels that have been exposed to bath solution with very low $[Ca^{2+}]_i$ (<5 nM) for a few minutes before the patch-clamp experiments, in various $[IP_3]$ as tabulated. The curves are least-square fits to the data using activation Hill equation $P_o = P_{Hill}\{1 + (K_{act}/[Ca^{2+}]_i)^{H_{act}}\}^{-1}$ with parameters as tabulated. Modified from Mak, McBride, Petrenko, et al. (2003). (See Color Insert.)

the data is not unique. Accordingly, no physical significance can be assigned to one particular set of Hill equation parameters, and conclusions drawn from them are questionable (Foskett et al., 2007).

Since IP_3 regulates IP_3R channels by modifying their sensitivity to Ca^{2+} inhibition (see later discussion), channels with broad, plateau-shaped P_o versus $[Ca^{2+}]_i$ curves at saturating $[IP_3]$ exhibit narrow bell-shaped P_o versus $[Ca^{2+}]_i$ curves at subsaturating $[IP_3]$ (see Fig. 2A with 10 or 20 nM IP_3), with P_{max} significantly less than observed in saturating $[IP_3]$. However, P_o at subsaturating $[IP_3]$ can still be well fitted by the same biphasic Hill equation (1) assuming that only K_{inh} was affected by $[IP_3]$ (Mak et al., 1998, 2001b).

Hill equations are empirical, and not based on any specific model for ligand regulation of channel gating. The equations enable the gating behaviors of the IP_3R to be characterized with a small number of parameters, provided that the Ca^{2+} dependence curve has a plateau shape with a well-defined P_{Hill} value. However, the empirical fit does not provide insights into the molecular mechanism(s) responsible for ligand regulation. For instance, the parameters H_{act} and H_{inh} do not have a direct model-independent relation with the number of activating or inhibitory Ca^{2+}-binding sites in the channel.

3. Ca^{2+} Activation of IP_3R Channels

Ca^{2+} activation of channel activity has been observed with comparable functional sensitivity to Ca^{2+} in most electrophysiological studies of the IP_3R, suggesting that the characteristics of Ca^{2+} activation are likely determined by intrinsic features of the IP_3R molecule that are conserved throughout evolution and among isoforms. This conservation probably exists because activation of IP_3R by a rise of Ca^{2+} from resting levels ($\sim$50 nM) to hundreds of nanomolar plays a key role in intracellular Ca^{2+} signaling. In the presence of activating levels of IP_3, elevations of $[Ca^{2+}]_i$ due to Ca^{2+} released by one IP_3R channel can in turn activate IP_3R channels nearby to release more Ca^{2+} in a positive feedback. CICR mediated by IP_3R channels acts as a critical communication mechanism between channels to coordinate their activities, generating large-scale Ca^{2+} signals (puffs and waves) from elementary Ca^{2+} release events emanating from individual IP_3R channels (Berridge, 1997).

Ca^{2+} activation of IP_3R channels is mostly positively cooperative (Fig. 2), enabling them to be activated sharply by Ca^{2+} within a narrow range. This property likely provides fine-tuning of the channel activity. Nevertheless, noncooperative Ca^{2+} activation has also been observed (Hagar et al., 1998; Mak et al., 2001b; Ramos-Franco, Fill, et al., 1998), suggesting that cooperativity is not an essential feature, but may rather provide diversity in Ca^{2+}

signals generated by different isoforms and under different conditions. For example, ATP regulation of the Ca^{2+} activation properties of the type 3 channel was associated with changes in the degree of cooperativity (Mak et al., 2001b).

IP_3R isoforms with higher sensitivity to Ca^{2+} activation (especially type 3 IP_3R; Fig. 2B) have higher P_o at resting $[Ca^{2+}]_i$ when IP_3 is present, than do channels formed by other isoforms. Thus, when an extracellular stimulus elicits production of IP_3 in a resting cell, such channels will have a higher probability of opening to release Ca^{2+} from the ER. They can therefore act as triggers to initiate IP_3-induced Ca^{2+} release (IICR; Mak et al., 2001b). Ca^{2+} released by these triggers may then raise local $[Ca^{2+}]_i$ sufficiently to cause nearby IP_3R channels to release Ca^{2+} through CICR, thus propagating the Ca^{2+} signals.

4. Ca^{2+} Inhibition of IP_3R Channels

Whereas Ca^{2+} activation of IP_3R is consistently observed in all single-channel experiments with comparable functional sensitivity to Ca^{2+}, the reported sensitivities of IP_3R channels to Ca^{2+} inhibition are highly variable even in saturating IP_3, so that a wide range of different shapes of channel P_o versus $[Ca^{2+}]_i$ curves has been reported (Foskett et al., 2007).

An understanding of the Ca^{2+} inhibition properties of the channels is important. Ca^{2+} inhibition can possibly serve as a negative feedback mechanism to either terminate or prevent Ca^{2+} release as the local $[Ca^{2+}]_i$ is raised by IP_3R-mediated Ca^{2+} release, even in the continuous presence of IP_3. This process may play a significant role in the generation of Ca^{2+} spikes and oscillations, as well as the generation of highly localized Ca^{2+} signals, by preventing runaway Ca^{2+} release due to the positive feedback of CICR. Furthermore, the range of $[Ca^{2+}]_i$ over which inhibition is exerted is also important, because it defines the spatial domain over which the channels may experience such regulation. For example, if the apparent affinity of the inhibitory Ca^{2+}-binding site is 20 μM, only channels quite close to an open Ca^{2+} channel will experience feedback inhibition. On the other hand, if the apparent affinity of the inhibitory Ca^{2+}-binding site is in the sub-micromolar range, then channels throughout the cytoplasm, even those that are quite far from individual release sites, will experience inhibitory $[Ca^{2+}]_i$.

The diversity of properties of Ca^{2+} inhibition of IP_3R channels observed in different studies may reflect a range of physiologically relevant modifications of Ca^{2+} inhibition by environmental factors that provide additional regulation of IP_3R-mediated Ca^{2+} signals. Alternately, such diversity may reflect a lack of control of important experimental variables during preparation or recording of the channels. There is no clear correlation between the sensitivity of an IP_3R channel to Ca^{2+} inhibition and the molecular structure of the IP_3R.

Nuclear patch-clamp studies revealed very similar Ca^{2+} inhibition characteristics (low Ca^{2+} sensitivity and high level of cooperativity) for homotetrameric channels formed by types 1 or 3 IP_3R isoforms (Mak et al., 1998, 2001b; Fig 2A and B). On the other hand, whereas similar Ca^{2+} inhibition characteristics for different splice variants of the same isoform of IP_3R (Ramos-Franco, Caenepeel, et al., 1998; Tu et al., 2002) or different isoforms (Tu, Wang, & Bezprozvanny, 2005) have been observed in some reconstitution studies, the same isoforms displayed very different Ca^{2+} inhibition properties in others (see Fig. 2 and cf. Mak et al., 1998; Tu, Wang, & Bezprozvanny, 2005). Ca^{2+} inhibition was absent for both type 2 (Ramos-Franco et al., 2000) and type 3 (Hagar et al., 1998) IP_3R channels in some planar bilayer reconstitution studies, but not in others (Tu, Wang, & Bezprozvanny, 2005). Furthermore, the Ca^{2+} inhibition properties are largely, but not completely, distinct between the reconstitution and nuclear patch-clamp studies. For example, similar high sensitivity to Ca^{2+} inhibition was observed for IP_3R channels formed by the same isoform studied using nuclear patching (Marchenko et al., 2005) or bilayer reconstitutions (Tu et al., 2002).

It is likely that various forms of regulation, both physiological and nonphysiological, impinge on the Ca^{2+} inhibition properties of the channel. The sensitivity of recombinant rat type 3 (IP_3R) to Ca^{2+} inhibition was altered by cytoplasmic ATP (Tu, Wang, & Bezprozvanny, 2005). Importantly, because Ca^{2+} inhibition is regulated by IP_3, as discussed later, different IP_3 sensitivities of channels in different studies may contribute to different Ca^{2+} inhibition properties. Different IP_3 sensitivities could in turn be generated by extrinsic factors such as membrane lipid composition (Cannon et al., 2003), covalent modifications of the channel (e.g., phosphorylation, redox), presence or absence of interacting proteins (Foskett et al., 2007), or to intrinsic differences in channel properties between species and/or isoforms. Even a simple nonphysiological maneuver can radically alter the channel's Ca^{2+} inhibition properties. Exposure of IP_3R channels to a very low $[Ca^{2+}]_i$ (< 10 nM) for a few minutes before they are exposed to IP_3 completely and reversibly relieves high $[Ca^{2+}]_i$ inhibition of types 1 and 3 IP_3R channels (Mak, McBride, Petrenko, et al., 2003; Fig. 2D). Thus, the Ca^{2+} inhibition properties of the channel are regulated by a distinct Ca^{2+}-binding site(s), which could be a locus for either regulation or disruption of Ca^{2+} inhibition. It is possible that this site was disrupted during protocols involved in channel reconstitution in those studies of the types 2 (Ramos-Franco et al., 2000) and 3 (Hagar et al., 1998) channel isoforms that failed to observe high-$[Ca^{2+}]$ inhibition.

One possible mechanism of regulating Ca^{2+} inhibition of IP_3R channels that has been repeatedly invoked is through interaction of the channel with calmodulin. However, single-channel studies of IP_3R are consistent with the absence of any regulatory role for calmodulin in Ca^{2+} inhibition of IP_3R. A fuller discussion of the calmodulin interaction with IP_3R is provided in Section III.A.

5. IP_3 Regulates IP_3R Channel Activity Through Modulation of its Sensitivity to Ca^{2+} Inhibition

Spontaneous IP_3R activity with very low P_o (approximately a few percentage) can be observed in the absence of IP_3 (Mak, McBride, & Foskett, 2003; Ramos-Franco, Fill, et al., 1998). However, both Ca^{2+} (in appropriate concentrations) and IP_3 must be present on the cytoplasmic side of the channel to activate it appreciably. IP_3 and Ca^{2+} are not simply equivalent coagonists of the channel, since the two ligands activate the IP_3R in very different manners. Systematic studies of the *Xenopus* type 1 (*X*-IP_3R-1), r-IP_3R-3, and Sf9 IP_3R channels under a broad range of [IP_3] and $[Ca^{2+}]_i$ in steady-state conditions revealed that the channel becomes more sensitive to inhibition by high $[Ca^{2+}]_i$ in the presence of subsaturating [IP_3]. That is, at lower [IP_3], the channel is inhibited by lower $[Ca^{2+}]_i$ (Fig. 2A–C). In contrast, the sensitivity to Ca^{2+} activation, and the cooperativity of Ca^{2+} activation and inhibition are not significantly affected by IP_3. At very low [IP_3], the maximum channel P_o observed and the range of $[Ca^{2+}]_i$ over which the channel is active are both reduced (Fig. 2A–C). Both effects can be fully accounted for by an increase in the sensitivity of the channel to Ca^{2+} inhibition. At very low [IP_3], the channel is so sensitive to Ca^{2+} inhibition that it begins to be inhibited before it is fully activated. This pattern of IP_3 regulation of channel activity by solely modulating the sensitivity of the channel to inhibition by Ca^{2+} is consistently observed for IP_3R channels of different isoforms (types 1 and 3) from very different species (*Xenopus*, rat, and insect; Ionescu et al., 2006; Mak et al., 1998, 2001b), suggesting that it is a common mechanism underlying ligand regulation of all IP_3R (although see Wagner et al., 2008).

Because IP_3 affects only the sensitivity to Ca^{2+} inhibition, the IP_3 sensitivity of the channel can be defined by the changes it causes in this parameter. A half-maximal effect was observed at ~50 nM IP_3 for both *X*-IP_3R-1 and r-IP_3R-3 (Fig. 2A and B). The response of these channels is fully saturated by 100 nM IP_3 (Mak et al., 1998, 2001b). In general, this functional sensitivity to IP_3 is in reasonable agreement with dissociation constants K_d derived from IP_3 binding assays, and EC_{50}s for IP_3 stimulation of Ca^{2+} release (Kaznacheyeva, Lupu, & Bezprozvanny, 1998; Mauger, Lièvremont, Piétri-Rouxel, Hilly, & Coquil, 1994; Meyer, Holowka, & Stryer, 1988; Miyakawa et al., 2001; Ramos-Franco et al., 2000; Taylor & Richardson, 1991). In contrast, the Sf9 IP_3R has lower sensitivity to IP_3, with half-maximal effect at about 400 nM, and its response is not saturated until $IP_3 \approx 1\ \mu M$ (Fig. 2C; Ionescu et al., 2006). Canine cerebellar IP_3R-1 apparently has a very low sensitivity to IP_3 such that the P_o versus $[Ca^{2+}]_i$ curve of the channel was still substantially changed when IP_3 was increased from 2 to 180 μM (Kaftan, Ehrlich, & Watras, 1997). This significantly lower IP_3 sensitivity agrees with the low IP_3 sensitivity of IP_3R-mediated Ca^{2+} release observed in cerebellar neurons (Khodakhah & Ogden, 1993, 1995), and may be due to the differences in IP_3R splice variants: IP_3R examined in Mak et al. (1998) was peripheral SII− form, whereas in Kaftan et al. (1997), the cerebellar SII+ form was examined, or due to the interference of IP_3R channel gating by phosphatidylinositol 4,5-bisphosphate bound to the cerebellar IP_3R (Lupu, Kaznacheyeva, Krishna, Falck, & Bezprozvanny, 1998).

a. Physiological Significance of IP_3 Modulation of Ca^{2+} Inhibition of IP_3R Channel. The modulation of IP_3R channel K_{inh} by IP_3 provides a possible mechanism intrinsic to the channel to generate graded Ca^{2+} release in response to different levels of extracellular stimulus intensity, instead of an all-or-nothing signal expected for the process of CICR (Berridge, 1997; Bootman & Berridge, 1995). Because IP_3R exposed to higher $[IP_3]$ has lower susceptibility to Ca^{2+} inhibition, $[Ca^{2+}]_i$ that can inhibit channel activity at low $[IP_3]$ will be insufficient to inhibit it when $[IP_3]$ is increased. Higher $[Ca^{2+}]_i$ at the vicinity of an open channel can therefore be achieved before reaching levels that will inhibit other closed channels in the same channel cluster before they can open (Berridge, 1997). This enables coordinated release of channels in the same local cluster, whereas in lower $[IP_3]$, Ca^{2+} released by the stochastic opening of one channel suppresses gating of the rest of the channels in the cluster. By enabling higher $[Ca^{2+}]_i$ to be achieved at the local level by coordinated Ca^{2+} release, higher $[IP_3]$ associated with more intense stimuli would promote greater diffusive spread of the local Ca^{2+} signal to other IP_3R channel clusters, thereby transforming highly localized signals at low levels of stimulation to more global coordinated Ca^{2+} release signals as the intensity of the stimulus is increased.

In addition, modulation by IP_3 of Ca^{2+} inhibition enables channel activity to be exquisitely sensitive to small changes in $[IP_3]$ within the subsaturating range. For *X*-IP_3R-1 in 10 nM IP_3, the channel is appreciably active only within a very narrow range of $[Ca^{2+}]_i$ (50–300 nM) with very low P_o ($P_{max} \sim 0.1$), whereas with IP_3 raised only to 100 nM, the channel is active over a very wide range of $[Ca^{2+}]_i$ (50 nM to 20 μM; Fig. 2A). Thus, the IP_3R-mediated Ca^{2+} signal can be finely controlled by small differences in the $[IP_3]$.

6. IP_3 and Ca^{2+} Regulate IP_3R Channel Activities Through Multiple Ca^{2+} Sensors

The most comprehensive studies of ligand regulation of IP_3R channel activity indicate that IP_3 regulates gating mainly by modulating its sensitivity to Ca^{2+} inhibition (Ionescu et al., 2006; Kaftan et al., 1997; Mak et al., 1998, 2001b). However, IP_3R channels that lack inhibition by Ca^{2+} (Ramos-Franco, Fill, et al., 1998), or with Ca^{2+} inhibition abolished by preexposure to very low levels of Ca^{2+} (Mak, McBride, & Petrenko et al., 2003), nevertheless still require IP_3 for activity (Fig. 2D). In these channels, increases in $[IP_3]$ increase the observed P_{max}, with little effect on the properties of Ca^{2+} activation. This suggests that there are at least two distinct kinds of Ca^{2+} sensors responsible for ligand regulation of IP_3R channel P_o. One kind of Ca^{2+} sensor is responsible for Ca^{2+} inhibition in saturating $[IP_3]$. This sensor has an apparent Ca^{2+} affinity of ~20–50 μM in most nuclear patch-clamp studies (Boehning et al., 2001; Ionescu et al., 2006; Mak et al., 1998, 2001b) and accounts for the descending phase of the P_o versus $[Ca^{2+}]_i$ curve in saturating $[IP_3]$. This Ca^{2+}-binding site is nonfunctional in the IP_3R channels that are not inhibited by high Ca^{2+}. The second kind of Ca^{2+} sensor is IP_3-dependent. It is responsible for the IP_3 sensitivity of the channel regardless of the functionality of the other inhibitory Ca^{2+} sensor. In the absence of IP_3, this site has very high Ca^{2+} affinity and inhibits channel opening. Thus, channels that lack high Ca^{2+} inhibition still require IP_3 for activation. In addition, comparisons of the sensitivities and levels of cooperativity for Ca^{2+} and IP_3 activation of different IP_3R isoform channels argues for the existence of a third Ca^{2+} sensor that is activating yet IP_3-independent. The following describes the properties of each of these sensors separately.

a. The IP_3-Independent Inhibitory Ca^{2+} Sensor. As noted earlier, single-channel studies of IP_3R have revealed diverse properties for Ca^{2+} inhibition of the channel, with very different sensitivities observed even for the same IP_3R isoform and splice variant in different studies. As a result, the P_o versus $[Ca^{2+}]_i$ curves observed in some studies are narrow and bell-shaped, whereas in others, they are broad and plateau-shaped. Ca^{2+} inhibition appeared to be totally absent in some studies. It has also been demonstrated that this inhibitory Ca^{2+} sensor can be reversibly rendered completely nonfunctional, thus abolishing high Ca^{2+} inhibition of the channel, by treating the channel with nanomolar $[Ca^{2+}]_i$ for a few minutes before exposure of the channel to IP_3 (Mak, McBride, & Petrenko et al., 2003). Channels that are not sensitive to Ca^{2+} inhibition are nevertheless fully IP_3-dependent (Ramos-Franco et al., 2000; Mak, McBride, & Petrenko et al., 2003; Fig. 2D). Together, these observations indicate that this inhibitory Ca^{2+} sensor is not responsible for IP_3 dependence of the channel and is therefore likely IP_3-independent.

Furthermore, the functional affinity for Ca^{2+} for this IP_3-independent inhibitory Ca^{2+} sensor is malleable, suggesting that it could possibly be under physiological regulation. Besides establishing the malleability of the inhibitory Ca^{2+} sensor, abolition of Ca^{2+} inhibition by preexposure of the channel to nanomolar $[Ca^{2+}]_i$ further indicates that its functionality is controlled by another, different type of Ca^{2+}-binding site(s) (Mak, McBride, & Petrenko et al., 2003).

b. The IP_3-Dependent Ca^{2+} Sensor. Another type of Ca^{2+} sensor is regulated by IP_3 binding to the channel. This IP_3-dependent Ca^{2+} sensor is extremely sensitive to small changes in $[IP_3]$ within the subsaturating range. For example, the sensitivity of *X*-IP_3R-1 channels to Ca^{2+} inhibition was reduced dramatically as $[IP_3]$ was raised from 10 to 100 nM. The channel was significantly inhibited by 160 nM Ca^{2+} in 10 nM IP_3, but it was not inhibited until cytoplasmic $[Ca^{2+}]_i$ was >30 μM in 100 nM IP_3 (Fig. 2A). To reconcile this exquisite sensitivity of the channel to subsaturating levels of IP_3 with the tetrameric structure of the channel consisting of four IP_3R molecules each with a single IP_3-binding site, an allosteric model was proposed in which the IP_3-dependent Ca^{2+} sensors in the channel (one per IP_3R molecule, total of four in each channel) act as inhibitory Ca^{2+}-binding sites to inhibit channel gating when bound to Ca^{2+} in the absence of IP_3 (Mak, McBride, & Foskett, 2003). However, as IP_3 concentration is raised and IP_3 binds to the channel, Ca^{2+} binding to the IP_3-dependent Ca^{2+} sensors starts to favor opening of the channel. In effect, this Ca^{2+} sensor becomes an activating Ca^{2+}-binding site. Thus, IP_3 regulates IP_3R channel activity with very high effectiveness by modifying not only the functional affinity of the IP_3-dependent Ca^{2+} sensors, but also their functional nature, changing them from inhibitory to activating sites.

The interplay of the two different types of Ca^{2+} sensors, one IP_3-sensitive, the other IP_3-insensitive, enables the response of IP_3R channel to IP_3 to saturate very abruptly despite its high sensitivity to subsaturating $[IP_3]$. Once IP_3 exceeds the saturating level of 100 nM, the P_o versus $[Ca^{2+}]_i$ curve of the *X*-IP_3R-1 exhibits no change even as IP_3 is further increased by over 3 orders of magnitude (Fig. 2A). This behavior results from the influence of the inhibitory Ca^{2+} sensor(s) that is IP_3-independent. Thus, at $IP_3 > 100$ nM, Ca^{2+} inhibition of the channel is caused by the IP_3-independent, purely inhibitory Ca^{2+} sensor. The abruptness in the saturation of the response of the channel to changes in IP_3 concentration is due to the insensitivity to IP_3 of these inhibitory Ca^{2+} sensors.

c. The IP_3-Independent Activating Ca^{2+} Sensor. In nuclear patch-clamp experiments, *X*-IP_3R-1 and r-IP_3R-3 channels exhibited similar sensitivities to activation by IP_3, even though the sensitivity and degree of cooperativity for Ca^{2+} activation of the two types of channels were very different (Fig. 2A and B; Mak et al., 1998, 2001b). On the other hand, the sensitivity and level of cooperativity for IP_3 activation of *X*-IP_3R-1 and Sf9 IP_3R channels are very different, even though the two types of channels have similar sensitivities and levels of cooperativity for Ca^{2+} activation. To quantitatively account for these different characteristics of Ca^{2+} and IP_3 activation, a third type of Ca^{2+} sensor must also play a role. This Ca^{2+} site is IP_3-independent and responsible for the consistent sensitivity to Ca^{2+} activation observed in various IP_3R channels despite their differences in IP_3 sensitivity or the presence or absence of Ca^{2+} inhibition.

7. A Model to Account for Multiple Ca^{2+} Sensor Regulation of IP_3R Steady-State Channel Gating

Numerical calculations (Mak, McBride, & Foskett, 2003) indicate that an allosteric model postulating the three types of Ca^{2+} sensors described above can account for all single-channel behaviors of various IP_3R channels studied by nuclear patch-clamp experiments (Foskett & Mak, 2004; Ionescu et al., 2006; Mak, McBride, & Foskett, 2003). In the simplest allosteric model that fits all observations in *Xenopus* oocyte nuclear patch-clamp studies of ligand regulation of steady-state channel gating behavior of *Xenopus* type 1 and rat type 3 IP_3R isoforms, the tetrameric channel can adopt six different conformations, the equilibria among which are controlled by one IP_3-binding site and three different functional Ca^{2+}-binding sites in each IP_3R monomer that directly affect the equilibria among active and closed conformations of the channel in a manner similar to the Monod–Wyman–Changeux (MWC) model (Fig. 3). One of the Ca^{2+}-binding sites is activating (so-called "F" site), whereas another is inhibitory (so-called "H" site), but both are independent of IP_3. In contrast, the third Ca^{2+}-binding site (so-called "G" site) is affected by IP_3, being inhibitory in the absence of IP_3 but becoming activating as $[IP_3]$ increases. Thus, two different Ca^{2+}-binding sites in each IP_3R monomer, the G and H sites, contribute to Ca^{2+} inhibition. Ca^{2+} binding to the IP_3-dependent G site inhibits IP_3R activity in the absence of IP_3, which is responsible for the lack of channel activity in the absence of IP_3 in normal physiological $[Ca^{2+}]_i$. IP_3 binding to the channel changes the effective affinities of the G site, and in so doing transforms it into an activating site. This IP_3-mediated transformation of the nature of this Ca^{2+}-binding site is responsible for all the IP_3 dependence of the channel, accounting for the extremely high sensitivity of channel gating to small changes in $[IP_3]$. The other Ca^{2+}-binding H site is strictly inhibitory with a lower Ca^{2+} affinity (10–30 μM) that is not modulated by IP_3 binding. The IP_3-independence of this site is responsible for the lack of further effect of IP_3 on types 1 and 3 channels in *Xenopus* oocytes once $[IP_3] > 1$ μM, accounting for the

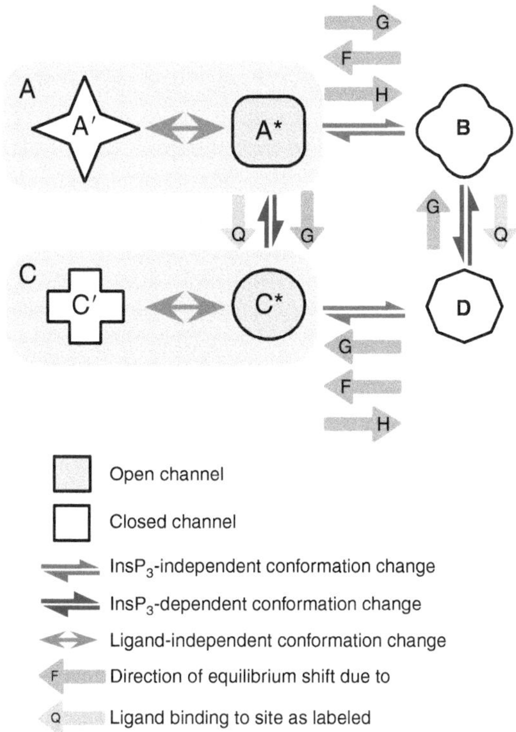

FIGURE 3 The MWC-based four + two-conformation model for IP_3R channel gating. Only conformation transitions are represented in the schemes. Reactions involving binding of IP_3 and Ca^{2+} to the IP_3R channel and the state of occupation of the various ligand-binding sites of the channel are omitted from the schemes for clarity. The shaded rectangles represent the grouping of the open A* and closed A′ conformations into the active A conformation, and the grouping of the C* and C′ conformations into the active C conformation. Modified from Mak, McBride, & Foskett (2003). (See Color Insert.)

observation that the effects of IP_3 abruptly saturate around this concentration. Furthermore, Ca^{2+} binding to the H site is also partly responsible for the observed inhibition of the channels even at lower $[Ca^{2+}]_i$ ($<$10–30 μM), when $[IP_3]$ is less than 1 μM. Whereas the properties of the H sites are insensitive to IP_3 binding, they are rendered nonfunctional by a nonphysiological protocol: exposure to an ultra-low bath $[Ca^{2+}]$. Nevertheless, the channel exposed to ultra-low bath $[Ca^{2+}]$ remains dependent on IP_3 because the other Ca^{2+}-binding sites, specifically the IP_3-dependent G site, are not affected by the low bath $[Ca^{2+}]$. Ca^{2+} binding to the G site inhibits IP_3R activity in the absence of IP_3.

8. Ligand-Dependent, IP_3-Induced IP_3R Channel Inactivation

An unusual aspect of IP_3R-mediated intracellular signaling is the phenomenon of "quantal release," defined (Meyer & Stryer, 1990; Muallem, Pandol, & Beeker, 1989) as the ability of cells to have graded release of Ca^{2+} from intracellular stores in response to incremental levels of extracellular agonist or IP_3 (reviewed in Bootman, 1994; Missiaen, Parys, De Smedt, Oike, & Casteels, 1994; Parys, Missiaen, Smedt, Sienaert, & Casteels, 1996; Taylor, 1998). This entails two processes: (i) an initial Ca^{2+} release whose rate is proportional to IP_3 concentration followed by (ii) a substantial reduction in rate or termination of Ca^{2+} release despite the presence of constant IP_3. Consequently, sustained exposure to submaximal levels of agonists, even over extensive periods, only mobilizes a fraction of total releasable Ca^{2+} in a cell. This is surprising because, with the steady-state ligand regulation of IP_3R channel P_o discussed so far, it might have been expected that all IP_3R channels should become activated in response to sufficient agonist stimulation, releasing all of the IP_3-sensitive Ca^{2+} stores, albeit at different rates depending on the agonist concentration.

In nuclear patch-clamp studies, abrupt termination of IP_3R channel activities despite the constant presence of agonist has been observed for all IP_3R investigated (Boehning et al., 2001; Ionescu et al., 2006; Mak & Foskett, 1994, 1997; Mak, McBride, & Foskett, 2001a; Mak et al., 2000). The mean duration of channel activity observed from its initial activation by IP_3 until the termination of activity in the presence of constant IP_3 (T_a) for vertebrate IP_3R is typically ~30 s (Mak & Foskett, 1997; Mak et al., 2000). A $[Ca^{2+}]_i$ dependence of T_a was qualitatively described for recombinant r-IP_3R-1 expressed in COS cell outer nuclear membrane (Boehning et al., 2001), but it was impossible, due to technical difficulties, to rule out the possibility that such abrupt termination of channel activity was a nonphysiological artifact associated with patching, for example, collisions of the channels with the walls of the glass pipette. In a nuclear patch-clamp study of the Sf9 IP_3R channel, it was demonstrated that T_a was dependent on the concentrations of both ligands (Ionescu et al., 2006). In optimal ligand conditions, T_a was ~120 s, substantially longer than the vertebrate channels. In optimal $[IP_3]$, T_a was reduced in $[Ca^{2+}]_i > 1\ \mu M$, with reduction by over 10-fold at 89 μM Ca^{2+}. In subsaturating IP_3, T_a already began to decrease in $[Ca^{2+}]_i \sim 300$ nM, substantially lower than that observed in saturating IP_3. The IP_3-induced termination of IP_3R channel activity was fully reversible upon ligand removal (Ionescu et al., 2006). These results suggest that the observed inevitable termination of channel activity is not an experimental artifact, and may be due to the entry of IP_3-liganded channels into a true inactivated state, driven by binding of Ca^{2+} to the channel at a relatively slow rate (Ionescu et al., 2006).

The inactivation kinetics observed in nuclear patch-clamp experiments: $T_a \sim$ 10–100 s in Sf9 IP_3R (Ionescu et al., 2006) and ~20–30 s in vertebrate IP_3R (Boehning et al., 2001; Mak & Foskett, 1997; Mak et al., 2000) are comparable to those estimated by ER permeability measurements in permeabilized hepatocytes (Hajnóczky & Thomas, 1994), as well as to the kinetics of IP_3-induced increases in IP_3 affinity of an apparently desensitized IP_3R in cerebellar microsomes (Coquil, Mauger, & Claret, 1996; half-life of channel activity ~15–45 s) and the kinetics of the transient fast phase of Ca^{2+} release in response to initial exposure to IP_3 in permeabilized and intact cells (see Meyer & Stryer, 1990; Muallem et al., 1989; Taylor & Potter, 1990). Of note, IP_3R inactivation observed in permeabilized hepatocytes (Hajnóczky & Thomas, 1994), which had kinetics similar to those observed in the single-channel studies, was shown to account for release termination associated with $[Ca^{2+}]_i$ oscillations (Hajnóczky & Thomas, 1997). These results suggest that the kinetics of inactivation observed in single-channel studies are of physiological relevance for $[Ca^{2+}]_i$ signaling in cells. However, it remains unclear whether distinct inactivation kinetics observed in different studies reflects methodological differences, distinct types of inactivation or inhibition, or a physiologically relevant range of kinetics of a common inactivation mechanism.

9. Ligand-Dependent IP_3R Channel Recruitment

In addition to termination of Ca^{2+} release in the presence of constant IP_3, quantal Ca^{2+} release requires that the initial rate of Ca^{2+} release from intracellular stores be proportional to IP_3 concentration. One mechanism to achieve this is by IP_3-tuning of Ca^{2+} inhibition of channel activity, as discussed earlier. Additional mechanisms were discovered to also play a role in recruiting channels into activity as a function of ligand stimulation.

A consistent rate of detection of IP_3R channel activity in nuclear patch-clamp experiments using isolated Sf9 nuclei enabled quantification of the average number of IP_3R channels detected in a nuclear patch-clamp experiment (N_A) under various ligand concentrations (Ionescu et al., 2006). In saturating IP_3, the number of active IP_3R in each patch was a biphasic function of $[Ca^{2+}]_i$. The observation that N_A was a function of stimulus strength is unexpected since it was anticipated that the entire channel population in a membrane patch would always become activated, albeit to different levels of activity (P_o) depending on the strength of ligand activation. Instead, these results indicate that suboptimal ligand concentrations are insufficient to activate all available IP_3R channels in a membrane patch that can be activated by optimal ligand concentrations. To account for these observations, a model was proposed in which Ca^{2+} binding at a very fast rate to a "sequestration" site, before these channel could actively gate open and be observed, sequestered some of the available IP_3R channels into a nonactive state when ligand conditions were suboptimal (Ionescu et al., 2006).

Thus, two mechanisms revealed by single-channel patch-clamp electrophysiology exist to grade Ca^{2+} release through a population of IP_3R channels: regulation of channel activity level (P_o) as well as recruitment of additional channels. Both mechanisms coexist and can occur even in the absence of crosstalk among channels by released Ca^{2+} (CICR). The importance of this channel recruitment mechanism can be appreciated by considering that the rate of Ca^{2+} flux through a population of IP_3R channels (J) is given by

$$J = \gamma N P_o, \tag{2}$$

where γ is the single-channel conductance, N is the number of activated channels, and P_o is the average single-channel open probability. Single-channel studies determined that the IP_3R channel conductance γ is largely IP_3-independent. Since both channel P_o and N_A are regulated by Ca^{2+} and IP_3, the ligand dependence of Ca^{2+} flux released through IP_3R channels is more accurately estimated in terms of $N_A P_o$. In saturating [IP_3], the dependence of $N_A P_o$ on $[Ca^{2+}]_i$ is biphasic: $N_A P_o$ increases by over 10-fold as Ca^{2+} is increased from 50 to 500 nM and then gradually decreases as $[Ca^{2+}]_i$ is further increased. $N_A P_o$ is also strongly dependent on IP_3 concentration. With IP_3 reduced to 33 nM, the dependence of $N_A P_o$ on $[Ca^{2+}]_i$ remains biphasic with peak $N_A P_o$ observed in $Ca^{2+} \approx 0.5$–$1\ \mu M$, but maximum $N_A P_o$ is an order of magnitude lower than that observed in saturating IP_3 (Ionescu et al., 2006).

B. *Insights from Kinetic Responses to Rapid Changes in Ligand Concentrations*

Recently, it became possible, with the development of the cytoplasmic side-out configuration in nuclear patch-clamp studies, to obtain kinetic information concerning single IP_3R channel responses to ligand concentration changes with millisecond resolution (Mak et al., 2007).

1. Ca^{2+} Activation and Deactivation

The kinetics of Ca^{2+} regulation were first studied by applying rapid $[Ca^{2+}]_i$ jumps in the presence of constant, saturating (10 μM) IP_3. Channel activation by a jump from low (< 10 nM) to optimal (2 μM) $[Ca^{2+}]_i$ had mean latency of ~40 ms (Fig. 4C), an order of magnitude faster than the response to IP_3 activation (350–500 ms; Fig. 4A). The latency for a $[Ca^{2+}]_i$ jump to

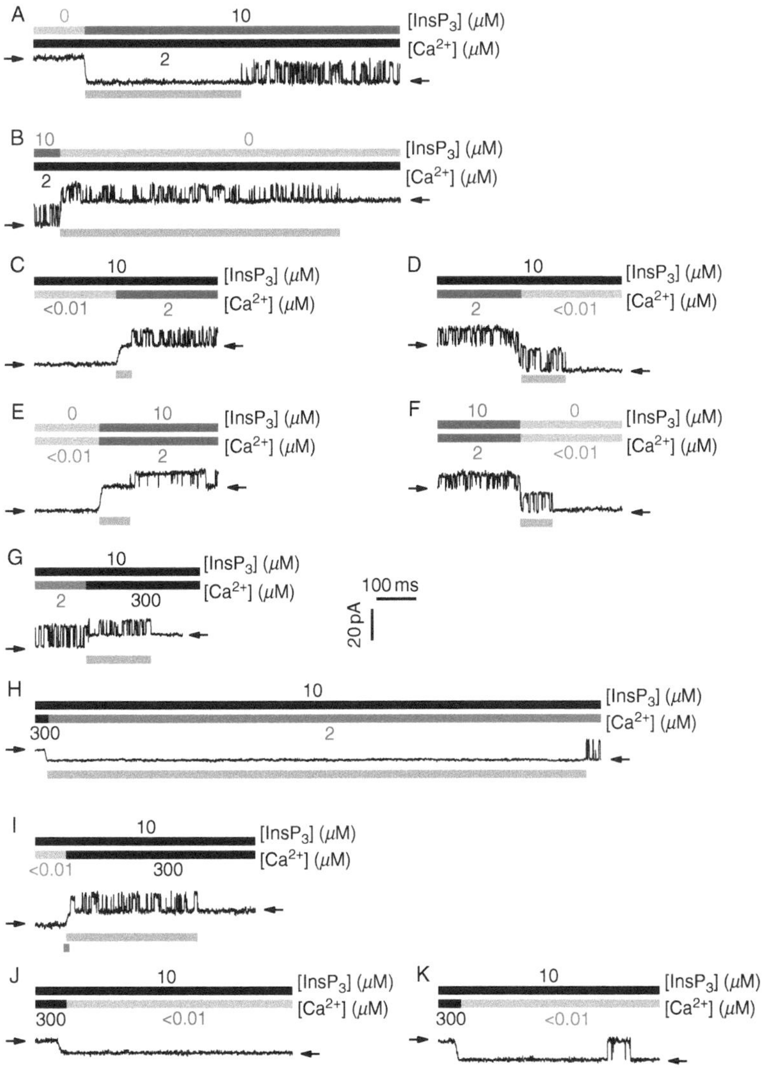

FIGURE 4 Nuclear patch-clamp studies of single IP_3R channel responses to rapid ligand concentration changes. (A–K) Single-channel current traces from cyto-out patches showing responses of Sf9 IP_3R channels to abrupt changes in ligand concentrations indicated by color bars above the traces. Ligand concentrations were changed by switching the perfusing bath solutions, which also caused a change in the closed channel baseline current level because of the different [KCl] in the perfusing solutions. This allowed the time course of the perfusing solution switch to be monitored. Response latencies are denoted by gray bars below the traces. Modified from Mak et al. (2007). (See Color Insert.)

300 μM was shorter but detectable (~24 ms; Fig. 4I), indicating that Ca^{2+} binding and subsequent conformation changes both contribute to the Ca^{2+} activation latency. Channel deactivation when Ca^{2+} was returned to <10 nM had a latency of ~160 ms (Fig. 4D), also significantly faster compared with channel deactivation due to removal of IP_3 (~690 ms; Fig. 4B). These results demonstrate that IP_3-bound channels respond rapidly to changes in $[Ca^{2+}]_i$, a property ideally suited for CICR to couple IP_3R channels in Ca^{2+} signaling.

Of note, the activation latency in response to simultaneous jumps of $[IP_3]$ (from 0 to 10 μM) and $[Ca^{2+}]_i$ (from <10 nM to 2 μM) was ~60 ms (Fig. 4E), much shorter than that for IP_3 activation in constant optimal $[Ca^{2+}]_i$ and comparable to the latency for Ca^{2+} activation in constant saturating $[IP_3]$. If IP_3 binding was required before Ca^{2+} can bind, then the activation latency for a simultaneous jump of the two ligands should be comparable to that for IP_3 in optimal $[Ca^{2+}]_i$. Thus, there is no requirement for strict sequential IP_3 and Ca^{2+} binding to activate the channel, and IP_3 and Ca^{2+} binding is not independent, since independence predicts that the activation latency for simultaneous ligand jumps should be longer than both activation latencies for $[IP_3]$ or $[Ca^{2+}]_i$ jumps individually. Instead, the kinetic data are consistent with a *negative* cooperative interference in which high-affinity Ca^{2+} binding to a channel not bound to IP_3 retards subsequent activation of the channel by IP_3. The deactivation latency for simultaneous removal of IP_3 (from 10 μM to 0) and Ca^{2+} (from 2 μM to <10 nM) of ~70 ms (Fig. 4F) was significantly shorter than deactivation latencies for removal of either IP_3 or Ca^{2+} alone. This implies that unbinding of IP_3 and Ca^{2+} from the IP_3R is also neither sequential nor independent. These results rule out many models for ligand regulation of IP_3R activity (Fraiman & Ponce Dawson, 2004; Hirose, Kadowaki, & Iino, 1998; Marchant & Taylor, 1997; Sneyd & Falcke, 2005; Swillens, Combettes, & Champeil, 1994; Swillens, Champeil, Combettes, & Dupont, 1998; Tang, Stephenson, & Othmer, 1996).

2. A Qualitative Model to Account for Rapid Ca^{2+} Regulation of Channel Activity

The observed response latencies point to a seemingly paradoxical interaction between IP_3 and Ca^{2+} in activation of the IP_3R. Longer activation latencies in response to jumps in $[IP_3]$ in constant optimal $[Ca^{2+}]_i$ compared to the activation latencies in response to simultaneous jumps in $[IP_3]$ and $[Ca^{2+}]_i$ indicate that Ca^{2+} can bind with high affinity to the channel in the absence of IP_3. However, unlike the high-affinity Ca^{2+} binding that activates the channel in concert with IP_3 binding established from steady-state studies (above), this Ca^{2+} binding interferes with IP_3 activation by predisposing the channel to activate through slower conformation transition pathways. On the other hand, the much more rapid deactivation of the channel in response to simultaneous drops in $[IP_3]$ and $[Ca^{2+}]_i$ compared to channel deactivation caused by drops in either ligand alone indicates that the IP_3 and the high-affinity Ca^{2+} sites have *positive* cooperative coupling in an activated channel, such that occupation of the IP_3 or Ca^{2+} sites causes the channel to deactivate through slower conformation transition pathways, thereby stabilizing the channel in active conformation(s).

A simplified version of the quantitative model proposed to account for steady-state ligand regulation of channel activity (above; Fig. 3) (Mak, McBride, & Foskett, 2003), which assumes only basic allosteric interactions between IP_3 and Ca^{2+} that arise solely from preferential binding of the two ligands to various channel conformations, is also able to account for these kinetic responses qualitatively. In its simplest form (Fig. 5), the model hypothesizes that the channel can assume four different conformations. In the absence of IP_3, it is energetically favorable for the channel to assume two conformations, one active (A) and one inactive (B), with Ca^{2+} binding to the high-affinity sites stabilizing the inactive B conformation. Thus, in the absence of IP_3, the high-affinity Ca^{2+} sites (G sites) are actually *inhibitory*. IP_3 binding to the channel energetically stabilizes the other two conformations, also one active (C) and one inactive (D). Ca^{2+} binding to the high-affinity sites stabilizes the active C conformation relative to the inactive D conformation. Consequently, the high-affinity G sites are *activating* when the channel is IP_3-bound. Thus, because of the different relative affinities of the high-affinity Ca^{2+} sites in the four channel conformations, IP_3 binding to the channel changes the nature of the high-affinity G Ca^{2+} sites from inhibitory to activating. This transformative effect of IP_3 on the nature of the high-affinity Ca^{2+} binding G sites can account for the seemingly paradoxical channel response latencies measured here. Because Ca^{2+} binding to the high-affinity G sites in the absence of IP_3 stabilizes the inactive B conformation which has low affinity for IP_3, preexposure of the channel to 2 μM Ca_i^{2+} slows binding of IP_3 to the channel and consequent activation kinetics compared with activation by simultaneous $[Ca^{2+}]_i$ and $[IP_3]$ jumps. Conversely, occupation of the IP_3 and high-affinity Ca^{2+} G sites stabilizes an active channel in the active C conformation with high affinity for IP_3 and Ca^{2+}. This accounts for the shorter mean deactivation latency by simultaneous removal of IP_3 and $Ca^{2+}{}_i$ compared with that due to removal of either ligand alone. The model hypothesizes that the channel can assume the different conformations in any ligand-binding state, with IP_3 and Ca^{2+} binding regulating channel activity by affecting the probability of the channel being in various conformations. Thus, active and inactive conformations are connected by multiple branched pathways through various ligand binding and conformation changes.

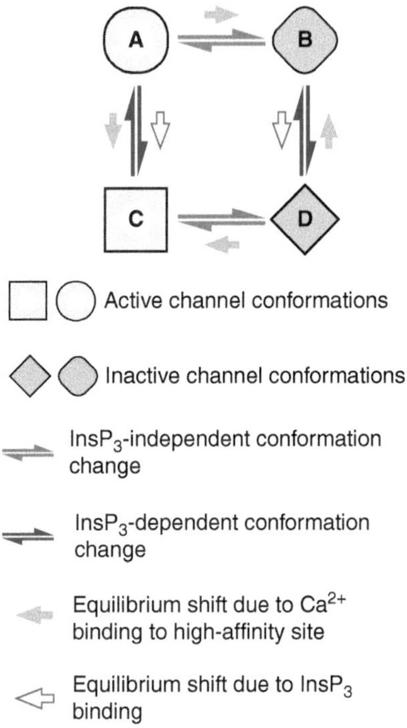

FIGURE 5 Kinetic scheme accounting for the observed IP_3R channel response latencies. A qualitative allosteric model stipulating one kind of high-affinity Ca^{2+} site in an IP_3R channel. Occupation of the IP_3-binding sites causes the channel to adopt the C and D conformations instead of the A and B conformations, so that Ca^{2+} binding to the high affinity Ca^{2+} site becomes activating instead of inhibitory. Only conformation changes are shown in this kinetic scheme. IP_3 and Ca^{2+} binding to the channel, and the ligand-binding state of the channel conformations, are omitted for clarity. Reproduced from Mak et al. (2007). (See Color Insert.)

3. Channel Inhibition and Recovery Kinetics

Switching $[Ca^{2+}]_i$ from 2 μM (optimal) to 300 μM (inhibitory) in the presence of 10 μM IP_3 inhibited channel gating (Fig. 4G), demonstrating that IP_3 binding does not render the low-affinity inhibitory (H) Ca^{2+} sites inaccessible for Ca^{2+} binding, disproving previous claims (Adkins & Taylor, 1999). The mean latency for Ca^{2+} inhibition ($\sim$300 ms) was substantially longer than that for Ca^{2+} activation. The mean latency for channel recovery from Ca^{2+} inhibition was $\sim$2.4 s (Fig. 4H). This remarkably long latency defines a refractory period between successive Ca^{2+} release events mediated by the same IP_3R channel or channel cluster. A refractory period has been invoked to account for mutual annihilation of colliding intracellular Ca^{2+} waves (Lechleiter, Girard, Peralta, & Clapham, 1991) and the periodicity of

Ca^{2+} oscillations (Parker & Ivorra, 1990), but its molecular basis remained unknown. The observed latency is in quantitative agreement with the mean refractory periods (~2.5 s) between two successive spontaneous IP_3R-mediated elementary Ca^{2+} release events (puffs) observed in the presence of saturating [IP_3] (Fraiman, Pando, Dargan, Parker, & Ponce Dawson, 2006). This suggests that the latency for channel recovery from Ca^{2+} inhibition, rather than other mechanisms such as lingering high $[Ca^{2+}]_i$ in the vicinity of the release sites after a Ca^{2+} release event, is responsible for the refractory behavior of Ca^{2+} release sites observed *in vivo* (Ilyin & Parker, 1994).

4. Kinetics of a Single-Channel Ca^{2+} Release Burst

In cells, Ca^{2+} released by an individual channel raises $[Ca^{2+}]_i$ to inhibitory (~60 μM) levels in the microdomain surrounding it. The kinetic results predict that this Ca^{2+} will feed back to inhibit it and terminate release after the burst of activity. $[Ca^{2+}]_i$ jumps from low subactivating (< 10 nM) directly to inhibitory (300 μM) levels in the presence of 10 μM IP_3 produced just such a burst of channel activity (Fig. 4I) as Ca^{2+} first bound to the high-affinity (F and G) to activate the channel before the low-affinity inhibitory (H) site(s) became occupied, with mean activation and inhibition latencies of channel activity bursts =~24 and ~470 ms, respectively. The mean single-channel burst duration was ~470 ms. The time course of these channel activity bursts represents the kinetics of the fundamental Ca^{2+} release event—the basic building block from which local concerted Ca^{2+} release by clusters of IP_3R channels (puffs) and global propagating IP_3R-mediated Ca^{2+} signals (waves) are generated through CICR.

A considerable fraction (~10%) of $[Ca^{2+}]_i$ jumps from < 10 nM to 300 μM failed to generate a burst. In comparison, < 1% of $[Ca^{2+}]_i$ jumps from < 10 nM to 2 μM failed to activate gating. Thus, Ca^{2+} binding to the high-affinity activating G sites and low-affinity inhibitory H sites are not sequential, and the failure of $[Ca^{2+}]_i$ jumps (< 10 nM to 300 μM) to generate an activity burst simply reflects the nonzero probability of Ca^{2+} binding to an inhibitory H site before the activating site(s) becomes occupied. Because the mean latency for recovery from Ca^{2+} inhibition is over an order of magnitude longer than the mean latency of deactivation by $Ca^{2+}{}_i$ removal, during most $[Ca^{2+}]_i$ drops (from 300 μM to < 10 nM), no channel activity was observed (Fig. 4J) as the channels were deactivated due to Ca^{2+} dissociation from the high-affinity activating F and G sites before they recovered from Ca^{2+} inhibition. However, in ~6% of $[Ca^{2+}]_i$ drops, the channel recovered from Ca^{2+} inhibition first, generating a brief burst of activity before it was deactivated by Ca^{2+} unbinding (Fig. 4K). This indicates that Ca^{2+} dissociation from the inhibitory and activating sites are also not sequential.

C. Insights from Modal Gating Analyses

Examination of long current recordings of single Sf9 IP_3R channels in the presence of constant [IP_3] and [Ca^{2+}]$_i$ revealed that the channels did not display steady-state gating behavior, but instead exhibited spontaneous changes in gating kinetics through distinct patterns of behavior, or modes (Ionescu et al., 2007). Three distinct gating modes of the IP_3R were identified: a mode in which the channel is mostly bursting, another with fast channel gating kinetics, and one with long quiescent periods, with channel P_o of 0.85, 0.24, and 0.007, respectively. In the gating mode with High P_o (H mode), the channel exhibits mainly bursting behavior with only brief gaps interrupting the long bursts of channel activity. In the mode with Intermediate P_o (I mode), the IP_3R gates frequently with mostly short openings and closings. In the Low P_o (L mode), the channel has long closed periods interrupted with brief, infrequent openings (Fig. 6).

Strikingly, the P_o within each mode (P_o^M, with M standing for H, I, or L) remains relatively consistent over all [Ca^{2+}]$_i$ (Ionescu et al., 2007). Accordingly, [Ca^{2+}]$_i$ regulates P_o predominantly by regulating the fraction of time the channel spends in the different gating modes (relative prevalence π^M; Fig. 7). The π^I exhibited relatively little [Ca^{2+}]$_i$ dependence, whereas both π^H and π^L showed strong ligand dependencies. In the presence of saturating IP_3 and optimal [Ca^{2+}]$_i$, the channel is found predominantly in the H mode and in the L mode for relatively little time, resulting in high overall P_o. The high π^H is due to the channel entering the H mode more frequently and staying there longer. In nonoptimal [Ca^{2+}]$_i$ (0.1 and 89 μM), the channel has low overall activity mainly because it enters the L mode more frequently.

Although the IP_3R channel P_o^M stayed within a narrow range over all [Ca^{2+}]$_i$ examined, the gating kinetics within a mode are not completely [Ca^{2+}]$_i$-independent. The mean open and closed channel durations and mean durations of channel activity bursts and gaps between those bursts in the three modes all exhibited some variations over the [Ca^{2+}]$_i$ range examined. Nevertheless, mode switching is the most relevant mechanism of ligand regulation of IP_3R gating.

The important role that modal gating plays in ligand regulation of IP_3R activity indicates that besides the time scales of channel openings and closings (approximately milliseconds), other, longer time scales in IP_3R gating kinetics are likely to be relevant for the kinetics of IP_3-mediated [Ca^{2+}]$_i$ signaling *in vivo*. One such time scale is associated with the channel burst and burst-terminating (interburst) gap durations. Whereas most of the short closings are the result of ligand-independent channel gating (Foskett & Mak, 2004; Mak, McBride, & Foskett, 2003), the time scales of the bursts and gaps

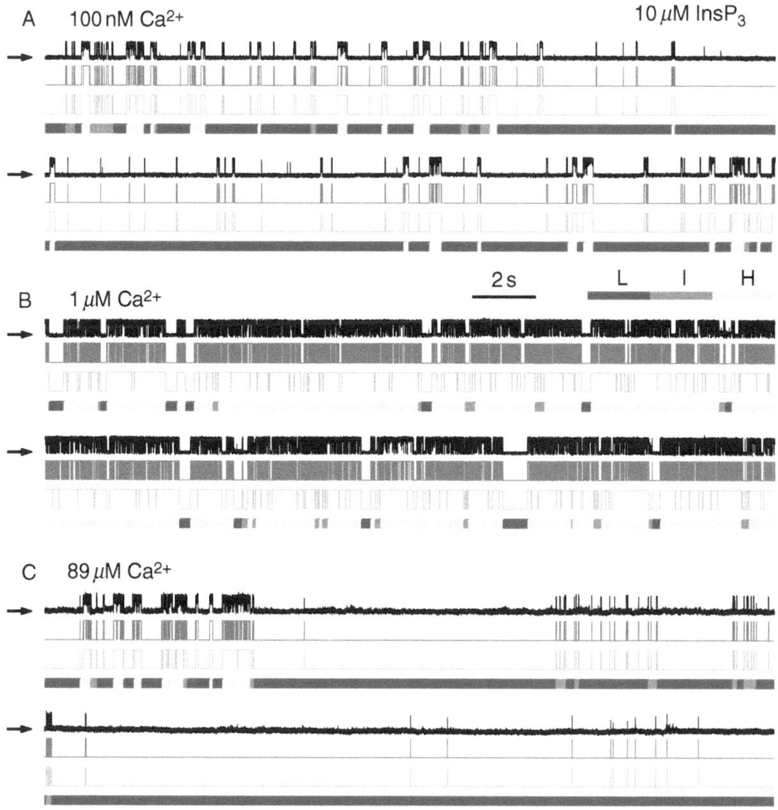

FIGURE 6 IP_3R channel modal gating. Single IP_3R channel gating records observed in saturating 10 μM IP_3 and suboptimal (100 nM), optimal (1 μM) and inhibiting (89 μM) $[Ca^{2+}]_i$ as tabulated. Each section consists of a set of four graphs derived from one 50-s long continuous single-channel record. From the top: raw single-channel current trace (in black with arrow indicating the closed-channel current level), idealized current trace (in blue), idealized current trace after burst analysis (in purple), and the gating mode of the channel with green, yellow, and red representing H, I, and L modes. The current traces were selected to show modal transitions among all three gating modes in one continuous recording under various $[Ca^{2+}]_i$. (See Color Insert.)

probably reflect the kinetics of ligand unbinding from and binding to, respectively, the channel and the associated IP_3R conformational changes. The studies of the kinetic responses of single IP_3R channels to rapid ligand concentration changes (above) observed that in the constant presence of saturating 10 μM IP_3, the mean lag times to termination of IP_3R channel activity from abrupt changes in $[Ca^{2+}]_i$ from optimal (2 μM) to subactivating (< 10 nM) or inhibitory (300 μM) levels were 160 and 290 ms, respectively.

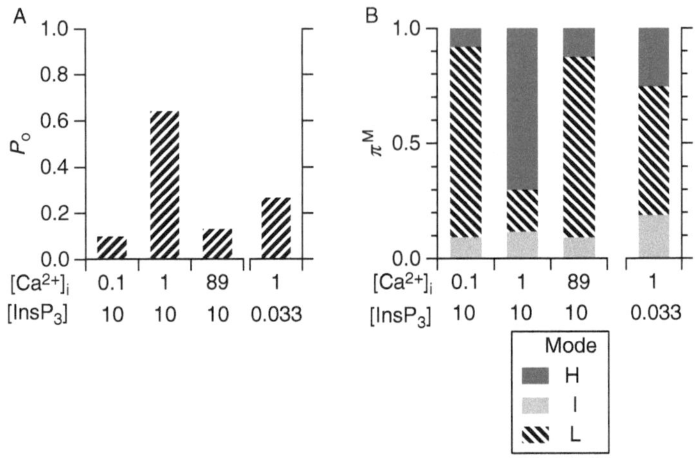

FIGURE 7 IP_3R channel modal gating parameters. Channel gating parameters observed in various $[Ca^{2+}]_i$ and $[IP_3]$ as tabulated (in μM). (A) Channel open probability P_o. (B) Relative prevalence π^M (amount of time the channel spent in a mode/total duration of all analyzed channel current records) for the three gating M (M stands for H, I, or L, depending on the mode involved). The H, I, and L modes are plotted in bar graphs in green, yellow, and red, respectively, as indicated by the key. Modified from Ionescu et al. (2007). (See Color Insert.)

In constant 2 μM $Ca^{2+}{}_i$, the mean lag time to channel activity termination from an abrupt drop in $[IP_3]$ from 10 μM to 0 was 700 ms. In those experiments, the channels were most likely in the H mode before the activity-terminating ligand concentration change, with burst duration of 200–600 ms. The similarity between the mean channel lag times observed in rapid perfusion experiments and the mean burst duration suggests that the kinetics of channel responses to changes in ligand concentrations are likely determined by how fast the channel can exit from a burst when the ligand concentration change occurs. If that is the case, then instead of a channel opening, a channel burst probably constitutes a stereotypical single-channel IP_3R Ca^{2+}-release event.

The weak dependence of channel burst duration on $[Ca^{2+}]_i$ may possibly be a mechanism by which an active IP_3R channel can avoid being prematurely inhibited by the Ca^{2+} that it releases, because an increase of $[Ca^{2+}]_i$ from 1 μM (optimal) to 89 μM (inhibitory) only reduces the burst duration from ~500 to ~300 ms when the channel is in H mode. Moreover, the stabilization by activating $[Ca^{2+}]_i$ of the H mode, which has significantly longer burst durations, may possibly play an important role in CICR. Thus, in the presence of sufficiently high $[IP_3]$, an increase in $[Ca^{2+}]_i$ above the resting level can encourage an IP_3R channel to enter the H mode from the

L mode. As a result, the burst duration of the channel increases from that for a L mode (~10 ms), which may not release enough Ca^{2+} to recruit nearby channels to propagate a Ca^{2+} signal (individual blips or puffs), to that of a H mode (~200 ms), which enables the IP_3R channel to continue releasing Ca^{2+} even when the local $[Ca^{2+}]_i$ is raised to a high level. Such long channel bursts can release sufficient Ca^{2+} to recruit neighboring IP_3Rs or IP_3R clusters for a regenerative Ca^{2+} signal.

D. Molecular Bases for Ca^{2+} Regulation of IP_3R Channel Activity

Single-channel studies of ligand regulation of IP_3R channel activity have revealed that Ca^{2+} is intimately involved in regulating many aspects of IP_3R channel activity, including activation (increasing steady-state channel P_o), inhibition (decreasing steady-state channel P_o), inactivation (termination of channel activity in the presence of constant ligand concentrations), and recruitment (changing the fraction of channels in a population that is activated). Moreover, other ligands exert their effect on IP_3R channel activity by altering the functional Ca^{2+} affinities of the channel: the main effect of IP_3 is modulating sensitivity of the channel to Ca^{2+} inhibition, and ATP potentiates channel activity by increasing its sensitivity to Ca^{2+} activation. However, despite the identification of multiple putative Ca^{2+}-binding sites in the IP_3R primary sequences (Sienaert et al., 1996, 1997), relatively little progress has been made in determining the amino acid sequences that constitute the functional Ca^{2+} sensors in the IP_3R.

1. Residues in the IP_3R Sequence

Eight GST-fused denatured peptide fragments of the IP_3R located throughout the linear sequence were found to bind Ca^{2+} in gel-overlays (Sienaert et al., 1996, 1997), and although the biochemical detection of several sites is consistent with the presence of multiple Ca^{2+}-binding sites inferred from studies of single-channel gating (above), the physiological implications of such data are unclear. Mutagenesis of a conserved glutamic acid residue in IP_3R (Glu2100) affected $[Ca^{2+}]_i$ signals (Miyakawa et al., 2001) and shifted the apparent Ca^{2+} dependence of reconstituted channel P_o by approximately fivefold, from 0.2 to 1 μM (Tu et al., 2003). A peptide spanning residues Glu1932–Arg2270 bound Ca^{2+} with an apparent affinity of 160 nM as measured by tryptophan fluorescence, which was decreased to approximately 1 μM when Glu2100 was mutated (Tu et al., 2003). Although it has been concluded that this residue and region are important for Ca^{2+} regulation (Miyakawa et al., 2001; Tu et al., 2003), it was not determined whether the observed effects on channel function were due to modification of one of the Ca^{2+}-binding sites discussed above, or whether they were secondary effects caused by long-range allosteric mechanisms. Second, metal-binding sites in proteins generally comprise several interacting residues that help to coordinate and stabilize the ion in a binding pocket. Ca^{2+}-binding sites usually consist of 6 or 7 coordinating oxygen atoms provided by side-chain carboxyls, main chain carbonyls, and water (Yang, Lee, Hellinga, & Yang, 2002). However, additional residues that might interact with Glu2100 to coordinate and bind Ca^{2+} have not been identified. With these caveats in mind, this region of the channel may play a role in Ca^{2+} sensing, but further experimentation is necessary to determine whether it is a Ca^{2+}-binding site, and which of the functional sites it represents. Another region of the receptor that has been considered to be involved in Ca^{2+} regulation of channel gating is the ligand-binding domain. Two surface acidic clusters were observed in the crystal structure of the IP_3-bound 225–604 fragment (Bosanac et al., 2002). Site 1 was contained completely within the β-trefoil domain, whereas site 2 was located across the two domains. Both sites had been shown previously to bind Ca^{2+} in gel overlay experiments (Sienaert et al., 1997). The residues that contribute to the acidic patches are highly conserved among isoforms. Nevertheless, mutagenesis studies failed to provide evidence for a role for residues in either site in Ca^{2+} activation since none of the mutations affected the ability of the channel to function at low cytoplasmic $[Ca^{2+}]_i$ in Ca^{2+} release assays (Joseph, Brownell, & Khan, 2005). However, only single point mutations were examined, and the relatively low resolution of Ca^{2+} release assays for measuring detailed channel properties may require multiple residues in a Ca^{2+} binding motif to be mutated to observe functional consequences. Single-channel studies of these mutant channels may reveal more subtle effects on the Ca^{2+} dependence of gating.

2. Summary of Functional Ca^{2+} Regulatory Sites

Detailed single-channel studies have revealed that Ca^{2+} regulates different aspects of IP_3R channel activity with very different functional affinities, binding kinetics, and ligand dependencies. Therefore, multiple different distinct Ca^{2+} sensors are likely to be involved. There is one Ca^{2+} sensor responsible for IP_3 modulation of Ca^{2+} inhibition of the channel. It changes from an inhibitory Ca^{2+}-binding site to an activating binding site depending on the concentration of IP_3. Another IP_3-independent Ca^{2+} sensor is responsible for the Ca^{2+} activation with consistent Ca^{2+} sensitivity observed in a wide variety of IP_3R channels. Yet another Ca^{2+} sensor is IP_3-independent but inhibitory, with variable functional Ca^{2+} affinity that can be changed by factors extrinsic to the IP_3R channel like membrane lipid composition and exposure to low $[Ca^{2+}]_i$. Elimination of the functionality of this purely inhibitory Ca^{2+}-binding site by exposure to very low bath Ca^{2+} concentrations implies that the channel possesses a fourth Ca^{2+}-binding site with very high Ca^{2+} affinity that must be occupied by Ca^{2+} for the IP_3-independent Ca^{2+} inhibitory site to be functional. There is a fifth Ca^{2+} sensor that binds Ca^{2+} at a low rate ($\sim$0.01 s^{-1}), but only when the channel is liganded with IP_3. This sensor is responsible for IP_3-induced inactivation of IP_3R channels. Ca^{2+} binding to this site causes the channel to enter into an inactive state from which it cannot emerge in the continuous presence of IP_3. A sixth Ca^{2+} sensor is responsible for regulation of IP_3R channel recruitment.

Thus, the evidence points to a complex picture with multiple functional Ca^{2+} sensors in the channel, and the molecular identities of these sensors remain to be properly elucidated. Regulation of IP_3R channel activity by these Ca^{2+} sensors is probably complicated, with various degrees of cooperativity and hierarchical organization so that Ca^{2+} binding to one sensor can regulate functionality of another Ca^{2+} sensor. Structurally, these Ca^{2+} sensors are most likely to be intrinsic to the channel, although some could reside in a protein(s) associated with the IP_3R channel complex.

IV. Ca^{2+} BINDING PROTEIN REGULATION OF IP_3R CHANNEL GATING

A. Calmodulin

The ubiquitous and conserved Ca^{2+}-binding protein calmodulin (CaM) confers Ca^{2+}-dependent regulation on many proteins, including ion channels (Saimi & Kung, 2002). Binding of Ca^{2+} to CaM (Ca^{2+}–CaM) triggers conformational changes that promote or modulate its interaction with binding partners. Ca^{2+}-free CaM (apoCaM) can also interact with target proteins, enabling CaM to function, in essence, as a Ca^{2+}-regulated subunit of the protein. In this way, CaM mediates Ca^{2+}-dependent and -independent regulation of many ion channels, including the ryanodine receptor (Balshaw, Yamaguchi, & Meissner, 2002). Although a substantial literature suggests that CaM interacts with the IP_3R, the *in vivo* association between the two proteins as well as the functional implications are far from unequivocal (recently reviewed in Foskett et al., 2007; Nadif Kasri et al., 2002; Rossi & Taylor, 2004; Taylor & Laude, 2002).

CaM was reported to bind within residues 1564–1585 (Fig. 2B) of the coupling domain of the mouse type 1 channel (Maeda et al., 1991). The interaction was Ca^{2+}-dependent in CaM-sepharose binding assays (Maeda et al., 1991; Yamada et al., 1995). In an early report, purified IP_3R-1 channels reconstituted into lipid bilayers lacked inhibition by high [Ca^{2+}], whereas channels reconstituted from cerebellar microsomes displayed high [Ca^{2+}] inhibition (Michikawa et al., 1999). This result was interpreted as indicating that a component of Ca^{2+}-dependent inhibition was lost during purification. The authors inferred that Ca^{2+} inhibition of IP_3R gating was therefore not mediated by Ca^{2+} binding to the IP_3R itself (Michikawa et al., 1999). Addition of CaM to the bath restored Ca^{2+} inhibition of purified receptors, suggesting that it mediated high $[Ca^{2+}]$ inhibition of channel gating (Michikawa et al., 1999). Addition of the CaM antagonists W7 or the CaM binding domain of CaM kinase II activated reconstituted cerebellar IP_3R, which was interpreted as indicating that microsomal IP_3R is constitutively associated with and inhibited by CaM. The conclusion that high $[Ca^{2+}]$ inhibition of IP_3R gating is mediated by CaM was reinforced by observations that IP_3R-3 did not bind CaM (Cardy & Taylor, 1998; Lin, Widjaja, & Joseph, 2000; Yamada et al., 1995) and its gating was not inhibited by high $[Ca^{2+}]_i$ (Hagar et al., 1998).

Michikawa et al. (1999) has been repeatedly cited as evidence that Ca^{2+} inhibition of IP_3R channel gating is mediated by CaM. However, many experimental results argue against this conclusion. The K_d for Ca^{2+}–CaM binding to the channel was reported to be 27 μM (Hirota, Michikawa, Natsume, Furuichi, & Mikoshiba, 1999), suggesting a low-affinity interaction, inconsistent with constitutive association with reconstituted channels. Indeed, high concentrations (10 μM) of CaM were required to inhibit Ca^{2+} release from purified IP_3R-1 (Hirota et al., 1999). Second, IP_3R-3 is not immune to Ca^{2+} inhibition as originally claimed. Biphasic $[Ca^{2+}]_i$ regulation of type 3 channels has been demonstrated in electrophysiological recordings by nuclear membrane patching (Mak et al., 2001b) and in planar bilayer reconstitutions (Tu, Wang, Nosyreva, De Smedt, & Bezprozvanny, 2005). Because the type 3 channel does not bind CaM, Ca^{2+} inhibition of channel gating can therefore occur in the absence of CaM binding (Cardy & Taylor, 1998). Third, in nuclear patch-clamp recordings of *X*-IP_3R-1, inhibition of channel activity by high $[Ca^{2+}]_i$ was observed in the presence of W7 and despite the overexpression of a dominant-negative, Ca^{2+}-insensitive CaM (Mak, McBride, Petrenko, et al., 2003). Fourth, mutation of the CaM-binding site eliminated CaM interaction with the IP_3R (Yamada et al., 1995), but it did not affect Ca^{2+} inhibition of channel activity (Nosyreva et al., 2002; Zhang & Joseph, 2001). The published data taken together provide little support for the proposition that high $[Ca^{2+}]_i$ inhibition of IP_3R gating is mediated by CaM.

In addition to a putative role in Ca^{2+} inhibition of channel gating, CaM has also been invoked to regulate other properties of the IP_3R. These studies have been reviewed elsewhere (Foskett et al., 2007). There are few biochemical data to suggest that CaM and IP_3R interact *in vivo*. The evidence of interactions relies almost exclusively on the various *in vitro* binding and pharmacological studies, while successful coimmunoprecipitation of the two proteins has not been demonstrated. Therefore, despite considerable effort to place CaM in the IP_3R signaling complex, there is little consensus on its functional role.

B. CaBP and CIB1

The IP_3R interacts with a family of CaM-related Ca^{2+}-binding proteins termed CaBPs, a subset of the neuronal Ca^{2+} sensor (NCS) family of EF-hand-containing proteins (Yang et al., 2002). Eight CaBP genes have been identified (CaBP1–8) with alternate splicing generating long and short forms of CaBP1 and CaBP2 (Haeseleer, Imanishi, Sokal, Filipek, & Palczewski, 2002; Haeseleer et al., 2000). CaBP1 binds with high affinity (apparent $K_d < 50$ nM) in a Ca^{2+}-dependent manner (apparent affinity $\sim 1\ \mu M$) to the amino-terminal 600 residues of all three rat IP_3R isoforms (Yang et al., 2002), a region that encompasses the IP_3-binding domain. The Ca^{2+} dependence of the interaction was mediated by the EF hands of CaBP (White, Yang, Monteiro, & Foskett, 2006; Yang et al., 2002). Truncation mutagenesis of the IP_3R ligand-binding domain localized CaBP1 binding to the amino-terminal 225 residues (Kasri et al., 2004), although the binding observed was Ca^{2+}-independent, whereas the high-affinity interaction of CaBP1 with the IP_3R is strongly Ca^{2+}-dependent (White et al., 2006; Yang et al., 2002). Thus, the significance of CaBP1 binding to this region is unclear.

In nuclear membrane patch-clamp recording of *X*-IP_3R-1, channel activity with high P_o and gating kinetics similar to those elicited by IP_3 was observed in the absence of IP_3 when recombinant CaBP1 was included in the pipette solution at optimal $[Ca^{2+}]_i$. In contrast, the triple-EF-hand mutant CaBP1 that lacked Ca^{2+} binding failed to activate the channel (Yang et al., 2002). Purified CaBP2 and CaBP5 each had similar effects (Yang et al., 2002). These results identified the CaBP Ca^{2+} sensors as a family of protein ligands of the IP_3R channel. It was proposed that this regulation might enable the channel to become activated in the absence of phosphoinositide metabolism, and that it may modulate the sensitivity of the channel to IP_3, which may play a role in spatially restricting Ca^{2+} release (Yang et al., 2002). Because IP_3R-mediated Ca^{2+} signals are shaped by messenger diffusion, degradation and removal, processes that will have distinct kinetics for IP_3 compared with CaBPs, the identification of novel ligands for the IP_3R provided new insights into the dimensions and versatility of this ubiquitous signaling pathway.

Members of the CaBP family are exclusively expressed in brain and retina (Haeseleer et al., 2000), but this mode of IP_3R regulation likely extends more widely to nonneuronal cell types as well. The ubiquitously expressed protein

CIB1 (calcium- and integrin-binding protein; also called calmyrin or KIP) was subsequently shown to interact with the IP_3R (White et al., 2006). CIB1 belongs to a distinct subfamily of NCS proteins (Gentry et al., 2005). Nevertheless, like CaBP1, CIB1 binds to the first 600-residue ligand-binding region of all isoforms of the IP_3R in a Ca^{2+}-sensitive fashion, with an affinity in the μM range, and has a similar dependence on functional EF-hands. Like CaBPs, CIB1 activated *Xenopus* and Sf9 IP_3R-1 channel gating in the absence of IP_3, establishing it as a novel protein agonist. CIB1 behaves as a partial agonist of the channel, since it was less efficacious than optimal [IP_3] in activating channel gating (White et al., 2006).

Although acute exposure of the IP_3R to purified recombinant CaBP/CIB1 proteins *activates* channel gating, overexpression of either CaBP or CIB1 was found to *inhibit* agonist-induced IP_3R-dependent Ca^{2+} release in intact cells (Haynes, Tepikin, & Burgoyne, 2004; Kasri et al., 2004; White et al., 2006). This apparent conflict was resolved by the demonstration that CaBP/CIB1 initially activates IP_3R channel activity, but then subsequently drives the channel into an inactivated state that cannot be activated by IP_3 (White et al., 2006). In CaBP1/CIB1 overexpression studies, much of the IP_3R channel population would be in the inactivated state, accounting for the muted IP_3-mediated responses observed in CaBP1 or CIB1-expressing cells (Haynes et al., 2004; Kasri et al., 2004; White et al., 2006). These results suggest that prolonged exposure to CIB1 can effectively remove functional channels from the total IP_3R population by a process of ligand-induced channel inactivation.

Phosphorylation or other covalent modifications, and other protein interactions may regulate the affinities of the interactions of CaBP1 and CIB1 with the IP_3R. Mutation of a conserved phosphorylation site in CaBP1 resulted in even greater $[Ca^{2+}]_i$ signaling inhibition (Kasri et al., 2004), although this site is not conserved in CIB1. Regulated unbinding of the protein ligands could enable the channel to escape from inactivation, allowing it to again be activated by protein ligand rebinding, in a mechanism of repeated channel activation that is independent of IP_3.

C. Chromogranins

Chromogranins are high-capacity, low-affinity Ca^{2+} binding proteins that are enriched in secretory granules but are also found in ER and nuclear compartments (Aunis & Metz-Boutigue, 2000). Both chromogranins A and B (CGA/CGB) interact within the third lumenal loop of IP_3R-1 within the stretch of residues that link the lumenal end of the pore selectivity filter to the beginning of TM6 (Yoo & Lewis, 1994, 2000; Yoo et al., 2000). The P_o of reconstituted IP_3R-1 incorporated into bilayers was dramatically increased by both CGA and CGB, from ~5%, the typical maximum P_o observed for the reconstituted channel, to 20–80% in different studies (Choe, Harrison, Grant, & Ehrlich, 2004; Thrower et al., 2003; Thrower, Park, So, Yoo, & Ehrlich, 2002). This remarkable level of stimulation was associated with the appearance of a new, long open state (Thrower et al., 2003, 2002). Remarkably, the chromogranin-activated channel lacked all cytoplasmic Ca^{2+} sensitivity, being equally active 10 nM, 1 μM, and 100 μM Ca^{2+} (Thrower et al., 2002). However, it is not known whether the interaction is modulated by physiological changes in luminal [Ca^{2+}].

Acknowledgments

The authors thank members of the lab who have contributed over the years to our understanding of IP_3R channel function. This work supported by GM/DK56328 and MH059937 to J. K. F., GM074999 to D.-O. D. M, and GM065830 to J. K. F. and D.-O. D. M.

References

Adkins, C. E., & Taylor, C. W. (1999). Lateral inhibition of inositol 1,4,5-trisphosphate receptors by cytosolic Ca^{2+}. *Current Biology, 9*, 1115–1118.

Aunis, D., & Metz-Boutigue, M. H. (2000). Chromogranins: Current concepts. Structural and functional aspects. *Advances in Experimental Medicine and Biology, 482*, 21–38.

Balshaw, D. M., Yamaguchi, N., & Meissner, G. (2002). Modulation of intracellular calcium-release channels by calmodulin. *The Journal of Membrane Biology, 185*, 1–8.

Berridge, M. J. (1997). Elementary and global aspects of calcium signalling. *Journal de Physiologie, 499*, 291–306.

Boehning, D., Joseph, S. K., Mak, D.-O. D., & Foskett, J. K. (2001). Single-channel recordings of recombinant inositol trisphosphate receptors in mammalian nuclear envelope. *Biophysical Journal, 81*, 117–124.

Bootman, M. D. (1994). Quantal Ca^{2+} release from $InsP_3$-sensitive intracellular Ca^{2+} stores. *Molecular and Cellular Endocrinology, 98*, 157–166.

Bootman, M. D., & Berridge, M. J. (1995). The elemental principles of calcium signaling. *Cell, 83*, 675–678.

Bosanac, I., Alattia, J. R., Mal, T. K., Chan, J., Talarico, S., Tong, F. K., et al. (2002). Structure of the inositol 1,4,5-trisphosphate receptor binding core in complex with its ligand. *Nature, 420*, 696–700.

Cannon, B., Hermansson, M., Gyorke, S., Somerharju, P., Virtanen, J. A., & Cheng, K. H. (2003). Regulation of calcium channel activity by lipid domain formation in planar lipid bilayers. *Biophysical Journal, 85*, 933–942.

Cardy, T. J., & Taylor, C. W. (1998). A novel role for calmodulin: Ca^{2+}-independent inhibition of type-1 inositol trisphosphate receptors. *The Biochemical Journal, 334*, 447–455.

Choe, C. U., Harrison, K. D., Grant, W., & Ehrlich, B. E. (2004). Functional coupling of chromogranin with the inositol 1,4,5-trisphosphate receptor shapes calcium signaling. *The Journal of Biological Chemistry, 279*, 35551–35556.

Coquil, J. F., Mauger, J. P., & Claret, M. (1996). Inositol 1,4,5-trisphosphate slowly converts its receptor to a state of higher affinity in sheep cerebellum membranes. *The Journal of Biological Chemistry, 271*, 3568–3574.

Foskett, J. K., & Mak, D.-O. D. (2004). Novel model of calcium and inositol 1,4,5-trisphosphate regulation of $InsP_3$ receptor channel gating in native endoplasmic reticulum. *Biological Research, 37*, 513–519.

Foskett, J. K., White, C., Cheung, K. H., & Mak, D.-O. D. (2007). Inositol trisphosphate receptor Ca^{2+} release channels. *Physiological Reviews, 87*, 593–658.

Fraiman, D., Pando, B., Dargan, S., Parker, I., & Ponce Dawson, S. (2006). Analysis of puff dynamics in oocytes: Interdependence of puff amplitude and interpuff interval. *Biophysical Journal, 90*, 3897–3907.

Fraiman, D., & Ponce Dawson, S. (2004). A model of IP_3 receptor with a luminal calcium binding site: stochastic simulations and analysis. *Cell Calcium, 35*, 403–413.

Gentry, H. R., Singer, A. U., Betts, L., Yang, C., Ferrara, J. D., Sondek, J., et al. (2005). Structural and biochemical characterization of CIB1 delineates a new family of EF-hand-containing proteins. *The Journal of Biological Chemistry, 280*, 8407–8415.

Gin, E., Falcke, M., Wagner, L. E., 2nd, Yule, D. I., & Sneyd, J. (2009). A kinetic model of the inositol trisphosphate receptor based on single-channel data. *Biophysical Journal, 96*, 4053–4062.

Haeseleer, F., Imanishi, Y., Sokal, I., Filipek, S., & Palczewski, K. (2002). Calcium-binding proteins: Intracellular sensors from the calmodulin superfamily. *Biochemical and Biophysical Research Communications, 290*, 615–623.

Haeseleer, F., Sokal, I., Verlinde, C. L., Erdjument-Bromage, H., Tempst, P., Pronin, A. N., et al. (2000). Five members of a novel Ca^{2+}-binding protein (CABP) subfamily with similarity to calmodulin. *The Journal of Biological Chemistry, 275*, 1247–1260.

Hagar, R. E., Burgstahler, A. D., Nathanson, M. H., & Ehrlich, B. E. (1998). Type III $InsP_3$ receptor channel stays open in the presence of increased calcium. *Nature, 396*, 81–84.

Hajnóczky, G., & Thomas, A. P. (1994). The inositol trisphosphate calcium channel is inactivated by inositol trisphosphate. *Nature, 370*, 474–477.

Hajnóczky, G., & Thomas, A. P. (1997). Minimal requirements for calcium oscillations driven by the IP_3 receptor. *The EMBO Journal, 16*, 3533–3543.

Haynes, L. P., Tepikin, A. V., & Burgoyne, R. D. (2004). Calcium-binding protein 1 is an inhibitor of agonist-evoked, inositol 1,4,5-trisphosphate-mediated calcium signaling. *The Journal of Biological Chemistry, 279*, 547–555.

Hirota, J., Michikawa, T., Natsume, T., Furuichi, T., & Mikoshiba, K. (1999). Calmodulin inhibits inositol 1,4,5-trisphosphate-induced calcium release through the purified and reconstituted inositol 1,4,5-trisphosphate receptor type 1. *FEBS Letters, 456*, 322–326.

Hirose, K., Kadowaki, S., & Iino, M. (1998). Allosteric regulation by cytoplasmic Ca^{2+} and IP_3 of the gating of IP_3 receptors in permeabilized guinea-pig vascular smooth muscle cells. *J Physiol, 506*, 407–414.

Ilyin, V., & Parker, I. (1994). Role of cytosolic Ca^{2+} in inhibition of $InsP_3$-evoked Ca^{2+} release in *Xenopus* oocytes. *Journal de Physiologie, 477*, 503–509.

Ionescu, L., Cheung, K. H., Vais, H., Mak, D.-O. D., White, C., & Foskett, J. K. (2006). Graded recruitment and inactivation of single $InsP_3$ receptor Ca^{2+}-release channels: Implications for quantal Ca^{2+} release. *Journal de Physiologie, 573*, 645–662.

Ionescu, L., White, C., Cheung, K. H., Shuai, J., Parker, I., Pearson, J. E., et al. (2007). Mode switching is the major mechanism of ligand regulation of $InsP_3$ receptor calcium release channels. *The Journal of General Physiology, 130*, 631–645.

Joseph, S. K., Brownell, S., & Khan, M. T. (2005). Calcium regulation of inositol 1,4,5-trisphosphate receptors. *Cell Calcium, 38*, 539–546.

Kaftan, E. J., Ehrlich, B. E., & Watras, J. (1997). Inositol 1,4,5-trisphosphate ($InsP_3$) and calcium interact to increase the dynamic range of $InsP_3$ receptor-dependent calcium signaling. *The Journal of General Physiology, 110*, 529–538.

Kasri, N. N., Holmes, A. M., Bultynck, G., Parys, J. B., Bootman, M. D., Rietdorf, K., et al. (2004). Regulation of $InsP_3$ receptor activity by neuronal Ca^{2+}-binding proteins. *The EMBO Journal, 23*, 312–321.

Kaznacheyeva, E., Lupu, V. D., & Bezprozvanny, I. (1998). Single-channel properties of inositol (1,4,5)-trisphosphate receptor heterologously expressed in HEK-293 cells. *The Journal of General Physiology, 111*, 847–856.

Khodakhah, K., & Ogden, D. (1993). Functional heterogeneity of calcium release by inositol trisphosphate in single Purkinje neurones, cultured cerebellar astrocytes, and peripheral tissues. *Proceedings of the National Academy of Sciences of the United States of America, 90*, 4976–4980.

Khodakhah, K., & Ogden, D. (1995). Fast activation and inactivation of inositol trisphosphate-evoked Ca^{2+} release in rat cerebellar Purkinje neurones. *Journal de Physiologie, 487*, 343–358.

Lechleiter, J., Girard, S., Peralta, E., & Clapham, D. (1991). Spiral calcium wave propagation and annihilation in *Xenopus laevis* oocytes. *Science, 252*, 123–126.

Lin, C., Widjaja, J., & Joseph, S. K. (2000). The interaction of calmodulin with alternatively spliced isoforms of the type-I inositol trisphosphate receptor. *The Journal of Biological Chemistry, 275*, 2305–2311.

Lupu, V. D., Kaznacheyeva, E., Krishna, U. M., Falck, J. R., & Bezprozvanny, I. (1998). Functional coupling of phosphatidylinositol 4,5-bisphosphate to inositol 1,4,5-trisphosphate receptor. *The Journal of Biological Chemistry, 273*, 14067–14070.

Maeda, N., Kawasaki, T., Nakade, S., Yokota, N., Taguchi, T., Kasai, M., et al. (1991). Structural and functional characterization of inositol 1,4,5-trisphosphate receptor channel from mouse cerebellum. *The Journal of Biological Chemistry, 266*, 1109–1116.

Mak, D.-O. D., & Foskett, J. K. (1994). Single-channel inositol 1,4,5-trisphosphate receptor currents revealed by patch clamp of isolated *Xenopus* oocyte nuclei. *The Journal of Biological Chemistry, 269*, 29375–29378.

Mak, D.-O. D., & Foskett, J. K. (1997). Single-channel kinetics, inactivation, and spatial distribution of inositol trisphosphate (IP_3) receptors in *Xenopus* oocyte nucleus. *The Journal of General Physiology, 109*, 571–587.

Mak, D.-O. D., McBride, S., & Foskett, J. K. (1998). Inositol 1,4,5-trisphosphate activation of inositol trisphosphate receptor Ca^{2+} channel by ligand tuning of Ca^{2+} inhibition. *Proceedings of the National Academy of Sciences of the United States of America, 95*, 15821–15825.

Mak, D.-O. D., McBride, S., & Foskett, J. K. (2001a). ATP-dependent adenophostin activation of inositol 1,4,5-trisphosphate receptor channel gating: Kinetic implications for the durations of calcium puffs in cells. *The Journal of General Physiology, 117*, 299–314.

Mak, D.-O. D., McBride, S., & Foskett, J. K. (2001b). Regulation by Ca^{2+} and inositol 1,4,5-trisphosphate ($InsP_3$) of single recombinant type 3 $InsP_3$ receptor channels. Ca^{2+} activation uniquely distinguishes types 1 and 3 $InsP_3$ receptors. *The Journal of General Physiology, 117*, 435–446.

Mak, D.-O. D., McBride, S., Raghuram, V., Yue, Y., Joseph, S. K., & Foskett, J. K. (2000). Single-channel properties in endoplasmic reticulum membrane of recombinant type 3 inositol trisphosphate receptor. *The Journal of General Physiology, 115*, 241–256.

Mak, D.-O. D., McBride, S. M., & Foskett, J. K. (2003). Spontaneous channel activity of the inositol 1,4,5-trisphosphate ($InsP_3$) receptor ($InsP_3R$). Application of allosteric modeling to calcium and $InsP_3$ regulation of $InsP_3R$ single-channel gating. *The Journal of General Physiology, 122*, 583–603.

Mak, D.-O. D., McBride, S. M., Petrenko, N. B., & Foskett, J. K. (2003). Novel regulation of calcium inhibition of the inositol 1,4,5-trisphosphate receptor calcium-release channel. *The Journal of General Physiology, 122*, 569–581.

Mak, D.-O. D., Pearson, J. E., Loong, K. P. C., Datta, S., Fernández-Mongil, M., & Foskett, J. K. (2007). Rapid ligand-regulated gating kinetics of single inositol 1,4,5-trisphosphate receptor Ca^{2+} release channels. *EMBO Reports, 8*, 1044–1051.

Marchenko, S. M., Yarotskyy, V. V., Kovalenko, T. N., Kostyuk, P. G., & Thomas, R. C. (2005). Spontaneously active and $InsP_3$-activated ion channels in cell nuclei from rat cerebellar Purkinje and granule neurones. *Journal de Physiologie, 565*, 897–910.

Marchant, J. S., & Taylor, C. W. (1997). Cooperative activation of IP_3 receptors by sequential binding of IP_3 and Ca^{2+} safeguards against spontaneous activity. *Curr Biol, 7*, 510–518.

Mauger, J.-P., Lièvremont, J.-P., Piétri-Rouxel, F., Hilly, M., & Coquil, J.-F. (1994). The inositol 1,4,5-trisphosphate receptor: Kinetic properties and regulation. *Molecular and Cellular Endocrinology, 98*, 133–139.

Meyer, T., Holowka, D., & Stryer, L. (1988). Highly cooperative opening of calcium channels by inositol 1,4,5-trisphosphate. *Science, 240*, 653–656.

Meyer, T., & Stryer, L. (1990). Transient calcium release induced by successive increments of inositol 1,4,5-trisphosphate. *Proceedings of the National Academy of Sciences of the United States of America, 87*, 3841–3845.

Michikawa, T., Hirota, J., Kawano, S., Hiraoka, M., Yamada, M., Furuichi, T., et al. (1999). Calmodulin mediates calcium-dependent inactivation of the cerebellar type 1 inositol 1,4,5-trisphosphate receptor. *Neuron, 23*, 799–808.

Missiaen, L., Parys, J. B., De Smedt, H., Oike, M., & Casteels, R. (1994). Partial calcium release in response to submaximal inositol 1,4,5-trisphosphate receptor activation. *Molecular and Cellular Endocrinology, 98*, 147–156.

Miyakawa, T., Mizushima, A., Hirose, K., Yamazawa, T., Bezprozvanny, I., Kurosaki, T., et al. (2001). Ca^{2+}-sensor region of IP_3 receptor controls intracellular Ca^{2+} signaling. *The EMBO Journal, 20*, 1674–1680.

Muallem, S., Pandol, S. J., & Beeker, T. G. (1989). Hormone-evoked calcium release from intracellular stores is a quantal process. *The Journal of Biological Chemistry, 264*, 205–212.

Nadif Kasri, N., Bultynck, G., Sienaert, I., Callewaert, G., Erneux, C., Missiaen, L., et al. (2002). The role of calmodulin for inositol 1,4,5-trisphosphate receptor function. *Biochimica et Biophysica Acta, 1600*, 19–31.

Nosyreva, E., Miyakawa, T., Wang, Z., Glouchankova, L., Mizushima, A., Iino, M., et al. (2002). The high-affinity calcium–calmodulin-binding site does not play a role in the modulation of type 1 inositol 1,4,5-trisphosphate receptor function by calcium and calmodulin. *The Biochemical Journal, 365*, 659–667.

Parker, I., & Ivorra, I. (1990). Inhibition by Ca^{2+} of inositol trisphosphate-mediated Ca^{2+} liberation: A possible mechanism for oscillatory release of Ca^{2+}. *Proceedings of the National Academy of Sciences of the United States of America, 87*, 260–264.

Parys, J. B., Missiaen, L., Smedt, H. D., Sienaert, I., & Casteels, R. (1996). Mechanisms responsible for quantal Ca^{2+} release from inositol trisphosphate-sensitive calcium stores. *Pflugers Archive, 432*, 359–367.

Ramos-Franco, J., Bare, D., Caenepeel, S., Nani, A., Fill, M., & Mignery, G. (2000). Single-channel function of recombinant type 2 inositol 1,4,5-trisphosphate receptor. *Biophysical Journal, 79*, 1388–1399.

Ramos-Franco, J., Caenepeel, S., Fill, M., & Mignery, G. (1998). Single channel function of recombinant type-1 inositol 1,4,5-trisphosphate receptor ligand binding domain splice variants. *Biophysical Journal, 75*, 2783–2793.

Ramos-Franco, J., Fill, M., & Mignery, G. A. (1998). Isoform-specific function of single inositol 1,4,5-trisphosphate receptor channels. *Biophysical Journal, 75*, 834–839.

Rossi, A. M., & Taylor, C. W. (2004). Ca^{2+} regulation of inositol 1,4,5-trisphosphate receptors: Can Ca^{2+} function without calmodulin? *Molecular Pharmacology, 66*, 199–203.

Saimi, Y., & Kung, C. (2002). Calmodulin as an ion channel subunit. *Annual Review of Physiology, 64*, 289–311.

Sienaert, I., De Smedt, H., Parys, J. B., Missiaen, L., Vanlingen, S., Sipma, H., et al. (1996). Characterization of a cytosolic and a luminal Ca^{2+} binding site in the type I inositol 1,4,5-trisphosphate receptor. *The Journal of Biological Chemistry, 271*, 27005–27012.

Sienaert, I., Missiaen, L., De Smedt, H., Parys, J. B., Sipma, H., & Casteels, R. (1997). Molecular and functional evidence for multiple Ca^{2+}-binding domains in the type 1 inositol 1,4,5-trisphosphate receptor. *The Journal of Biological Chemistry, 272*, 25899–25906.

Striggow, F., & Ehrlich, B. E. (1996). The inositol 1,4,5-trisphosphate receptor of cerebellum. Mn^{2+} permeability and regulation by cytosolic Mn^{2+}. *The Journal of General Physiology, 108*, 115–124.

Sneyd, J., & Falcke, M. (2005). Models of the inositol trisphosphate receptor. *Prog Biophys Mol Biol, 89*, 207–245.

Swillens, S., Champeil, P., Combettes, L., & Dupont, G. (1998). Stochastic simulation of a single inositol 1, 4, 5-trisphosphate-sensitive Ca^{2+} channel reveals repetitive openings during 'blip-like' Ca^{2+} transients. *Cell Calcium, 23*, 291–302.

Swillens, S., Combettes, L., & Champeil, P. (1994). Transient inositol 1,4,5-trisphosphate-induced Ca^{2+} release: A model based on regulatory Ca^{2+}-binding sites along the permeation pathway. *Proc Natl Acad Sci USA, 91*, 10074–10078.

Tang, Y., Stephenson, J. L., & Othmer, H. G. (1996). Simplification and analysis of models of calcium dynamics based on IP_3-sensitive calcium channel kinetics. *Biophys J, 70*, 246–263.

Taylor, C. W. (1998). Inositol trisphosphate receptors: Ca^{2+}-modulated intracellular Ca^{2+} channels. *Biochimica et Biophysica Acta, 1436*, 19–33.

Taylor, C. W., & Laude, A. J. (2002). IP_3 receptors and their regulation by calmodulin and cytosolic Ca^{2+}. *Cell Calcium, 32*, 321–334.

Taylor, C. W., & Potter, B. V. (1990). The size of inositol 1,4,5-trisphosphate-sensitive Ca^{2+} stores depends on inositol 1,4,5-trisphosphate concentration. *The Biochemical Journal, 266*, 189–194.

Taylor, C. W., & Richardson, A. (1991). Structure and function of inositol trisphosphate receptors. *Pharmacology & Therapeutics, 51*, 97–137.

Thrower, E. C., Choe, C. U., So, S. H., Jeon, S. H., Ehrlich, B. E., & Yoo, S. H. (2003). A functional interaction between chromogranin B and the inositol 1,4,5-trisphosphate receptor/Ca^{2+} channel. *The Journal of Biological Chemistry, 278*, 49699–49706.

Thrower, E. C., Park, H. Y., So, S. H., Yoo, S. H., & Ehrlich, B. E. (2002). Activation of the inositol 1,4,5-trisphosphate receptor by the calcium storage protein chromogranin A. *The Journal of Biological Chemistry, 277*, 15801–15806.

Tu, H., Miyakawa, T., Wang, Z., Glouchankova, L., Iino, M., & Bezprozvanny, I. (2002). Functional characterization of the type 1 inositol 1,4,5-trisphosphate receptor coupling domain SII(+/−) splice variants and the Opisthotonos Mutant Form. *Biophysical Journal, 82*, 1995–2004.

Tu, H., Nosyreva, E., Miyakawa, T., Wang, Z., Mizushima, A., Iino, M., et al. (2003). Functional and biochemical analysis of the type 1 inositol (1,4,5)-trisphosphate receptor calcium sensor. *Biophysical Journal, 85*, 290–299.

Tu, H., Wang, Z., & Bezprozvanny, I. (2005). Modulation of mammalian inositol 1,4,5-trisphosphate receptor isoforms by calcium: A role of calcium sensor region. *Biophysical Journal, 88*, 1056–1069.

Tu, H., Wang, Z., Nosyreva, E., De Smedt, H., & Bezprozvanny, I. (2005). Functional characterization of mammalian inositol 1,4,5-trisphosphate receptor isoforms. *Biophysical Journal, 88*, 1046–1055.

Wagner, L. E., 2nd, Joseph, S. K., & Yule, D. I. (2008). Regulation of single inositol 1,4,5-trisphosphate receptor channel activity by protein kinase A phosphorylation. *Journal de Physiologie, 586*, 3577–3596.

White, C., Yang, J., Monteiro, M. J., & Foskett, J. K. (2006). CIB1, a ubiquitously-expressed Ca^{2+}-binding protein-ligand of the $InsP_3$ receptor Ca^{2+}-release channel. *The Journal of Biological Chemistry, 281*, 20825–20833.

Yamada, M., Miyawaki, A., Saito, K., Nakajima, T., Yamamoto-Hino, M., Ryo, Y., et al. (1995). The calmodulin-binding domain in the mouse type 1 inositol 1,4,5-trisphosphate receptor. *The Biochemical Journal, 308*, 83–88.

Yang, J., McBride, S., Mak, D.-O. D., Vardi, N., Palczewski, K., Haeseleer, F., et al. (2002). Identification of a family of calcium sensors as protein ligands of inositol trisphosphate receptor Ca^{2+} release channels. *Proceedings of the National Academy of Sciences of the United States of America, 99*, 7711–7716.

Yang, W., Lee, H. W., Hellinga, H., & Yang, J. J. (2002). Structural analysis, identification, and design of calcium-binding sites in proteins. *Proteins, 47*, 344–356.

Yoo, S. H., & Lewis, M. S. (1994). pH-dependent interaction of an intraluminal loop of inositol 1,4,5-trisphosphate receptor with chromogranin A. *FEBS Letters, 341*, 28–32.

Yoo, S. H., & Lewis, M. S. (2000). Interaction of chromogranin B and the near N-terminal region of chromogranin B with an intraluminal loop peptide of the inositol 1,4,5-trisphosphate receptor. *The Journal of Biological Chemistry, 275*, 30293–30300.

Yoo, S. H., So, S. H., Kweon, H. S., Lee, J. S., Kang, M. K., & Jeon, C. J. (2000). Coupling of the inositol 1,4,5-trisphosphate receptor and chromogranins A and B in secretory granules. *The Journal of Biological Chemistry, 275*, 12553–12559.

Zhang, X., & Joseph, S. K. (2001). Effect of mutation of a calmodulin binding site on Ca^{2+} regulation of inositol trisphosphate receptors. *The Biochemical Journal, 360*, 395–400.

CHAPTER 12

Regulation of Inositol 1,4,5-Trisphosphate Receptors by Phosphorylation and Adenine Nucleotides

Matthew J. Betzenhauser* and David I. Yule†

*Department of Physiology and Cellular Biophysics, Columbia University Medical School, New York City, New York, USA

†Department of Pharmacology and Physiology, University of Rochester Medical School, Rochester, New York, USA

I. OVERVIEW

A wide range of cytosolic factors regulate the activity of the three members of the inositol 1,4,5-trisphosphate receptor (IP_3R) family (IP_3R1, IP_3R2, IP_3R3; Patel, Joseph, & Thomas, 1999; Patterson, Boehning, & Snyder, 2004). IP_3 and Ca^{2+} are the primary activators of the receptors whereas phosphorylation events, nucleotides, and protein-binding partners are

Current Topics in Membranes, Volume 66
Copyright 2010, Elsevier Inc. All right reserved.
1063-5823/10 $35.00
DOI: 10.1016/S1063-5823(10)66012-7

important allosteric modulators of IP_3R function (Bezprozvanny, 2005; Foskett, White, Cheung, & Mak, 2007). The rich diversity of IP_3R regulatory inputs allows for rapid, reversible, and cell-type specific forms of modulation. Furthermore, the three IP_3R isoforms are differentially regulated by many of these cytosolic factors. Such isoform diversity likely facilitates the generation of cell-type-specific Ca^{2+} signals, which in turn are thought to underlie the impressive array of physiological functions controlled by cytosolic Ca^{2+} (Berridge, Lipp, & Bootman, 2000). Here, we describe the current state of knowledge regarding the mechanisms and physiological ramifications of IP_3R regulation by phosphorylation and ATP binding.

II. REGULATION OF IP_3R BY PHOSPHORYLATION

Phosphorylation by various kinases represents a common mode of ion channel regulation (Dai, Hall, & Hell, 2009). The transfer of phosphate allows rapid, reversible modification of channel function not provided by changes in gene transcription which, given the large size of IP_3R proteins, are more energetically demanding. Similar to other ion channels, IP_3R are phosphorylated by several ser/thr kinases and tyr kinases, which comprise the two major classes of mammalian protein kinases (Manning, Whyte, Martinez, Hunter, & Sudarsanam, 2002). Kinases are directed to substrate residues by recognition sequences surrounding the target ser/thr or tyr residues. Analysis of amino acid sequences therefore can provide information about the possibility of a protein constituting a kinase substrate. Given the large number of serine, threonine, and tyrosine residues in the primary sequence of IP_3R, these channels have a high potential for acting as kinase substrates. In fact, computer-based searches such as Net Phos (http://www.cbs.dtu.dk/services/NetPhosK/) of the rat IP_3R1 sequence indicate the presence of dozens of possible serine, threonine, or tyrosine-kinase consensus sequences. The actual number of substrate residues is undoubtedly smaller since a large number of the greater than 2749 residues in IP_3R1 are most likely buried in the 3D structure of the receptor.

A. Second Messenger Regulated Kinases

1. Phosphorylation by PKA

Second messenger regulated protein kinases couple signaling cascades to effector function. The prototypical second messenger activated protein kinase, the cAMP-dependent protein kinase (PKA), was the first kinase demonstrated to phosphorylate IP_3R (Supattapone et al., 1988). PKA is activated by increases in the second messenger cAMP, and phosphorylation of IP_3R allows interplay or crosstalk between the ubiquitous cAMP and Ca^{2+} signaling pathways (Bruce, Straub, & Yule, 2003). IP_3R1 was known as a major brain PKA substrate prior to its identification as the receptor for IP_3 (Walaas, Nairn, & Greengard, 1986). The molecular cloning of the cDNA sequence for IP_3R1 revealed two strong consensus motifs for PKA with serine residues S1589 and S1755 preceded by two arginine residues at the -3 and -2 positions (R-R-X-S; numbering based on the rat IP_3R1 sequence; Mignery, Newton, Archer, & Sudhof, 1990). The incorporation of phosphate into these two serines was specifically demonstrated by phosphopeptide mapping studies (Ferris, Cameron, Bredt, Huganir, & Snyder, 1991). Both sites are efficiently phosphorylated to stoichiometry levels generally reaching greater than 1 mol of phosphate per IP_3R monomer (Ferris, Cameron, et al., 1991; Nakade, Rhee, Hamanaka, & Mikoshiba, 1994; Supattapone et al., 1988; Wojcikiewicz, & Luo, 1998).

IP_3R2 and IP_3R3 do not contain PKA phosphorylation sites equivalent to S1589 and S1755, but are nevertheless substrates for the kinase (Blondel, Takeda, Janssen, Seino, & Bell, 1993; Sudhof, Newton, Archer, Ushkaryov, & Mignery, 1991; Wojcikiewicz & Luo, 1998). Wojcikiewicz and Luo (1998) demonstrated variable stoichiometry for PKA phosphorylation of IP_3R1, IP_3R2, and IP_3R3 using receptor isolated from cells predominantly expressing one of the three isoforms (SHSY-5Y, AR4-2J, and RinM5F cells, respectively). While IP_3R1 was a robust substrate for PKA, incorporating 0.65 mol phosphates per monomer, IP_3R3 incorporated 0.14 mol of phosphate, and IP_3R2 incorporated 0.04 mol (Wojcikiewicz & Luo). In later experiments using a recombinant phosphorylation site mutant IP_3R1, the same group demonstrated that activation of PKA in intact cells induced a similar level of phosphorylation at S1589 and S1755 (Soulsby, Alzayady, Xu, & Wojcikiewicz, 2004). These workers also identified three serine residues in IP_3R3 that were PKA substrates of variable stoichiometry (Soulsby & Wojcikiewicz, 2005). We have recently identified S937 of IP_3R2 as a PKA substrate, but it is not known whether PKA phosphorylates any additional residues in this isoform (Betzenhauser, Fike, Wagner, & Yule, 2009).

While biochemical evidence of IP_3R phosphorylation by PKA was readily apparent, the intracellular localization and the presence of multiple IP_3R isoforms in virtually all cell types made it difficult to link the biochemical events to IP_3R function. Initial experiments suggested that PKA phosphorylation reduced IP_3R1-mediated Ca^{2+} release (Supattapone et al., 1988). These results were based on studies examining IP_3-induced Ca^{2+} release (IICR) from brain microsomes. In these experiments, SERCA activity is required to load $^{45}Ca^{2+}$ into the microsomes and the loss of Ca^{2+} in response to suitable IP_3 and Ca^{2+} provides measure of IP_3R activity. As such, the effects of PKA could arise as a result of increased SERCA activity or alternatively decreased IP_3R activity. Numerous subsequent studies have demonstrated that the direct effect of PKA phosphorylation on IP_3R1 is to increase its activity (DeSouza et al., 2002; Tang, Tu, Wang, & Bezprozvanny, 2003; Wagner, Li, & Yule, 2003). This was demonstrated using immunopurified IP_3R1 incorporated into lipid vesicles and in later studies examining the effects of PKA on IP_3R1 single-channel activity in lipid bilayers (Tang et al., 2003). In these studies, exogenous PKA caused an increase in single-channel open-probability (P_o) that was reversible by treatment with the phosphatase PP1-α.

The generation of the DT40 triple knockout (DT40-3KO) cell line provided the first null background to facilitate the mutational analysis of specific phosphorylated residues. Taking advantage of the high degree of homologous recombination in DT40 cells, Kurosaki and colleagues systematically deleted the genes for all three IP_3R isoforms (Sugawara, Kurosaki, Takata, & Kurosaki, 1997). We and others have exploited this system to study the effects of mutating phosphorylation sites on IP_3R Ca^{2+} release and single-channel activities (Betzenhauser, Fike, et al., 2009; Betzenhauser, Wagner, Won, & Yule, 2008; Khan, Wagner, Yule, Bhanumathy, & Joseph, 2006; Soulsby & Wojcikiewicz, 2007; Wagner, Joseph, & Yule, 2008; Wagner, Li, Joseph, & Yule, 2004; Wagner et al., 2003). Since the phosphorylation sites on IP_3R1 were already identified, the examination of the relative contributions of the phosphorylation sites was relatively straightforward. Elimination of both S1589 and S1755 by substitution of these residues with alanines in the rat IP_3R1 completely prevented the enhancing effects of PKA activation on IP_3R1 activity in both the peripheral $S2^-$ and the neuronal $S2^+$ splice isoforms. Substitutions of S1589A and S1755A individually established that the result of phosphorylating these residues was dependent on the splicing of the receptor at the S2 site (Wagner et al., 2003). While phosphorylation of S1755 was necessary and sufficient for enhancing the activity of the neuronal $S2^+$ spice variant, phosphorylation of either S1589 or S1755 could enhance the activity of the peripheral $S2^-$ isoform. These results were obtained using the muscarinic receptor-induced activation of IP_3R1 variants transiently coexpressed with muscarinic receptors in DT40-3KO cells. Similarly, forskolin treatment enhanced Ca^{2+} signals in response to endogenous IP_3-mobilizing receptors in DT40 cells stably expressing mouse IP_3R1 (Soulsby & Wojcikiewicz, 2007). Analyzing the effects of the mutations on the activation of IP_3R1 using a cell-permeable "caged" IP_3 corroborated these results (Wagner et al., 2003).

Further insights into the mechanisms by which PKA enhances IP_3R1 activity were obtained by using "phosphomimetic" glutamate substitutions at S1589 and S1755 (Wagner et al., 2004). These mutant receptors exhibited increased sensitivity toward muscarinic stimulation and activation by photorelease of caged IP_3. Once again, the S1755E mutation was sufficient for this effect in the $S2^+$ isoform and both S1589E and S1755E were required for the maximal increase in sensitivity to activation in the context of the $S2^-$ splice variant. These results suggested that phosphorylation resulted in an increase in the sensitivity of the receptor to activation by IP_3 (Wagner et al., 2004). This was consistent with results from bilayer experiments where active PKA in the cytosolic face of the chamber resulted in increased P_o at lower [IP_3] (Tang et al., 2003).

Recently, additional mechanistic information about the enhancing effects of phosphorylation was provided by single-channel measurements of plasma membrane IP_3R1 in stable DT40 cell lines (Fig. 1; Wagner et al., 2008). This technique, first described by Taylor and colleagues, allows direct channel measurements to be made of small numbers (typically <4) of channels in the plasma membrane while in the whole-cell patch-clamp configuration (Dellis et al., 2006; Dellis, Rossi, Dedos, & Taylor, 2008). Similar to intact Ca^{2+} imaging experiments and in lipid bilayer channel measurements, the primary effect of forskolin treatment or phosphomimetic mutations was to cause an increase in P_o at subsaturating [IP_3] (Fig. 1A and B). Interestingly, there was no apparent effect on IP_3 binding or the biphasic effects of Ca^{2+} on the receptor. There were, however, clear effects on the kinetics of channel opening and closing with the phosphomimetic mutations increasing the likelihood that the channel would exhibit high P_o burst activity when compared to nonphosphorylatable constructs (Fig. 1C and D; Wagner et al., 2008).

Given the large amount of work aimed at understanding the functional effects of PKA activation on IP_3R1 channel activity, it is surprising that relatively little is understood regarding the physiological implications of PKA on IP_3R1 activity in neurons, where this isoform is enriched over the other two subtypes (Taylor, Genazzani, & Morris, 1999; Wojcikiewicz, 1995). Bezprozvanny and colleagues have proposed a model whereby regulating the phosphorylation state of IP_3R1 by PKA and the phosphatase PP1-α regulates basal intracellular Ca^{2+} levels in spiny neurons of the striatum (Tang & Bezprozvanny, 2004; Tang et al., 2003). Specifically, they propose that activation of PKA or inhibition of PP1 both results in elevated basal Ca^{2+} levels, possibly due to the maintained phosphorylation of IP_3R1. This elevated basal Ca^{2+} was also associated with suppressed dopamine-induced Ca^{2+} oscillations (Tang & Bezprozvanny, 2004). Excessive phosphorylation of IP_3R1 at S1755 may also promote a reduction in ER Ca^{2+} store content by inducing a chronic Ca^{2+} leak, promoting apoptosis in a Bcl-2-dependent manner (Oakes et al., 2005). Further insights into the physiological consequences of IP_3R1 phosphorylation may be facilitated by the generation of knockin mouse models that express IP_3R1 containing S1589/S1755 alanine or glutamate substitutions.

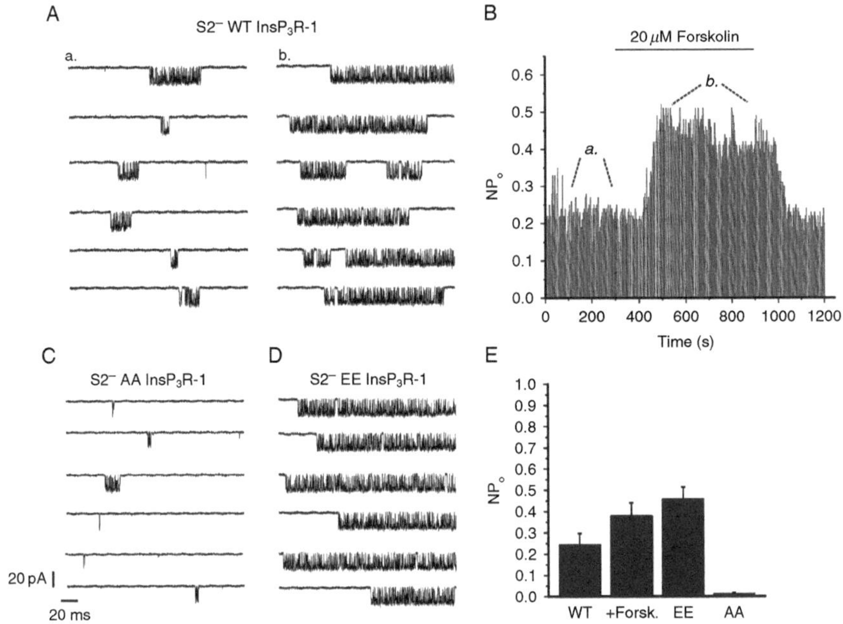

FIGURE 1 PKA-activation enhances the open-probability of IP_3R1. In (A), S2$^-$ IP_3R1 activity stably expressed in the plasma membrane of DT40-3KO cells monitored in the whole cell mode of the patch-clamp technique induced by 0.5 μM adenophostin A and activating concentrations of Ca^{2+} and ATP (200 nM and 0.5 mM, respectively). In the left hand panel, representative sweeps are shown under control conditions. In the right panel. the increase in S2$^-$ IP_3R1 single-channel activity is shown in the same cell after exposure to forskolin, an agent which leads to an increase in cAMP and PKA activation. (B). Represents a diary plot detailing the changes in open-probability throughout the experiment. Traces shown in (A); left panel are representative sweeps from the section indicated as "*a*." Traces shown in (A); right panel are representative sweeps from the section indicated as "*b*." Shown in (C) is single-channel activity of a nonphosphorylatable S2$^-$ IP_3R1 mutant construct (Ser 1589-Ala and Ser 1755-Ala; named "AA"). (D). Shows a "phosphomimetic" S2$^-$ IP_3R1 construct (Ser 1589-Glu and Ser 1755-Glu; named "EE") under identical stimulating conditions (as in A) and demonstrates the enhanced P_o of the phosphomimetic construct. (E). Shows relative activity of the S2$^-$ IP_3R1 under these conditions. Data adapted from Wagner et al. (2008) by permission.

The effects of PKA on IP_3R2 and IP_3R3 have also been examined on these isoforms expressed in isolation in DT40 cells. Activation of PKA by forskolin or the cAMP analog, cBIMPs resulted in increased muscarinic receptor-induced Ca^{2+} signals (Betzenhauser, Fike, et al., 2009). Forskolin also increased the P_o of IP_3R2 single channels measured in the plasma membrane of DT40 cells. The enhancing effect of cBIMPs was eliminated when serine 937 was replaced with an alanine. The functional consequences of PKA on IP_3R2 in these experiments were consistent with reports of PKA activity enhancing Ca^{2+} release in AR4-2J cells, which predominantly express IP_3R2 (Regimbald-Dumas, Arguin, Fregeau, & Guillemette, 2007; Wojcikiewicz & Luo, 1998). In contrast to IP_3R1 and IP_3R2, no enhancing effects of PKA activation were observed in DT40 cells expressing IP_3R3 (Soulsby & Wojcikiewicz, 2007). A slight inhibitory effect was found, but this occurred even in cells expressing IP_3R3 devoid of all known PKA phosphorylation sites. These experiments were performed in populations of cells and strong enhancing effects of PKA were observed in parallel experiments using cells expressing IP_3R1. It is possible that a weak effect of phosphorylation on IP_3R3 could be masked under these conditions. In this regard, single-channel analysis or the examination of Ca^{2+} signals in individual cells is warranted.

2. Phosphorylation by Cyclic GMP-Dependent Kinases

Other second messenger activated serine/threonine kinases, including cyclic-GMP (cGMP)-dependent protein kinase (PKG; Soulsby et al., 2004), protein kinase C (PKC), and Ca^{2+}–calmodulin-activated kinase II (CaMKII; Ferris, Huganir, Bredt, Cameron, & Snyder, 1991), have been shown to phosphorylate IP_3R1. PKG phosphorylates IP_3R1 at the same residues as PKA (Soulsby et al., 2004), but the functional effects of direct PKG phosphorylation were distinct from those of PKA phosphorylation. Whereas PKA phosphorylation enhances the activity of both the $S2^-$ and $S2^+$ splice isoforms, PKG activation increased the activity of only the $S2^+$ splice isoform (Wagner et al., 2003). Furthermore, this effect on the $S2^+$ splice variant was completely eliminated by mutating serine 1755 to alanine. There have been no reports of functional effects of direct PKG phosphorylation of IP_3R2 or IP_3R3. Of note, the functional effects of PKG activation can also result in diminished IP_3R activity as a consequence of the phosphorylation of IP_3R-associated PKG substrate (IRAG) (Schlossmann et al., 2000). This situation occurs in cells where IRAG is expressed including smooth muscle cells and platelets where PKG activation results in IP_3R inhibition (Antl et al., 2007; Geiselhoringer et al., 2004). A similar inhibition of Ca^{2+} release by PKA activation was also demonstrated in these studies, but was not dependent on the presence of IRAG. This suggests that some other factor, possibly an unidentified protein-binding partner, prevents the direct effects of PKA phosphorylation on IP_3R1 in smooth muscle.

3. Phosphorylation by Protein Kinase C and Ca^{2+}–Calmodulin-Dependent Kinases

Initial reports suggested that PKC and Ca^{2+}–calmodulin-dependent kinase type 2 (CaMKII) were not able to phosphorylate cerebellar IP_3R (Supattapone et al., 1988). Snyder and colleagues demonstrated in a subsequent study, this time using a detergent free IP_3R preparation, that cerebellar IP_3R is a substrate for both kinases *in vitro* (Ferris, Huganir, et al., 1991). The phosphorylation sites for PKC and CaMKII on IP_3R1 were found to be distinct from the PKA phosphorylation sites. PKC phosphorylation of IP_3R1 *in vitro* was also enhanced by prior PKA phosphorylation and by Ca^{2+}–CaM binding to the receptor (Vermassen et al., 2004).

Given that some PKC isoforms and all CaMK isoforms are activated by a rise in intracellular Ca^{2+}, the ability of these kinases to phosphorylate IP_3R may provide Ca^{2+}-dependent feedback on IP_3R activity distinct from direct effects of Ca^{2+} and CaM binding. It is, however, unclear what effects PKC or CaMKII phosphorylation events have on IP_3R activity. In particular, any effects of these kinases on IP_3R2 and IP_3R3 are also poorly understood. PKC activation has been shown to enhance IP_3R activity in hepatocytes (Matter, Ritz, Freyermuth, Rogue, & Malviya, 1993), but inhibit IP_3R activity in AR4-2J cells (Arguin, Regimbald-Dumas, Fregeau, Caron, & Guillemette, 2007) even though both cell types predominantly express IP_3R2 (Wojcikiewicz, 1995). PKC activation also decreased IP_3R activity in RinM5F cells which express mostly IP_3R3 (Caron, Chaloux, Arguin, & Guillemette, 2007). CaMKII can also phosphorylate IP_3R2 *in vitro* and phosphorylation was retained in an N-terminal fragment consisting of the first 1053 amino acids of rat IP_3R2 (Bare, Kettlun, Liang, Bers, & Mignery, 2005). The exact residue or the phosphorylation of additional regions of IP_3R2 are unknown, but treatment of cardiac ventricular IP_3R (mostly IP_3R2) with purified CaMKII dramatically reduced the P_o of IP_3R in lipid bilayers (Bare et al., 2005).

B. Phosphorylation by Other Ser/Thr Kinases

Further work has identified additional ser/thr kinases that phosphorylate IP_3R that are coupled to cell-cycle changes, oocyte maturation, or to the cellular metabolic state. These kinases include cdc2/CyB, mitogen-activated protein kinases (MAPKs), and Akt (Bai, Yang, Huang, & Sun, 2006; Ito et al., 2008; Khan et al., 2006; Lee et al., 2006; Malathi et al., 2003; Sun, Haun, Jones, Edmondson, & Machaca, 2009; Szado et al., 2008; Vanderheyden et al., 2009; Yang, Bai, Huang, & Sun, 2006). Again, while the biochemical evidence for phosphorylation has become readily apparent, the functional consequences of many of these events have yet to be firmly established.

1. Phosphorylation by Cyclin-Dependent Kinases

Cyclin-dependent kinases exert control of eukaryotic cell division by regulating cell-cycle stages through the phosphorylation of various substrates (Morgan, 1997). A sequence motif search for the cell-cycle cdc2/CyB recognition motif (S/T)-P-X-(K/R) yielded two candidate residues (serine 421 and threonine 799 in the rat) conserved in IP_3R1 sequences. Phosphospecific antibodies raised against these epitopes demonstrated phosphorylation by cdc2 *in vitro*. Both sites were also phosphorylated in nocodozole-arrested DT40 cells. Phosphorylation by cdc2 is associated with an increase in IP_3 affinity and sensitized brain microsomes to IP_3-mediated Ca^{2+} release (Malathi et al., 2003, 2005). These results suggest that Ca^{2+} release through IP_3R can be differentially regulated during different stages of the cell cycle.

2. Phosphorylation by MAP Kinases

MAPK cascades represent a highly conserved signaling paradigm represented in eukaryotes from yeast to man (Widmann, Gibson, Jarpe, & Johnson, 1999). MAPK pathways are involved in regulating a multitude of cellular events in response to environmental changes (Cuevas, Abell, & Johnson, 2007; Turjanski, Vaque, & Gutkind, 2007). The canonical MAPKs, the extracellular signal-regulated kinases 1 and 2 (ERK1/2) are the terminal kinases activated by the sequential activation of the ras GTPase, the MAPK-kinase-kinase, raf-1, and MAPK-kinase 1 (Avruch, 2007). ERK1/2 are known to phosphorylate numerous cytosolic and nuclear proteins in response to growth factor receptor activation and other stimuli. Sequence analysis of IP_3R1 revealed three canonical MAPK recognition motifs (P-X-S/T-P) surrounding residues S436, T945, S1765 (Bai et al., 2006). None of these motifs are conserved in all three mammalian isoforms, but S436 is conserved in IP_3R1 sequences in *Xenopus* and mammals. Recombinant ERK2 phosphorylates S436 and S1765 in mouse IP_3R1 and ERK1/2 physically interacts IP_3R1 likely as a result of a consensus ERK docking domain (D-domain) contained in a C-terminal region of the receptor (residues 2078-2087; Bai et al., 2006; Yang et al., 2006). The functional effects of phosphorylating S436 have been examined for IP_3-binding to the purified IP_3-binding domain of IP_3R1 and in Ca^{2+} release assays in DT40 cells. Phosphorylation of S436 by ERK2 was associated with a reduction in IP_3 binding possibly as a result of enhancing the interactions between the IP_3-binding domain and the suppressor domain (Bai et al., 2006). Consistent with the inhibitory effects of phosphorylating S436, IICR was also inhibited by conditions that promoted this phosphorylation event (Yang et al., 2006).

3. Phosphorylation by M-Phase Protein Kinases

Xenopus oocyte maturation is accompanied by an increased sensitivity of IICR. The phosphorylation state of *Xenopus* IP_3R1 also changes with oocyte maturation leading to incorporation of phosphate into at least three different residues (T-931, T-1136, and S-1145; Sun et al., 2009). A phosphor-specific antibody generated against the T-1136 site demonstrated phosphorylation downstream of maturation promoting factor activation or the MAPK signaling cascade. The altered phosphorylation state of IP_3R1 may play a role in the dramatic increase in IICR observed in oocytes upon maturation. Consistent with this idea, IP_3R1 is also phosphorylated during oocyte maturation in mouse and *Xenopus* eggs by M-phase protein kinases (Ito et al., 2008; Lee et al., 2006; Vanderheyden et al., 2009). Initial results were based on results using an MPM-2 phosphoepitope antibody (Ito et al., 2008). More detailed mutagenic strategies identified the phosphorylation of T2656 by the M-phase kinase, Polo-like kinase 1 (Plk-1; Vanderheyden et al., 2009). The phosphorylation of IP_3R1 at specific residues during oocyte maturation has been firmly established. What remains to be determined is what contribution these phosphorylation events play in promoting the dramatic increase in IICR sensitivity that occurs during oocyte maturation.

4. Phosphorylation by Akt

Akt (also known as protein kinase B) is a widely expressed protein kinase involved in regulating cellular proliferation by phosphorylating substrates at the consensus sequence R-X-R-X-X-S/T. A consensus Akt site is conserved in all three mammalian IP_3R isoforms in the C-terminus of the receptors and mutational analyses have examined the effects of phosphorylating this site (S2681) in IP_3R1 (Khan et al., 2006; Szado et al., 2008). Two different approaches to examining the effects of Akt phosphorylation on IP_3R1 activity resulted in opposing conclusions. Initial studies reported no apparent effect on IICR in $^{45}Ca^{2+}$ release studies and in single-channel studies examining S2681A and a putative phosphomimetic mutant (S2681E; Khan et al., 2006). Even though no discernable effects on channel activity were reported, caspase activation following staurosporine treatment occurred in DT40 cells expressing the S2681A mutant. A subsequent study using overexpression of constitutively active Akt demonstrated reduced IICR in HeLa cells that correlated with a reduction in apoptosis (Szado et al., 2008).

C. Phosphorylation by Tyrosine Kinases

Mammalian tyrosine kinases are classified into two main families based on whether or not the kinase activity is associated with a cell surface receptor (Hunter, 2009). Growth factor receptors, lymphocyte receptors, integrin receptors, and many cytokine receptors are receptor tyrosine kinases while soluble cytosolic enzymes of the src family are nonreceptor tyrosine kinases. Src and the related kinases Fyn and Lyn all phosphorylate IP_3R1 *in vitro* and in lymphocyte cell lines (Cui et al., 2004; deSouza, Cui, Dura, McDonald, & Marks, 2007; Harnick et al., 1995; Jayaraman, Ondrias, Ondriasova, &

Marks, 1996). Src also likely phosphorylates IP_3R2 following treatment of cells with the Na–K-ATPase inhibitor ouabain (Cui et al., 2004; deSouza et al., 2007; Harnick et al., 1995; Jayaraman et al., 1996). Tyrosine phosphorylation of IP_3R1 was first observed in activated Jurkat T-cells (Harnick et al., 1995). T-cell activation involves the interaction of the major histocompatability complex with the T-cell receptor (TCR) and TCR activation subsequently recruits a number of signaling molecules to the T-cell synapse, including phospholipase Cγ and Fyn (Salmond, Filby, Qureshi, Caserta, & Zamoyska, 2009). Purified Src and Fyn directly phosphorylated both neuronal and peripheral IP_3R1 splice variants *in vitro* which provided evidence that one of the Src family of kinases was responsible for the tyr phosphorylation of IP_3R1 upon TCR activation (Jayaraman et al., 1996). Later studies identified T353 of IP_3R1 as the target for Lyn phosphorylation downstream of both T- and B-cell receptor (BcR) activation (Cui et al., 2004). Tyr353 is located in the IP_3-binding core of the receptor and phosphorylation of this residue leads to an increase in the affinity of IP_3R1 for IP_3 (Cui et al., 2004). Single-channel measurements demonstrated that Lyn-phosphorylated IP_3R1 is more sensitive to IP_3 at optimal Ca^{2+} concentrations and also leads to a reduction in the Ca^{2+}-mediated inhibition of IP_3R1 (deSouza et al., 2007). The combined effects of an increased sensitivity to IP_3 activation and a reduced sensitivity to Ca^{2+}-mediated inhibition ensure that the receptor is open under a wide range of IP_3 and Ca^{2+}. These effects may facilitate the maintenance of Ca^{2+} release by allowing IP_3R1 to remain open in the face of the elevated $[Ca^{2+}]_i$ and reduced IP_3 that occurs during prolonged receptor signaling (Cui et al., 2004; deSouza et al., 2007).

III. REGULATION OF IP_3R BY ADENINE NUCLEOTIDES

Early studies also indicated that ATP was an important modulator of IICR from permeabilized vascular smooth muscle cells (Smith, Smith, & Higgins, 1985) and peritoneal macrophages (Hirata et al., 1985). Submillimolar ATP was reported to augment the activity of IP_3-activated channels isolated from aortic smooth muscle when reconstituted in planer lipid bilayers (Ehrlich & Watras, 1988). It was further demonstrated that the photoaffinity label α-^{32}P azidoadenosine 5′-triphosphate could be cross-linked stoichiometrically to the purified cerebellar IP_3R (Maeda et al., 1991) and that ATP was a similarly potent modulator of Ca^{2+} release (Ferris, Huganir, & Snyder, 1990) and IP_3-activated channel activity of the cerebellar receptor (Maeda et al., 1991; Bezprozvanny & Ehrlich, 1993). Higher concentrations of ATP (greater than millimolar) were reported in some studies to inhibit IICR (Ferris et al., 1990; Iino, 1991;

Missiaen et al., 1997). While no consensus has been reached on the mechanism underlying this inhibition, it appears to occur through interaction with the IP_3R at a distinct, low affinity site. Since high concentrations of ATP have been shown to inhibit IP_3 binding to microsomal membranes (Guillemette, Balla, Baukal, & Catt, 1987; Maeda et al., 1991), it is possible that the inhibition of Ca^{2+} release by ATP reflects competitive inhibition of IP_3 binding.

ATP regulation of Ca^{2+} release has subsequently been reported in a diverse array of cell types, including primary neurons, smooth muscle, and exocrine cells (Iino, 1991; Kaplin, Snyder, & Linden, 1996; Maes et al., 1999; Mak, McBride, & Foskett, 1999; Park, Betzenhauser, Won, Chen, & Yule, 2008) and this is consistent with all IP_3R subtypes when studied in isolation being subject to nucleotide modulation, albeit as noted below, with somewhat different properties and nucleotide sensitivity (Fig. 2; Betzenhauser et al., 2008; Tu, Wang, Nosyreva, De Smedt, & Bezprozvanny, 2005; Miyakawa et al., 1999). ATP modulation of Ca^{2+} release attributed to effects on IP_3R1 and IP_3R3 (Maes et al., 2000; Miyakawa et al., 1999; Tu et al., 2005) is not contentious; however, the IP_3R2 has been reported not to be regulated by ATP (Miyakawa et al., 1999; Tu et al., 2005). A recent study has shown that, in contrast to IP_3R1 and IP_3R3 (Betzenhauser et al., 2008; Betzenhauser, Wagner, Park, & Yule, 2009), this is indeed the case at saturating $[IP_3]$ (Fig. 2A–C; Betzenhauser et al., 2008). However, IICR and single-channel activity of IP_3R2 was markedly augmented at subsaturating $[IP_3]$ (Betzenhauser et al., 2008; Park et al., 2008). Under identical conditions (submaximal IP_3 at basal intracellular Ca^{2+}), the affinity for ATP regulation of Ca^{2+} release also differs; the IP_3R2 has been reported most sensitive followed by IP_3R1 and IP_3R3 (Fig. 2D; $K_d \sim 40$, 100, and 500 μM, respectively).

A diverse array of nucleotides have been reported to influence Ca^{2+} release. These include ADP (Ferris et al., 1990; Iino, 1991; Maes et al., 2000), AMP (Ferris et al., 1990; Iino, 1991; Missiaen et al., 1997; Mak et al., 1999), adenosine (Missiaen et al., 1997), and GTP (Bezprozvanny & Ehrlich, 1993; Iino, 1991; Maes et al., 2000; Mak et al., 1999), as well as adenine containing pharmacologically relevant agents such as caffeine (Maes et al., 1999). The reported discrepancies in regard to the presence or absence of regulation by individual nucleotides and differences in the absolute degree of modulation described may reflect differing experimental methodologies utilized, together with the complement of IP_3R subtypes expressed in the particular experimental system. Furthermore, ATP itself may not be the most effective or potent nucleotide (Iino, 1991; Missiaen et al., 1997) and thus in a cellular context, regulation may reflect the competition at binding sites by various nucleotides.

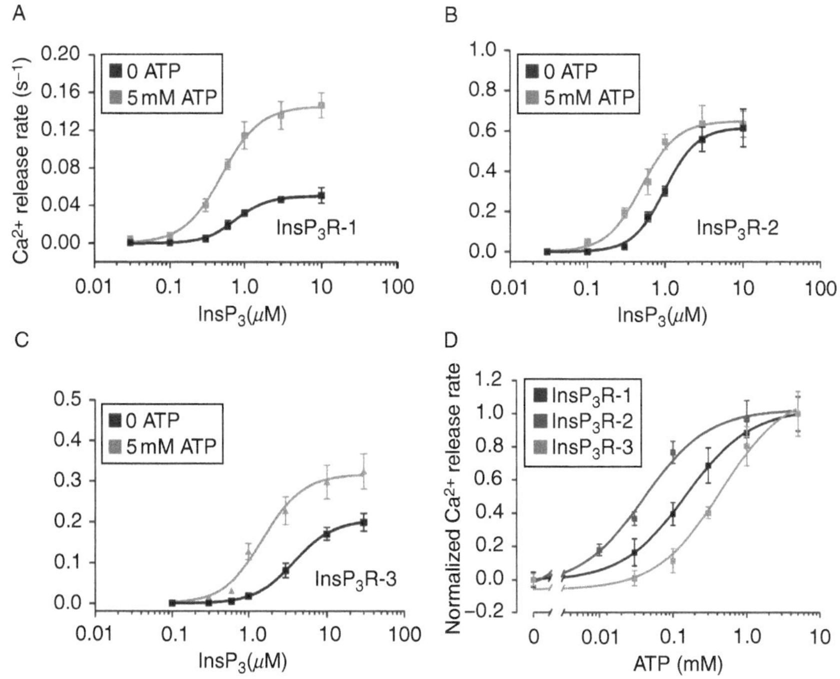

FIGURE 2 ATP regulation of IP_3R family members. The rate of Ca^{2+} release was monitored in permeabilized DT40-3KO cells expressing the indicated IP_3R family member. For details of methods see Betzenhauser, Wagner, Won, & Yule (2008). The effects of a saturating [ATP] on Ca^{2+} release at various [IP_3] at ~basal cytosolic Ca^{2+} is shown for the $S2^+$ IP_3R1 in (A), IP_3R2 in (B), and IP_3R3 in (C). Note that Ca^{2+} release via the IP_3R2 in contrast to IP_3R1 and IP_3R3 is only subject to modulation at submaximal [IP_3], under these conditions. In (D), comparison of the concentration–response relationship for the modulation of Ca^{2+} release by ATP through IP_3R family members. Adapted from Betzenhauser et al. (2008) and Betzenhauser, Wagner, et al. (2009) with permission. (See Color Insert.)

A. Binding of ATP to IP_3R

Considerable evidence indicates that the functional effects of adenine nucleotides are mediated through direct binding of ATP to IP_3R. For example, IP_3R purified from cerebellum and incorporated in proteoliposomes was shown to bind ^{35}S-ATPβS (Ferris et al., 1990). Similarly, a microsomal fraction from cerebellum was photoaffinity labeled with α-^{32}P azidoadenosine 5′-triphosphate (Maeda et al., 1991). Scatchard analysis of ^{32}P-ATP binding to purified IP_3R revealed a single site with affinity of 5–17 μM and B_{max} almost identical to that reported for IP_3 binding (Ferris et al., 1990; Maeda et al., 1991). IP_3 itself does not influence ATP binding, indicating a distinct interaction site (Maeda et al., 1991). The binding specificity was indicated as ATP effectively competed for ^{32}P-ATP binding, with ADP > AMP > GTP being progressively less effective (Ferris et al., 1990; Maeda et al., 1991). Subsequently, ATP binding has been reported to purified recombinant IP_3R1 and IP_3R3 from microsomes prepared from bacculovirus-infected insect SF9 cells (Maes et al., 2000; Maes et al., 2001). The affinity for competition by unlabelled ATP to inhibit crosslinking of the photoaffinity label was 10-fold higher for IP_3R3 when compared to IP_3R1, consistent with the differing ATP requirement to enhance Ca^{2+} release in cells expressing predominately IP_3R1 or IP_3R3 (Maes et al., 2000).

B. Molecular Recognition of ATP Binding

All IP_3R family members harbor putative nucleotide-binding sequences in their primary structure. These consist of glycine-rich domains (Gly-X-Gly-X-X-Gly) reminiscent of Walker type A motifs present in proteins which utilize ATP in a catalytic manner (Fig. 3; Walker, Saraste, Runswick, & Gay, 1982). It has been generally assumed that these domains are responsible for ATP binding and subsequent downstream functional effects (Maeda et al., 1991; Maes et al., 2000, 2001, 1999; Patel et al., 1999). Three such motifs are present in the primary sequence of IP_3R family members (Fig. 3B–E). A region termed the ATPB site is conserved in all three family members (aa 2016–2021 in IP_3R1, aa 1943–2013 in IP_3R2, and aa 1849–1944 in IP_3R3). The so-called ATPA site is unique to the IP_3R1 (aa 1773–1780) and an additional motif termed ATPC (aa 1574–1765) is introduced by excision of the S2 splice site in $S2^-$ IP_3R1. There is convincing evidence that these sites indeed bind ATP. Glutathione *S*-transferase (GST) fusion proteins of fragments of $S2^+$ IP_3R1 containing either the ATPA and ATPB sites were shown to bind α-^{32}P-ATP specifically (Maes et al., 1999). In addition, fluorescent TNP–ATP has been shown to bind to GST-fragments containing the ATPA and ATBC sites of the IP_3R1 (Wagner, Betzenhauser, & Yule, 2006) and the ATPB motif in the context of each IP_3R family member (Betzenhauser et al., 2008; Wagner et al., 2006). In each case, binding of TNP–ATP was abolished in the GST-fragments where the Walker motif was disrupted by substitution of the central glycine residue for alanine or by excess "cold" unlabelled ATP (Betzenhauser et al., 2008; Wagner et al., 2006). ATP binding has also been demonstrated in fragments of purified recombinant full-length IP_3R subjected to limited tryptic digestion (Maes et al., 2001). In the IP_3R1, ATP binding was limited to two fragments corresponding to regions harboring the ATPA and ATPB sites while in IP_3R3 ATP bound to a single fragment predicted to contain the ATPB site (Maes et al., 2001). The presence of two

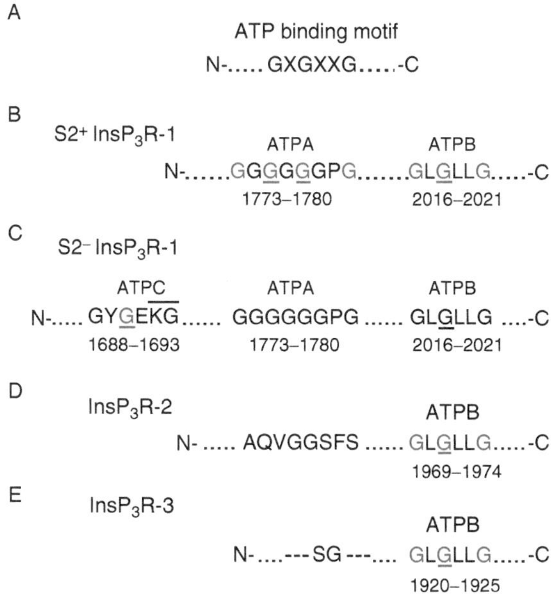

FIGURE 3 Walker type A nucleotide-binding motifs in the IP_3R isoforms. (A) Shows the canonical sequence for "Walker-type" nucleotide-binding motifs. (B) Walker motifs in $S2^+$ IP_3R1. (C) Walker motifs in $S2^-$ IP_3R1. (D) Walker type motif in IP_3R2. (E) Walker type motif in IP_3R3. Glycine residues in red indicate those which have been mutated to alanine in functional studies. Underlined residues indicate glycine residues mutated to alanine in the context of GST-fragments encompassing those sites. Over-lined residues indicate the two amino acids constituting the N- and C-terminal residues demarking the S2 splice site in $S2^-$ IP_3R1. (See Color Insert.)

putative-binding sites in IP_3R1 has also been suggested to explain the higher functional sensitivity of the IP_3R1 to ATP when compared with IP_3R3 (Maes et al., 2001). Moreover, this idea is also consistent with the reduced functional sensitivity to ATP modulation of the *Opisthotonos* (*opt*) mutation in IP_3R1 which results in an in-frame deletion of 107 amino acids which includes the ATPA site (Tu et al., 2002).

Despite this wealth of data demonstrating ATP binding to these motifs in IP_3R, the functional link between these binding data and the modulation of Ca^{2+} release is nonetheless purely correlative. For example, the binding studies described above do not necessarily preclude additional-binding sites within the regions expressing the Walker A type motifs. Functional roles for the specific sites can only be definitely established by functional experiments combined with mutagenesis of the corresponding sites. A series of studies from our laboratory has systematically studied the functional consequences of mutation of all known Walker A type motifs in the context of the full-length IP_3R family members (Fig. 3; Betzenhauser et al., 2008; Betzenhauser, Wagner, et al., 2009; Wagner et al., 2006). Given the binding data described previously, these studies have resulted in somewhat surprising conclusions.

Initially a mutant full-length $S2^-$ IP_3R1 was generated in which the ATPC site was disrupted by an analogous mutation shown to abolish TNP–ATP binding to a GST-fragment containing the ATPC motif (G1690A; termed ΔATPC IP_3R1). The ATP sensitivity of Ca^{2+} release of ΔATPC IP_3R measured following transient expression in DT40 3KO cells was essentially unaltered when compared to WT $S2^-$ IP_3R1. Given that ΔATPC IP_3R1 contains intact ATPA and ATPB sites and that the $S2^+$ IP_3R, lacking ATPC is regulated by ATP (Betzenhauser, Fike, et al., 2009; Ferris, Cameron, et al., 1991; Tu et al., 2002, 2005) these data were not unexpected. However, the mutation of ΔATPC IP_3R1 abolished phosphorylation by PKA and the subsequent potentiated Ca^{2+} release (Wagner et al., 2006). These data suggest that while ATP binding to the ATPC site does not modulate Ca^{2+} release directly, it appears to permit the $S2^-$ IP_3R to adopt a conformation amenable to phosphorylation by PKA and leads indirectly to enhanced IP_3R1 channel activity (Wagner et al., 2004). Although the ATP requirement for this effect was not reported, these data raise the possibility that the functional effects of PKA phosphorylation in cells expressing the nonneuronal $S2^-$ IP_3R1 may be tied to cellular ATP levels and thus to the metabolic status of the cell.

Because the ATPB site is the sole Walker A type motif present in the IP_3R2 and IP_3R3, this site was reasonably predicted to mediate the functional effects of ATP via these receptors. The ATPB site in mouse IP_3R2 was eliminated by substituting alanine residues for each glycine in the motif (G1969A, G1971A, and G1974A; termed ΔATPB). This construct eliminated ATP regulation of IICR over a range of [IP_3]. In addition, both the oscillation frequency and amplitude of Ca^{2+} signals following BcR activation were reduced in cells expressing ΔATPB IP_3R2 when compared to WT IP_3R2. These data confirm that in the context of IP_3R2, the ATPB site is solely responsible for mediating the effects of ATP binding. Importantly, despite the high affinity for the effect of ATP at IP_3R2 (Betzenhauser et al., 2008; Park et al., 2008), these data also support the idea that nucleotide binding has functional consequences following stimulation of an endogenous signaling pathway (Betzenhauser et al., 2008).

In contrast, a similar mutation in the IP_3R3 (G1920A, G1922A, G1925A; termed ΔATPB IP_3R3) resulted in identical ATP regulation of Ca^{2+} release to WT IP_3R3, and moreover, BcR activation evoked Ca^{2+} oscillations in cells expressing ΔATPC which were indistinguishable from WT IP_3R3. These data indicate that despite evidence of ATP binding in a region that contains the ATPB site that the site itself is functionally irrelevant for regulation of Ca^{2+} release via IP_3R3. This finding although surprising, nevertheless appears in hindsight, consistent with IP_3R2 and IP_3R3 having somewhat different characteristics of ATP modulation (Betzenhauser et al., 2008).

Similar approaches were taken to study the role of the ATPB motif in the $S2^+$ IP_3R1. Expression of ΔATPB IP_3R1 (G2016A, G2018A, G2021A IP_3R1) resulted in no obvious alteration in the sensitivity of Ca^{2+} release to ATP (Betzenhauser, Wagner, et al., 2009). Because this construct also contains the ATPA site, with a presumed higher affinity for nucleotide, these data were not unexpected. Remarkably, however, when $S2^+$ IP_3R1 was constructed devoid of any known ATP-binding sites (mutation of G1773, G1775A, G1777A, G1780A in ATPA and G2016A, G2018A, G2021A in ATPB; $S2^+$ IP_3R1 ΔATPA/B), the modulation of Ca^{2+} release by ATP was essentially identical to WT IP_3R1 (Betzenhauser, Wagner, et al., 2009). Indeed, at the single-channel level, monitored in the "on-nucleus patch" configuration, modulation of P_o by ATP, the single-channel conductance and the predominant mode of gating of the mutant channel were indistinguishable from native $S2^+$ IP_3R1. The inescapable conclusion from these data are that the known ATP-binding sites in IP_3R1, similar to IP_3R3, are not responsible for mediating the functional effects of ATP binding on Ca^{2+} release. Given the absence of any further predicted nucleotide-binding sites, a possibility is that the interaction is dependent on the tertiary structure of the IP_3R. A further possibility is that a cryptic ATP-binding site may be exposed following the large conformational changes the receptor undergoes when binding Ca^{2+}. Conceivably, the functional effects of ATP could also be mediated through a tightly bound accessory protein. Further work is obviously necessary to define the molecular determinants of nucleotide function in the IP_3R1 and IP_3R3.

C. Mechanism of ATP Action on IP_3R

Unlike the ryanodine receptor (RyR) where P_o is augmented by ATP in the absence of its primary ligand Ca^{2+} (Fill & Copello, 2002; Meissner, 1984; Meissner, Darling, & Eveleth, 1986; Smith, Coronado, & Meissner, 1985), modulation of IP_3R gating by ATP requires IP_3 binding (Iino, 1991; Taylor & Putney, 1985; Smith, Smith, et al., 1985). Some characteristics of regulation by adenine nucleotides are shared by both channels; for example, hydrolysis of ATP is not required, as poorly hydrolysable analogs of ATP effectively substitute for ATP itself in augmenting the activity of both families of channel (Bezprozvanny & Ehrlich, 1993; Ferris et al., 1990; Iino, 1991; Meissner, 1984; Ogawa & Ebashi, 1976; Smith, Smith, et al., 1985).

Consistent with no requirement for a high-energy acyl phosphate intermediate or kinase reaction in modulation of IP_3R channel activity, the potentiation does not require magnesium bound ATP (Betzenhauser et al., 2008; Betzenhauser, Wagner, et al., 2009; Iino, 1991; Miyakawa et al., 1999). Nevertheless, numerous fluorescence imaging and Ca^{2+} flux studies have suggested that Mg^{2+}ATP, the major cellular molecular species, is as effective as "free" ATP in modulating IP_3R channel activity (Maes et al., 2000). A similar conclusion was reached in a study of cerebellar IP_3R incorporated in lipid bilayers (Bezprozvanny & Ehrlich, 1993). In contrast, studies measuring IP_3R single-channel activity in isolated *Xenopus* oocytes nuclei have offered evidence that channel activity is exclusively activated by micromolar levels of ATP-free acid (presumably ATP^{3-} or ATP^{4-}) in a similar fashion to RyR (Copello et al., 2002; Sonnleitner, Fleischer, & Schindler, 1997). Specifically, channel open-probability (P_o) was unaffected by ATP if the [Mg^{2+}] was increased to reduce the calculated free ATP to negligible levels (Mak et al., 1999; Mak, McBride, & Foskett, 2001a); however, elevated free ATP levels increased channel P_o whether Mg^{2+} was present or absent. The reasons for these disparate reports are not clear; however, the interpretation of data in cellular Ca^{2+} flux experiments may be confounded because MgATP may impact multiple effectors, including SERCA pumps, IP_3 metabolizing enzymes, and kinases which alter channel function. Given the relative concentrations of free cellular ATP (10–100 μM) and Mg^{2+} complexed ATP (0.5–5 mM) and the reported K_d for ATP regulation of release (50–500 μM ATP), the species responsible for regulation of IP_3R is nevertheless, fundamentally important in regard to whether the channel is physiologically regulated by fluctuating nucleotide levels. Further work is necessary to firmly establish that free ATP is the relevant molecular species responsible for regulation of all mammalian IP_3R subtypes.

The most detailed evaluation of the biophysical basis for the action of ATP has been performed using "on-nucleus" single-channel patch-clamp recordings (Betzenhauser, Wagner, et al., 2009; Boehning, Mak, Foskett, & Joseph, 2001; Mak & Foskett, 1994, 1997; Taufiq Ur, Skupin, Falcke, & Taylor, 2009) of the endogenous *Xenopus* oocyte IP_3R (X-IP_3R; Mak et al., 1999). Since the X-IP_3R is most closely related to mammalian IP_3R1 (Patel et al., 1999), these properties are presumed to reflect the characteristics of IP_3R1. Similar to earlier bilayer and Ca^{2+} release studies, ATP failed to gate the channel in the absence of IP_3. The primary effect of ATP was to increase channel P_o by allosterically modulating the sensitivity of the X-IP_3R to activation by cytoplasmic Ca^{2+}. The predominant kinetic effect observed was a shortening of mean channel closed time (τ_c) and lengthening of mean open times (τ_o) at low cytoplasmic [Ca^{2+}]. These data indicate that ATP appears to both stabilize an open state(s) and concurrently destabilize a closed channel state(s) leading to enhanced P_o. An additional study using "on-nucleus" patch-clamp of rat IP_3R1 stably expressed in DT40-3KO chicken lymphocytes at approximately basal cytoplasmic [Ca^{2+}] reported that the predominant effect of ATP was to markedly decrease τ_c without significant effects on τ_o (Betzenhauser, Wagner, et al., 2009). This ultimately resulted in enhanced bursting activity of the channel manifested as extended periods of fast channel opening and closings. ATP is not an obligate ligand for IP_3R1 opening since at saturating [IP_3], but low cytosolic Ca^{2+}, ATP did not increase the absolute maximum P_o of X-IP_3R (Mak et al., 1999, 2001a). In contrast, in the absence of ATP, simply increasing Ca^{2+} was sufficient to ultimately result in maximum channel activity. Similar kinetic effects at the single-channel level have been reported for the action of ATP on recombinant rat IP_3R3 expressed in *Xenopus* oocytes (Mak et al., 2001a). Binding of ATP to the IP_3R3 also appears to establish the distinct cytoplasmic Ca^{2+} sensitivity of this receptor when compared with X-IP_3R, since in the absence of ATP the Ca^{2+} sensitivities of the two receptors are essentially identical (Mak, McBride, & Foskett, 2001b). ATP has been reported to increase the single-channel activity of IP_3R2 when recorded in the plasma membrane of DT40 cells (Betzenhauser et al., 2008), a detailed kinetic evaluation of its effects has however not been reported. Because IP_3 binding has been reported to enhance P_o primarily by modifying the Ca^{2+} sensitivity of the IP_3R (Foskett et al., 2007; Mak & Foskett, 1994; Mak, McBride, & Foskett, 1998; Mak et al., 2001b), a model based on these data suggests that the ultimate *functional* cellular IP_3R sensitivity is governed dynamically in an interdependent fashion by the levels of these three ligands.

D. Physiological Relevance

The molecular species of ATP responsible for the effects, clearly will have impact on any physiological rather than purely pathological relevance for the regulation of the channel. The levels of free ATP are much lower than MgATP and fall within a range which might dynamically regulate the channel in specific cellular microenvironments. Regulation by adenine nucleotides may reflect a ubiquitous mechanism whereby the degree of Ca^{2+} release can be dynamically tuned to the metabolic status of the cell in both physiological and pathological situations. Furthermore, nucleotide modulation may selectively modulate Ca^{2+} release at particular subcellular locales. At ER–mitochondrial tethering points, ATP levels are predicted to be locally high and nucleotide modulation would facilitate optimal Ca^{2+} release and effective ER–mitochondrial coupling (Csordas et al., 2006; de Brito & Scorrano, 2008). The converse situation could be envisioned to occur in subcellular regions of high ATP utilization in membrane domains where pumps are selectively localized close to Ca^{2+} release sites (Lee et al., 1997; Lee et al., 1997).

Acknowledgments

The authors would like to thank Keigan Park and Jill Thompson for proof reading of the manuscript. The work was supported by NIH grants DK 054568 and DE14756 to DIY.

References

Antl, M., von Bruhl, M. L., Eiglsperger, C., Werner, M., Konrad, I., Kocher, T., et al. (2007). IRAG mediates NO/cGMP-dependent inhibition of platelet aggregation and thrombus formation. *Blood, 109*, 552–559.

Arguin, G., Regimbald-Dumas, Y., Fregeau, M. O., Caron, A. Z., & Guillemette, G. (2007). Protein kinase C phosphorylates the inositol 1,4,5-trisphosphate receptor type 2 and decreases the mobilization of Ca^{2+} in pancreatoma AR4-2J cells. *The Journal of Endocrinology, 192*, 659–668.

Avruch, J. (2007). MAP kinase pathways: The first twenty years. *Biochimica et Biophysica Acta, 1773*, 1150–1160.

Bai, G. R., Yang, L. H., Huang, X. Y., & Sun, F. Z. (2006). Inositol 1,4,5-trisphosphate receptor type 1 phosphorylation and regulation by extracellular signal-regulated kinase. *Biochemical and Biophysical Research Communications, 348*, 1319–1327.

Bare, D. J., Kettlun, C. S., Liang, M., Bers, D. M., & Mignery, G. A. (2005). Cardiac type 2 inositol 1,4,5-trisphosphate receptor: Interaction and modulation by calcium/calmodulin-dependent protein kinase II. *The Journal of Biological Chemistry, 280*, 15912–15920.

Berridge, M. J., Lipp, P., & Bootman, M. D. (2000). The versatility and universality of calcium signalling. *Nature Reviews. Molecular Cell Biology, 1*, 11–21.

Betzenhauser, M. J., Fike, J. L., Wagner, L. E., 2nd, Yule, D. I. (2009). Protein kinase A increases type-2 inositol 1,4,5-trisphosphate receptor activity by phosphorylation of serine 937. *The Journal of Biological Chemistry, 284*, 25116–25125.

Betzenhauser, M. J., Wagner, L. E., 2nd, Iwai, M., Michikawa, T., Mikoshiba, K., & Yule, D. I. (2008). ATP modulation of Ca^{2+} release by type-2 and type-3 inositol (1,4,5)-triphosphate receptors. Differing ATP sensitivities and molecular determinants of action. *The Journal of Biological Chemistry, 283*, 21579–21587.

Betzenhauser, M. J., Wagner, L. E., 2nd, Park, H. S., & Yule, D. I. (2009). ATP regulation of type-1 inositol 1,4,5-trisphosphate receptor activity does not require walker A-type ATP-binding motifs. *The Journal of Biological Chemistry, 284*, 16156–16163.

Betzenhauser, M. J., Wagner, L. E., 2nd, Won, J. H., & Yule, D. I. (2008). Studying isoform-specific inositol 1,4,5-trisphosphate receptor function and regulation. *Methods, 46*, 177–182.

Bezprozvanny, I. (2005). The inositol 1,4,5-trisphosphate receptors. *Cell Calcium, 38*, 261–272.

Bezprozvanny, I., & Ehrlich, B. E. (1993). ATP modulates the function of inositol 1,4,5-trisphosphate-gated channels at two sites. *Neuron, 10*, 1175–1184.

Blondel, O., Takeda, J., Janssen, H., Seino, S., & Bell, G. I. (1993). Sequence and functional characterization of a third inositol trisphosphate receptor subtype, IP3R-3, expressed in pancreatic islets, kidney, gastrointestinal tract, and other tissues. *The Journal of Biological Chemistry, 268*, 11356–11363.

Boehning, D., Mak, D. O., Foskett, J. K., & Joseph, S. K. (2001). Molecular determinants of ion permeation and selectivity in inositol 1,4,5-trisphosphate receptor Ca^{2+} channels. *The Journal of Biological Chemistry, 276*, 13509–13512.

Bruce, J. I., Straub, S. V., & Yule, D. I. (2003). Crosstalk between cAMP and Ca^{2+} signaling in non-excitable cells. *Cell Calcium, 34*, 431–444.

Caron, A. Z., Chaloux, B., Arguin, G., & Guillemette, G. (2007). Protein kinase C decreases the apparent affinity of the inositol 1,4,5-trisphosphate receptor type 3 in RINm5F cells. *Cell Calcium, 42*, 323–331.

Copello, J. A., Barg, S., Sonnleitner, A., Porta, M., Diaz-Sylvester, P., Fill, M., et al. (2002). Differential activation by Ca^{2+}, ATP and caffeine of cardiac and skeletal muscle ryanodine receptors after block by Mg^{2+}. *The Journal of Membrane Biology, 187*, 51–64.

Csordas, G., Renken, C., Varnai, P., Walter, L., Weaver, D., Buttle, K. F., et al. (2006). Structural and functional features and significance of the physical linkage between ER and mitochondria. *The Journal of Cell Biology, 174*, 915–921.

Cuevas, B. D., Abell, A. N., & Johnson, G. L. (2007). Role of mitogen-activated protein kinase kinase kinases in signal integration. *Oncogene, 26*, 3159–3171.

Cui, J., Matkovich, S. J., deSouza, N., Li, S., Rosemblit, N., & Marks, A. R. (2004). Regulation of the type 1 inositol 1,4,5-trisphosphate receptor by phosphorylation at tyrosine 353. *The Journal of Biological Chemistry, 279*, 16311–16316.

Dai, S., Hall, D. D., & Hell, J. W. (2009). Supramolecular assemblies and localized regulation of voltage-gated ion channels. *Physiological Reviews, 89*, 411–452.

de Brito, O. M., & Scorrano, L. (2008). Mitofusin 2 tethers endoplasmic reticulum to mitochondria. *Nature, 456*, 605–610.

Dellis, O., Dedos, S. G., Tovey, S. C., Taufiq Ur, R., Dubel, S. J., & Taylor, C. W. (2006). Ca^{2+} entry through plasma membrane IP3 receptors. *Science, 313*, 229–233.

Dellis, O., Rossi, A. M., Dedos, S. G., & Taylor, C. W. (2008). Counting functional inositol 1,4,5-trisphosphate receptors into the plasma membrane. *The Journal of Biological Chemistry, 283*, 751–755.

deSouza, N., Cui, J., Dura, M., McDonald, T. V., & Marks, A. R. (2007). A function for tyrosine phosphorylation of type 1 inositol 1,4,5-trisphosphate receptor in lymphocyte activation. *The Journal of Cell Biology, 179*, 923–934.

DeSouza, N., Reiken, S., Ondrias, K., Yang, Y. M., Matkovich, S., & Marks, A. R. (2002). Protein kinase A and two phosphatases are components of the inositol 1,4,5-trisphosphate receptor macromolecular signaling complex. *The Journal of Biological Chemistry, 277*, 39397–39400.

Ehrlich, B. E., & Watras, J. (1988). Inositol 1,4,5-trisphosphate activates a channel from smooth muscle sarcoplasmic reticulum. *Nature, 336*, 583–586.

Ferris, C. D., Cameron, A. M., Bredt, D. S., Huganir, R. L., & Snyder, S. H. (1991). Inositol 1,4,5-trisphosphate receptor is phosphorylated by cyclic AMP-dependent protein kinase at serines 1755 and 1589. *Biochemical and Biophysical Research Communications, 175*, 192–198.

Ferris, C. D., Huganir, R. L., Bredt, D. S., Cameron, A. M., & Snyder, S. H. (1991). Inositol trisphosphate receptor: Phosphorylation by protein kinase C and calcium calmodulin-dependent protein kinases in reconstituted lipid vesicles. *Proceedings of the National Academy of Sciences of the United States of America, 88*, 2232–2235.

Ferris, C. D., Huganir, R. L., & Snyder, S. H. (1990). Calcium flux mediated by purified inositol 1,4,5-trisphosphate receptor in reconstituted lipid vesicles is allosterically regulated by adenine nucleotides. *Proceedings of the National Academy of Sciences of the United States of America, 87*, 2147–2151.

Fill, M., & Copello, J. A. (2002). Ryanodine receptor calcium release channels. *Physiological Reviews, 82*, 893–922.

Foskett, J. K., White, C., Cheung, K. H., & Mak, D. O. (2007). Inositol trisphosphate receptor Ca^{2+} release channels. *Physiological Reviews, 87*, 593–658.

Geiselhoringer, A., Werner, M., Sigl, K., Smital, P., Worner, R., Acheo, L., et al. (2004). IRAG is essential for relaxation of receptor-triggered smooth muscle contraction by cGMP kinase. *The EMBO Journal, 23*, 4222–4231.

Guillemette, G., Balla, T., Baukal, A. J., & Catt, K. J. (1987). Inositol 1,4,5-trisphosphate binds to a specific receptor and releases microsomal calcium in the anterior pituitary gland. *Proceedings of the National Academy of Sciences of the United States of America, 84*, 8195–8199.

Harnick, D. J., Jayaraman, T., Ma, Y., Mulieri, P., Go, L. O., & Marks, A. R. (1995). The human type 1 inositol 1,4,5-trisphosphate receptor from T lymphocytes. Structure, localization, and tyrosine phosphorylation. *The Journal of Biological Chemistry, 270*, 2833–2840.

Hirata, M., Kukita, M., Sasaguri, T., Suematsu, E., Hashimoto, T., & Koga, T. (1985). Increase in Ca^{2+} permeability of intracellular Ca^{2+} store membrane of saponin-treated guinea pig peritoneal macrophages by inositol 1,4,5-trisphosphate. *Journal of Biochemistry, 97*, 1575–1582.

Hunter, T. (2009). Tyrosine phosphorylation: Thirty years and counting. *Current Opinion in Cell Biology, 21*, 140–146.

Iino, M. (1991). Effects of adenine nucleotides on inositol 1,4,5-trisphosphate-induced calcium release in vascular smooth muscle cells. *The Journal of General Physiology, 98*, 681–698.

Ito, J., Yoon, S. Y., Lee, B., Vanderheyden, V., Vermassen, E., Wojcikiewicz, R., et al. (2008). Inositol 1,4,5-trisphosphate receptor 1, a widespread Ca^{2+} channel, is a novel substrate of polo-like kinase 1 in eggs. *Developmental Biology, 320*, 402–413.

Jayaraman, T., Ondrias, K., Ondriasova, E., & Marks, A. R. (1996). Regulation of the inositol 1,4,5-trisphosphate receptor by tyrosine phosphorylation. *Science, 272*, 1492–1494.

Kaplin, A. I., Snyder, S. H., & Linden, D. J. (1996). Reduced nicotinamide adenine dinucleotide-selective stimulation of inositol 1,4,5-trisphosphate receptors mediates hypoxic mobilization of calcium. *The Journal of Neuroscience, 16*, 2002–2011.

Khan, M. T., Wagner, L., 2nd, Yule, D. I., Bhanumathy, C., & Joseph, S. K. (2006). Akt kinase phosphorylation of inositol 1,4,5-trisphosphate receptors. *The Journal of Biological Chemistry, 281*, 3731–3737.

Lee, B., Vermassen, E., Yoon, S. Y., Vanderheyden, V., Ito, J., Alfandari, D., et al. (2006). Phosphorylation of IP3R1 and the regulation of $[Ca^{2+}]_i$ responses at fertilization: A role for the MAP kinase pathway. *Development, 133*, 4355–4365.

Lee, M. G., Xu, X., Zeng, W., Diaz, J., Kuo, T. H., Wuytack, F., et al. (1997). Polarized expression of Ca^{2+} pumps in pancreatic and salivary gland cells. Role in initiation and propagation of $[Ca^{2+}]_i$ waves. *The Journal of Biological Chemistry, 272*, 15771–15776.

Lee, M. G., Xu, X., Zeng, W., Diaz, J., Wojcikiewicz, R. J., Kuo, T. H., et al. (1997). Polarized expression of Ca^{2+} channels in pancreatic and salivary gland cells. Correlation with initiation and propagation of $[Ca^{2+}]_i$ waves. *The Journal of Biological Chemistry, 272*, 15765–15770.

Maeda, N., Kawasaki, T., Nakade, S., Yokota, N., Taguchi, T., Kasai, M., et al. (1991). Structural and functional characterization of inositol 1,4,5-trisphosphate receptor channel from mouse cerebellum. *The Journal of Biological Chemistry, 266*, 1109–1116.

Maes, K., Missiaen, L., De Smet, P., Vanlingen, S., Callewaert, G., Parys, J. B., et al. (2000). Differential modulation of inositol 1,4,5-trisphosphate receptor type 1 and type 3 by ATP. *Cell Calcium, 27*, 257–267.

Maes, K., Missiaen, L., Parys, J. B., De Smet, P., Sienaert, I., Waelkens, E., et al. (2001). Mapping of the ATP-binding sites on inositol 1,4,5-trisphosphate receptor type 1 and type 3 homotetramers by controlled proteolysis and photoaffinity labeling. *The Journal of Biological Chemistry, 276*, 3492–3497.

Maes, K., Missiaen, L., Parys, J. B., Sienaert, I., Bultynck, G., Zizi, M., et al. (1999). Adenine-nucleotide binding sites on the inositol 1,4,5-trisphosphate receptor bind caffeine, but not adenophostin A or cyclic ADP-ribose. *Cell Calcium, 25*, 143–152.

Mak, D. O., & Foskett, J. K. (1994). Single-channel inositol 1,4,5-trisphosphate receptor currents revealed by patch clamp of isolated *Xenopus* oocyte nuclei. *The Journal of Biological Chemistry, 269*, 29375–29378.

Mak, D. O., & Foskett, J. K. (1997). Single-channel kinetics, inactivation, and spatial distribution of inositol trisphosphate (IP3) receptors in *Xenopus* oocyte nucleus. *The Journal of General Physiology, 109*, 571–587.

Mak, D. O., McBride, S., & Foskett, J. K. (1998). Inositol 1,4,5-trisphosphate [correction of tris-phosphate] activation of inositol trisphosphate [correction of tris-phosphate] receptor Ca^{2+} channel by ligand tuning of Ca^{2+} inhibition. *Proceedings of the National Academy of Sciences of the United States of America, 95*, 15821–15825.

Mak, D. O., McBride, S., & Foskett, J. K. (1999). ATP regulation of type 1 inositol 1,4,5-trisphosphate receptor channel gating by allosteric tuning of Ca(2+) activation. *The Journal of Biological Chemistry, 274*, 22231–22237.

Mak, D. O., McBride, S., & Foskett, J. K. (2001a). ATP regulation of recombinant type 3 inositol 1,4,5-trisphosphate receptor gating. *The Journal of General Physiology, 117*, 447–456.

Mak, D. O., McBride, S., & Foskett, J. K. (2001b). Regulation by Ca^{2+} and inositol 1,4,5-trisphosphate (IP3) of single recombinant type 3 IP3 receptor channels. Ca^{2+} activation uniquely distinguishes types 1 and 3 insp3 receptors. *The Journal of General Physiology, 117*, 435–446.

Malathi, K., Kohyama, S., Ho, M., Soghoian, D., Li, X., Silane, M., et al. (2003). Inositol 1,4,5-trisphosphate receptor (type 1) phosphorylation and modulation by Cdc2. *Journal of Cellular Biochemistry, 90*, 1186–1196.

Malathi, K., Li, X., Krizanova, O., Ondrias, K., Sperber, K., Ablamunits, V., et al. (2005). Cdc2/cyclin B1 interacts with and modulates inositol 1,4,5-trisphosphate receptor (type 1) functions. *Journal of Immunology, 175*, 6205–6210.

Manning, G., Whyte, D. B., Martinez, R., Hunter, T., & Sudarsanam, S. (2002). The protein kinase complement of the human genome. *Science, 298*, 1912–1934.

Matter, N., Ritz, M. F., Freyermuth, S., Rogue, P., & Malviya, A. N. (1993). Stimulation of nuclear protein kinase C leads to phosphorylation of nuclear inositol 1,4,5-trisphosphate receptor and accelerated calcium release by inositol 1,4,5-trisphosphate from isolated rat liver nuclei. *The Journal of Biological Chemistry, 268*, 732–736.

Meissner, G. (1984). Adenine nucleotide stimulation of Ca^{2+}-induced Ca^{2+} release in sarcoplasmic reticulum. *The Journal of Biological Chemistry, 259*, 2365–2374.

Meissner, G., Darling, E., & Eveleth, J. (1986). Kinetics of rapid Ca^{2+} release by sarcoplasmic reticulum. Effects of Ca^{2+}, Mg^{2+}, and adenine nucleotides. *Biochemistry, 25*, 236–244.

Mignery, G. A., Newton, C. L., Archer, B. T., 3rd, Sudhof, T. C. (1990). Structure and expression of the rat inositol 1,4,5-trisphosphate receptor. *The Journal of Biological Chemistry, 265*, 12679–12685.

Missiaen, L., Parys, J. B., Smedt, H. D., Sienaert, I., Sipma, H., Vanlingen, S., et al. (1997). Effect of adenine nucleotides on myo-inositol-1,4,5-trisphosphate-induced calcium release. *The Biochemical Journal, 325*(Pt 3), 661–666.

Miyakawa, T., Maeda, A., Yamazawa, T., Hirose, K., Kurosaki, T., & Iino, M. (1999). Encoding of Ca^{2+} signals by differential expression of IP3 receptor subtypes. *The EMBO Journal, 18*, 1303–1308.

Morgan, D. O. (1997). Cyclin-dependent kinases: Engines, clocks, and microprocessors. *Annual Review of Cell and Developmental Biology, 13*, 261–291.

Nakade, S., Rhee, S. K., Hamanaka, H., & Mikoshiba, K. (1994). Cyclic AMP-dependent phosphorylation of an immunoaffinity-purified homotetrameric inositol 1,4,5-trisphosphate receptor (type I) increases Ca^{2+} flux in reconstituted lipid vesicles. *The Journal of Biological Chemistry, 269*, 6735–6742.

Oakes, S. A., Scorrano, L., Opferman, J. T., Bassik, M. C., Nishino, M., Pozzan, T., et al. (2005). Proapoptotic BAX and BAK regulate the type 1 inositol trisphosphate receptor and calcium leak from the endoplasmic reticulum. *Proceedings of the National Academy of Sciences of the United States of America, 102*, 105–110.

Ogawa, Y., & Ebashi, S. (1976). Ca-releasing action of beta, gamma-methylene adenosine triphosphate on fragmented sarcoplasmic reticulum. *Journal of Biochemistry, 80*, 1149–1157.

Park, H. S., Betzenhauser, M. J., Won, J. H., Chen, J., & Yule, D. I. (2008). The type 2 inositol (1,4,5)-trisphosphate (InsP3) receptor determines the sensitivity of InsP3-induced Ca^{2+} release to ATP in pancreatic acinar cells. *The Journal of Biological Chemistry, 283*, 26081–26088.

Patel, S., Joseph, S. K., & Thomas, A. P. (1999). Molecular properties of inositol 1,4,5-trisphosphate receptors. *Cell Calcium, 25*, 247–264.

Patterson, R. L., Boehning, D., & Snyder, S. H. (2004). Inositol 1,4,5-trisphosphate receptors as signal integrators. *Annual Review of Biochemistry, 73*, 437–465.

Regimbald-Dumas, Y., Arguin, G., Fregeau, M. O., & Guillemette, G. (2007). cAMP-dependent protein kinase enhances inositol 1,4,5-trisphosphate-induced Ca^{2+} release in AR4-2J cells. *Journal of Cellular Biochemistry, 101*, 609–618.

Salmond, R. J., Filby, A., Qureshi, I., Caserta, S., & Zamoyska, R. (2009). T-cell receptor proximal signaling via the Src-family kinases, Lck and Fyn, influences T-cell activation, differentiation, and tolerance. *Immunological Reviews, 228*, 9–22.

Schlossmann, J., Ammendola, A., Ashman, K., Zong, X., Huber, A., Neubauer, G., et al. (2000). Regulation of intracellular calcium by a signalling complex of IRAG, IP3 receptor and cGMP kinase Ibeta. *Nature, 404*, 197–201.

Smith, J. B., Smith, L., & Higgins, B. L. (1985). Temperature and nucleotide dependence of calcium release by myo-inositol 1,4,5-trisphosphate in cultured vascular smooth muscle cells. *The Journal of Biological Chemistry, 260*, 14413–14416.

Smith, J. S., Coronado, R., & Meissner, G. (1985). Sarcoplasmic reticulum contains adenine nucleotide-activated calcium channels. *Nature, 316*, 446–449.

Sonnleitner, A., Fleischer, S., & Schindler, H. (1997). Gating of the skeletal calcium release channel by ATP is inhibited by protein phosphatase 1 but not by Mg^{2+}. *Cell Calcium, 21*, 283–290.

Soulsby, M. D., Alzayady, K., Xu, Q., & Wojcikiewicz, R. J. (2004). The contribution of serine residues 1588 and 1755 to phosphorylation of the type I inositol 1,4,5-trisphosphate receptor by PKA and PKG. *FEBS Letters, 557*, 181–184.

Soulsby, M. D., & Wojcikiewicz, R. J. (2005). The type III inositol 1,4,5-trisphosphate receptor is phosphorylated by cAMP-dependent protein kinase at three sites. *The Biochemical Journal, 392*, 493–497.

Soulsby, M. D., & Wojcikiewicz, R. J. (2007). Calcium mobilization via type III inositol 1,4,5-trisphosphate receptors is not altered by PKA-mediated phosphorylation of serines 916, 934, and 1832. *Cell Calcium, 42*, 261–270.

Sudhof, T. C., Newton, C. L., Archer, B. T., 3rd, Ushkaryov, Y. A., & Mignery, G. A. (1991). Structure of a novel InsP3 receptor. *The EMBO Journal, 10*, 3199–3206.

Sugawara, H., Kurosaki, M., Takata, M., & Kurosaki, T. (1997). Genetic evidence for involvement of type 1, type 2 and type 3 inositol 1,4,5-trisphosphate receptors in signal transduction through the B-cell antigen receptor. *The EMBO Journal, 16*, 3078–3088.

Sun, L., Haun, S., Jones, R. C., Edmondson, R. D., & Machaca, K. (2009). Kinase-dependent regulation of inositol 1,4,5-trisphosphate-dependent Ca^{2+} release during oocyte maturation. *The Journal of Biological Chemistry, 284*, 20184–20196.

Supattapone, S., Danoff, S. K., Theibert, A., Joseph, S. K., Steiner, J., & Snyder, S. H. (1988). Cyclic AMP-dependent phosphorylation of a brain inositol trisphosphate receptor decreases its release of calcium. *Proceedings of the National Academy of Sciences of the United States of America, 85*, 8747–8750.

Szado, T., Vanderheyden, V., Parys, J. B., De Smedt, H., Rietdorf, K., Kotelevets, L., et al. (2008). Phosphorylation of inositol 1,4,5-trisphosphate receptors by protein kinase B/Akt inhibits Ca^{2+} release and apoptosis. *Proceedings of the National Academy of Sciences of the United States of America, 105*, 2427–2432.

Tang, T. S., & Bezprozvanny, I. (2004). Dopamine receptor-mediated Ca(2+) signaling in striatal medium spiny neurons. *The Journal of Biological Chemistry, 279*, 42082–42094.

Tang, T. S., Tu, H., Wang, Z., & Bezprozvanny, I. (2003). Modulation of type 1 inositol (1,4,5)-trisphosphate receptor function by protein kinase a and protein phosphatase 1alpha. *The Journal of Neuroscience, 23*, 403–415.

Taufiq Ur, R., Skupin, A., Falcke, M., & Taylor, C. W. (2009). Clustering of InsP3 receptors by InsP3 retunes their regulation by InsP3 and Ca^{2+}. *Nature, 458*, 655–659.

Taylor, C. W., Genazzani, A. A., & Morris, S. A. (1999). Expression of inositol trisphosphate receptors. *Cell Calcium, 26*, 237–251.

Taylor, C. W., & Putney, J. W., Jr. (1985). Size of the inositol 1,4,5-trisphosphate-sensitive calcium pool in guinea-pig hepatocytes. *The Biochemical Journal, 232*, 435–438.

Tu, H., Miyakawa, T., Wang, Z., Glouchankova, L., Iino, M., & Bezprozvanny, I. (2002). Functional characterization of the type 1 inositol 1,4,5-trisphosphate receptor coupling domain SII(+/−) splice variants and the Opisthotonos mutant form. *Biophysical Journal, 82*, 1995–2004.

Tu, H., Wang, Z., Nosyreva, E., De Smedt, H., & Bezprozvanny, I. (2005). Functional characterization of mammalian inositol 1,4,5-trisphosphate receptor isoforms. *Biophysical Journal, 88*, 1046–1055.

Turjanski, A. G., Vaque, J. P., & Gutkind, J. S. (2007). MAP kinases and the control of nuclear events. *Oncogene, 26*, 3240–3253.

Vanderheyden, V., Wakai, T., Bultynck, G., De Smedt, H., Parys, J. B., & Fissore, R. A. (2009). Regulation of inositol 1,4,5-trisphosphate receptor type 1 function during oocyte maturation by MPM-2 phosphorylation. *Cell Calcium, 46*, 56–64.

Vermassen, E., Fissore, R. A., Nadif Kasri, N., Vanderheyden, V., Callewaert, G., Missiaen, L., et al. (2004). Regulation of the phosphorylation of the inositol 1,4,5-trisphosphate receptor by protein kinase C. *Biochemical and Biophysical Research Communications, 319*, 888–893.

Wagner, L. E., 2nd, Betzenhauser, M. J., & Yule, D. I. (2006). ATP binding to a unique site in the type-1 S2- inositol 1,4,5-trisphosphate receptor defines susceptibility to phosphorylation by protein kinase A. *The Journal of Biological Chemistry, 281*, 17410–17419.

Wagner, L. E., 2nd, Joseph, S. K., & Yule, D. I. (2008). Regulation of single inositol 1,4,5-trisphosphate receptor channel activity by protein kinase A phosphorylation. *Journal de Physiologie, 586*, 3577–3596.

Wagner, L. E., 2nd, Li, W. H., Joseph, S. K., & Yule, D. I. (2004). Functional consequences of phosphomimetic mutations at key cAMP-dependent protein kinase phosphorylation sites in the type 1 inositol 1,4,5-trisphosphate receptor. *The Journal of Biological Chemistry, 279*, 46242–46252.

Wagner, L. E., 2nd, Li, W. H., & Yule, D. I. (2003). Phosphorylation of type-1 inositol 1,4,5-trisphosphate receptors by cyclic nucleotide-dependent protein kinases: A mutational analysis of the functionally important sites in the S2+ and S2− splice variants. *The Journal of Biological Chemistry, 278*, 45811–45817.

Walaas, S. I., Nairn, A. C., & Greengard, P. (1986). PCPP-260, a Purkinje cell-specific cyclic AMP-regulated membrane phosphoprotein of Mr 260, 000. *The Journal of Neuroscience, 6*, 954–961.

Walker, J. E., Saraste, M., Runswick, M. J., & Gay, N. J. (1982). Distantly related sequences in the alpha- and beta-subunits of ATP synthase, myosin, kinases and other ATP-requiring enzymes and a common nucleotide binding fold. *The EMBO Journal, 1*, 945–951.

Widmann, C., Gibson, S., Jarpe, M. B., & Johnson, G. L. (1999). Mitogen-activated protein kinase: Conservation of a three-kinase module from yeast to human. *Physiological Reviews, 79*, 143–180.

Wojcikiewicz, R. J. (1995). Type I, II, and III inositol 1,4,5-trisphosphate receptors are unequally susceptible to down-regulation and are expressed in markedly different proportions in different cell types. *The Journal of Biological Chemistry, 270*, 11678–11683.

Wojcikiewicz, R. J., & Luo, S. G. (1998). Phosphorylation of inositol 1,4,5-trisphosphate receptors by cAMP-dependent protein kinase. Type I, II, and III receptors are differentially susceptible to phosphorylation and are phosphorylated in intact cells. *The Journal of Biological Chemistry, 273*, 5670–5677.

Yang, L. H., Bai, G. R., Huang, X. Y., & Sun, F. Z. (2006). ERK binds, phosphorylates InsP3 type 1 receptor and regulates intracellular calcium dynamics in DT40 cells. *Biochemical and Biophysical Research Communications, 349*, 1339–1344.

CHAPTER 13

Role of Thiols in the Structure and Function of Inositol Trisphosphate Receptors

Suresh K. Joseph
Department of Pathology and Cell Biology, Thomas Jefferson University, Philadelphia, Pennsylvania, USA

I. OVERVIEW

IP_3 receptors, like ryanodine receptors, are redox regulated Ca^{2+} channels. Both pharmacological and pathophysiological conditions mediating oxidative stress have been reported to have potentiating or inhibitory effects on channel function. However, the precise cysteine residues involved have not been identified. This review summarizes our knowledge of redox regulation of IP_3R isoforms and describes what is known regarding the reactivity of key

Current Topics in Membranes, Volume 66
1063-5823/10 $35.00
DOI: 10.1016/S1063-5823(10)66013-9
Copyright 2010, Elsevier Inc. All right reserved.

cysteine residues. This information is compared with similar studies of ryanodine receptors. The potential utility of using the reactivity of endogenous and mutant cysteines to study the structure and conformational changes of the receptor in native membranes is also described.

II. INTRODUCTION

Inositol trisphosphate receptors are Ca^{2+} channels that function to release Ca^{2+} from the endoplasmic reticulum in response to a wide array of hormones, growth factors, and neurotransmitters (Berridge, 2009; Mikoshiba, 2007). At least one or more of the three different IP_3R isoforms are expressed in nearly every cell type in the body. Many fundamental biological processes that are activated or regulated by Ca^{2+} signals require IP_3R function. These include such critical functions as secretion (Petersen & Tepikin, 2008), smooth muscle contraction (Sanderson, Delmotte, Bai, & Perez-Zogbhi, 2008), gene transcription (Lewis, 2001), and fertilization (Malcuit, Kurokawa, & Fissore, 2006). It is increasingly being recognized that the aberrant function of IP_3Rs may contribute to the pathogenesis of several diseases (Chapter 14). The principal regulators of IP_3R channel activity are IP_3 and Ca^{2+} (Foskett, White, Cheung, & Mak, 2007), although the channel is also modulated by phosphorylation (Vanderheyden, Devogelaere, et al., 2009), ATP (Yule et al., 2010), and numerous interacting proteins (Choe & Ehrlich, 2006). This review focuses on the regulation of the channel by redox agents that target the endogenous thiol residues on the protein. In addition, we summarize studies on the accessibility and reactivity of specific cysteine residues in the receptor and discuss the implications of these studies for the structure of the protein.

III. REGULATION OF IP_3R FUNCTION BY CHANGES IN THIOL REDOX STATE

A. Pharmacological Effects of Thiol Reagents

A large number of thiol oxidizing agents have been shown to cause an elevation of cytosolic Ca^{2+} in intact cells. These agents include thimerosal (reviewed in Elferink, 1999), *t*-butylhydroperoxide (Henschke & Elliott, 1995; Zhang, Hisatsune, Matsu-ura, & Mikoshiba, 2009), menadione (Pruijn, Sibeijn, & Bast, 1990), H_2O_2 (Redondo, Salido, Rosado, & Pariente, 2004), 2,2′-dithiodipyridine (Islam, Kindmark, Larsson, & Berggren, 1997), and superoxide generating systems such as hypoxanthine/xanthine oxidase

(Madesh et al., 2005; Suzuki & Ford, 1992; Wesson & Elliott, 1995). Even when these experiments are conducted in the absence of extracellular Ca^{2+}, the attribution of these effects solely to modulation of IP_3R channels is by no means established. Nevertheless, there is considerable evidence that the changes of the thiol redox state of the cell can directly influence the activity of the IP_3R channel. The pharmacological manipulations cited above would be expected to move the cytoplasmic ratio of reduced to oxidized glutathione (GSH/GSSG) to lower (more oxidized) values. There is evidence that oxidized glutathione sensitizes the IP_3R to respond to lower concentrations of IP_3 (Lopez-Colome & Lee, 1996; Renard, Seitz, & Thomas, 1992; Renard-Rooney, Joseph, Seitz, & Thomas, 1995; Zammit & Caldwell, 1992). The physiological range of cytosolic GSH/GSSG ratio is >30 (Hwang, Sinskey, & Lodish, 1992). A systematic study of the relationship between the glutathione redox potential and IP_3R channel activity has not been undertaken, but experiments which have monitored $[^3H]$-IP_3 binding in rat liver show that IP_3Rs become sensitive only when the ratio falls below 0.5 (Renard-Rooney et al., 1995). These values more resemble those found within the ER lumen and may indicate redox regulation of the receptor from within this compartments (see below). Overall, the results suggest that thiol agents that are capable of markedly decreasing the glutathione redox ratio may exert some of their effects by increasing the concentration of GSSG (Henschke & Elliott, 1995; Zhang et al., 2009). However, changes in glutathione are unlikely to be the sole basis for the effects of all the redox agents that modulate intracellular Ca^{2+} signaling. It should be noted that buthionine sulfoximine, an inhibitor of glutathione biosynthesis which depletes cellular glutathione and renders the cells more prone to oxidative stress, has little effect on Ca^{2+} transients elicited from internal stores (Hawkins et al., 2009; Missiaen et al., 1993).

By far the strongest evidence of a direct modulation of thiol groups on the receptor comes from experiments with the drug thimerosal. This organomercurial compound, which has gained notoriety from its widespread use as a preservative in vaccines, has been repeatedly shown to elicit the release of Ca^{2+} from intracellular stores of many different cell types (Adunyah & Dean, 1986; Bird, Burgess, & Putney, 1993; Bootman, Taylor, & Berridge, 1992; Hecker, Brune, Decker, & Ullrich, 1989; Miyazaki et al., 1992; Thorn, Brady, Llopis, Gallacher, & Petersen, 1992). The involvement of IP_3Rs as a potential target of the drug was indicated by the finding that microinjection of an antibody to the IP_3R was able to block the Ca^{2+} oscillations evoked by thimerosal in hamster eggs (Miyazaki et al., 1992). Several studies have since shown that thimerosal promotes an enhancement of $[^3H]$-IP_3 binding (Bultynck et al., 2004; Lopez-Colome & Lee, 1996; Vanlingen et al., 2001) and shifts the IP_3 concentration dependence of Ca^{2+} release to lower concentrations (Bootman et al., 1992; Bultynck et al., 2004; Missiaen et al., 1996).

Thus, thimerosal and other potentiating thiol reagents are viewed as sensitizing the IP_3R to the ambient levels of endogenous IP_3. The observation that these effects can be observed even after purification and functionally reconstitution of the receptor supports the view that thimerosal interacts directly with thiol groups on the channel protein (Kaplin, Ferris, Voglmaier, & Snyder, 1994; Thrower, Duclohier, Lea, Molle, & Dawson, 1996). The effects of the drug on IP_3R single channel activity in planar lipid membranes are complex with effects on both channel opening kinetics and conductance (Thrower et al., 1996). Two features of thimerosal action deserve further comment. First, the effects of thimerosal are biphasic with higher concentrations inhibiting IP_3R channel function (Bultynck et al., 2004; Kaplin et al., 1994; Parys, Missiaen, De Smedt, Droogmans, & Casteels, 1993; Poirier, Poitras, Laflamme, & Guillemette, 2001). Kinetic evidence suggests that the stimulatory effect of thimerosal involves reaction with multiple thiols (Mezna & Michelangeli, 1997). At higher thimerosal concentrations additional thiols that inhibit IP_3 binding and/or channel function may become reactive. Thus, these data point to the IP_3R containing sets of both potentiating and inhibitory thiols. A second observation is that not all organomercurials have the same effects as thimerosal. Related drugs such as *p*-chloromercurobenzoate (Palade, Dettbarn, Alderson, & Volpe, 1989) or mersalyl (Joseph, Ryan, Pierson, Renard-Rooney, & Thomas, 1995) are inhibitors of channel function. This indicates that the chemical and steric environments of the potentiating and inhibitory thiols are different; a conclusion that has implications for the mechanistic role of thiols in channel regulation and function (see below).

In contrast to the pharmacological interventions discussed above, there is less firm evidence that physiological (and pathophysiological) conditions of oxidative stress alter IP_3R-mediated Ca^{2+} signaling by changing the thiol redox state of the receptor. For example, in hepatocytes (Higashi & Hoek, 1991; Hoek, Nomura, & Higashi, 1994) and other cell types (Bokkala, Rubin, & Joseph, 1999; Briner, Tsai, Wang, & Schrier, 1993), acute ethanol treatment leads to an inhibition of intracellular Ca^{2+} release whereas chronic ethanol treatment has the opposite effect and stimulates Ca^{2+} release. The latter conditions, in particular, are well established to be associated with oxidative stress (Cederbaum, Lu, & Wu, 2009). Although IP_3-mediated Ca^{2+} release is more sensitive to IP_3 in permeabilized hepatocytes from ethanol-fed animals, no alterations in the potentiating effect of GSSG were observed (Nomura et al., 1996). Hence, it was concluded that alcohol treatment was unlikely to have changed the thiol redox state of the receptor (Nomura et al., 1996). In other systems, the sensitization of the IP_3R seen in ethanol-treated cells was attributed to increased receptor levels (Bokkala et al., 1999) or increased generation of IP_3 (Briner et al., 1993). IP_3R sensitization mediated by increased ROS generation has been linked to aberrant Ca^{2+}-dependent phosphorylation of cytoskeletal proteins in brain microvascular endothelial cells and has been suggested to underlie the impairment of the blood–brain barrier in chronic alcoholics (Haorah, Knipe, Gorantla, Zheng, & Persidsky, 2007). Similarly, in cells expressing the activated form of the small G-protein Rac1 there is an enhancement of IP_3R-mediated intracellular Ca^{2+} signaling which has been attributed in part to the enhanced production of ROS generated by Rac1 expression (Zhang & Forscher, 2009). Ischemia/reperfusion injury is another system in which oxidative stress may be linked to perturbed IP_3R signaling. There is evidence of a cytosolic Ca^{2+} elevation during both sustained ischemia and the reperfusion phase of the injury that, in part, reflects IP_3R channel activity (Peters & Piper, 2007). Several studies have observed a selective decrease of the type-I IP_3R mRNA and protein levels in response to prolonged cerebral ischemia (Nagata et al., 1994) and reperfusion (Xia, Simonyi, & Sun, 1998). In a recent report, it was noted that H_2O_2 at low concentrations can induce the degradation of IP_3Rs by the ubiquitin/proteasomal pathway (Martin-Garrido et al., 2009), raising the possibility that oxidative modifications of the IP_3R may not only induce changes in channel function but also initiate a slower proteolytic disposal of the protein.

B. Identification and Functional Role of Critical Thiols in IP_3Rs

The type-I IP_3R is the only isoform that has been investigated in any detail with regard to the functional role of its thiol groups and the discussion is therefore focused on this isoform. Its primary sequence contains 60 cysteine residues. Of these, approximately 70% appear free to react with the small fluorescent thiol reagent monobromobimane (Joseph, Nakao, & Sukumvanich, 2005). These measurements were made in the absence of redox buffers using denatured immunoprecipitated IP_3R. Therefore, it is possible that many of the 30% nonreactive thiol residues may have autooxidized during experimental manipulations. Although an accurate estimate is not available of the number of free thiols under the redox conditions prevailing in the cytosol, it is likely that a major proportion of cytosol-exposed IP_3R thiols are maintained in a reduced condition.

The structure of the IP_3R can be divided into an N-terminal suppressor domain (SD; aa1–225), a core IP_3-binding domain (IBC; aa226–604), a central regulatory domain (RD; aa605–1931), and the C-terminal channel domain (CD; aa1932–2749). The SD functions to inhibit IP_3 binding to the IBC and is also thought to be critical for relaying the conformational changes in the IBC to the channel domain (Rossi et al., 2009). Using a chicken DT40 cell lacking all three IP_3R isoforms as a background, Bultynck et al. (2004) found that stable transfection of a deletion mutant lacking the SD resulted in loss of the thimerosal enhancement of IP_3 binding and Ca^{2+} release. In addition, thimerosal stimulated the *in vitro* interaction between fusion proteins encoding the SD and IBC (Bultynck et al., 2004). It was concluded that the potentiating thiols modified by thimerosal may be located within the SD. However, loss of the SD itself causes a marked increase in affinity of the IP_3R which may not be susceptible to further increases by thimerosal. Furthermore, deletion of the SD completely inactivates the channel (Uchida, Miyauchi, Furuichi, Michikawa, & Mikoshiba, 2003) making it difficult to draw any conclusions regarding a lack of sensitization of Ca^{2+} release. The systematic mutation of each of the six cysteines in the SD could provide information on whether potentiating thiols are located in this domain. Initial mutagenesis of two of these cysteines (C56 and C61) did not alter thimerosal sensitivity (Bultynck et al., 2004).

Two cysteines at positions 2610 and 2613 in the channel domain have been identified as inhibitory for channel function based on mutagenesis and stable transfection in the DT40 cell system (Uchida et al., 2003). These cysteines are conserved in all IP_3R and RyR isoforms and are located at the cytosolic emergence of the S6 pore-lining helix. Their critical position has led to the suggestion that these residues play a critical role in the channel gating mechanism (Uchida et al., 2003). The C2610S mutation, unlike the C2613S mutation, showed marked structural changes in the receptor, as detected by an aberrant trypsin digestion profile (Uchida et al., 2003). Previous studies have established that the cytosolic 160 amino acid C-terminal tail contains several determinants that aid in the proper assembly of channel dimers and tetramers (Galvan & Mignery, 2002; Schug & Joseph, 2006). The recently solved crystal structure of the full-length KcsA channel shows the C-terminal tail to have twofold symmetry immediately distal to the pore-lining helix and fourfold symmetry further into the C-tail (Uysal et al., 2009). Thus, it cannot be presently excluded that the cysteine mutants interfere with the complex assembly of the C-tail structure of IP_3Rs.

The functionally inhibitory cysteines are located within the sequence context CXXC which is a well-established motif found at the catalytic site of thiol oxido-reductases (Wouters, Fan, & Haworth, 2010). The XX residues in IP_3Rs are phenylalanine and isoleucine which are also highly conserved in IP_3Rs and RyRs. Surprisingly, mutation of these intervening residues has no effect on channel function (Schug & Joseph, 2006). Under the prevailing reducing environment of the cytosol it would be anticipated that the 2610/2613 thiols are in a reduced state. However, it is possible that a change in the redox environment could induce these thiols to form intra- or intermolecular disulphides and thereby alter channel function. The CXXC sequence in the

C-tail is followed by two histidine residues at positions 2630 and 2635 which are in the sequence context $CXXC(X)_{16}HXXXXH$. This arrangement is conserved in all three IP_3Rs and RyRs. Although the spacing between the cysteines and histidines are atypical, the arrangement is characteristic of Zn finger motifs (e.g., the transcription factor regulator Prickle-1; Katoh & Katoh, 2003). The histidine at position 2630 is essential since the mutation H2630A completely eliminates channel function whereas mutation of adjacent amino acids are without effect (Bhanumathy & Joseph, unpublished observations). The mutation H2635A also strongly inhibits channel function but the result is more difficult to interpret since expression of the protein in COS-7 cells is also impaired (Bhanumathy & Joseph, unpublished observations). Zn finger motifs are commonly found in transcription factors but can also serve as protein recognition modules (Gamsjaeger, Liew, Loughlin, Crossley, & Mackay, 2007). Whether the C-tail of IP_3Rs has the ability to coordinate Zn^{2+} has not been determined. Nevertheless, the possibility that the CXXC motif in the C-tail may be part of a larger motif and subserve multiple functions in the receptor needs to be tested experimentally.

There are three other CXXC motifs that are unique to the primary sequence of the type-I IP_3R. These are not accompanied by any nearby paired histidines. Early studies indicated that NADH has a direct stimulatory effect on IP_3-mediated Ca^{2+} release (Kaplin, Snyder, & Linden, 1996; Patterson, van Rossum, Kaplin, Barrow, & Snyder, 2005). This regulatory modulation appears to be important in the protection of neurons against hypoxic injury when subjected to a short preconditioning period of hypoxia (Bickler, Fahlman, Gray, & McKleroy, 2009). Patterson et al. (2005) also found that glyceraldehyde phosphate dehydrogenase was associated with IP_3Rs and it was suggested that locally produced NADH may mediate the potentiating effects on channel function. They also showed by mutagenesis that cysteines 992 and 995 were critical for association with glyceraldehyde phosphate dehydrogenase (Patterson et al., 2005). The pair of cysteines involved are one of the unique CXXC sequences present in the type-I isoform. In addition to being possible sites of protein association, the CXXC residues in IP_3Rs could potentially act as reactive redox centers to shuttle reducing equivalents between tightly associated oxidoreductases and the receptor. Depending on the cysteines involved, this may have potentiating or inhibitory effects on channel function.

C. ER Luminal Redox Regulation

Based on the assigned topology of the IP_3R there are four ER intraluminal cysteines that are present in all three IP_3R isoforms which are located in the relatively long intraluminal loop between TM domains 5 and 6 (located in type-I at positions 2496, 2504, 2527, and 2533). The type-I isoform contains an additional unique cysteine in the intraluminal loop between TM domains 2 and 3. Each of the four intraluminal cysteines in the type-I isoform has been mutated. Mutations C2496S and C2504S have no effect on channel function (Higo et al., 2005). However, these residues are important for interaction with ERp44, an oxidoreductase of the thioredoxin family, which suppresses IP_3-mediated Ca^{2+} release (Higo et al., 2005). Association with ERp44 is favored by a more reduced intraluminal oxidation state and by depletion of intraluminal Ca^{2+} (Higo et al., 2005). The ability of C2496 and C2504 to form a disulphide bond has been demonstrated in a fusion protein encoding the intraluminal loop (Kang et al., 2008). By contrast, cysteine 2527 remains in a reduced state even when C2596 and 2504 are disulphide linked. The C2527S and C2527A mutants result in complete or partial inactivation of channel function, respectively (Higo et al., 2005; Schug et al., 2007). The cysteine at position 2533 is predicted to be at the luminal aspect of the small pore helix that precedes the selectivity filter of the channel. The mutation C2533A inactivates the channel (Schug et al., 2007). Thus, of the four conserved intraluminal cysteines, two have indirect effects on channel function by altering interaction with the ERp44 regulatory protein (C2496 and C2504), and two appear to be essential for channel activity (C2527 and C2533). The contribution of changes in the intraluminal redox state to the physiological (or pathophysiological) regulation of the receptor remains to be explored. A recent study has suggested that macrophage apoptosis mediated by stimulation of the unfolded protein response may be linked to enhanced IP_3R-mediated Ca^{2+} release (Li et al., 2009). The proposed mechanism implicates an increased activity of ER-1α oxidase, a hyperoxidation of the ER lumen, and a decreased association of the IP_3R with ERp44.

D. *Isoform Specificity of Thiol Regulatory Effects*

An interesting question is whether all three IP_3R isoforms are equally sensitive to redox regulation. The total number of cysteines present in the rat type-II and -III isoform are 56 and 51, respectively, compared to 60 for the rat type-I isoform. Of these, 41 cysteines are conserved in all three isoforms (Fig. 1). Despite the large number of conserved cysteines, differences in the redox behavior of the isoforms have been noted. Thus, thimerosal is ineffective in potentiating IP_3 binding or IP_3-mediated Ca^{2+} release mediated by the type-III isoform in several experimental systems. This includes cells which predominantly express the type-III isoform (Missiaen et al., 1998), recombinant full-length type-III receptor expressed in Sf9 cells (Vanlingen et al., 2001), or DT40 cells expressing the rat type-III

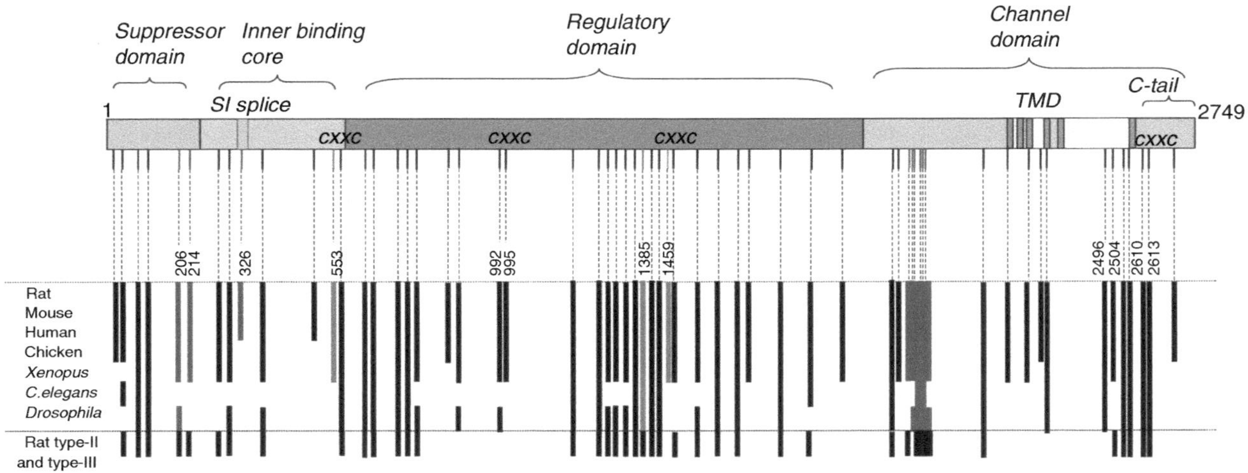

FIGURE 1 Domain organization and distribution of cysteine residues in IP_3Rs. The figure shows the division of the type-I isoform into its constituent functional domains. The location of CXXC domains, the SI splice site, and the six transmembrane domains (TMD) are shown. The location of three intraluminal segments is indicated in yellow. The distribution of 60 cysteine residues in the rat type-I sequence (accession P29994) is indicated as tick marks and homology across species and other isoforms is shown as vertical bars. The analysis was carried out using the CLUSTAL-W program. Cysteines specifically referred to in the text are numbered. The red bars indicate cysteines that are proposed to be located at the receptor's surface and highly reactive with PEG-maleimides. The brown bars are cysteines predicted to be highly homologous to highly reactive cysteines identified in type-I ryanodine receptors. For additional details see text. (See Color Insert.)

isoform (Bultynck et al., 2004). Superoxide generation triggers intracellular Ca^{2+} transients in DT40 cells containing the chicken type-I, type-II but not type-III IP_3Rs (Madesh & Hajnoczky, unpublished observations). Type-III IP_3Rs have a greatly reduced sensitivity to IP_3 when compared to the type-I or -II isoform (Missiaen et al., 1998; Tu, Wang, Nosyreva, De Smedt, & Bezprozvanny, 2005). Since thiol oxidants appear to potentiate IP_3 responses by enhancing the IP_3 sensitivity of the receptor, it is possible that the absence of effects on the type-III isoform reflects sensitization that is insufficient to trigger responses to the low endogenous concentrations of IP_3 that can induce channel opening in the other isoforms. It is also possible that these differences reflect the presence of cysteines that are conserved in the type-I and -II but not the type-III isoforms. However, other factors cannot be discounted. It has been proposed that thiol oxidants may interfere with the conformational interactions between the SD and IBC (Bultynck et al., 2004). All IP_3R isoforms appear to have the same high affinity for IP_3 in their IBC (Iwai, Michikawa, Bosanac, Ikura, & Mikoshiba, 2007). Therefore, differences in the SD–IBC interaction may be responsible for the lower affinity of the type-III isoform. Thus, reaction with even a highly conserved cysteine could potentially have a different structural consequence in the type-III IP_3R than the other isoforms.

Selectivity of responses is also seen at the level of intraluminal redox regulation since only the type-I isoform is sensitive to regulation by ERp44 (Higo et al., 2005). The two cysteines involved in ERp44 interaction (C2496 and C2505) were stated to be highly conserved in all three IP_3R isoforms based on alignment of the mouse IP_3R sequences. However, gaps had to be introduced between the cysteines in order to optimize the alignment and the rat sequences show only the C2505 position to be conserved (Fig. 1). Based on the experimental data it would be predicted that there are structural differences in this region of the protein in the different isoforms that account for the specificity of ERp44 interaction. Finally, it should be noted that there are differences in the thiol reactivity of the alternatively spliced variants of the type-I receptor that are described in more detail in Section V.B.

IV. COMPARISON OF THIOL REGULATION OF IP_3Rs AND RYRs

The regulation of RyRs by alterations in thiol redox state is well established and is described elsewhere in this volume. It is instructive to make some comparisons of the molecular mechanisms of thiol modulation in these two intracellular Ca^{2+} release channels. The skeletal muscle RyR1 channel contains 100 cysteines of which ~25–50 are in the free reduced form under redox conditions approximating the cytosol (Dulhunty, Haarmann, Green, & Hart,

2000; Sun, Xu, Eu, Stamler, & Meissner, 2001). Many of the redox modulators that affect RyR channel activity also regulate IP_3Rs in a similar manner. This includes agents such as thimerosal and glutathione (Hidalgo, Bull, Marengo, Perez, & Donoso, 2000; Zable, Favero, & Abramson, 1997). However, there are some exceptions. The activity of RyR1 is potentiated at low physiological pO_2 by agents that act as donors of NO (Sun et al., 2001). There is evidence that multiple cysteines in RyR1 are nitrosylated including C3635 which is in the calmodulin-binding domain (Porter, Zhang, & Hamilton, 1999; Zhang, Zhang, Danila, & Hamilton, 2003). Guanylate cyclase is a target of NO and a demonstration of a direct effect of NO on IP_3Rs is complicated by the fact that IP_3Rs are also regulated by cGMP-dependent protein kinase (Fritsch et al., 2004). However, there are several instances where such a direct effect has been invoked. For example, it has been reported that the enhancement of IP_3 binding in neuronal nuclei caused by hypoxia can be prevented by treatment with an NO synthase inhibitor (Mishra, Qayyum, & Delivoria-Papadopoulos, 2003). An NO donor has also been reported to promote intracellular Ca^{2+} release from intact neutrophils by a mechanism that is inhibited by IP_3R inhibitors (Pan, Zhang, Song, Wu, & Xu, 2008). However, it has also been reported that an NO donor does not alter IP_3R-mediated Ca^{2+} release in a permeabilized cell system (Bultynck et al., 2004). Therefore, the evidence for a direct, functionally relevant, *S*-nitrosylation of cysteine residues in IP_3Rs is not well established. Similarly, specific residues that undergo glutathionylation in RyR1 have been identified (Aracena-Parks et al., 2006) but analogous evidence of direct glutathionylation of IP_3Rs is lacking.

Voss, Lango, Ernst-Russell, Morin, and Pessah (2004) have used a small fluorescent maleimide derivative in combination with mass spectrometry to identify the most rapidly reactive cysteines in the closed state of RyR1. They identified seven residues including C3635. In a separate study, Aracena-Parks et al. (2006) used different proteomic approaches to identify nine residues that are either endogenously modified by *S*-nitrosylation, *S*-glutathionylation, or form disulphide-bonds in response to H_2O_2. A total of 16 residues were identified in the two studies as being subject to modification. Interestingly, none of the residues fell within the channel domain and reactivity of the CXXC domain in the C-tail was not detected in either study. The limited crystal structures available suggest that RyRs and IP_3Rs may share a similar structural organization, at least in specific regions of the protein (Bosanac et al., 2005; Lobo & Van Petegem, 2009). Alignment programs (ClustalW) indicate that three of the highly reactive cysteines in RyR1 can be aligned with similar sequences in type-I IP_3R. Two of these residues are part of CXXC domains that are unique to the type-I IP_3R (C553, C1459) whereas the third is a cysteine in the RD that is conserved in all three isoforms (C1385; shown in brown in Fig. 1). Identification of highly reactive or modified cysteines does not necessarily prove that the particular cysteine mediates functional responses to redox changes. Potentially, cysteine mutagenesis could be informative but it is more than likely that the redox sensing capabilities of this family of channels are mediated by multiple cysteines. The available proteomic information and analogies to RyRs may provide valuable guides to limit the exhaustive mutagenesis that may be required to identify the critical redox-sensitive cysteines in IP_3Rs. Finally, a mode of regulation that is observed with RyRs is that the thiol redox state interferes with the association of the channel protein with accessory proteins such as calmodulin (Porter et al., 1999; Zhang et al., 2003), FKBP12 (Zissimopoulos, Docrat, & Lai, 2007), junctophilin (Phimister et al., 2007), or the type-II chloride channel (Jalilian, Gallant, Board, & Dulhunty, 2008). Despite interaction with a plethora of ancillary proteins, this mode of redox regulation has not yet been documented for IP_3Rs.

V. CYSTEINE RESIDUES AS PROBES OF IP_3R STRUCTURE

The relatively large size of the IP_3R tetramer (>1 MDa) represents a challenge in obtaining structural information on the protein. Although small segments of the monomer have been utilized to obtain crystal structure (Bosanac et al., 2002, 2005), and progress has been made in obtaining low-resolution EM images of the purified tetramer (da Fonseca, Morris, Nerou, Taylor, & Morris, 2003; Sato et al., 2004; Serysheva et al., 2003), we are far from obtaining a full picture of the receptor's appearance and dynamic behavior in its native lipid environment (reviewed in Chapter VIII). One methodology that provides this type of information utilizes probes that are directed at reactive thiol residues in the receptor. The basic rationale is that the different conformational states of the protein may expose or occlude reactive thiols. Many small fluorescent thiol reactive probes are available but usually this type of experiments requires the use of purified proteins. An exception is the use of junctional SR membranes in which RyRs make up the predominant protein. In this system it has been shown that CPM, a small fluorescent maleimide, reacts 12 times faster with RyRs in the closed state than in the open state (Phimister et al., 2007). This level of enrichment of RyRs in junctional SR is not easy to achieve with IP_3Rs in naturally occurring membranes. An alternative strategy is to use much larger maleimides and to monitor reactivity of the receptor in crude membranes using gel-shifts detected by immunoblotting. IP_3Rs are cleaved by trypsin into a limited number of well-defined fragments (five in the case of the type-I isoform; Yoshikawa, Iwasaki, Michikawa, Furuichi, & Mikoshiba, 1999). Therefore, by interposing a trypsin cleavage step after reaction with the maleimide, it is possible to obtain information on the location of the reactive sites. Potentially, this approach can help to identify the most highly accessible surface-reactive endogenous cysteines present in IP_3Rs. In addition, cysteine substitution mutants can provide information on the accessibility of selected regions of the protein. Finally, this approach can be used to monitor conformational changes occurring in the receptor in its native membrane environment.

A. Location of Highly Reactive Cysteines

Maleimides coupled to polyethylene glycol of either 5 or 20 kDa have been reacted for short periods (5 min) with cerebellum microsomes, which are enriched in the type-I IP_3R isoform. Surprisingly few gel-shifted bands are observed after trypsin digestion and immunoblot analysis. A single shift is observed for the N-terminal trypsin fragment-1 (aa1–345) and two shifted bands occur in the centrally located trypsin fragment-3 (aa923–1581) (Joseph et al., 2005). With MPEG-20 the reaction of the fragment-3 cysteines is very rapid (complete within 15 s) suggesting that at least two cysteines are highly accessible in this domain. The N-terminal fragment-1 appears in immunoblots as a doublet of bands that are known to reflect the presence or absence of a 15 amino acid splice variant referred to as SI. Interestingly, only the SI(+) form of fragment-1 is reactive with either MPEG-5 or MPEG-20. The splice site contains a single cysteine in its sequence. However, mutation of the SI(+) cysteine (C326S) does not prevent the MPEG-shifts, and neither does individual mutation of the eight cysteines in fragment-1 (Joseph et al., 2005). However, mutation of C326S in combination with C206S/C214S completely prevented the MPEG-shifts. The results were interpreted to suggest that there are multiple and independently reactive cysteines that are present in a closely spaced cluster. Reaction of the large MPEG molecule with a single cysteine would sterically prevents further reactions from occurring, thereby generating only a single shifted band. The observation that only the SI(+) form is reactive suggests that the presence of the splice site has a major impact on the conformation of the protein and the formation of the cysteine cluster. The precise functional role of the SI splice site is not known. The SI splice site is expressed in different regions of the brain and is developmentally regulated (Choi, Beaman-Hall, & Vallano, 2004; Nakagawa, Okano, Furuichi, Aruga, & Mikoshiba, 1991). The site appears to influence Ca^{2+} regulation of IP_3 binding (Regan, Lin, Emerick, & Agnew, 2005) and susceptibility to degradation by the ubiquitin proteasome pathway (Bhanumathy, Nakao, & Joseph, 2006). The type-I IP_3R present in echinoderms (Iwasaki et al., 2002) is lacking the SI site but this species still shows a thimerosal sensitization of IP_3 responses (Tanaka & Tashjian, 1994). Whether the presence of the SI splice site has an influence on the redox regulation of vertebrate type-I IP_3Rs remains to be ascertained.

B. Accessibility of the N-Terminal Domain

The introduction of site-specific cysteines into proteins is an established approach used to study the structure and function of ion-channels and membrane receptors (Karlin & Akabas, 1998; Liapakis, Simpson, & Javitch, 2001; Loo & Clarke, 1999; Lu & Deutsch, 2001). Usually, the method requires the removal by mutagenesis of all endogenous cysteines in order to limit reaction to only the cysteine of interest. Mutation of the 60 endogenous cysteines from the full-length type-I IP_3R would be a formidable cloning task and is unlikely to be functionally innocuous. It is fortuitous that the large MPEG molecules have only a limited reactivity with selected regions of the receptor. In the case of the SI(−) N-terminal domain, the absence of MPEG reaction allows this splice variant to be used as a null background for the introduction of cysteine mutants. In a recent study, 10 endogenous serines or alanine residues were mutated to cysteine within the SD and IBC of the SI(−) type-I IP_3R and examined for MPEG reactivity (Anyatonwu & Joseph, 2008). Two main findings emerged from these studies. First, only the S2C mutant at the N-terminus was accessible to MPEG-20 under control conditions, although MPEG-5 did react with the other three cysteine substitution mutants introduced in the SD (S6C, S7C, A189C; Anyatonwu & Joseph, 2008). Accessibility to the ~50 Å MPEG-5 suggests that these residues are oriented at the surface of the receptor. Therefore, it was proposed that the regions of the SD containing these residues would be unlikely to make direct interactions with interior segments of the channel domain (Anyatonwu & Joseph, 2008). The second main finding was that physiological concentrations of Ca^{2+} markedly influenced accessibility of specific cysteines. Some cysteine mutants that were previously inaccessible became accessible to MPEG-5. Other mutants that were accessible to MPEG-5 also became accessible to the ~130 Å MPEG-20. This implies that there are very large structural changes induced by Ca^{2+} in the IP_3R in native membranes. The results were interpreted in the context of a model in which the Ca^{2+}-induced movement of the RDs away from the central density allows increased access to MPEGs (Fig. 2). The conformation transition being observed in native membranes in MPEG accessibility studies may be identical to the square/windmill conformational transitions observed in EM analysis of purified IP_3R preparations (Hamada, Terauchi, & Mikoshiba, 2003). No changes in MPEG reactivity were observed with IP_3 suggesting

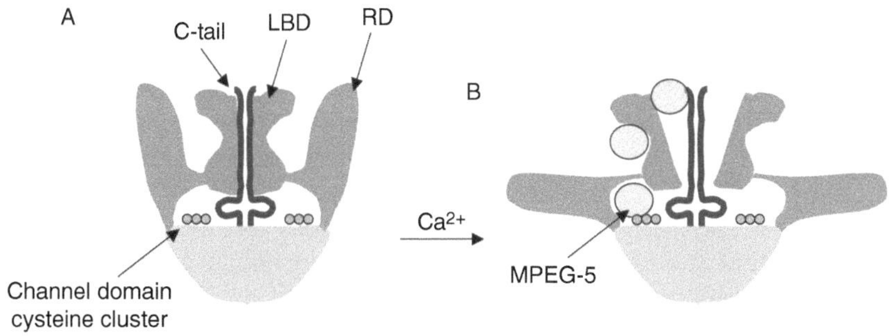

FIGURE 2 Model of Ca^{2+} induced conformational changes in the IP_3R deduced from cysteine accessibility studies using PEG-maleimides. (A) The model shows a representation of the IP_3R based on the EM structure reported by da Fonseca et al. (2003) in which the densities corresponding to the ligand-binding domain (LBD; pink), regulatory domain (RD; gray), and channel domains (blue) are as indicated. A cross-sectional side view with just two subunits is shown for clarity. The C-terminal tail is proposed to project perpendicularly from the membrane and run in a central channel formed by the ligand-binding domain. (B) The model depicts a small vestibule above the channel domain which is proposed to enlarge as a result of a Ca^{2+}-driven conformational change arising from the outward movement of ligand-binding and regulatory domains. This results in the experimentally observed increases in MPEG-5 accessibility to sites in the LBD and to an endogenous cluster of cysteines located in the channel domain. The outward movement of the ligand-binding domain also allows increased accessibility to the C-tail. For additional details see text. (See Color Insert.)

that the conformational transitions occurring upon IP_3 binding are smaller and/or the selected substitution mutants are not appropriately located to monitor the transitions.

C. Accessibility of the C-Tail

The trypsin fragment containing the channel domain (fragment 5) is also nonreactive with MPEG-5 and MPEG-20 and therefore can be used as a null background for cysteine substitution. So far these studies have been confined to the 160 amino acid C-terminal tail. Cysteine substitution of the most C-terminal residue (A2749C) indicates that this portion of the C-tail is highly accessible based on reactivity with both MPEG-5 and MPEG-20. However, three other cysteine mutants placed progressively further into the tail show no reactivity. This is surprising because the C-tail is the site of interaction with at least seven proteins of different sizes ranging from 12 to 98 kDa (Boehning, van Rossum, Patterson, & Snyder, 2005; Fukatsu, Bannai, Inoue, & Mikoshiba, 2006; Kawaai et al., 2009; Tang et al., 2003; Tang, Tu, Wang, & Bezprozvanny, 2003; White et al., 2005; Zhang et al., 2009). The C-tail is also the site of phosphorylation by Akt (Khan, Wagner, Yule,

Bhanumathy, & Joseph, 2006; Szado et al., 2008) and Polo kinases (Vanderheyden, Wakai, et al., 2009). This apparent discrepancy may in part be explained by the observation that Ca^{2+} also has marked effects on accessibility of cysteine residues in the channel domain fragment. In the presence of Ca^{2+} an endogenous cysteine becomes reactive with MPEG-5 (but not MPEG-20) and the MPEG-5 reactivity of the cysteine substitution mutants in the C-tail is also increased (Anyatonwu et al., 2010). Thus, it is possible that Ca^{2+} plays a major role in determining accessibility of the regulatory proteins to the C-tail. The location of the endogenous cysteine exposed by Ca^{2+} has been investigated by mutagenesis. As observed when similar studies were carried out in the N-terminal domain, the individual mutation of endogenous cysteines was not sufficient to remove the Ca^{2+}-dependent reactivity. This required the simultaneous mutation of six cysteines which are present in the primary sequence as two closely spaced clusters (Anyatonwu et al., 2010). Whether this cluster plays a functional role in the receptor's response to redox agents and whether Ca^{2+} influences this reactivity remains to be examined. There are also a number of structural implications of these data. First, the changes induced by Ca^{2+} must provide space to accommodate a molecule of the size of MPEG-5 in the vicinity of the channel domain. Presumably, the confined environment also limits MPEG reactivity to only one cysteine in the cluster at any one time. One hypothesis is that the outward movement of the RDs enlarges a vestibule/space present on the cytosolic side of the channel domain seen in some EM models of the channel (da Fonseca et al., 2003; Sato et al., 2004; Serysheva et al., 2003; Fig. 2). Second, we suggest that the C-tail is present as a perpendicular bundle that runs in a central channel formed within the ligand-binding domain. This would account for the relative inaccessibility of the C-tail cysteines and also permit the close associations required for dimerization and tetramerization of the C-tail. Ca^{2+}-dependent conformational changes in the ligand-binding domain could indirectly influence accessibility to the C-tail (Fig. 2). Indeed, the deletion of the SD was found to entirely eliminate Ca^{2+}-dependent reactivity of MPEG-5 to the channel domain cysteine cluster (Anyatonwu et al., 2010).

VI. FUTURE DIRECTIONS

IP_3Rs in their natural environment are exposed to changes in their redox environment caused by endogenously produced ROS or by external stimuli that cause oxidative stress. The functional consequences on IP_3R channel activity are mediated at the level of critical potentiating or inhibitory thiol residues that have yet to be identified. Modulation of channel function may occur at several levels including an alteration in IP_3-binding affinity, or an altered efficiency of the coupling between IP_3 binding and the channel, or direct effects on the channel gating mechanism itself. A combination of proteomic approaches and cysteine mutagenesis studies is required to identify the location of the functionally relevant thiols in the IP_3R. Although clusters of highly reactive, surface-exposed thiols are present in both the N-terminal and channel segments of the receptor, their functional relevance remains to be ascertained. In addition, the interplay between regulation of the IP_3R channel by changes in thiol redox state and regulation by other key factors such as Ca^{2+} or phosphorylation has yet to be explored. From a structural perspective, the cysteines (either endogenous or introduced by mutation) can potentially be exploited to provide useful information. Cysteine-directed cross-linking reagents can be used to obtain information on nearest neighbor relationships between domains in the native protein. In addition, PEG-maleimides can be utilized to build up a picture of the surface disposition of specific regions of the protein. Since the larger PEG-maleimides can be observed under EM, site-specific labeling can be used to localize domains in the EM structure. The observed changes in thiol reactivity in the presence of Ca^{2+} can also be used to provide further information on the conformational changes occurring in the receptor in its native membrane environment. Additional studies on the modification of cysteines in the IP_3R have the potential to provide many further insights into the structure, function, and regulation of the IP_3R channel.

References

Adunyah, S. E., & Dean, W. L. (1986). Effects of sulfhydryl reagents and other inhibitors on Ca^{2+} transport and inositol trisphosphate-induced Ca^{2+} release from human platelet membranes. *The Journal of Biological Chemistry, 261*, 13071–13075.

Anyatonwu, G., & Joseph, S. K. (2008). Mapping surface accessibility and conformational changes of type-I Inositol trisphosphate receptors using cysteine substitution mutagenesis. *The Journal of Biological Chemistry, 284*, 8093–8102.

Anyatonwu, G., Khan, M. T., Schug, Z. T., da Fonseca, P. C., Morris, E. P., & Joseph, S.K. (2010). Calcium dependent conformational changes in Inositol trisphosphate receptors. *The Journal of Biological Chemistry*, in press, PMID: 20530483.

Aracena-Parks, P., Goonasekera, S. A., Gilman, C. P., Dirksen, R. T., Hidalgo, C., & Hamilton, S. L. (2006). Identification of cysteines involved in S-nitrosylation, S-glutathionylation, and oxidation to disulfides in ryanodine receptor type 1. *The Journal of Biological Chemistry, 281*, 40354–40368.

Berridge, M. J. (2009). Inositol trisphosphate and calcium signalling mechanisms. *Biochimica et Biophysica Acta, 1793*, 933–940.

Bhanumathy, C. D., Nakao, S. K., & Joseph, S. K. (2006). Mechanism of proteasomal degradation of inositol trisphosphate receptors in CHO-K1 cells. *The Journal of Biological Chemistry, 281*, 3722–3730.

Bickler, P. E., Fahlman, C. S., Gray, J., & McKleroy, W. (2009). Inositol 1,4,5-triphosphate receptors and NAD(P)H mediate Ca^{2+} signaling required for hypoxic preconditioning of hippocampal neurons. *Neuroscience, 160*, 51–60.

Bird, G., Burgess, G., & Putney, J. W., Jr. (1993). Sulfhydryl reagents and cAMP-dependent kinase increase the sensitivity of the inositol 1,4,5-trisphosphate receptor in hepatocytes. *The Journal of Biological Chemistry, 268*, 17917–17923.

Boehning, D., van Rossum, D. B., Patterson, R. L., & Snyder, S. H. (2005). A peptide inhibitor of cytochrome c/inositol 1,4,5-trisphosphate receptor binding blocks intrinsic and extrinsic cell death pathways. *Proceedings of the National Academy of Sciences of the United States of America, 102*, 1466–1471.

Bokkala, S., Rubin, E., & Joseph, S. K. (1999). The effect of chronic ethanol exposure on Inositol Trisphosphate receptors in WB rat liver epithelial cells. *Alcoholism, Clinical and Experimental Research, 23*, 1875–1883.

Bootman, M. D., Taylor, C. W., & Berridge, M. J. (1992). The thiol reagent, thimerosal, evokes Ca^{2+} spikes in HeLa cells by sensitizing the inositol 1,4,5-trisphosphate receptor. *The Journal of Biological Chemistry, 267*, 25113–25119.

Bosanac, I., Alattia, J. R., Mal, T. K., Chan, J., Talarico, S., Tong, F. K., et al. (2002). Structure of the inositol 1,4,5-trisphosphate receptor binding core in complex with its ligand. *Nature, 420*, 696–700.

Bosanac, I., Yamazaki, H., Matsu-ura, T., Michikawa, T., Mikoshiba, K., & Ikura, M. (2005). Crystal structure of the ligand binding suppressor domain of type 1 inositol 1,4,5-trisphosphate receptor. *Molecular Cell, 17*, 193–203.

Briner, V. A., Tsai, P., Wang, X., & Schrier, R. W. (1993). Divergent effects of acute and chronic ethanol exposure on contraction and Ca^{2+} mobilization in cultured vascular smooth muscle cells. *American Journal of Hypertension, 6*, 268–275.

Bultynck, G., Szlufcik, K., Kasri, N. N., Assefa, Z., Callewaert, G., Missiaen, L., et al. (2004). Thimerosal stimulates Ca2+ flux through inositol 1,4,5-trisphosphate receptor type 1, but not type 3, via modulation of an isoform-specific Ca2+-dependent intramolecular interaction. *The Biochemical Journal, 381*, 87–96.

Cederbaum, A. I., Lu, Y., & Wu, D. (2009). Role of oxidative stress in alcohol-induced liver injury. *Archives of Toxicology, 83*, 519–548.

Choe, C. U., & Ehrlich, B. E. (2006). The inositol 1,4,5-trisphosphate receptor (IP3R) and its regulators: Sometimes good and sometimes bad teamwork. *Science STKE, 2006*, re15.

Choi, J. Y., Beaman-Hall, C. M., & Vallano, M. L. (2004). Granule neurons in cerebellum express distinct splice variants of the inositol trisphosphate receptor that are modulated by calcium. *American Journal of Physiology. Cell Physiology, 287*, C971–980.

da Fonseca, P. C., Morris, S. A., Nerou, E. P., Taylor, C. W., & Morris, E. P. (2003). Domain organization of the type 1 inositol 1, 4, 5-trisphosphate receptor as revealed by single-particle analysis. *Proceedings of the National Academy of Sciences of the United States of America, 100*, 3936–3941.

Dulhunty, A., Haarmann, C., Green, D., & Hart, J. (2000). How many cysteine residues regulate ryanodine receptor channel activity? *Antioxidants and Redox Signaling, 2*, 27–34.

Elferink, J. G. (1999). Thimerosal: A versatile sulfhydryl reagent, calcium mobilizer, and cell function-modulating agent. *General Pharmacology, 33*, 1–6.

Foskett, J. K., White, C., Cheung, K. H., & Mak, D. O. (2007). Inositol trisphosphate receptor Ca^{2+} release channels. *Physiological Reviews, 87*, 593–658.

Fritsch, R. M., Saur, D., Kurjak, M., Oesterle, D., Schlossmann, J., Geiselhoringer, A., et al. (2004). InsP3R-associated cGMP kinase substrate (IRAG) is essential for nitric oxide-induced inhibition of calcium signaling in human colonic smooth muscle. *The Journal of Biological Chemistry, 279*, 12551–12559.

Fukatsu, K., Bannai, H., Inoue, T., & Mikoshiba, K. (2006). 4.1N binding regions of inositol 1,4,5-trisphosphate receptor type 1. *Biochemical and Biophysical Research Communications, 342*, 573–576.

Galvan, D. L., & Mignery, G. A. (2002). Carboxyl-terminal sequences critical for inositol 1,4,5-trisphosphate receptor subunit assembly. *The Journal of Biological Chemistry, 277*, 48248–48260.

Gamsjaeger, R., Liew, C. K., Loughlin, F. E., Crossley, M., & Mackay, J. P. (2007). Sticky fingers: Zinc-fingers as protein-recognition motifs. *Trends in Biochemical Sciences, 32*, 63–70.

Hamada, K., Terauchi, A., & Mikoshiba, K. (2003). Three-dimensional rearrangements within inositol 1,4,5-trisphosphate receptor by calcium. *The Journal of Biological Chemistry, 278*, 52881–52889.

Haorah, J., Knipe, B., Gorantla, S., Zheng, J., & Persidsky, Y. (2007). Alcohol-induced blood-brain barrier dysfunction is mediated via inositol 1,4,5-triphosphate receptor (IP3R)-gated intracellular calcium release. *Journal of Neurochemistry, 100*, 324–336.

Hawkins, B. J., Lien, Y., Irrinki, K., Wang, Y., Bhanumathy, C., Ramasamy, S., et al. (2009). S-Glutathionylation activates STIM1 and alters mitochondrial homeostasis. *Molecular Cell*, (Unpublished observations).

Hecker, M., Brune, B., Decker, K., & Ullrich, V. (1989). The sulfhydryl reagent thimerosal elicits human platelet aggregation by mobilization of intracellular calcium and secondary prostaglandin endoperoxide formation. *Biochemical and Biophysical Research Communications, 159*, 961–968.

Henschke, P. N., & Elliott, S. J. (1995). Oxidized glutathione decreases luminal Ca^{2+} content of the endothelial cell ins(1,4,5)P3-sensitive Ca^{2+} store. *The Biochemical Journal, 312*(Pt 2), 485–489.

Hidalgo, C., Bull, R., Marengo, J. J., Perez, C. F., & Donoso, P. (2000). SH oxidation stimulates calcium release channels (ryanodine receptors) from excitable cells. *Biological Research, 33*, 113–124.

Higashi, K., & Hoek, J. B. (1991). Ethanol causes desensitization of receptor-mediated phospholipase C activation in isolated hepatocytes. *The Journal of Biological Chemistry, 266*, 2178–2190.

Higo, T., Hattori, M., Nakamura, T., Natsume, T., Michikawa, T., & Mikoshiba, K. (2005). Subtype-specific and ER lumenal environment-dependent regulation of inositol 1,4,5-trisphosphate receptor type 1 by ERp44. *Cell, 120*, 85–98.

Hoek, J. B., Nomura, T., & Higashi, K. (1994). Ethanol and phospholipid dependent signal transduction: The view from the liver. In C. Alling (Ed.), *Alcohol, cell membranes and signal transduction in brain* (pp. 219–233). New York: Plenum Press.

Hwang, C., Sinskey, A., & Lodish, H. F. (1992). Oxidized redox state of glutathione in the endoplasmic reticulum. *Science, 257*, 1496–1501.

Islam, M. S., Kindmark, H., Larsson, O., & Berggren, P. O. (1997). Thiol oxidation by 2, 2'-dithiodipyridine causes a reversible increase in cytoplasmic free Ca2+ concentration in pancreatic beta-cells. Role for inositol 1,4,5-trisphosphate-sensitive Ca2+ stores. *The Biochemical Journal, 321*(Pt 2), 347–354.

Iwai, M., Michikawa, T., Bosanac, I., Ikura, M., & Mikoshiba, K. (2007). Molecular basis of the isoform-specific ligand-binding affinity of inositol 1,4,5-trisphosphate receptors. *The Journal of Biological Chemistry, 282*, 12755–12764.

Iwasaki, H., Chiba, K., Uchiyama, T., Yoshikawa, F., Suzuki, F., Ikeda, M., et al. (2002). Molecular characterization of the starfish inositol 1,4,5-trisphosphate receptor and its role during oocyte maturation and fertilization. *The Journal of Biological Chemistry, 277*, 2763–2772.

Jalilian, C., Gallant, E. M., Board, P. G., & Dulhunty, A. F. (2008). Redox potential and the response of cardiac ryanodine receptors to CLIC-2, a member of the glutathione S-transferase structural family. *Antioxidants and Redox Signaling, 10*, 1675–1686.

Joseph, S. K., Nakao, S. K., & Sukumvanich, S. (2005). Reactivity of free thiol groups in type-I inositol trisphosphate receptors. *The Biochemical Journal, 393*, 575–582.

Joseph, S. K., Ryan, S. V., Pierson, S., Renard-Rooney, D., & Thomas, A. P. (1995). The effect of mersalyl on inositol 1,4,5-trisphosphate receptor binding and ion channel function. *The Journal of Biological Chemistry, 270*, 3588–3593.

Kang, S., Kang, J., Kwon, H., Frueh, D., Yoo, S. H., Wagner, G., et al. (2008). Effects of redox potential and Ca^{2+} on inositol 1,4,5-trisphosphate receptor L3–1 loop region: Implications for receptor regulation. *The Journal of Biological Chemistry, 283*, 25567–25575.

Kaplin, A. I., Ferris, C. D., Voglmaier, S. M., & Snyder, S. H. (1994). Purified reconstituted inositol 1,4,5-trisphosphate receptors. Thiol reagents act directly on receptor protein. *The Journal of Biological Chemistry, 269*, 28972–28978.

Kaplin, A. I., Snyder, S. H., & Linden, D. J. (1996). Reduced nicotinamide adenine dinucleotide-selective stimulation of inositol 1,4,5-trisphosphate receptors mediates hypoxic mobilization of calcium. *The Journal of Neuroscience, 16*, 2002–2011.

Karlin, A., & Akabas, M. H. (1998). Substituted-cysteine accessibility method. *Methods in Enzymology, 293*, 123–145.

Katoh, M., & Katoh, M. (2003). Identification and characterization of human PRICKLE1 and PRICKLE2 genes as well as mouse Prickle1 and Prickle2 genes homologous to Drosophila tissue polarity gene prickle. *International Journal of Molecular Medicine, 11*, 249–256.

Kawaai, K., Hisatsune, C., Kuroda, Y., Mizutani, A., Tashiro, T., & Mikoshiba, K. (2009). 80K-H interacts with inositol 1,4,5-trisphosphate (IP3) receptors and regulates IP3-induced calcium release activity. *The Journal of Biological Chemistry, 284*, 372–380.

Khan, M. T., Wagner, L., Yule, D. I., Bhanumathy, C. D., & Joseph, S. K. (2006). Akt kinase phosphorylation of inositol 1,4,5-trisphosphate receptors. *The Journal of Biological Chemistry, 281*, 3731–3737.

Lewis, R. S. (2001). Calcium signaling mechanisms in T lymphocytes. *Annual Review of Immunology, 19*, 497–521.

Li, G., Mongillo, M., Chin, K. T., Harding, H., Ron, D., Marks, A. R., et al. (2009). Role of ERO1-alpha-mediated stimulation of inositol 1,4,5-triphosphate receptor activity in endoplasmic reticulum stress-induced apoptosis. *The Journal of Cell Biology, 186*, 783–792.

Liapakis, G., Simpson, M. M., & Javitch, J. A. (2001). The substituted-cysteine accessibility method (SCAM) to elucidate membrane protein structure. *Current Protocols in Neuroscience, Chapter 4, Unit 4.15.*

Lobo, P. A., & Van Petegem, F. (2009). Crystal structures of the N-terminal domains of cardiac and skeletal muscle ryanodine receptors: Insights into disease mutations. *Structure, 17*, 1505–1514.

Loo, T. W., & Clarke, D. M. (1999). Determining the structure and mechanism of the human multidrug resistance P-glycoprotein using cysteine-scanning mutagenesis and thiol-modification techniques. *Biochimica et Biophysica Acta, 1461*, 315–325.

Lopez-Colome, A. M., & Lee, I. (1996). Pharmacological characterization of inositol-1,4,5,-trisphosphate binding to membranes from retina and retinal cultures. *Journal of Neuroscience Research, 44*, 149–156.

Lu, J., & Deutsch, C. (2001). Pegylation: A method for assessing topological accessibilities in Kv1.3. *Biochemistry, 40*, 13288–13301.

Madesh, M., Hawkins, B. J., Milovanova, T., Bhanumathy, C. D., Joseph, S. K., Ramachandrarao, S. P., et al. (2005). Selective role for superoxide in InsP3 receptor-mediated mitochondrial dysfunction and endothelial apoptosis. *The Journal of Cell Biology, 170*, 1079–1090.

Malcuit, C., Kurokawa, M., & Fissore, R. A. (2006). Calcium oscillations and mammalian egg activation. *Journal of Cellular Physiology, 206*, 565–573.

Martin-Garrido, A., Boyano-Adanez, M. C., Alique, M., Calleros, L., Serrano, I., Griera, M., et al. (2009). Hydrogen peroxide down-regulates inositol 1,4,5-trisphosphate receptor content through proteasome activation. *Free Radical Biology and Medicine, 47*, 1362–1370.

Mezna, M., & Michelangeli, F. (1997). Effects of thimerosal on the transient kinetics of inositol 1,4,5-trisphosphate-induced Ca^{2+} release from cerebellar microsomes. *The Biochemical Journal, 325*(Pt 1), 177–182.

Mikoshiba, K. (2007). IP3 receptor/Ca^{2+} channel: From discovery to new signaling concepts. *Journal of Neurochemistry, 102*, 1426–1446.

Mishra, O. P., Qayyum, I., & Delivoria-Papadopoulos, M. (2003). Hypoxia-induced modification of the inositol triphosphate receptor in neuronal nuclei of newborn piglets: Role of nitric oxide. *Journal of Neuroscience Research, 74*, 333–338.

Missiaen, L., De Smedt, H., Parys, J. B., Sienaert, I., Valingen, S., & Casteels, R. (1996). Threshold for inositol 1,4,5-trisphosphate action. *The Journal of Biological Chemistry, 271*, 12287–12293.

Missiaen, L., Parys, J. B., Bootman, M. D., Smedt, H. D., Droogmans, G., & Casteels, R. (1993). Normal Ca^{++} signalling in glutathione-depleted and dithiothrietol-treated HeLa cells. *Pflügers Archiv, 423*, 480–484.

Missiaen, L., Parys, J. B., Sienaert, I., Maes, K., Kunzelmann, K., Takahashi, M., et al. (1998). Functional properties of the type-3 InsP3 receptor in 16HBE14o- bronchial mucosal cells. *The Journal of Biological Chemistry, 273*, 8983–8986.

Miyazaki, S., Shirakawa, H., Nakada, K., Honda, Y., Yuzaki, M., Nkade, S., et al. (1992). Antibody to the inositol trisphosphate receptor blocks thimerosal-enhanced Ca^{++}-induced calcium release. *FEBS Letters, 309*, 180–184.

Nagata, E., Tanaka, K., Gomi, S., Mihara, B., Shirai, T., Nogawa, S., et al. (1994). Alteration of inositol 1,4,5-trisphosphate receptor after six-hour hemispheric ischemia in the gerbil brain. *Neuroscience, 61*, 983–990.

Nakagawa, T., Okano, H., Furuichi, T., Aruga, J., & Mikoshiba, K. (1991). The subtypes of the mouse inositol 1,4,5-trisphosphate receptor are expressed in a tissue-specific and developmentally specific manner. *Proceedings of the National Academy of Sciences of the United States of America, 88*, 6244–6248.

Nomura, T., Higashi, K., Hoshino, M., Saso, K., Itou, M., & Hoek, J. B. (1996). Effect of glutathione on Inositol 1,4,5-trisphophate induced Ca2+ release in permeabilized hepatocytes from control and chronic ethanol-fed rats. *Alcoholism, Clinical and Experimental Research, 20*, 325–329.

Palade, P., Dettbarn, C., Alderson, B., & Volpe, P. (1989). Pharmacologic differentiation between inositol-1,4,5- trisphosphate-induced Ca^{2+} release and Ca^{2+}- or caffeine-induced Ca^{2+} release from intracellular membrane systems. *Molecular Pharmacology, 36*, 673–680.

Pan, L., Zhang, X., Song, K., Wu, X., & Xu, J. (2008). Exogenous nitric oxide-induced release of calcium from intracellular IP3 receptor-sensitive stores via S-nitrosylation in respiratory burst-dependent neutrophils. *Biochemical and Biophysical Research Communications, 377*, 1320–1325.

Parys, J., Missiaen, L., De Smedt, H., Droogmans, G., & Casteels, R. (1993). Bell-shaped activation of inositol-1,4,5-trisphosphate-induced Ca^{2+} release by thimerosal in permeabilized A7r5 smooth-muscle cells. *Pflugers Archiv, 424*, 516–522.

Patterson, R. L., van Rossum, D. B., Kaplin, A. I., Barrow, R. K., & Snyder, S. H. (2005). Inositol 1,4,5-trisphosphate receptor/GAPDH complex augments Ca2+ release via locally derived NADH. *Proceedings of the National Academy of Sciences of the United States of America, 102*, 1357–1359.

Peters, S. C., & Piper, H. M. (2007). Reoxygenation-induced Ca^{2+} rise is mediated via Ca^{2+} influx and Ca^{2+} release from the endoplasmic reticulum in cardiac endothelial cells. *Cardiovascular Research, 73*, 164–171.

Petersen, O. H., & Tepikin, A. V. (2008). Polarized calcium signaling in exocrine gland cells. *Annual Review of Physiology, 70*, 273–299.

Phimister, A. J., Lango, J., Lee, E. H., Ernst-Russell, M. A., Takeshima, H., Ma, J., et al. (2007). Conformation-dependent stability of junctophilin 1 (JP1) and ryanodine receptor type 1 (RyR1) channel complex is mediated by their hyper-reactive thiols. *The Journal of Biological Chemistry, 282*, 8667–8677.

Poirier, S. N., Poitras, M., Laflamme, K., & Guillemette, G. (2001). Thiol-reactive agents biphasically regulate inositol 1,4,5-trisphosphate binding and Ca^{2+} release activities in bovine adrenal cortex microsomes. *Endocrinology, 142*, 2614–2621.

Porter, M. C., Zhang, J. Z., & Hamilton, S. L. (1999). A role for cysteine 3635 of RYR1 in redox modulation and calmodulin binding. *The Journal of Biological Chemistry, 274*, 36831–36834.

Pruijn, F., Sibeijn, J., & Bast, A. (1990). Changes in inositol-1,4,5-trisphosphate binding to hepatic plasma membranes caused by temperature, N-ethylmaleimide and menadione. *Biochemical Pharmacology, 40*, 1947–1952.

Redondo, P. C., Salido, G. M., Rosado, J. A., & Pariente, J. A. (2004). Effect of hydrogen peroxide on Ca^{2+} mobilisation in human platelets through sulphydryl oxidation dependent and independent mechanisms. *Biochemical Pharmacology, 67*, 491–502.

Regan, M. R., Lin, D. D., Emerick, M. C., & Agnew, W. S. (2005). The effect of higher order RNA processes on changing patterns of protein domain selection: A developmentally regulated transcriptome of type 1 inositol 1,4,5-trisphosphate receptors. *Proteins, 59*, 312–331.

Renard, D., Seitz, M., & Thomas, A. (1992). Oxidized glutathione causes sensitization of calcium release to inositol 1,4,5-trisphosphate in permeabilized hepatocytes. *The Biochemical Journal, 284*, 507–512.

Renard-Rooney, D. C., Joseph, S. K., Seitz, M. B., & Thomas, A. P. (1995). Effect of oxidized glutathione and temperature on inositol 1,4,5-trisphosphate binding in permeabilized hepatocytes. *The Biochemical Journal, 310*(Pt 1), 185–192.

Rossi, A. M., Riley, A. M., Tovey, S. C., Rahman, T., Dellis, O., Taylor, E. J., et al. (2009). Synthetic partial agonists reveal key steps in IP3 receptor activation. *Nature Chemical Biology, 5*, 631–639.

Sanderson, M. J., Delmotte, P., Bai, Y., & Perez-Zogbhi, J. F. (2008). Regulation of airway smooth muscle cell contractility by Ca^{2+} signaling and sensitivity. *Proceedings of the American Thoracic Society, 5*, 23–31.

Sato, C., Hamada, K., Ogura, T., Miyazawa, A., Iwasaki, K., Hiroaki, Y., et al. (2004). Inositol 1,4,5-trisphosphate receptor contains multiple cavities and L-shaped ligand-binding domains. *Journal of Molecular Biology, 336*, 155–164.

Schug, Z. T., da Fonseca, P. C., Bhanumathy, C. D., Wagner, L., Zhang, X., Bailey, B., et al. (2007). Molecular characterization of the inositol 1,4,5-trisphosphate pore-forming segment. *The Journal of Biological Chemistry, 283*, 2939–2948.

Schug, Z. T., & Joseph, S. K. (2006). The role of the S4–S5 linker and C-terminal tail in inositol 1,4,5-trisphosphate receptor function. *The Journal of Biological Chemistry, 281*, 24431–24440.

Serysheva, I. I., Bare, D. J., Ludtke, S. J., Kettlun, C. S., Chiu, W., & Mignery, G. A. (2003). Structure of the type 1 inositol 1,4,5-trisphosphate receptor revealed by electron cryomicroscopy. *The Journal of Biological Chemistry, 278*, 21319–21322.

Sun, J., Xu, L., Eu, J. P., Stamler, J. S., & Meissner, G. (2001). Classes of thiols that influence the activity of the skeletal muscle calcium release channel. *The Journal of Biological Chemistry, 276*, 15625–15630.

Suzuki, Y. J., & Ford, G. D. (1992). Superoxide stimulates IP3-induced Ca^{2+} release from vascular smooth muscle sarcoplasmic reticulum. *American Journal de Physiologie, 262*, H114–H116.

Szado, T., Vanderheyden, V., Parys, J. B., De Smedt, H., Rietdorf, K., Kotelevets, L., et al. (2008). Phosphorylation of inositol 1,4,5-trisphosphate receptors by protein kinase B/Akt inhibits Ca^{2+} release and apoptosis. *Proceedings of the National Academy of Sciences of the United States of America, 105*, 2427–2432.

Tanaka, Y., & Tashjian, A. H. Jr., (1994). Thimerosal potentiates Ca^{2+} release mediated by both the inositol 1,4,5-trisphosphate and the ryanodine receptors in sea urchin eggs. Implications for mechanistic studies on Ca^{2+} signaling. *The Journal of Biological Chemistry, 269*, 11247–11253.

Tang, T. S., Tu, H., Chan, E. Y., Maximov, A., Wang, Z., Wellington, C. L., et al. (2003). Huntingtin and huntingtin-associated protein 1 influence neuronal calcium signaling mediated by inositol-(1,4,5) triphosphate receptor type 1. *Neuron, 39*, 227–239.

Tang, T. S., Tu, H., Wang, Z., & Bezprozvanny, I. (2003). Modulation of type 1 inositol (1,4,5)-trisphosphate receptor function by protein kinase a and protein phosphatase 1alpha. *The Journal of Neuroscience, 23*, 403–415.

Thorn, P., Brady, P., Llopis, J., Gallacher, D., & Petersen, O. (1992). Cytosolic Ca^{2+} spikes evoked by the thiol reagent thimerosal in both intact and internally perfused single pancreatic acinar cells. *Pflugers Archiv, 422*, 173–178.

Thrower, E. C., Duclohier, H., Lea, E. J., Molle, G., & Dawson, A. P. (1996). The inositol 1,4,5-trisphosphate-gated Ca^{2+} channel: Effect of the protein thiol reagent thimerosal on channel activity. *The Biochemical Journal, 318*, 61–66.

Tu, H., Wang, Z., Nosyreva, E., De Smedt, H., & Bezprozvanny, I. (2005). Functional characterization of mammalian inositol 1,4,5-trisphosphate receptor isoforms. *Biophysical Journal, 88*, 1046–1055.

Uchida, K., Miyauchi, H., Furuichi, T., Michikawa, T., & Mikoshiba, K. (2003). Critical regions for activation gating of the inositol 1,4,5-trisphosphate receptor. *The Journal of Biological Chemistry, 278*, 16551–16560.

Uysal, S., Vasquez, V., Tereshko, V., Esaki, K., Fellouse, F. A., Sidhu, S. S., et al. (2009). Crystal structure of full-length KcsA in its closed conformation. *Proceedings of the National Academy of Sciences of the United States of America, 106*, 6644–6649.

Vanderheyden, V., Devogelaere, B., Missiaen, L., De Smedt, H., Bultynck, G., & Parys, J. B. (2009). Regulation of inositol 1,4,5-trisphosphate-induced Ca2+ release by reversible phosphorylation and dephosphorylation. *Biochimica et Biophysica Acta, 1793*, 959–970.

Vanderheyden, V., Wakai, T., Bultynck, G., De Smedt, H., Parys, J. B., & Fissore, R. A. (2009). Regulation of inositol 1,4,5-trisphosphate receptor type 1 function during oocyte maturation by MPM-2 phosphorylation. *Cell Calcium, 46*, 56–64.

Vanlingen, S., Sipma, H., De Smet, P., Callewaert, G., Missiaen, L., De Smedt, H., et al. (2001). Modulation of inositol 1,4,5-trisphosphate binding to the various inositol 1,4,5-trisphosphate receptor isoforms by thimerosal and cyclic ADP-ribose. *Biochemical Pharmacology, 61*, 803–809.

Voss, A. A., Lango, J., Ernst-Russell, M., Morin, D., & Pessah, I. N. (2004). Identification of hyperreactive cysteines within ryanodine receptor type 1 by mass spectrometry. *The Journal of Biological Chemistry, 279*, 34514–34520.

Wesson, D. E., & Elliott, S. J. (1995). The H2O2-generating enzyme, xanthine oxidase, decreases luminal Ca^{2+} content of the IP3-sensitive Ca^{2+} store in vascular endothelial cells. *Microcirculation, 2*, 195–203.

White, C., Li, C., Yang, J., Petrenko, N. B., Madesh, M., Thompson, C. B., et al. (2005). The endoplasmic reticulum gateway to apoptosis by Bcl-X(L) modulation of the InsP3R. *Nature Cell Biology, 7*, 1021–1028.

Wouters, M. A., Fan, S. W., & Haworth, N. L. (2010). Disulfides as redox switches: From molecular mechanisms to functional significance. *Antioxidants and Redox Signaling, 12*, 53–91.

Xia, J., Simonyi, A., & Sun, G. Y. (1998). Changes in IP3R1 and SERCA2b mRNA levels in the gerbil brain after chronic ethanol administration and transient cerebral ischemia-reperfusion. *Brain Research. Molecular Brain Research, 56*, 22–28.

Yoshikawa, F., Iwasaki, H., Michikawa, T., Furuichi, T., & Mikoshiba, K. (1999). Trypsinized cerebellar inositol 1,4,5-trisphosphate receptor—structural and functional coupling of cleaved ligand binding and channel domains. *The Journal of Biological Chemistry, 274*, 316–327.

Yule, D. I., Betzenhauser, M. J., & Joseph, S. K. (2010). Linking structure to function: Recent lessons from inositol 1,4,5-trisphosphate receptor mutagenesis. *Cell Calcium,* 2010 May 24. [Epub ahead of print].

Zable, A. C., Favero, T. G., & Abramson, J. J. (1997). Glutathione modulates ryanodine receptor from skeletal muscle sarcoplasmic reticulum. Evidence for redox regulation of the Ca^{2+} release mechanism. *The Journal of Biological Chemistry, 272*, 7069–7077.

Zammit, V. A., & Caldwell, A. M. (1992). Direct demosnstration that increased phosphorylation of 3-hydroxy-3-methylglutaryl-CoA reductase does not increase its rate of degredation in isolated rat hepatocytes. *The Biochemical Journal, 284*, 901–904.

Zhang, H., Zhang, J. Z., Danila, C. I., & Hamilton, S. L. (2003). A noncontiguous, intersubunit binding site for calmodulin on the skeletal muscle Ca^{2+} release channel. *The Journal of Biological Chemistry, 278*, 8348–8355.

Zhang, S., Hisatsune, C., Matsu-ura, T., & Mikoshiba, K. (2009). G-protein-coupled receptor kinase-interacting proteins inhibit apoptosis by inositol 1,4,5-triphosphate receptor-mediated Ca^{2+} signal regulation. *The Journal of Biological Chemistry, 284*, 29158–29169.

Zhang, X. F., & Forscher, P. (2009). Rac1 modulates stimulus-evoked Ca(2+) release in neuronal growth cones via parallel effects on microtubule/endoplasmic reticulum dynamics and reactive oxygen species production. *Molecular Biology of the Cell, 20*, 3700–3712.

Zissimopoulos, S., Docrat, N., & Lai, F. A. (2007). Redox sensitivity of the ryanodine receptor interaction with FK506-binding protein. *The Journal of Biological Chemistry, 282*, 6976–6983.

CHAPTER 14

Inositol 1,4,5-Tripshosphate Receptor, Calcium Signaling, and Polyglutamine Expansion Disorders

Ilya Bezprozvanny
Department of Physiology, University of Texas Southwestern Medical Center at Dallas, Dallas, Texas, USA

I. OVERVIEW

Huntington's disease (HD), spinocerebellar ataxia type 2 (SCA2), and type 3 (SCA3) are autosomal-dominant neurodegenerative disorders. HD is caused by polyglutamine (polyQ) expansion (exp) in the amino-terminal region of a protein huntingtin (Htt) and primarily affects medium spiny striatal neurons (MSNs). SCA2 is caused by a polyQ expansion in the amino-terminal region of a cytosolic protein ataxin-2 (Atxn2) and primarily affects cerebellar Purkinje neurons. SCA3 is caused by a polyQ expansion in the carboxy-terminal portion of a cytosolic protein ataxin-3 (Atxn3) and

Current Topics in Membranes, Volume 66
Copyright 2010, Elsevier Inc. All right reserved.
1063-5823/10 $35.00
DOI: 10.1016/S1063-5823(10)66014-0

primarily affects dentate and pontine nuclei and substantia nigra. The reasons for neuronal dysfunction and death in cases of HD, SCA2, and SCA3 remain poorly understood and no cure is available for the patients. Our laboratory discovered that mutant Huntingtin, ataxin-2, and ataxin-3 proteins specifically bind to the carboxy-terminal region of the type 1 inositol 1,4,5-trisphosphate receptor (IP_3R1), an intracellular Ca^{2+} release channel. Moreover, we found that association with IP_3R1 causes sensitization of IP_3R1 to activation by IP_3 in planar lipid bilayers and in neuronal cells. These results suggested that deranged neuronal Ca^{2+} signaling might play an important role in pathogenesis of HD, SCA2, and SCA3. In support of this idea, we demonstrated a connection between abnormal Ca^{2+} signaling and neuronal cell death in experiments with HD, SCA2, and SCA3 transgenic mouse models. Based on these results, I propose that IP_3R and other Ca^{2+} signaling proteins should be considered as potential therapeutic targets for treatment of HD, SCA2, and SCA3.

II. HUNTINGTON'S DISEASE, SPINOCEREBELLAR ATAXIA TYPE 2, AND SPINOCEREBELLAR ATAXIA TYPE 3

HD is an autosomal-dominant neurodegenerative disorder with the age of onset between 35 and 50 years and inexorable progression to death 15–20 years after onset. The symptoms include motor abnormalities, including chorea and psychiatric disturbance with gradual dementia (Vonsattel & DiFiglia, 1998). Neuropathological analysis reveals selective and progressive neuronal loss in the striatum (caudate nucleus, putamen, and globus pallidus) (Vonsattel & DiFiglia, 1998; Vonsattel et al., 1985). GABAergic MSNs are the most sensitive to neuronal degeneration in HD (Vonsattel & DiFiglia, 1998; Vonsattel et al., 1985). Positional cloning efforts demonstrated that at the molecular level, the cause of HD is polyQ expansion in the amino-terminal of a 350 kDa evolutionary conserved cytosolic protein called Htt (The Huntington's Disease Collaborative Research Group, 1993). Clinical signs of HD develop if the length of polyQ track in Htt exceeds a pathological threshold of 35Q. The CAG repeat length is inversely correlated with age of onset (Langbehn, Brinkman, Falush, Paulsen, & Hayden, 2004). Htt is widely expressed in the brain and in nonneuronal tissues and not particularly enriched in the striatum (Li et al., 1993; Sharp et al., 1995; Strong et al., 1993). Htt plays an essential function in development, as deletion of the Htt gene in mice is embryonic lethal (Duyao et al., 1995; Nasir et al., 1995). Analysis of Htt primary sequence suggests that Htt is likely to function as a signaling scaffold (MacDonald, 2003), but the precise function of Htt in cells is not known. In order to elucidate the pathogenesis of HD, a number of transgenic HD mouse models have been generated (Menalled & Chesselet, 2002; Rubinsztein, 2002).

Spinocerebellar ataxias (SCAs) constitute a heterogeneous group of autosomal-dominant genetic and neurodegenerative disorders. SCAs are generally characterized by cerebellar atrophy and a progressive incoordination of movement known as ataxia (Filla et al., 1999; Lastres-Becker, Rub, & Auburger, 2008; Schols, Bauer, Schmidt, Schulte, & Riess, 2004). A total of 29 SCAs have been identified and named in the chronological order of their discovery from SCA1 to SCA30. SCA15 and SCA16 are allelic (Hara et al., 2008; Iwaki et al., 2008). Although all SCA patients present with the phenotypic overlap of cerebellar atrophy and ataxia, other brain regions are differentially affected in each SCA. Seventeen genes have been associated with these SCAs and it is not understood how mutations in those SCA-associated genes lead to the SCA pathogenesis (Paulson, 2009). The pathogenesis of SCAs is not fully understood; however, several different pathogenic mechanisms have been studied in SCAs such as dysregulation of transcription and gene expression, alterations in calcium homeostasis and synaptic neurotransmission, and mitochondrial stress and apoptosis (reviewed in Bezprozvanny & Klockgether, 2010; Carlson, Andresen, & Orr, 2009; Matilla-Duenas et al., 2009). Currently, therapy for SCA patients is mainly supportive and directed at treating individual symptoms in each patient using methods such as deep brain stimulation and levodopa supplements (Bezprozvanny & Klockgether, 2010; Pirker, Back, Gerschlager, Laccone, & Alesch, 2003). No disease modifying therapy exists for any of the SCAs. In order to develop a successful treatment of SCAs, it will be important that a valid therapeutic target and the pathogenic pathways are identified.

In this chapter, I discuss recently emerging results that support the concept that excessive activity of inositol 1,4,5-triphosphate receptors and abnormal Ca^{2+} signaling is playing a major role in pathogenesis of HD, SCA2, and SCA3.

III. MUTANT HUNTINGTIN SPECIFICALLY SENSITIZES IP_3R1 TO IP_3

The inositol 1,4,5-triphosphate receptor (IP_3R) is an intracellular calcium (Ca^{2+}) release channel that plays an important role in neuronal Ca^{2+} signaling (Berridge, 1998). Three isoforms of IP_3R have been identified (Furuichi, Kohda, Miyawaki, & Mikoshiba, 1994). The type 1 receptor (IP_3R1) is the predominant neuronal isoform. Mice lacking IP_3R1 display severe ataxic behavior (Matsumoto et al., 1996), and mice with a spontaneous mutation in the *IP_3R1* gene experience convulsions and ataxia (Street et al., 1997), suggesting a major role of IP_3R1 in neuronal function. Our laboratory has been heavily invested in elucidating the interaction between mHtt and IP_3, as well as other Ca^{2+}-related mechanisms relevant to HD (for reviews see Bezprozvanny, 2009; Bezprozvanny & Hayden, 2004). We first discovered that mHtt binds directly and specifically to the C-terminal region of the type 1 IP_3 receptor (IP_3R1) (Tang et al., 2003). Recently, unbiased high-throughput screening assays confirmed mHtt binding to IP_3R1 (Kaltenbach et al., 2007). Interestingly the affinity of mHtt to IP_3R1 increases when mHtt is associated with HAP1a (Tang et al., 2003). Moreover, mHtt, but not normal Htt, augmented IP_3R1 activity in planar lipid bilayers (Tang et al., 2003). Similarly, application of subthreshold concentrations of DHPG, an mGluR1/5 agonist, sensitized Ca^{2+} release in YAC128 primary MSN cultures (Tang et al., 2003). This is consistent with the fact that glutamate-induced apoptosis of MSNs from YAC128 mice is mediated by mGluR1/5 and NR2B receptors (Tang et al., 2005). In fact, specific blockade of IP_3R1 with 2-APB and Enoxaparin is neuroprotective in the same model (Tang et al., 2005). In a follow-up study, we found disturbed Ca^{2+} signaling, enhanced glutamate-induced apoptosis, and augmented NR2B activity in cultured MSNs from YAC128 mice, which express full-length human Htt, but these effects were completely absent in shortstop mice that express an amino-terminal fragment of mHtt (exons 1 and 2), suggesting a critical role for full-length mHtt in altered MSN Ca^{2+} signaling (Zhang et al., 2008). In more recent experiments, we demonstrated that viral delivery of a peptide that disrupts mHtt association with IP_3R1 protected YAC128 MSNs *in vitro* and *in vivo* (Tang, Guo, Wang, Chen, & Bezprozvanny, 2009). Augmented IP_3R1 activity further implicates mGluR5 receptor signaling in HD pathology. Importantly, inhibitors of IP_3R1 may impede intracellular Ca^{2+} overload early in the disease state and protect MSNs from glutamate-induce excitotoxicity. MSN are highly enriched for mGluR5, a member of the group I mGluRs (Kerner, Standaert, Penney, Young, & Landwehrmeyer, 1997; Mao & Wang, 2001, 2002; Tallaksen-Greene, Kaatz, Romano, & Albin, 1998; Testa, Standaert, Landwehrmeyer, Penney, & Young, 1995). Stimulation of group I mGluR in MSN leads to the generation of IP_3 and release of Ca^{2+}. The alterations in endoplasmic reticulum (ER) enzymes that have been observed in HD postmortem brains (Cross, Crow, Johnson, Dawson, & Peters, 1985) are consistent with malfunction of ER Ca^{2+} handling in HD MSN neurons.

IV. MUTANT HUNTINGTIN ACTIVATES NR2B-CONTAINING NMDA RECEPTORS

MSN abundantly express NR2B subtype of NMDA receptors (Landwehrmeyer, Standaert, Testa, Penney, & Young, 1995; Monyer, Burnashev, Laurie, Sakmann, & Seeburg, 1994; Portera-Cailliau, Price, &

Martin, 1996). In contrast to NMDA receptors containing NR2A subtype, NR2B-containing NMDA receptors have significant permeability for Ca^{2+} and activation of these receptors may have a dramatic effect on intracellular Ca^{2+} signals in MSN. Importantly, studies from Lynn Raymond's and Michael Hayden's laboratories suggested that expression of mutant Htt^{exp} protein facilitates activity of NR2B subtype of NMDA receptors in a heterologous HEK293 cells expression system (Chen et al., 1999). Interestingly, the potentiating effect of Htt^{exp} was specific for the NR1/NR2B NMDAR subtype and not for the NR1/NR2A NMDAR subtype. Using the same HEK293 cells expression system it was also demonstrated that cells cotransfected with NMDAR and Htt-138Q plasmids were more sensitive to NMDA-induced apoptosis than the cells cotransfected with NMDAR and Htt-15Q or GFP (control) plasmids (Zeron et al., 2001). Similar to effects on NMDAR currents, potentiating effects of Htt-138Q on excitotoxic cell death were more pronounced in the presence of the NR1/NR2B NMDAR subunit combination than in the presence of the NR1/NR2A subunit combination (Zeron et al., 2001). These initial findings were corroborated in the YAC mouse, whereby facilitation of NMDA-evoked current amplitude in MSNs from YAC72 and YAC128 mice compared to wild type (WT) was mediated by the NR1/NR2B receptor (Zeron et al., 2002; Zhang et al., 2008). Similarly, cultured MSNs from YAC72 mice showed enhanced NMDA-mediated cell death relative to WTs. Furthermore, treatment with ifenprodil, an NR2B-type selective antagonist, blocked most NMDA-mediated cell death in both YAC72 and WT mice (Zeron et al., 2002). The neuroprotective effect of ifenprodil was later recapitulated in YAC128 MSN cultures (Tang et al., 2005). Findings from dissociated MSNs from presymptomatic R6/2 mice have revealed that NR2A, and not NR1 or NR2B, mRNA levels are reduced, which is concomitant with larger NR2B-mediated NMDAR currents compared to WTs (Ali & Levine, 2006).

The studies that indicated NR2A-type NMDARs are targeted primarily to the synapse and NR2B-type receptors are preferentially expressed at extrasynaptic sites (outside the synapse) is intriguing (Li et al., 1998; Stocca & Vicini, 1998; Tovar & Westbrook, 1999). In fact, dual roles for synaptic and extrasynaptic NMDARs have been proposed (for details see Papadia & Hardingham, 2007; Vanhoutte & Bading, 2003). Whereas synaptic NMDARs activate prosurvival pathways (e.g., CREB phosphorylation, BDNF transcription, ERK activation, antioxidant mechanisms), extrasynaptic NMDARs activate pathways mediate cell death (e.g., loss of mitochondrial membrane potential, inhibition of ERK and BDNF, dendritic blebbing) (Gouix et al., 2009; Hardingham, Fukunaga, & Bading, 2002; Papadia et al., 2008). A recent study by Milnerwood and colleagues has shown that, indeed, mHtt results in increased extrasynaptic NMDAR expression and currents (Milnerwood et al., 2010; also see Li, Murphy, Hayden, & Raymond, 2004). Examination of extrasynaptic activation, by mimicking glutamate spillover into the synaptic cleft with either strong-evoked excitatory postsynaptic currents (EPSCs) or paired-pulse stimuli, revealed increased EPSC charge and elevated NMDA peak current and charge in presymptomatic (1 month) YAC128 MSNs relative to YAC18 (Milnerwood et al., 2010). Accordingly, application of TBOA, a GLT1 transport inhibitor that enhances glutamate spillover to extrasynaptic receptors, resulted in slowing of NMDAR EPSC kinetics in WT and YAC18 mice, but the slowing was consistently more augmented in YAC72 and YAC128. Moreover, in the presence of TBOA, NMDA currents and peak charge were larger in YAC128 compared to WT and YAC18, while YAC72 exhibited intermediate values. Altered TBOA-induced NMDA currents in YAC128 mice were ameliorated upon application of ifenprodil, which blocks NR2B extrasynaptic signaling. Spontaneous NMDA activity was not altered by mHtt, further suggesting involvement of NR2B-type and not NR2A-type NMDARs in altered MSN function. Surprisingly, treatment with low dose memantine, which is thought to preferentially antagonize extrasynaptic over synaptic NMDARs, in 2–4 month YAC128 mice resulted in increased CREB activation and improved motor learning (number of falls when learning the rotarod) relative to WTs (Milnerwood et al., 2010). Direct *in vivo* evidence for NR2B-mediated excitotoxicity in HD comes from recent findings demonstrating exacerbated striatal neurodegeneration in double-mutant mice that express 150Q and overexpress NR2B subunits (Heng, Detloff, Wang, Tsien, & Albin, 2009).

How does mHtt augment NR2B activity? Numerous studies have investigated mRNA and protein expression of NDMARs in human postmortem brain and in HD mouse models, yet no conclusive correlation exists between CAG repeat number and receptor expression levels, nor receptor sensitivity (for discussion see Milnerwood & Raymond, 2006). Some studies suggested that Htt^{exp} affects NMDAR function via PSD95-dependent mechanism (Song, Zhang, Parsons, & Liu, 2003; Sun, Savanenin, Reddy, & Liu, 2001). Indeed, treatment of YAC72 and YAC128 MSN cultures with the Tat-NR2B9C peptide (200 nM), which blocks binding of NR2B with PSD95 and reduces surface expression of NMDARs, resulted in a marked reduction in NMDA-induced apoptosis (Fan, Cowan, Zhang, Hayden, & Raymond, 2009). Knockdown of PSD95 with siRNA resulted in the same effect in YAC128 MSNs (Fan et al., 2009). Another possibility is related to changes in NMDAR trafficking to the plasma membrane in the presence of Htt^{exp} mutation (Fan, Fernandes, Zhang, Hayden, & Raymond, 2007). Future experiments will be needed to clarify exact molecular mechanisms responsible for increase in NMDAR activity in HD MSNs.

From above discussion is clear that hyperfunction of extrasynaptic NMDARs (NR2B) is a major problem in HD and these receptors warrant targeting. The NR2B-specific inhibitors (e.g., ifenprodil, memantine) are likely candidate targets. Indeed, low dose memantine is neuroprotective *in vitro* (Okamoto et al., 2009; Wu, Tang, & Bezprozvanny, 2006) and *in vivo* (Milnerwood et al., 2010; Okamoto et al., 2009). Memantine demonstrated some beneficial effects in small scale pilot evaluation in HD patients (Ondo, Mejia, & Hunter, 2007) and large phase IV clinical trial of memantine in HD patients is currently in progress. Furthermore, a continued search for compounds that positively regulate synaptic NMDARs (NR2A) and/or negatively regulate extrasynaptic NMDARs (NR2B) is warranted.

V. DERANGED Ca^{2+} SIGNALING AND APOPTOSIS OF HD MSN

Several lines of evidence indicate that glutamate-mediated excitotoxicity plays a role in neurodegeneration of HD MSN. Striatal injection of kainic acid induced death of MSN and yielded one of the first animal models of HD (Coyle & Schwarcz, 1976; McGeer & McGeer, 1976). Importantly, effects of kainate required presence of corticostriatal neurons (McGeer, McGeer, & Singh, 1978), suggesting that glutamate release is required for kainate-induced MSN cell death. More direct evidence for an involvement of NMDAR was obtained when HD-like lesions were observed following striatal injection of the NMDAR agonist quinolinic acid (Beal, Ferrante, Swartz, & Kowall, 1991; Beal et al., 1986; Hantraye, Riche, Maziere, & Isacson, 1990). Consistent with the excitotoxicity hypothesis, striatal neurons from YAC72 mouse were sensitized to neuronal death induced by quinolinic acid and NMDA (Zeron et al., 2002). Moreover, excitotoxic cell death of YAC72 MSN was blocked by ifenprodil (Zeron et al., 2002), supporting a direct involvement of NR1/NR2B NMDAR subtypes in HD.

Based on the results described above (Sections III and IV), we previously suggested that overactivation of IP_3R1-mediated Ca^{2+} release and NR2B-mediated Ca^{2+} influx in HD MSN may lead to Ca^{2+} overload and apoptosis of these neurons (Bezprozvanny & Hayden, 2004). To test this "Ca^{2+} hypothesis of HD" my laboratory used TUNEL assay to compare glutamate-induced apoptosis of MSN cultured from WT mice and mice expressing mutant human Htt-128Q gene (YAC128 mouse; Slow et al., 2003). The mice expressing normal human Htt-18Q gene (YAC18; Hodgson et al., 1999) was used as a control in these experiments. At 14 DIV all three groups of MSN were challenged by an 8 h application of glutamate (from 0 to 250 μM) to mimic physiological stimulation. Following exposure to glutamate, MSN were fixed, permeabilized, and scored for apoptotic cell death using

TUNEL staining. We determined that in basal conditions (no glutamate added) approximately 10% of MSN in all three experimental groups were apoptotic (TUNEL-positive). Addition of 25 or 50 μM glutamate increased the number of apoptotic cells to 15–20% in all three experimental groups. Addition of 100 or 250 μM glutamate increased apoptotic death to 60–70% for YAC128 MSN, but only to 25–30% for WT and YAC18 MSN. Thus, we reasoned that exposure to glutamate concentrations in 100–250 μM range leads to selective apoptosis of YAC128 MSN (Tang et al., 2005).

This "*in vitro* HD" model enabled us to test a connection between abnormal Ca^{2+} signaling and apoptosis of HD MSN. We found that inhibition of mGluR1/5 receptors (by a mixture of MPEP and CPCCOEt) reduced the glutamate-induced apoptosis of YAC128 MSN to WT MSN levels. NMDAR-inhibitor (+) MK801 or NR2B-specific antagonist ifenprodil had similar neuroprotective effects. Consistent with direct involvement of IP_3R1, preincubation of the MSN cultures with a membrane-permeable IP_3R blocker 2-APB (Maruyama, Kanaji, Nakade, Kanno, & Mikoshiba, 1997) protected YAC128 MSN from glutamate-induced apoptosis. All these results supported an idea that glutamate-induced Ca^{2+} overload plays a key role in induction of apoptotic cell death of HD MSN. In more recent experiments, we demonstrated that viral delivery of a peptide that disrupts mHtt association with IP_3R1 protected YAC128 MSNs *in vitro* and *in vivo* (Tang et al., 2009).

VI. IP_3R AND ABNORMAL Ca^{2+} SIGNALING IN SCA2 NEURONS

SCA2 patients suffer from a progressive cerebellar syndrome with ataxia of gait and stance, ataxia of limb movements, and dysarthria (Filla et al., 1999; Lastres-Becker et al., 2008; Schols et al., 2004). Similar to HD, the SCA2 is caused by an expansion and translation of unstable CAG repeats in the gene encoding ataxin-2 from the normal 22 to more than 31 extra glutamine repeats (Imbert et al., 1996; Pulst et al., 1996; Sanpei et al., 1996). The pathogenesis of SCA2 is currently not understood. Similar to WT ataxin-2, polyQ-expanded ataxin-2 protein is ubiquitously expressed in cells without severe aggregation and formation of inclusion bodies (Huynh, Figueroa, Hoang, & Pulst, 2000), but the cerebellar PCs in SCA2 patients are mostly affected with a loss of over 75% (Schols et al., 2004). Genetic knockouts of ATXN2 orthologs in fly and worm resulted in embryonic lethality (Kiehl, Shibata, & Pulst, 2000; Satterfield, Jackson, & Pallanck, 2002). Atxn2 knockout mice were viable but displayed a late-onset obesity phenotype (Kiehl et al., 2006). Mice deficient in Atxn2 did not show Purkinje cell loss or marked changes in the Purkinje cell dendritic tree (Kiehl et al., 2006).

Nonessential role of Atxn2 in rodents is most likely related to the presence of orthologs and redundancy in its function (Kiehl et al., 2006). Thus, the polyQ expansion in ataxin-2 likely does not cause a loss of function nor a dominant negative effect but a gain of toxic function (Kiehl et al., 2006). The role of calcium signaling in the pathogenesis of SCA2 is supported by the genetic association between polymorphisms in the CACNA1A gene and the age of disease onset in patients diagnosed with SCA2. The CACNA1A gene encodes the pore-forming α1A subunit of $Ca_V2.1$, a P/Q-type voltage-gated calcium channel. Patients with a prematurely early age of onset of SCA2 tended to have a longer CAG-repeat length in the CACNA1A gene (Pulst et al., 2005). Longer CAG-repeat lengths in this gene have been genetically linked to SCA6.

The role of aberrant neuronal calcium signaling in SCA2 pathogenesis was strengthened further by a finding in our lab that $ATX2^{exp}$ but not WT ATX2 ($ATX2^{wt}$) specifically binds IP_3R1 (Liu et al., 2009). This suggested that its association would result in either a sensitization or desensitization of IP_3R1 to activation by inositol 1,4,5-triphosphate (IP_3) during glutamate signaling in PCs. PCs express extremely high levels of intracellular IP_3R1 (Furuichi et al., 1993; Sharp et al., 1999), which are present on ER membranes. $ATX2^{exp}$ and $ATX2^{wt}$ were found to localize and associate with ER membranes (van de Loo, Eich, Nonis, Auburger, & Nowock, 2009). In a lipid bilayer reconstitution experiment, we examined the effect of $ATX2^{exp}$ expression on IP_3R1 activation in single channel recordings of IP_3R1 coexpressed with $ATX2^{exp}$. We found the presence of $ATX2^{exp}$ substantially sensitized IP_3R1 to activation by inositol 1,4,5 triphosphate (IP_3) (Liu et al., 2009).

To test the importance of Ca^{2+} signaling in SCA2 pathogenesis we performed a series of experiments with SCA2 transgenic mouse model (SCA2-58Q, generated by Huynh et al., 2000) expressing 58 glutamine repeats ($ATX2^{exp}$) in the ataxin-2 gene under the control of the L7/Pcp2 PC-specific promoter. These mice exhibit behavioral deficits, loss of Purkinje cell dendritic arborization and Purkinje cell death, which is progressive and akin to SCA2 patient pathology (Huynh et al., 2000). We found that there was a significant increase in calcium release from ER stores via ITPR1s in calcium imaging experiments in primary PCs cultured from SCA2-58Q transgenic mice, which was not observed in PCs cultured from WT littermates (Liu et al., 2009). When ryanodine or dantrolene were added to block ryanodine receptors and ER calcium release in PC cultures, the effect of $ATX2^{exp}$ expression was immediately reversed as ER calcium release returned to WT levels (Liu et al., 2009). In a TUNEL assay of PC death, we found that the addition of dantrolene to block the excessive ER calcium release caused by $ATX2^{exp}$ expression attenuated exogenous glutamate-induced PC death

(Liu et al., 2009). Furthermore, long-term feeding of SCA2-58Q mice with a calcium stabilizer dantrolene alleviated the age-dependent motor coordination deficits in these mice quantified by rotarod and beam-walk behavioral assays (Liu et al., 2009). Stereological counting of PCs showed a rescue of PC death in 12-month-old SCA2-58Q mice fed with dantrolene (Liu et al., 2009). The above lines of evidence supported the hypothesis that deranged calcium signaling plays a role in SCA2 pathogenesis.

VII. IP_3R AND ABNORMAL Ca^{2+} SIGNALING IN SCA3 NEURONS

SCA3 is a multisystemic disorder characterized by degeneration of spinocerebellar tracts, dentate nucleus, pontine, and other brainstem nuclei, substantia nigra, and pallidum (Stevanin, Durr, & Brice, 2000). In contrast to most other SCA, the cerebellar cortex and the inferior olives are widely spared. MRIs show cerebellar atrophy, which is less pronounced than in SCA2 in combination with brainstem atrophy. Nuclear inclusions containing expanded ataxin-3 have been found in neurons of affected brain regions. The clinical picture of SCA3 is characterized by a wide range of clinical manifestations, the precise nature of which partly depends on repeat length. All SCA3 patients suffer from a progressive syndrome with ataxia of gait and stance, ataxia of limb movements, and dysarthria (Coutinho & Andrade, 1978). At molecular level the cause of SCA3 is a polyQ expansion in the carboxy-terminal of ataxin-3 (Atxn3) protein (Paulson et al., 1997). The ataxin-3 is a 43 kDa cytosolic protein of 316 amino acids which contains amino-terminal Josephin domain and 3 ubiquitin-interactions motifs (UIMs) (Albrecht, Golatta, Wullner, & Lengauer, 2004; Scheel, Tomiuk, & Hofmann, 2003). Consistent with the presence of Josephin domain Atxn3 functions as deubiquitinating enzyme (DUB) (Burnett, Li, & Pittman, 2003). The polyQ-expansion has no effect on DUB activity of Atxn3 (Burnett et al., 2003; Winborn et al., 2008), so it is not likely that neurodegeneration in SCA3 develops due to impaired normal function of Atxn3. Ataxin-3 has been shown to act as a potent transcriptional repressor, a function that is probably related to its DUB activity (Evert et al., 2006). Knockout of the *ATXN3* gene in mice increases protein ubiquitination but does not result in an obvious behavioral phenotype (Schmitt et al., 2007).

Abnormal neuronal Ca^{2+} signaling has also been implicated in SCA3 pathogenesis. It was discovered that inhibition of Ca^{2+}-dependent protease calpain suppressed aggregation of polyQ-expanded Atxn3 in transfected cells (Haacke, Hartl, & Breuer, 2007). Similar to mutant Htt and mutant Atxn2, mutant Atxn3 but not WT Atxn3 binds to and activates IP_3R1, an intracellular Ca^{2+} release channel (Chen et al., 2008). Moreover, long-term feeding of SCA3-YAC-84Q transgenic mouse with Ca^{2+} stabilizer dantrolene alleviated age-dependent motor coordination deficits in this mice and prevented neuronal loss in SNc and PN (Chen et al., 2008). These results provided strong experimental support to the "Ca^{2+} hypothesis of SCA3 pathogenesis" (Chen et al., 2008).

Abnormal neuronal Ca^{2+} signaling is not restricted to polyQ-expanded ataxias and may be important in other ataxias as well. For example, recent genetic evidence indicated that the cause of SCA15 is deletion of a fragment of a gene encoding IP_3R1 (van de Leemput et al., 2007).

VIII. Ca^{2+} BLOCKERS AND PERSPECTIVES FOR CLINICAL INTERVENTION IN HD AND SCA PATIENTS

Despite efforts by many laboratories and cloning of Huntingtin and ataxins more than 15 years ago, there is still no cure for these disorders. The most obvious targets are mutant Huntingtin, Atxn2 and Atxn3 proteins themselves. Expression of mutant Htt and mutant ataxin-1 (Atxn1) was reduced by injection of RNAi-encoding viruses, resulting in beneficial effects in mouse models of HD and SCA1 (Harper et al., 2005; Xia et al., 2004). Similar RNAi knockdown approach could also be expanded to SCA2 and SCA3. Selective knockdown of mRNA encoding mutant Htt and mutant Atxn3 proteins was recently demonstrated using peptide nucleic acid (PNA) and locked nucleic acid (LNA) antisense oligomers targeting expanded CAG repeats (Hu et al., 2009). These are very attractive approaches, but their utility at the moment is limited by the absence of RNAi or antisense brain delivery system that can be used in humans.

In addition to reducing expression of polyQ-expanded proteins at the mRNA level, another potential therapeutic strategy is based on developing agents that selectively bind to mutant forms of these proteins. A polyQ-containing protein can exist in multiple conformations, some of which are nontoxic and some of which are aggregation-prone and toxic (Shao & Diamond, 2007). One possibility to reduce the amount of toxic conformations of Atxn2 and Atxn3 proteins is to increase the levels of chaperons, which control protein folding. Although these approaches are being developed for polyQ-expansion disorders (Wacker, Zareie, Fong, Sarikaya, & Muchowski, 2004; Warrick et al., 1999; Wyttenbach et al., 2002), they have not produced clinical leads yet. Several small molecules were isolated in screens as inhibitors of polyQ aggregation. Some of these molecules were able to reduce polyQ aggregation and toxicity in cellular and animal models (Chopra et al., 2007; Ehrnhoefer et al., 2006; Heiser et al., 2002; Sanchez, Mahlke, & Yuan, 2003; Wang, Gines, MacDonald, & Gusella, 2005; Zhang et al., 2005). A specific polyQ-binding peptide (QBP1) was isolated in the phage library screen (Nagai et al., 2000). It has been reported that QBP1 peptide prevented toxic conformational transition within polyQ-expanded proteins (Nagai et al., 2000) and exerted neuroprotective effects in cellular and animal models of polyQ toxicity (Nagai et al., 2003; Popiel, Nagai, Fujikake, & Toda, 2007, 2009). Despite excellent scientific rationals, so far none of these candidates has successfully advanced into clinical trials due to problems related to finding an appropriate mode of delivery, poor pharmacokinetics, and low efficacy *in vivo*. Recent solution of the crystal structure of Htt exon 1 containing polyQ region (Kim, Chelliah, Kim, Otwinowski, & Bezprozvanny, 2009) opens perspective for rational design of novel polyQ-binding therapeutic agents.

Our laboratory demonstrated beneficial effects of Ca^{2+} stabilizer dantrolene in both SCA2 and SCA3 mouse models (Chen et al., 2008; Liu et al., 2009). Dantrolene also has a long history of human use for treatment of malignant hyperthermia and for neurological indications. In considering design of the clinical trial potential side effects of long-term exposure to lithium or dantrolene must be taken into consideration. NMDAR inhibitor memantine and antiglutamate agent riluzole were neuroprotective in experiments with primary MSN cultures from YAC128 HD mouse model, with memantine being more effective (Wu et al., 2006). Memantine was also effective in 3-NP model of HD (Lee et al., 2006) and in YAC128 genetic HD model (Okamoto et al., 2009). Memantine demonstrated some beneficial effects in small scale pilot evaluation in HD patients (Ondo et al., 2007) and is currently being tested soon in phase IV HD clinical trial. Riluzole was tested in phase III HD clinical trial, but it was not successful (Landwehrmeyer et al., 2007). In addition to NMDAR, voltage-gated Ca^{2+} channels constitute another potential target for SCA treatment. In recent studies, L-type Ca^{2+} channel inhibitor isradipine significantly protected SNc neurons in animal models of Parkinson's disease (Chan et al., 2007). Moreover, a recent retrospective epidemiological study has found that treatment of hypertension with Ca^{2+} channel antagonists significantly diminished the risk of developing PD (Becker, Jick, & Meier, 2008). It is possible that isradipine and other dihydropyridines (DHP) may also be useful for treatment of HD, SCA2, and SCA3 patients. More conservative potential strategy involves use of mitochondrial stabilizers and energizers, such as creatine, CoQ10, and MitoQ. In previous neurodegenerative trials these compounds resulted in modest benefit (Chaturvedi & Beal, 2008) and it is likely that similar result will be obtained for SCA2 and SCA3 patients as well. Controlled clinical evaluation of Ca^{2+} inhibitors in HD, SCA2, and SCA3 patients will provide an ultimate test to the proposed "Ca^{2+} hypothesis" of these disorders.

Acknowledgments

I thank members of my laboratory for their hard work. I am a holder of Carla Cocke Francis Professorship in Alzheimer's Research and supported by the McKnight Neuroscience of Brain Disorders Award. Our work on HD, SCA2, and SCA3 was supported by CHDI, Hereditary Disease Foundation, the National Organization for Rare Disorders, National Ataxia Foundation, Ataxia MJD Research Project, and the National Institutes of Health grants R01NS38082 and R01NS056224.

References

Albrecht, M., Golatta, M., Wullner, U., & Lengauer, T. (2004). Structural and functional analysis of ataxin-2 and ataxin-3. *European Journal of Biochemistry, 271*, 3155–3170.

Ali, N. J., & Levine, M. S. (2006). Changes in expression of N-methyl-D-aspartate receptor subunits occur early in the R6/2 mouse model of Huntington's disease. *Developmental Neuroscience, 28*, 230–238.

Beal, M. F., Ferrante, R. J., Swartz, K. J., & Kowall, N. W. (1991). Chronic quinolinic acid lesions in rats closely resemble Huntington's disease. *The Journal of Neuroscience, 11*, 1649–1659.

Beal, M. F., Kowall, N. W., Ellison, D. W., Mazurek, M. F., Swartz, K. J., & Martin, J. B. (1986). Replication of the neurochemical characteristics of Huntington's disease by quinolinic acid. *Nature, 321*, 168–171.

Becker, C., Jick, S. S., & Meier, C. R. (2008). Use of antihypertensives and the risk of Parkinson disease. *Neurology, 70*, 1438–1444.

Berridge, M. J. (1998). Neuronal calcium signaling. *Neuron, 21*, 13–26.

Bezprozvanny, I. (2009). Calcium signaling and neurodegenerative diseases. *Trends in Molecular Medicine, 15*, 89–100.

Bezprozvanny, I., & Hayden, M. R. (2004). Deranged neuronal calcium signaling and Huntington disease. *Biochemical and Biophysical Research Communications, 322*, 1310–1317.

Bezprozvanny, I., & Klockgether, T. (2010). Therapeutic prospects for spinocerebellar ataxia type 2 and 3. *Drugs of the Future, 34*, 991–999.

Burnett, B., Li, F., & Pittman, R. N. (2003). The polyglutamine neurodegenerative protein ataxin-3 binds polyubiquitylated proteins and has ubiquitin protease activity. *Human Molecular Genetics, 12*, 3195–3205.

Carlson, K. M., Andresen, J. M., & Orr, H. T. (2009). Emerging pathogenic pathways in the spinocerebellar ataxias. *Current Opinion in Genetics and Development, 19*, 247–253.

Chan, C. S., Guzman, J. N., Ilijic, E., Mercer, J. N., Rick, C., Tkatch, T., et al. (2007). 'Rejuvenation' protects neurons in mouse models of Parkinson's disease. *Nature, 447*, 1081–1086.

Chaturvedi, R. K., & Beal, M. F. (2008). Mitochondrial approaches for neuroprotection. *Annals of the New York Academy of Sciences, 1147*, 395–412.

Chen, N., Luo, T., Wellington, C., Metzler, M., McCutcheon, K., Hayden, M. R., et al. (1999). Subtype-specific enhancement of NMDA receptor currents by mutant huntingtin. *Journal of Neurochemistry, 72*, 1890–1898.

Chen, X., Tang, T.-S., Tu, H., Nelson, O., Pook, M. A., Hammer, R. E., et al. (2008). Deranged calcium signaling and neurodegeneration in spinocerebellar ataxia type 3. *The Journal of Neuroscience, 28*, 12713–12724.

Chopra, V., Fox, J. H., Lieberman, G., Dorsey, K., Matson, W., Waldmeier, P., et al. (2007). A small-molecule therapeutic lead for Huntington's disease: Preclinical pharmacology and efficacy of C2-8 in the R6/2 transgenic mouse. *Proceedings of the National Academy of Sciences of the United States of America, 104*, 16685–16689.

Coutinho, P., & Andrade, C. (1978). Autosomal dominant system degeneration in Portuguese families of the Azores Islands. A new genetic disorder involving cerebellar, pyramidal, extrapyramidal and spinal cord motor functions. *Neurology, 28*, 703–709.

Coyle, J. T., & Schwarcz, R. (1976). Lesion of striatal neurones with kainic acid provides a model for Huntington's chorea. *Nature, 263*, 244–246.

Cross, A. J., Crow, T. J., Johnson, J. A., Dawson, J. M., & Peters, T. J. (1985). Loss of endoplasmic reticulum-associated enzymes in affected brain regions in Huntington's disease and Alzheimer-type dementia. *Journal of the Neurological Sciences, 71*, 137–143.

Duyao, M. P., Auerbach, A. B., Ryan, A., Persichetti, F., Barnes, G. T., McNeil, S. M., et al. (1995). Inactivation of the mouse Huntington's disease gene homolog Hdh. *Science, 269*, 407–410.

Ehrnhoefer, D. E., Duennwald, M., Markovic, P., Wacker, J. L., Engemann, S., Roark, M., et al. (2006). Green tea (-)-epigallocatechin-gallate modulates early events in huntingtin misfolding and reduces toxicity in Huntington's disease models. *Human Molecular Genetics, 15*, 2743–2751.

Evert, B. O., Araujo, J., Vieira-Saecker, A. M., de Vos, R. A., Harendza, S., Klockgether, T., et al. (2006). Ataxin-3 represses transcription via chromatin binding, interaction with histone deacetylase 3, and histone deacetylation. *The Journal of Neuroscience, 26*, 11474–11486.

Fan, J., Cowan, C. M., Zhang, L. Y., Hayden, M. R., & Raymond, L. A. (2009). Interaction of postsynaptic density protein-95 with NMDA receptors influences excitotoxicity in the yeast artificial chromosome mouse model of Huntington's disease. *The Journal of Neuroscience, 29*, 10928–10938.

Fan, M. M., Fernandes, H. B., Zhang, L. Y., Hayden, M. R., & Raymond, L. A. (2007). Altered NMDA receptor trafficking in a yeast artificial chromosome transgenic mouse model of Huntington's disease. *The Journal of Neuroscience, 27*, 3768–3779.

Filla, A., De Michele, G., Santoro, L., Calabrese, O., Castaldo, I., Giuffrida, S., et al. (1999). Spinocerebellar ataxia type 2 in southern Italy: A clinical and molecular study of 30 families. *Journal of Neurology, 246*, 467–471.

Furuichi, T., Kohda, K., Miyawaki, A., & Mikoshiba, K. (1994). Intracellular channels. *Current Opinion in Neurobiology, 4*, 294–303.

Furuichi, T., Simon-Chazottes, D., Fujino, I., Yamada, N., Hasegawa, M., Miyawaki, A., et al. (1993). Widespread expression of inositol 1,4,5-trisphosphate receptor type 1 gene (IP_3R1) in the mouse central nervous system. *Receptors and Channels, 1*, 11–24.

Gouix, E., Leveille, F., Nicole, O., Melon, C., Had-Aissouni, L., & Buisson, A. (2009). Reverse glial glutamate uptake triggers neuronal cell death through extrasynaptic NMDA receptor activation. *Molecular and Cellular Neurosciences, 40*, 463–473.

Haacke, A., Hartl, F. U., & Breuer, P. (2007). Calpain inhibition is sufficient to suppress aggregation of polyglutamine-expanded ataxin-3. *The Journal of Biological Chemistry, 282*, 18851–18856.

Hantraye, P., Riche, D., Maziere, M., & Isacson, O. (1990). A primate model of Huntington's disease: Behavioral and anatomical studies of unilateral excitotoxic lesions of the caudate-putamen in the baboon. *Experimental Neurology, 108*, 91–104.

Hara, K., Shiga, A., Nozaki, H., Mitsui, J., Takahashi, Y., Ishiguro, H., et al. (2008). Total deletion and a missense mutation of ITPR1 in Japanese SCA15 families. *Neurology, 71*, 547–551.

Hardingham, G. E., Fukunaga, Y., & Bading, H. (2002). Extrasynaptic NMDARs oppose synaptic NMDARs by triggering CREB shut-off and cell death pathways. *Nature Neuroscience, 5*, 405–414.

Harper, S. Q., Staber, P. D., He, X., Eliason, S. L., Martins, I. H., Mao, Q., et al. (2005). RNA interference improves motor and neuropathological abnormalities in a Huntington's disease mouse model. *Proceedings of the National Academy of Sciences of the United States of America, 102*, 5820–5825.

Heiser, V., Engemann, S., Brocker, W., Dunkel, I., Boeddrich, A., Waelter, S., et al. (2002). Identification of benzothiazoles as potential polyglutamine aggregation inhibitors of Huntington's disease by using an automated filter retardation assay. *Proceedings of the National Academy of Sciences of the United States of America, 99*(Suppl. 4), 16400–16406.

Heng, M. Y., Detloff, P. J., Wang, P. L., Tsien, J. Z., & Albin, R. L. (2009). In vivo evidence for NMDA receptor-mediated excitotoxicity in a murine genetic model of Huntington disease. *The Journal of Neuroscience, 29*, 3200–3205.

Hodgson, J. G., Agopyan, N., Gutekunst, C. A., Leavitt, B. R., LePiane, F., Singaraja, R., et al. (1999). A YAC mouse model for Huntington's disease with full-length mutant huntingtin, cytoplasmic toxicity, and selective striatal neurodegeneration. *Neuron, 23*, 181–192.

Hu, J., Matsui, M., Gagnon, K. T., Schwartz, J. C., Gabillet, S., Arar, K., et al. (2009). Allele-specific silencing of mutant huntingtin and ataxin-3 genes by targeting expanded CAG repeats in mRNAs. *Nature Biotechnology, 27*, 478–484.

Huynh, D. P., Figueroa, K., Hoang, N., & Pulst, S. M. (2000). Nuclear localization or inclusion body formation of ataxin-2 are not necessary for SCA2 pathogenesis in mouse or human. *Nature Genetics, 26*, 44–50.

Imbert, G., Saudou, F., Yvert, G., Devys, D., Trottier, Y., Garnier, J. M., et al. (1996). Cloning of the gene for spinocerebellar ataxia 2 reveals a locus with high sensitivity to expanded CAG/glutamine repeats. *Nature Genetics, 14*, 285–291.

Iwaki, A., Kawano, Y., Miura, S., Shibata, H., Matsuse, D., Li, W., et al. (2008). Heterozygous deletion of ITPR1, but not SUMF1, in spinocerebellar ataxia type 16. *Journal of Medical Genetics, 45*, 32–35.

Kaltenbach, L. S., Romero, E., Becklin, R. R., Chettier, R., Bell, R., Phansalkar, A., et al. (2007). Huntingtin interacting proteins are genetic modifiers of neurodegeneration. *PLoS Genetics, 3*, e82.

Kerner, J. A., Standaert, D. G., Penney, J. B., Jr., Young, A. B., & Landwehrmeyer, G. B. (1997). Expression of group one metabotropic glutamate receptor subunit mRNAs in neurochemically identified neurons in the rat neostriatum, neocortex, and hippocampus. *Brain Research. Molecular Brain Research, 48*, 259–269.

Kiehl, T. R., Nechiporuk, A., Figueroa, K. P., Keating, M. T., Huynh, D. P., & Pulst, S. M. (2006). Generation and characterization of Sca2 (ataxin-2) knockout mice. *Biochemical and Biophysical Research Communications, 339*, 17–24.

Kiehl, T. R., Shibata, H., & Pulst, S. M. (2000). The ortholog of human ataxin-2 is essential for early embryonic patterning in *C. elegans*. *Journal of Molecular Neuroscience, 15*, 231–241.

Kim, M. W., Chelliah, Y., Kim, S. W., Otwinowski, Z., & Bezprozvanny, I. (2009). Secondary structure of Huntingtin amino-terminal region. *Structure, 17*, 1205–1212.

Landwehrmeyer, G. B., Dubois, B., de Yebenes, J. G., Kremer, B., Gaus, W., Kraus, P. H., et al. (2007). Riluzole in Huntington's disease: A 3-year, randomized controlled study. *Annals of Neurology, 62*, 262–272.

Landwehrmeyer, G. B., Standaert, D. G., Testa, C. M., Penney, J. B., Jr., & Young, A. B. (1995). NMDA receptor subunit mRNA expression by projection neurons and interneurons in rat striatum. *The Journal of Neuroscience, 15*, 5297–5307.

Langbehn, D. R., Brinkman, R. R., Falush, D., Paulsen, J. S., & Hayden, M. R. (2004). A new model for prediction of the age of onset and penetrance for Huntington's disease based on CAG length. *Clinical Genetics, 65*, 267–277.

Lastres-Becker, I., Rub, U., & Auburger, G. (2008). Spinocerebellar ataxia 2 (SCA2). *Cerebellum, 7*, 115–124.

Lee, S. T., Chu, K., Park, J. E., Kang, L., Ko, S. Y., Jung, K. H., et al. (2006). Memantine reduces striatal cell death with decreasing calpain level in 3-nitropropionic model of Huntington's disease. *Brain Research, 1118*, 199–207.

Li, L., Murphy, T. H., Hayden, M. R., & Raymond, L. A. (2004). Enhanced striatal NR2B-containing N-methyl-D-aspartate receptor-mediated synaptic currents in a mouse model of Huntington disease. *Journal of Neurophysiology, 92*, 2738–2746.

Li, S. H., Schilling, G., Young, W. S., 3rd, Li, X. J., Margolis, R. L., Stine, O. C., et al. (1993). Huntington's disease gene (IT15) is widely expressed in human and rat tissues. *Neuron, 11*, 985–993.

Li, J. H., Wang, Y. H., Wolfe, B. B., Krueger, K. E., Corsi, L., Stocca, G., et al. (1998). Developmental changes in localization of NMDA receptor subunits in primary cultures of cortical neurons. *The European Journal of Neuroscience, 10*, 1704–1715.

Liu, J., Tang, T. S., Tu, H., Nelson, O., Herndon, E., Huynh, D. P., et al. (2009). Deranged calcium signaling and neurodegeneration in spinocerebellar ataxia type 2. *The Journal of Neuroscience, 29*, 9148–9162.

MacDonald, M. E. (2003). Huntingtin: Alive and well and working in middle management. *Science STKE, 2003*, pe48.

Mao, L., & Wang, J. Q. (2001). Upregulation of preprodynorphin and preproenkephalin mRNA expression by selective activation of group I metabotropic glutamate receptors in characterized primary cultures of rat striatal neurons. *Brain Research. Molecular Brain Research, 86*, 125–137.

Mao, L., & Wang, J. Q. (2002). Glutamate cascade to cAMP response element-binding protein phosphorylation in cultured striatal neurons through calcium-coupled group I metabotropic glutamate receptors. *Molecular Pharmacology, 62*, 473–484.

Maruyama, T., Kanaji, T., Nakade, S., Kanno, T., & Mikoshiba, K. (1997). 2APB, 2-aminoethoxydiphenyl borate, a membrane-penetrable modulator of Ins(1,4,5)P3-induced Ca^{2+} release. *Journal of Biochemistry (Tokyo), 122*, 498–505.

Matilla-Duenas, A., Sanchez, I., Corral-Juan, M., Davalos, A., Alvarez, R., & Latorre, P. (2009). Cellular and molecular pathways triggering neurodegeneration in the spinocerebellar ataxias. *Cerebellum, 9*, 148–166.

Matsumoto, M., Nakagawa, T., Inoue, T., Nagata, E., Tanaka, K., Takano, H., et al. (1996). Ataxia and epileptic seizures in mice lacking type 1 inositol 1,4,5-trisphosphate receptor. *Nature, 379*, 168–171.

McGeer, E. G., & McGeer, P. L. (1976). Duplication of biochemical changes of Huntington's chorea by intrastriatal injections of glutamic and kainic acids. *Nature, 263*, 517–519.

McGeer, E. G., McGeer, P. L., & Singh, K. (1978). Kainate-induced degeneration of neostriatal neurons: Dependency upon corticostriatal tract. *Brain Research, 139*, 381–383.

Menalled, L. B., & Chesselet, M. F. (2002). Mouse models of Huntington's disease. *Trends in Pharmacological Sciences, 23*, 32–39.

Milnerwood, A. J., Gladding, C. M., Pouladi, M. A., Kaufman, A. M., Hines, R. M., Boyd, J. D., et al. (2010). Early increase in extrasynaptic NMDA receptor signaling and expression contributes to phenotype onset in Huntington's disease mice. *Neuron, 65*, 178–190.

Milnerwood, A. J., & Raymond, L. A. (2006). Synaptic abnormalities associated with Huntington's disease. In A. El-Husseini & A. Dityatev (Eds.), *In Molecular Mechanisms of Synaotogenesis* (pp. 457–469). New York: Springer.

Monyer, H., Burnashev, N., Laurie, D. J., Sakmann, B., & Seeburg, P. H. (1994). Developmental and regional expression in the rat brain and functional properties of four NMDA receptors. *Neuron, 12*, 529–540.

Nagai, Y., Fujikake, N., Ohno, K., Higashiyama, H., Popiel, H. A., Rahadian, J., et al. (2003). Prevention of polyglutamine oligomerization and neurodegeneration by the peptide inhibitor QBP1 in *Drosophila. Human Molecular Genetics, 12*, 1253–1259.

Nagai, Y., Tucker, T., Ren, H., Kenan, D. J., Henderson, B. S., Keene, J. D., et al. (2000). Inhibition of polyglutamine protein aggregation and cell death by novel peptides identified by phage display screening. *The Journal of Biological Chemistry, 275*, 10437–10442.

Nasir, J., Floresco, S. B., O'Kusky, J. R., Diewert, V. M., Richman, J. M., Zeisler, J., et al. (1995). Targeted disruption of the Huntington's disease gene results in embryonic lethality and behavioral and morphological changes in heterozygotes. *Cell, 81*, 811–823.

Okamoto, S., Pouladi, M. A., Talantova, M., Yao, D., Xia, P., Ehrnhoefer, D. E., et al. (2009). Balance between synaptic versus extrasynaptic NMDA receptor activity influences inclusions and neurotoxicity of mutant huntingtin. *Natural Medicines, 15*, 1407–1413.

Ondo, W. G., Mejia, N. I., & Hunter, C. B. (2007). A pilot study of the clinical efficacy and safety of memantine for Huntington's disease. *Parkinsonism and Related Disorders, 13*, 453–454.

Papadia, S., & Hardingham, G. E. (2007). The dichotomy of NMDA receptor signaling. *The Neuroscientist, 13*, 572–579.

Papadia, S., Soriano, F. X., Leveille, F., Martel, M. A., Dakin, K. A., Hansen, H. H., et al. (2008). Synaptic NMDA receptor activity boosts intrinsic antioxidant defenses. *Nature Neuroscience, 11*, 476–487.

Paulson, H. L. (2009). The spinocerebellar ataxias. *Journal of Neuroophthalmology, 29*, 227–237.

Paulson, H. L., Das, S. S., Crino, P. B., Perez, M. K., Patel, S. C., Gotsdiner, D., et al. (1997). Machado-Joseph disease gene product is a cytoplasmic protein widely expressed in brain. *Annals of Neurology, 41*, 453–462.

Pirker, W., Back, C., Gerschlager, W., Laccone, F., & Alesch, F. (2003). Chronic thalamic stimulation in a patient with spinocerebellar ataxia type 2. *Movement Disorders, 18*, 222–225.

Popiel, H. A., Nagai, Y., Fujikake, N., & Toda, T. (2007). Protein transduction domain-mediated delivery of QBP1 suppresses polyglutamine-induced neurodegeneration in vivo. *Molecular Therapy, 15*, 303–309.

Popiel, H. A., Nagai, Y., Fujikake, N., & Toda, T. (2009). Delivery of the aggregate inhibitor peptide QBP1 into the mouse brain using PTDs and its therapeutic effect on polyglutamine disease mice. *Neuroscience Letters, 449*, 87–92.

Portera-Cailliau, C., Price, D. L., & Martin, L. J. (1996). N-methyl-D-aspartate receptor proteins NR2A and NR2B are differentially distributed in the developing rat central nervous system as revealed by subunit-specific antibodies. *Journal of Neurochemistry, 66*, 692–700.

Pulst, S. M., Nechiporuk, A., Nechiporuk, T., Gispert, S., Chen, X. N., Lopes-Cendes, I., et al. (1996). Moderate expansion of a normally biallelic trinucleotide repeat in spinocerebellar ataxia type 2. *Nature Genetics, 14*, 269–276.

Pulst, S. M., Santos, N., Wang, D., Yang, H., Huynh, D., Velazquez, L., et al. (2005). Spinocerebellar ataxia type 2: PolyQ repeat variation in the CACNA1A calcium channel modifies age of onset. *Brain, 128*, 2297–2303.

Rubinsztein, D. C. (2002). Lessons from animal models of Huntington's disease. *Trends in Genetics, 18*, 202–209.

Sanchez, I., Mahlke, C., & Yuan, J. (2003). Pivotal role of oligomerization in expanded polyglutamine neurodegenerative disorders. *Nature, 421*, 373–379.

Sanpei, K., Takano, H., Igarashi, S., Sato, T., Oyake, M., Sasaki, H., et al. (1996). Identification of the spinocerebellar ataxia type 2 gene using a direct identification of repeat expansion and cloning technique, DIRECT. *Nature Genetics, 14*, 277–284.

Satterfield, T. F., Jackson, S. M., & Pallanck, L. J. (2002). A *Drosophila* homolog of the polyglutamine disease gene SCA2 is a dosage-sensitive regulator of actin filament formation. *Genetics, 162*, 1687–1702.

Scheel, H., Tomiuk, S., & Hofmann, K. (2003). Elucidation of ataxin-3 and ataxin-7 function by integrative bioinformatics. *Human Molecular Genetics, 12*, 2845–2852.

Schmitt, I., Linden, M., Khazneh, H., Evert, B. O., Breuer, P., Klockgether, T., et al. (2007). Inactivation of the mouse Atxn3 (ataxin-3) gene increases protein ubiquitination. *Biochemical and Biophysical Research Communications, 362*, 734–739.

Schols, L., Bauer, P., Schmidt, T., Schulte, T., & Riess, O. (2004). Autosomal dominant cerebellar ataxias: Clinical features, genetics, and pathogenesis. *Lancet Neurology, 3*, 291–304.

Shao, J., & Diamond, M. I. (2007). Polyglutamine diseases: Emerging concepts in pathogenesis and therapy. *Human Molecular Genetics, 16*, R115–R123. Spec No. 2.

Sharp, A. H., Loev, S. J., Schilling, G., Li, S. H., Li, X. J., Bao, J., et al. (1995). Widespread expression of Huntington's disease gene (IT15) protein product. *Neuron, 14*, 1065–1074.

Sharp, A. H., Nucifora, F. C., Jr., Blondel, O., Sheppard, C. A., Zhang, C., Snyder, S. H., et al. (1999). Differential cellular expression of isoforms of inositol 1,4,5-triphosphate receptors in neurons and glia in brain. *The Journal of Comparative Neurology, 406*, 207–220.

Slow, E. J., van Raamsdonk, J., Rogers, D., Coleman, S. H., Graham, R. K., Deng, Y., et al. (2003). Selective striatal neuronal loss in a YAC128 mouse model of Huntington disease. *Human Molecular Genetics, 12*, 1555–1567.

Song, C., Zhang, Y., Parsons, C. G., & Liu, Y. F. (2003). Expression of polyglutamine-expanded huntingtin induces tyrosine phosphorylation of N-methyl-D-aspartate receptors. *The Journal of Biological Chemistry, 278*, 33364–33369.

Stevanin, G., Durr, A., & Brice, A. (2000). Clinical and molecular advances in autosomal dominant cerebellar ataxias: From genotype to phenotype and physiopathology. *European Journal of Human Genetics, 8*, 4–18.

Stocca, G., & Vicini, S. (1998). Increased contribution of NR2A subunit to synaptic NMDA receptors in developing rat cortical neurons. *Journal de Physiologie, 507*(Pt 1), 13–24.

Street, V. A., Bosma, M. M., Demas, V. P., Regan, M. R., Lin, D. D., Robinson, L. C., et al. (1997). The type 1 inositol 1, 4, 5-trisphosphate receptor gene is altered in the opisthotonos mouse. *The Journal of Neuroscience, 17*, 635–645.

Strong, T. V., Tagle, D. A., Valdes, J. M., Elmer, L. W., Boehm, K., Swaroop, M., et al. (1993). Widespread expression of the human and rat Huntington's disease gene in brain and nonneural tissues. *Nature Genetics, 5*, 259–265.

Sun, Y., Savanenin, A., Reddy, P. H., & Liu, Y. F. (2001). Polyglutamine-expanded huntingtin promotes sensitization of N-methyl-D- aspartate receptors via post-synaptic density 95. *The Journal of Biological Chemistry, 276*, 24713–24718.

Tallaksen-Greene, S. J., Kaatz, K. W., Romano, C., & Albin, R. L. (1998). Localization of mGluR1a-like immunoreactivity and mGluR5-like immunoreactivity in identified populations of striatal neurons. *Brain Research, 780*, 210–217.

Tang, T. S., Guo, C., Wang, H., Chen, X., & Bezprozvanny, I. (2009). Neuroprotective effects of inositol 1,4,5-trisphosphate receptor C-terminal fragment in a Huntington's disease mouse model. *The Journal of Neuroscience, 29*, 1257–1266.

Tang, T. S., Slow, E., Lupu, V., Stavrovskaya, I. G., Sugimori, M., Llinas, R., et al. (2005). Disturbed Ca^{2+} signaling and apoptosis of medium spiny neurons in Huntington's disease. *Proceedings of the National Academy of Sciences of the United States of America, 102*, 2602–2607.

Tang, T. S., Tu, H., Chan, E. Y., Maximov, A., Wang, Z., Wellington, C. L., et al. (2003). Huntingtin and huntingtin-associated protein 1 influence neuronal calcium signaling mediated by inositol-(1,4,5) triphosphate receptor type 1. *Neuron, 39*, 227–239.

Testa, C. M., Standaert, D. G., Landwehrmeyer, G. B., Penney, J. B., Jr., & Young, A. B. (1995). Differential expression of mGluR5 metabotropic glutamate receptor mRNA by rat striatal neurons. *The Journal of Comparative Neurology, 354*, 241–252.

The Huntington's Disease Collaborative Research Group. (1993). A novel gene containing a trinucleotide repeat that is expanded and unstable on Huntington's disease chromosomes. *Cell, 72*, 971–983.

Tovar, K. R., & Westbrook, G. L. (1999). The incorporation of NMDA receptors with a distinct subunit composition at nascent hippocampal synapses in vitro. *The Journal of Neuroscience, 19*, 4180–4188.
van de Leemput, J., Chandran, J., Knight, M. A., Holtzclaw, L. A., Scholz, S., Cookson, M. R., et al. (2007). Deletion at ITPR1 underlies ataxia in mice and spinocerebellar ataxia 15 in humans. *PLoS Genetics, 3*, e108.
van de Loo, S., Eich, F., Nonis, D., Auburger, G., & Nowock, J. (2009). Ataxin-2 associates with rough endoplasmic reticulum. *Experimental Neurology, 215*, 110–118.
Vanhoutte, P., & Bading, H. (2003). Opposing roles of synaptic and extrasynaptic NMDA receptors in neuronal calcium signalling and BDNF gene regulation. *Current Opinion in Neurobiology, 13*, 366–371.
Vonsattel, J. P., & DiFiglia, M. (1998). Huntington disease. *Journal of Neuropathology and Experimental Neurology, 57*, 369–384.
Vonsattel, J. P., Myers, R. H., Stevens, T. J., Ferrante, R. J., Bird, E. D., & Richardson, E. P. Jr. (1985). Neuropathological classification of Huntington's disease. *Journal of Neuropathology and Experimental Neurology, 44*, 559–577.
Wacker, J. L., Zareie, M. H., Fong, H., Sarikaya, M., & Muchowski, P. J. (2004). Hsp70 and Hsp40 attenuate formation of spherical and annular polyglutamine oligomers by partitioning monomer. *Nature Structural and Molecular Biology, 11*, 1215–1222.
Wang, J., Gines, S., MacDonald, M. E., & Gusella, J. F. (2005). Reversal of a full-length mutant huntingtin neuronal cell phenotype by chemical inhibitors of polyglutamine-mediated aggregation. *BMC Neuroscience, 6*, 1.
Warrick, J. M., Chan, H. Y., Gray-Board, G. L., Chai, Y., Paulson, H. L., & Bonini, N. M. (1999). Suppression of polyglutamine-mediated neurodegeneration in *Drosophila* by the molecular chaperone HSP70. *Nature Genetics, 23*, 425–428.
Winborn, B. J., Travis, S. M., Todi, S. V., Scaglione, K. M., Xu, P., Williams, A. J., et al. (2008). The deubiquitinating enzyme ataxin-3, a polyglutamine disease protein, edits Lys63 linkages in mixed linkage ubiquitin chains. *The Journal of Biological Chemistry, 283*, 26436–26443.
Wu, J., Tang, T., & Bezprozvanny, I. (2006). Evaluation of clinically relevant glutamate pathway inhibitors in in vitro model of Huntington's disease. *Neuroscience Letters, 407*, 219–223.
Wyttenbach, A., Sauvageot, O., Carmichael, J., Diaz-Latoud, C., Arrigo, A. P., & Rubinsztein, D. C. (2002). Heat shock protein 27 prevents cellular polyglutamine toxicity and suppresses the increase of reactive oxygen species caused by huntingtin. *Human Molecular Genetics, 11*, 1137–1151.
Xia, H., Mao, Q., Eliason, S. L., Harper, S. Q., Martins, I. H., Orr, H. T., et al. (2004). RNAi suppresses polyglutamine-induced neurodegeneration in a model of spinocerebellar ataxia. *Natural Medicines, 10*, 816–820.
Zeron, M. M., Chen, N., Moshaver, A., Lee, A. T., Wellington, C. L., Hayden, M. R., et al. (2001). Mutant huntingtin enhances excitotoxic cell death. *Molecular and Cellular Neurosciences, 17*, 41–53.
Zeron, M. M., Hansson, O., Chen, N., Wellington, C. L., Leavitt, B. R., Brundin, P., et al. (2002). Increased sensitivity to N-methyl-D-aspartate receptor-mediated excitotoxicity in a mouse model of Huntington's disease. *Neuron, 33*, 849–860.
Zhang, H., Li, Q., Graham, R. K., Slow, E., Hayden, M. R., & Bezprozvanny, I. (2008). Full length mutant huntingtin is required for altered Ca^{2+} signaling and apoptosis of striatal neurons in the YAC mouse model of Huntington's disease. *Neurobiology of Disease, 31*, 80–88.
Zhang, X., Smith, D. L., Meriin, A. B., Engemann, S., Russel, D. E., Roark, M., et al. (2005). A potent small molecule inhibits polyglutamine aggregation in Huntington's disease neurons and suppresses neurodegeneration in vivo. *Proceedings of the National Academy of Sciences of the United States of America, 102*, 892–897.

Index

J

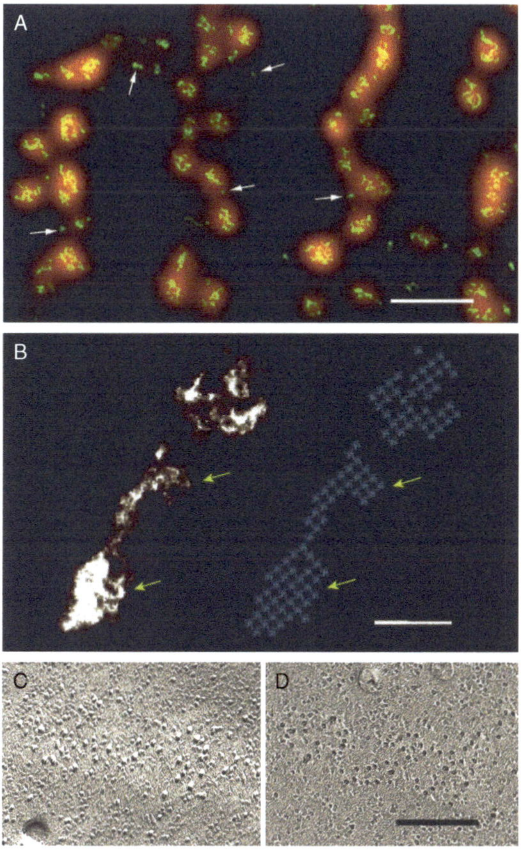

Please refer to Fig. 1.1 in text for figure legend.

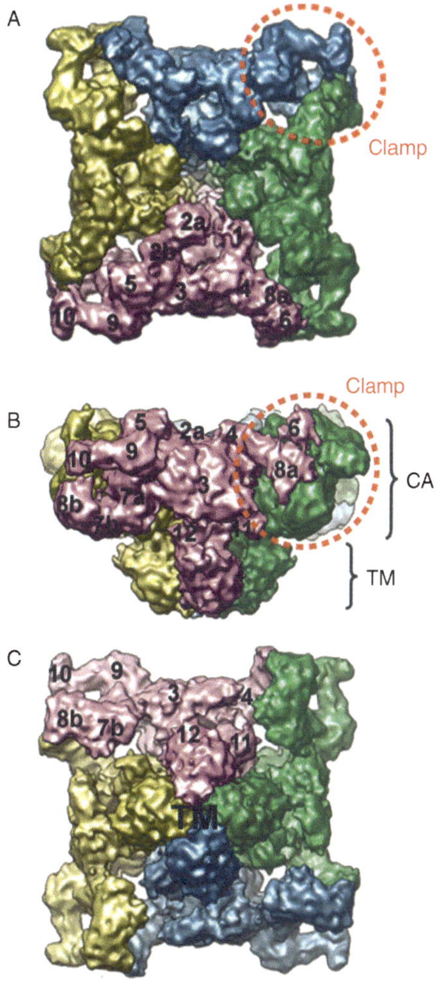

Please refer to Fig. 2.1 in text for figure legend.

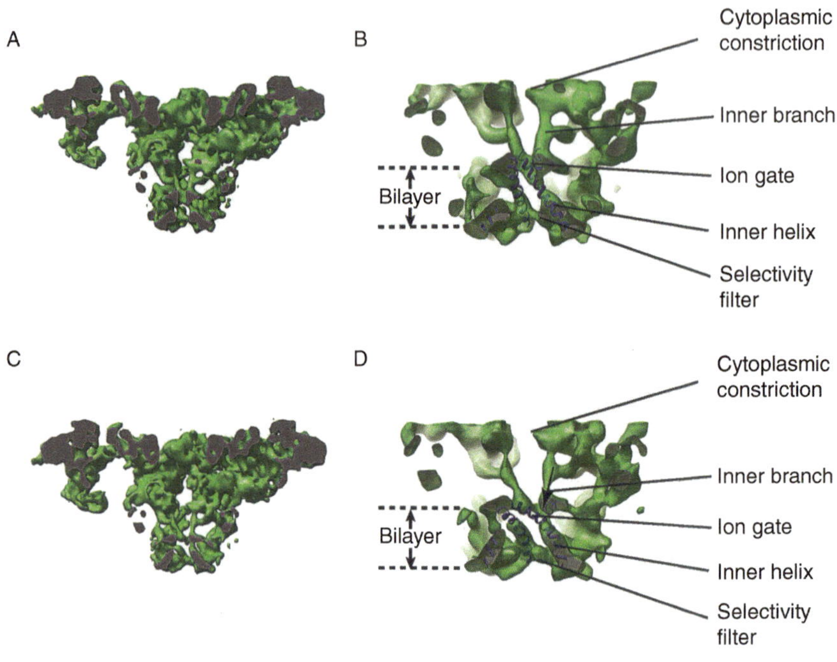

Please refer to Fig. 2.2 in text for figure legend.

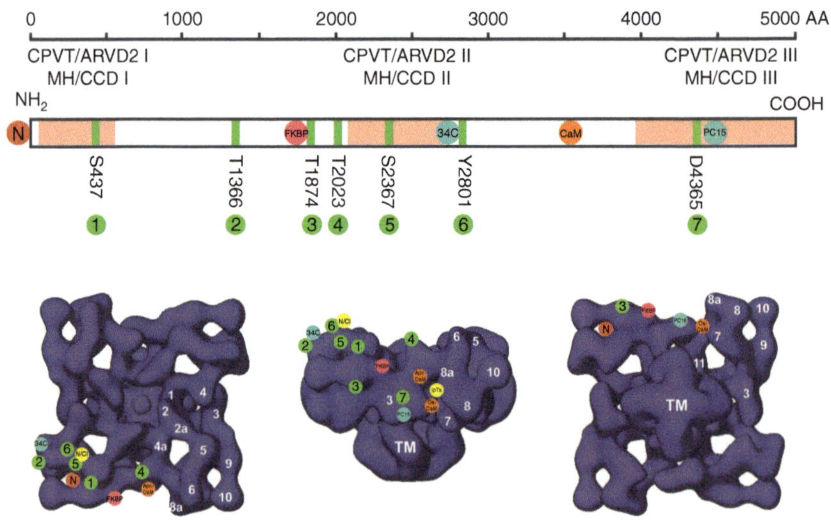

Please refer to Fig. 2.3 in text for figure legend.

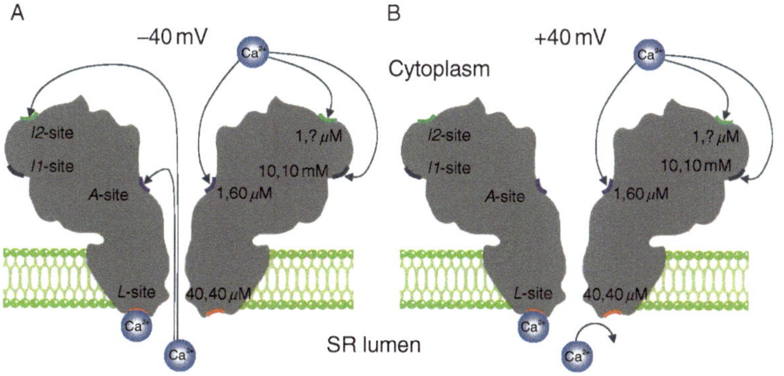

Please refer to Fig. 4.1 in text for figure legend.

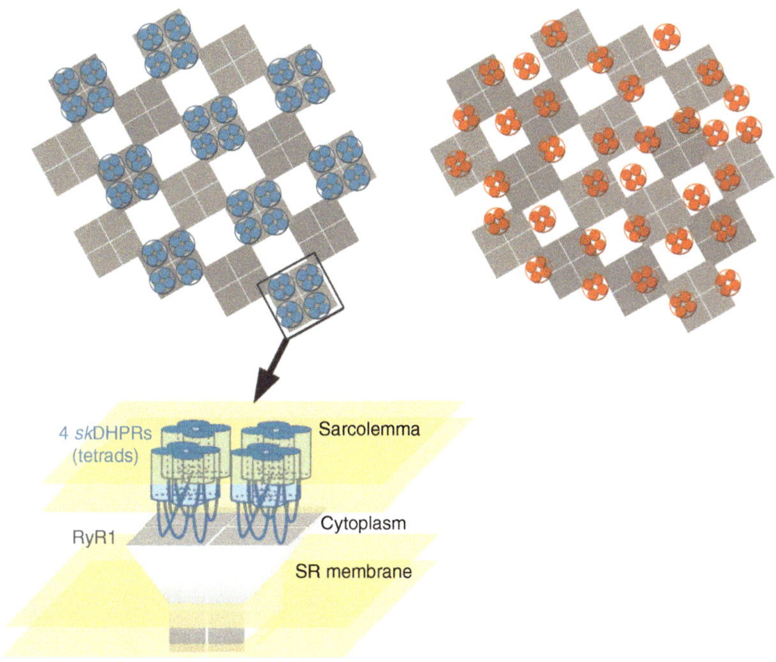

Please refer to Fig. 6.1 in text for figure legend.

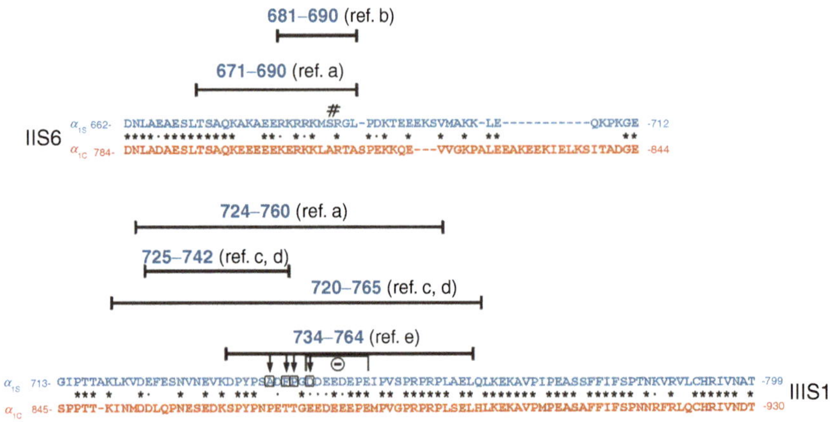

Please refer to Fig. 6.2 in text for figure legend.

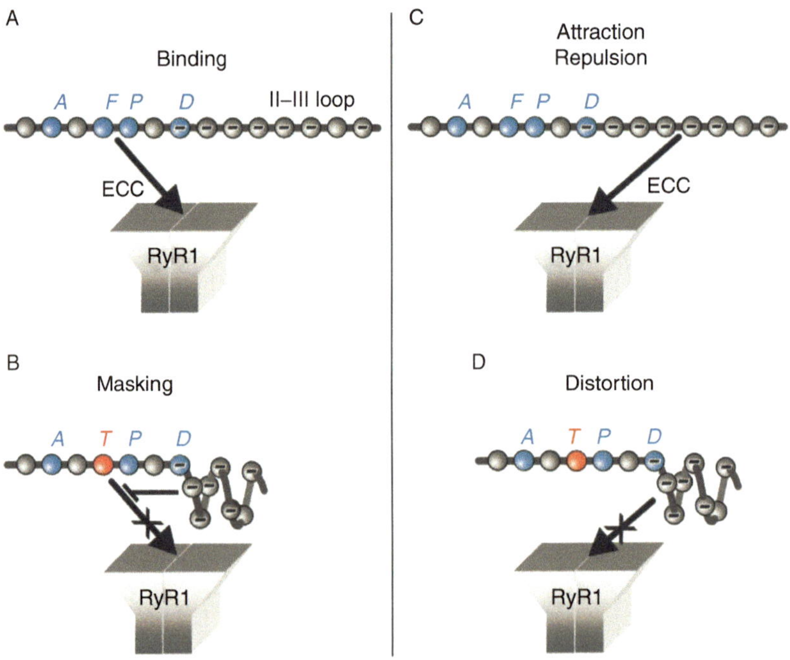

Please refer to Fig. 6.3 in text for figure legend.

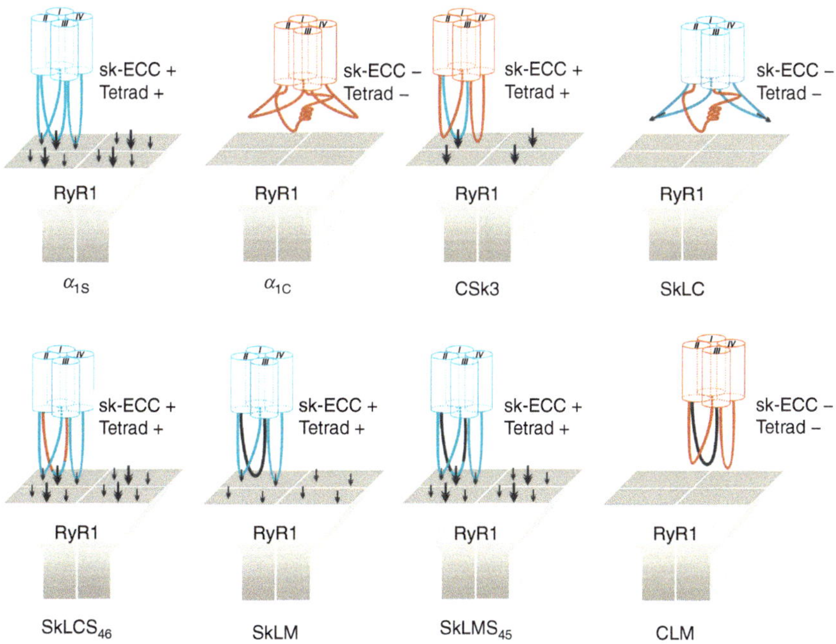

Please refer to Fig. 6.5 in text for figure legend.

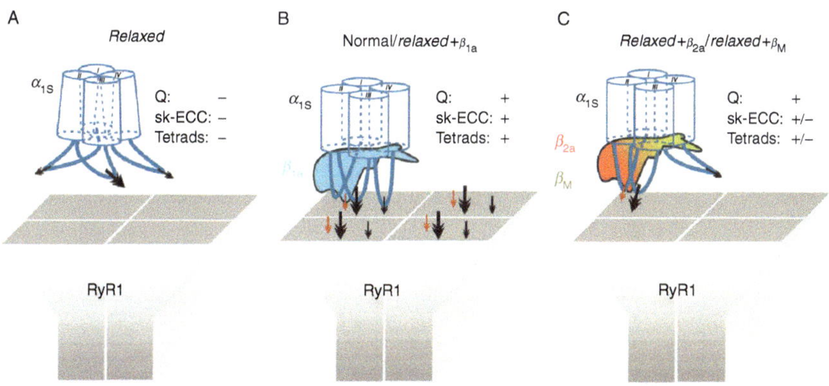

Please refer to Fig. 6.6 in text for figure legend.

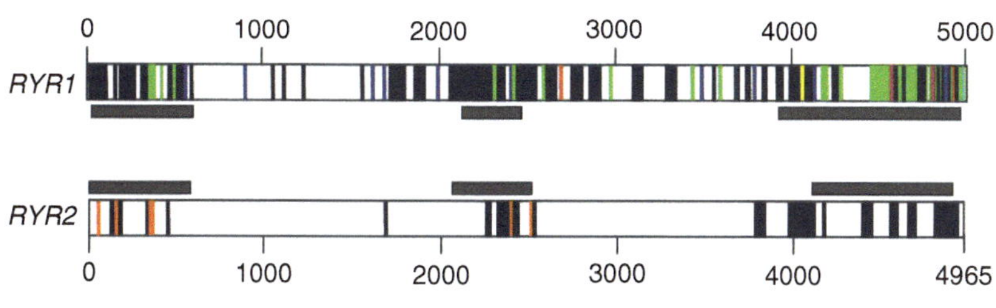

Please refer to Fig. 7.1 in text for figure legend.

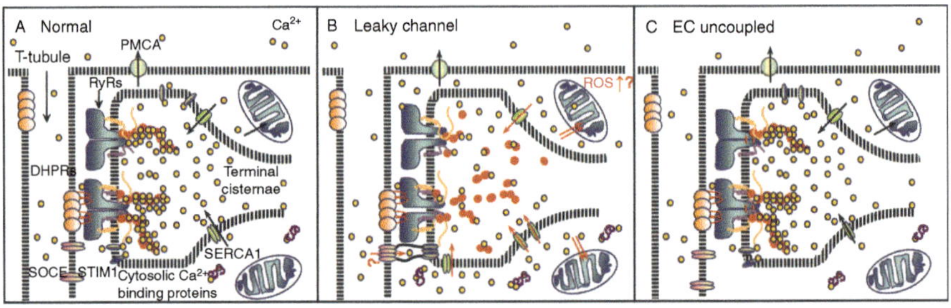

Please refer to Fig. 7.2 in text for figure legend.

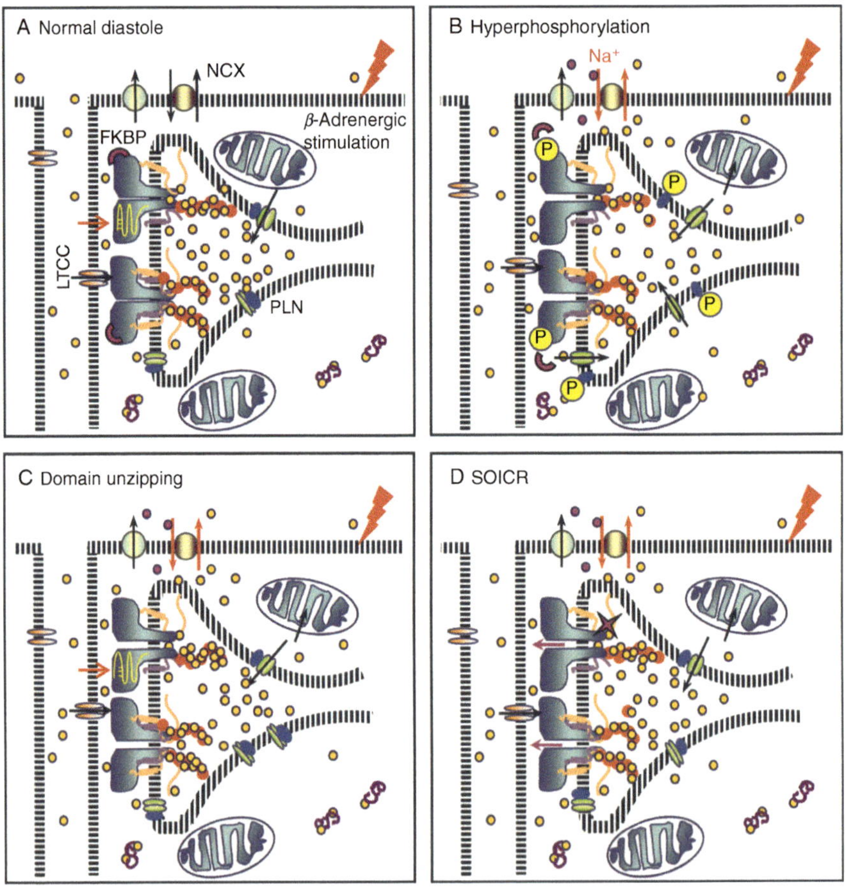

Please refer to Fig. 7.3 in text for figure legend.

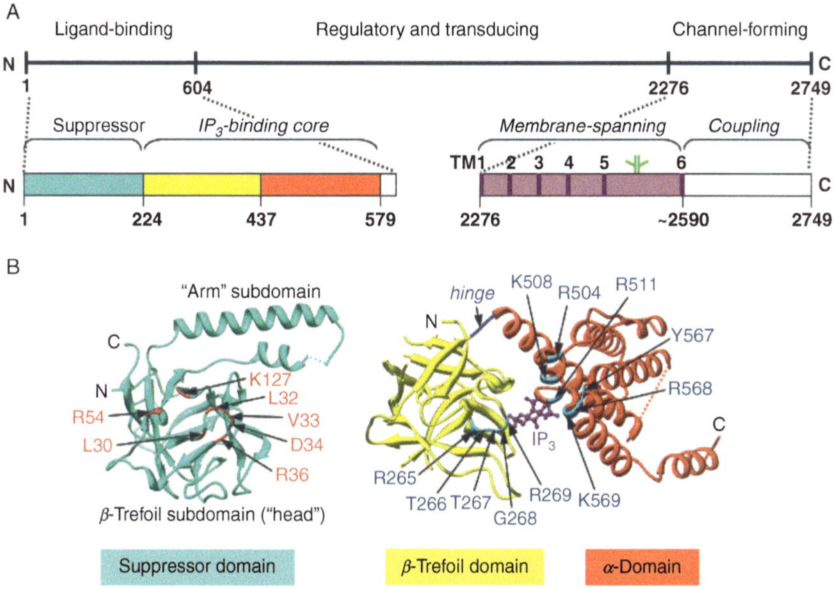

Please refer to Fig. 8.1 in text for figure legend.

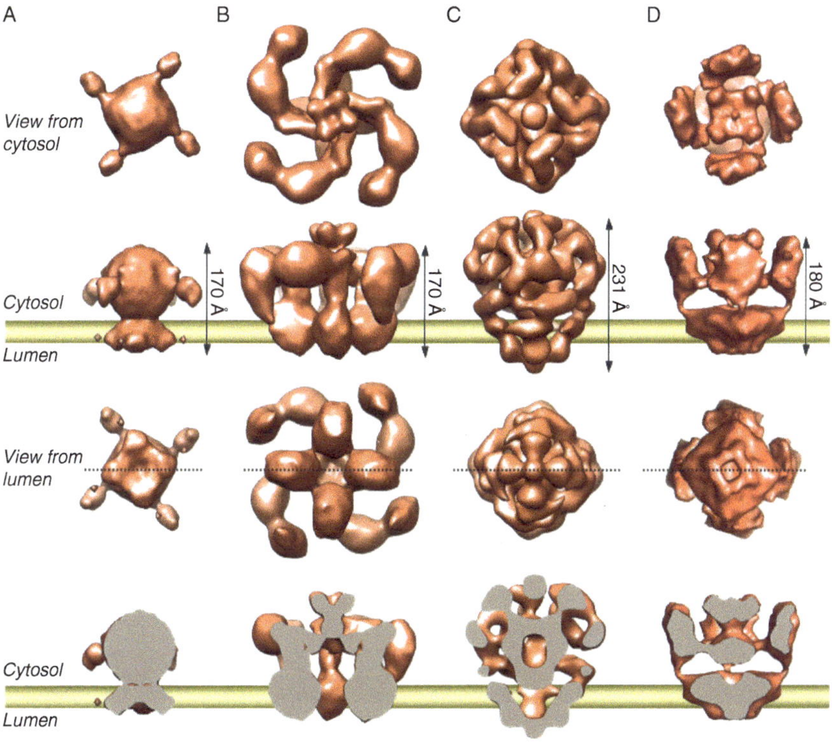

Please refer to Fig. 8.3 in text for figure legend.

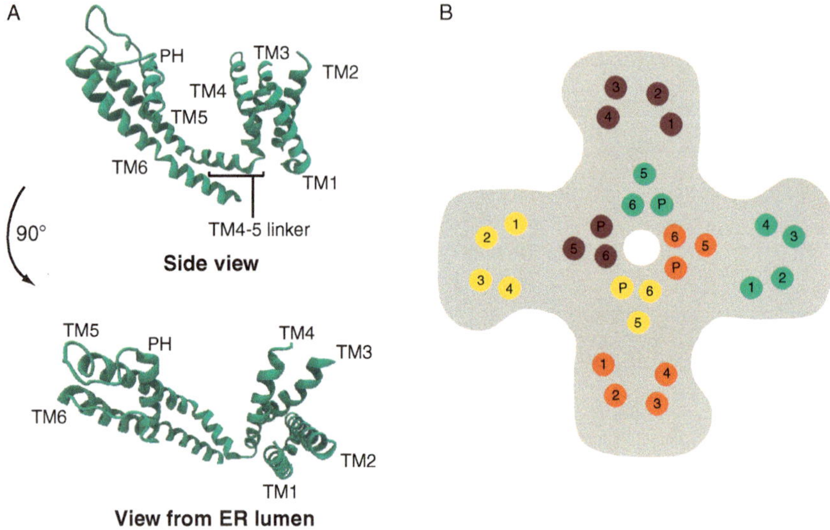

Please refer to Fig. 9.1 in text for figure legend.

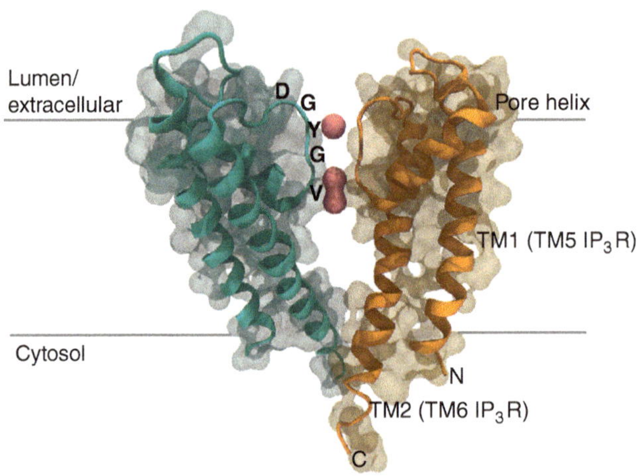

Please refer to Fig. 9.2 in text for figure legend.

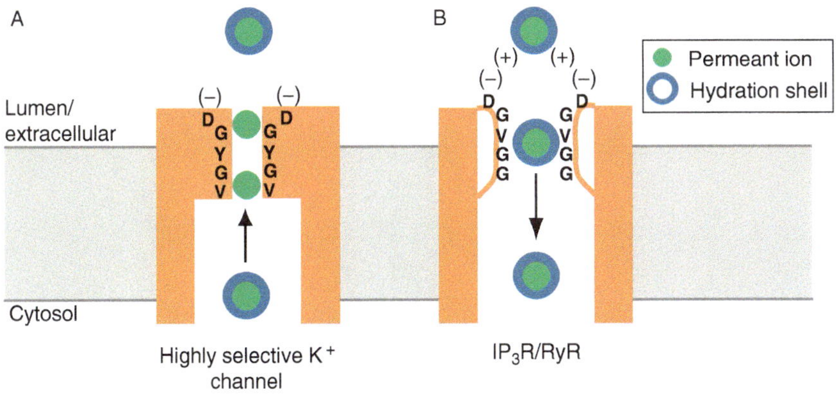

Please refer to Fig. 9.3 in text for figure legend.

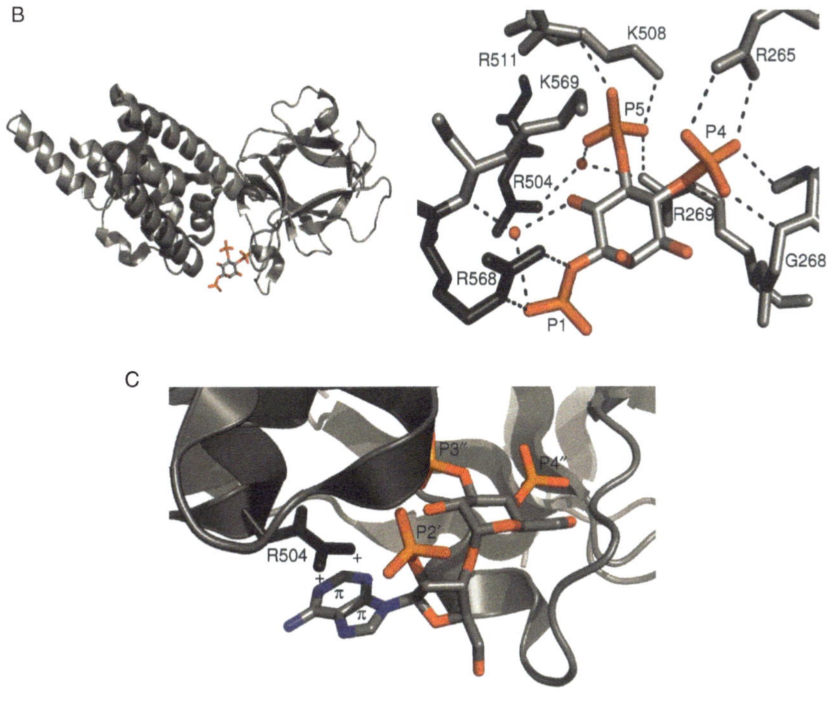

Please refer to Fig. 10.5 in text for figure legend.

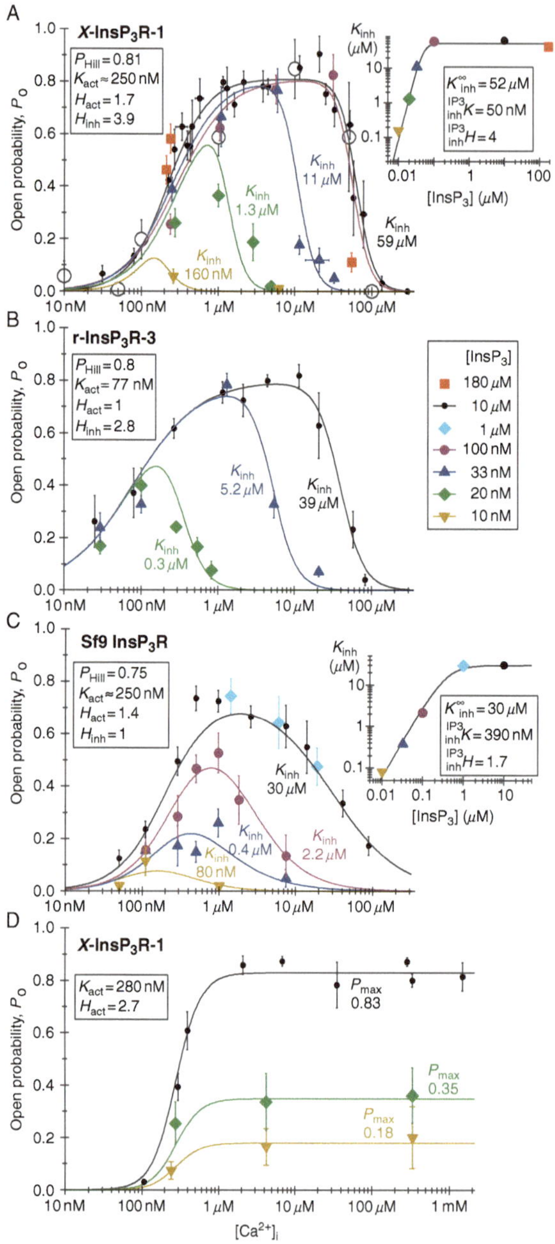

Please refer to Fig. 11.2 in text for figure legend.

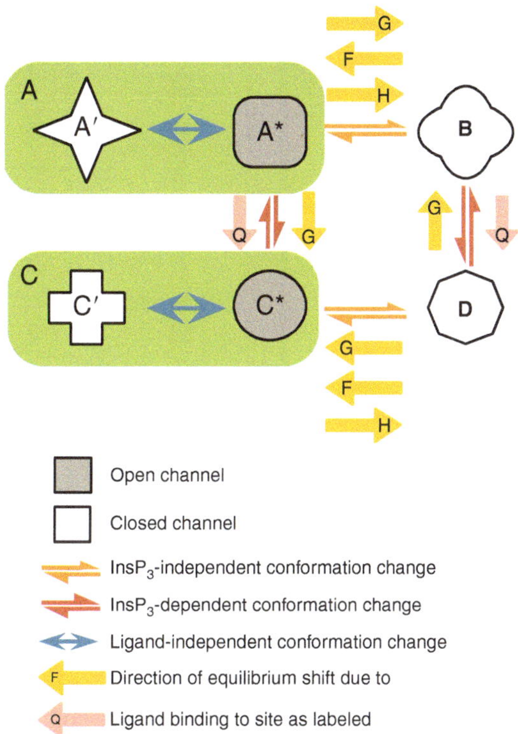

Please refer to Fig. 11.3 in text for figure legend.

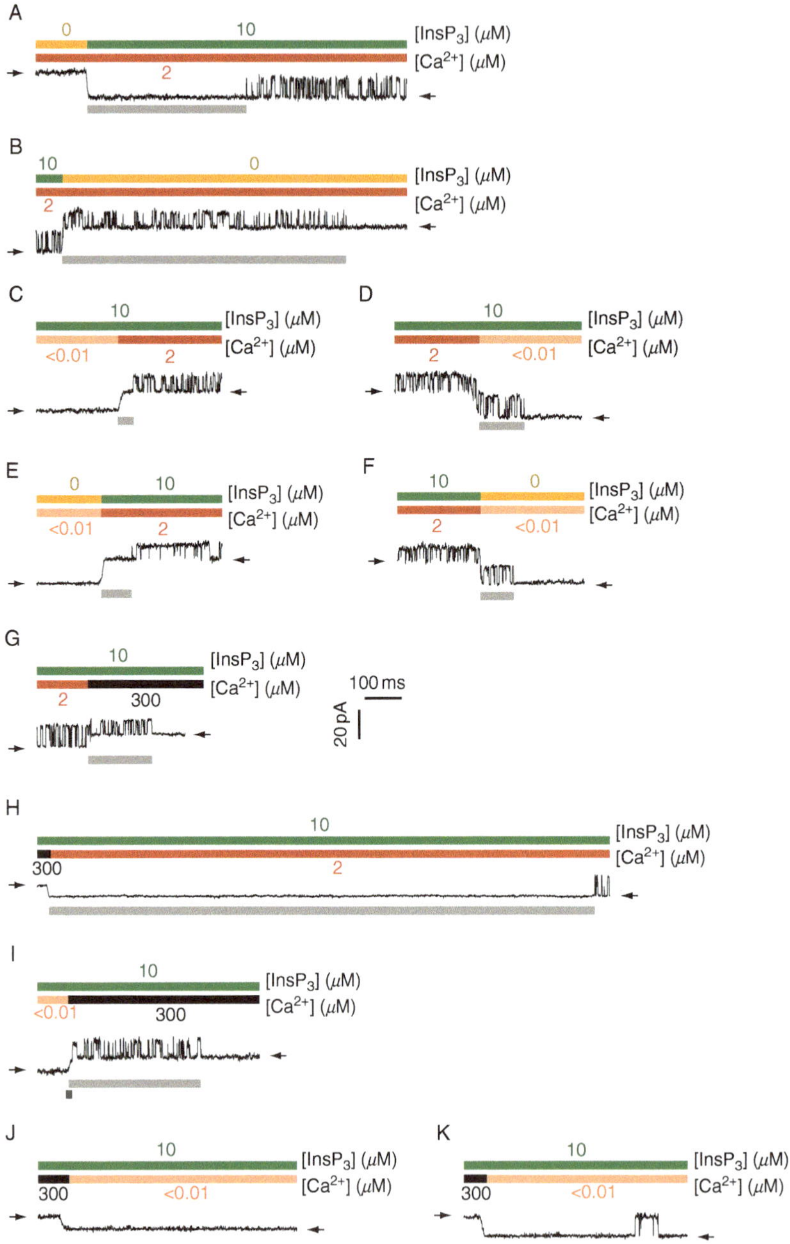

Please refer to Fig. 11.4 in text for figure legend.

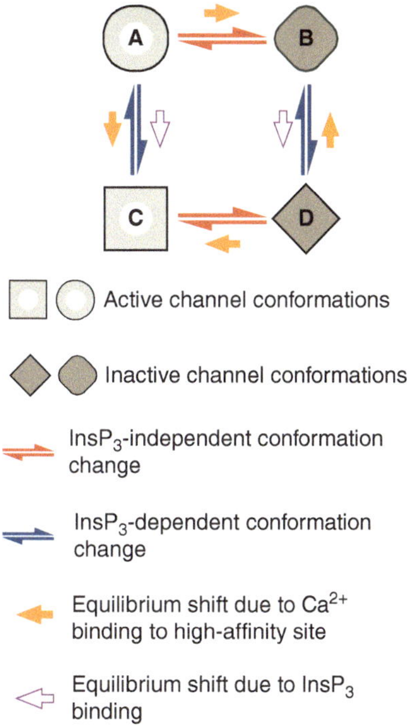

Please refer to Fig. 11.5 in text for figure legend.

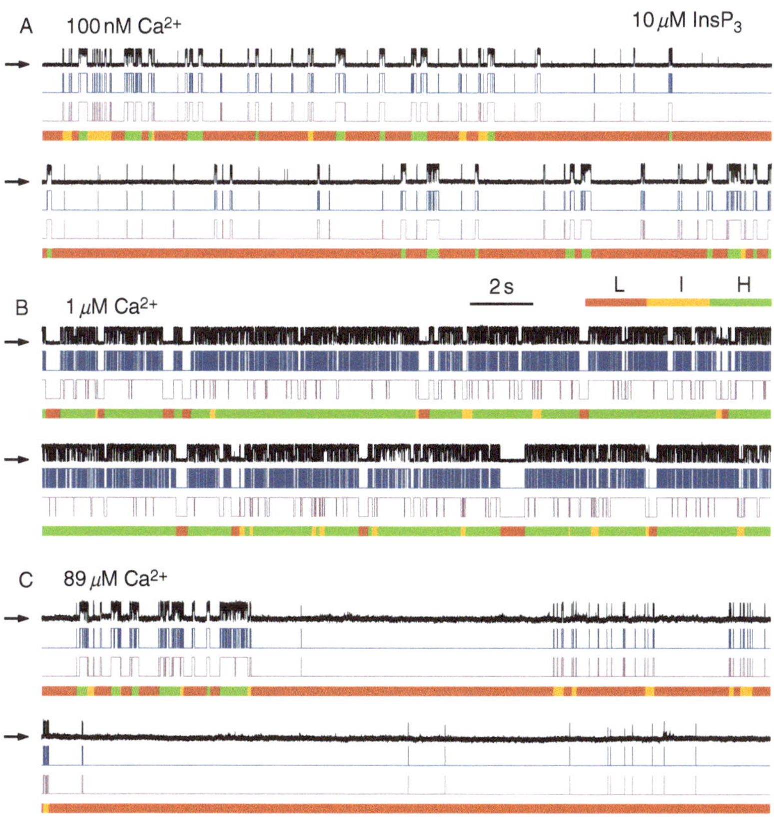

Please refer to Fig. 11.6 in text for figure legend.

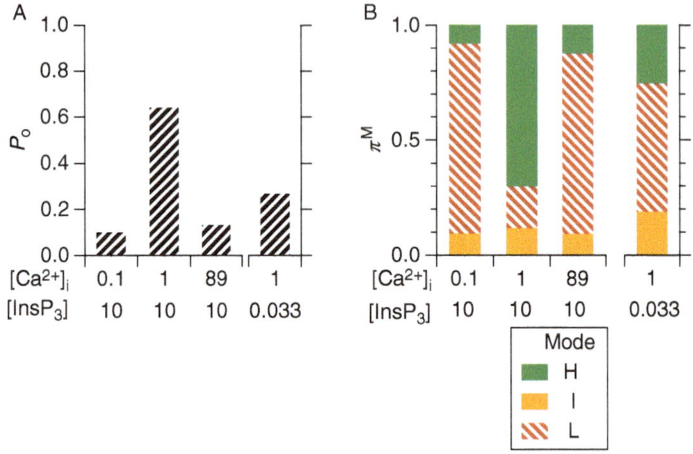

Please refer to Fig. 11.7 in text for figure legend.

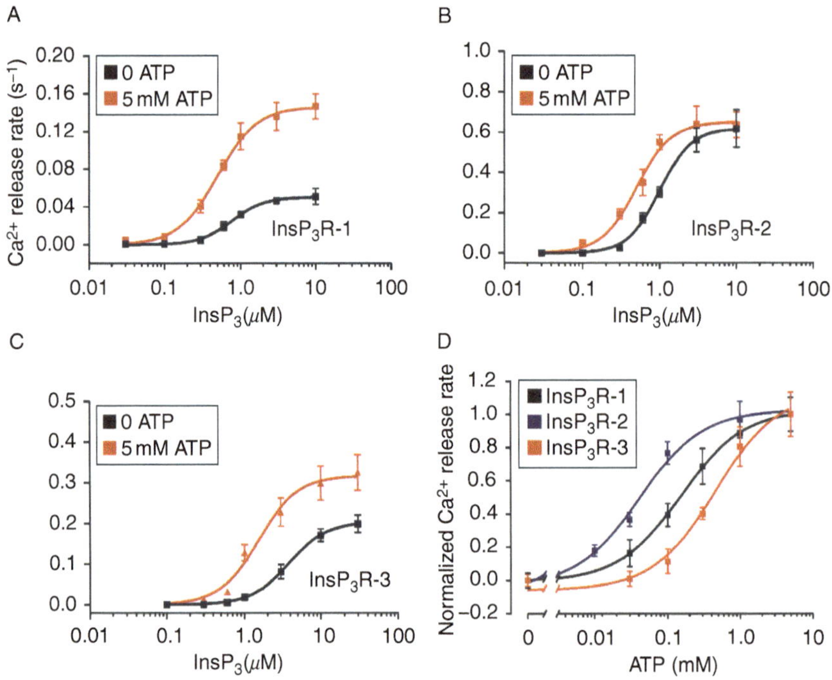

Please refer to Fig. 12.2 in text for figure legend.

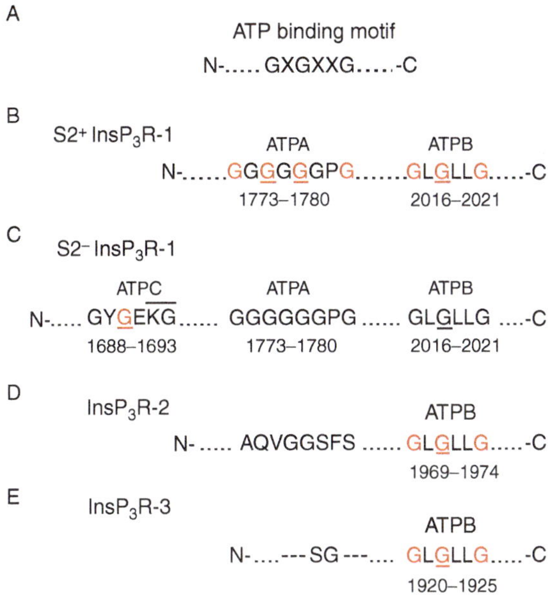

Please refer to Fig. 12.3 in text for figure legend.

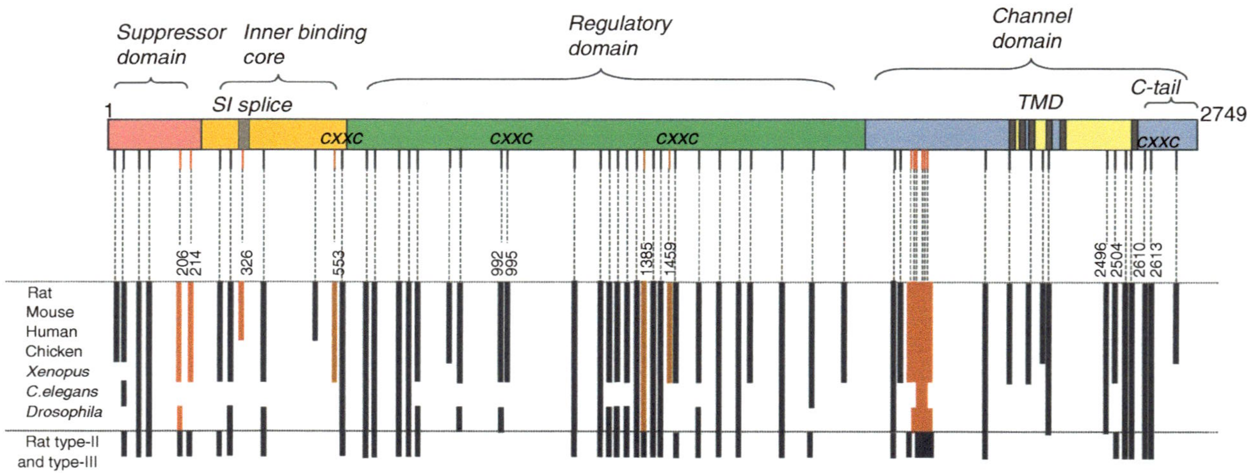

Please refer to Fig. 13.1 in text for figure legend.

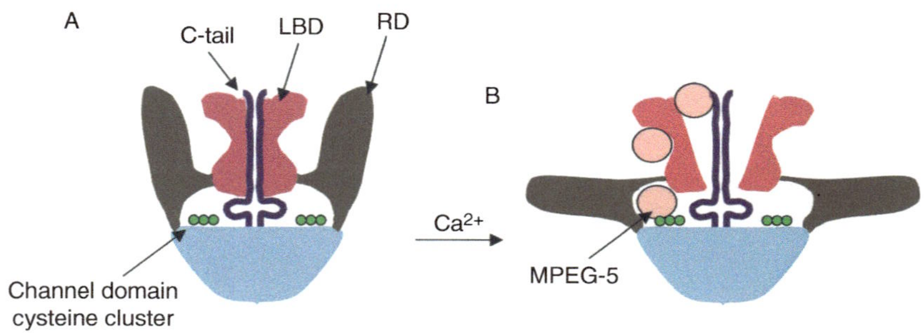

Please refer to Fig. 13.2 in text for figure legend.

FSC
www.fsc.org
MIX
Papier | Fördert
gute Waldnutzung
FSC® C083411

Kolibri 360 GmbH
Ferdinand-Jühlke-Straße 7
99095 Erfurt, Deutschland
produktsicherheit@kolibri360.de